Unraveling Lipid Metabolism with Microarrays

Unraveling Lipid Metabolism with Microarrays

EDITED BY

ALVIN BERGER

Icoria, Inc.
Research Triangle Park, North Carolina, U.S.A.

MATTHEW A. ROBERTS

Nestlé Global Strategic Business Unit
and Research and Development
St. Louis, Missouri, U.S.A.

 MARCEL DEKKER   NEW YORK

Although great care has been taken to provide accurate and current information, neither the author(s) nor the publisher, nor anyone else associated with this publication, shall be liable for any loss, damage, or liability directly or indirectly caused or alleged to be caused by this book. The material contained herein is not intended to provide specific advice or recommendations for any specific situation.

Trademark notice: Product or corporate names may be trademarks or registered trademarks and are used only for identification and explanation without intent to infringe.

Library of Congress Cataloging-in-Publication Data
A catalog record for this book is available from the Library of Congress.

ISBN: 0-8247-5811-0

This book is printed on acid-free paper.

Headquarters
Marcel Dekker, 270 Madison Avenue, New York, NY 10016, U.S.A.
tel: 212-696-9000; fax: 212-685-4540

Distribution and Customer Service
Marcel Dekker, Cimarron Road, Monticello, New York 12701, U.S.A.
tel: 800-228-1160; fax: 845-796-1772

World Wide Web
http://www.dekker.com

Current printing (last digit):

10 9 8 7 6 5 4 3 2 1

PRINTED IN THE UNITED STATES OF AMERICA

Preface

Lipids are known to affect a plethora of physiological functions in numerous organisms. Lipids have known structural roles as components of biological membranes, where they associate with proteins and affect membrane biophysical properties. Lipids are also known to be converted to numerous bioactive eicosanoids, affecting physiological function of neighboring cells via receptor interactions.[1]

Despite almost 100 years of targeted lipid research, our understanding of how lipids affect such diverse functions, has remained largely a black box.[2] Furthermore, since very specific hypotheses were evaluated in a given experiment, each new experiment could only answer whether the narrowly focused hypothesis under consideration was true or false.[3] In essence the conventional approach encouraged good hypothesis testing but limited serendipitous discovery. With the advent of more global analytical technologies and an examination of all experimental endpoints, we can rapidly expand our knowledge of lipids, moving beyond the initial hypothesis—some would say beyond our wildest dreams.

"Omic" technologies have clearly begun to shed light on this aforementioned black box by providing a truly comprehensive picture of metabolism. These technologies span genomics, transcriptomics, proteomics, and metabolomics (including lipidomics), to name a few. In this text, an omic approach will be viewed as any massively parallel analytical platform that covers a significant portion, if not all, of the set of molecules that a researcher wishes to determine (i.e., all lipids, all amino acids, all mRNA transcripts, etc.). The collected mass of data should be evaluated using modern rigorous statistical and bioinformatic approaches, including clustering and pattern recognition. Good use of the emerging data analysis systems and an intuitive, clear presentation of the data are prerequisites for useable omic experimental observations. Lastly, the text touches on systems biology, which combines multiple omic data streams.

The technology which is most mature and which has been most widely dispersed in the research community is the microarray. Massively parallel quantification of mRNA transcripts across treatments has been enabled by the microarray and is now known as transcriptomics. Prior to the advent of the microarray, techniques such as Northern blot analysis allowed researchers to compare upto 10 transcripts across treatments in a convenient manner, whereas reverse transcriptase–polymerase chain reaction (RT–PCR) allowed up to 100 or so transcripts to be compared conveniently. With microarrays, transcripts representing all coding sequences from an entire genome can now be evaluated.

In *Understanding Lipid Metabolism with Microarrays and Other "Omic" Approaches* our aim was to demonstrate the power of microarrays and transcriptomic approaches to dramatically increase our understanding of how lipids function, and to expand the known biological functions of lipids. Furthermore, the focus on microarray technology provides guidance on how to handle omic datasets, which will benefit the lipid and scientific community. Herein, several leading laboratories throughout the world present reviews of selected fields including some previously unpublished work. Highlights and summations of each book section and chapter are presented.

DATA ANALYSIS AND STATISTICAL INTERPRETATION IN MICROARRAY EXPERIMENTS

As previously described, the challenge in microarray-based transcriptomics and all omic approaches, is to manage the deluge of data in a robust, statistically sound manner. In Chapter 1, Goldstein and Delorenzi present an extensive overview of the statistical approaches in current usage to evaluate microarray data.

EXPLORING FATTY ACID AND CHOLESTEROL METABOLISM USING MICROARRAYS

Berger and Roberts were one of the first groups to use microarray-based transcriptomic approaches to elucidate the process of lipid signaling through known transcriptional signatures.[1,4–6] In Chapters 2 and 3, the authors present newly published findings showing that the well-known long chain polyunsaturated fatty acids (LC-PUFA), arachidonic acid and docosahexaenoic acid (DHA, as a component of fish oil), affect numerous pathways and potential physiological functions not previously associated with these LC-PUFA. Chapter 2 focuses on the liver, whereas Chapter 3 focuses on the hippocampus. Similarities and differences between the two organs and between their own work and that of others published in the literature and this book are described. The authors also update interpretations of their original findings based on literature published since 2002. Unlike their earlier works, the authors do not focus exclusively on known transcriptional signatures.

In Chapter 4, Fujiwara and Masumoto describe how LC-PUFA affects gene expression in hepatic cells. Transcriptomic experiments with cells are vital because of the ease, rapidity, lack of ethical issues, and cost savings associated with cell work when compared with animal work. One of the goals of transcriptomics is to more fully validate different *in vitro* and *in vivo* models. By comparing Fujiwara and Masumoto's work to the hepatic murine work of Berger and Roberts, differences between the *in vitro* and *in vivo* models can be evaluated, even if the starting sources of fatty acids were different.

In Chapter 5, Farkas et al. provide a review of their recent work describing how several diverse LC-PUFA affect transcriptomics in several brain regions. They are the only group to have studied how lipids affect transcriptomics in multiple brain regions at present. Cross-talk lipid signaling between brain regions is a particularly important and challenging area of investigation. Dr. Farkas was a renowned scientist who died while completing the above chapter.

In Chapter 6, Gaines et al. manipulate endogenous levels of DHA in the cerebrum by varying amounts of the dietary precursor fatty acid linolenic acid (18:3n-3). The authors then evaluate how gene transcripts are affected during murine development, and uncover some potential signaling cascades that may be DHA-regulated, and related to behavior. This work is extremely important since infants in North America and elsewhere may not receive sufficient n-3 LC-PUFA and DHA in utero and in the diet.

Leptin is a well-known and extremely important hormone, most studied for its role in adiposity, and its hypothalamic neuronal effects on appetite. Recently, leptin has been established to affect adiposity via fatty acid metabolism. Specifically, leptin decreases Δ9 desaturation of stearic acid, leading to reduced adiposity,[7] a salient feature of obesity, diabetes, and metabolic syndrome. Leptin signaling may also affect hypothalamic levels of LC-PUFA-derived endocannabinoids,[8] which in turn, affect appetite. For these reasons, it was important to include a chapter on the effects of leptin on transcriptomics, and to evaluate whether LC-PUFA and leptin affected similar transcripts. In Chapter 7, Kaszubska et al. studied transcriptomic effects of leptin in a hypothalamic cell line. Berger and Roberts (Chapter 3) and Kaszubska et al. found that some of the same general signaling cascades may be affected by LC-PUFA and leptin (e.g., nucleotide binding), but specific transcripts that changed were not equivalent with the two experimental approaches.

Cholesterol feeding is known to affect not only levels of cholesterol, but also other biological processes. For example, cholesterol may affect cholesterylation of proteins, biophysical properties of membranes, and physiological imprinting.[9] Thus, to expand our knowledge of dietary cholesterol on broad metabolism, Vergnes et al. fed cholesterol to atherosclerosis-susceptible and -resistant mice (Chapter 8). Since cholesterol and LC-PUFA both signal through sterol regulatory element binding protein (SREBP), it was not surprising to observe some similar transcripts affected in the works of Vergnes et al. and Berger and Roberts with LC-PUFA (see Chapter 2 for a comparative discussion of the two chapters).

Dietary lipids have been known for many years to affect behavior, via mostly unknown mechanisms. More recently, lipids have been shown to benefit various neurological and psychiatric disorders.[1] Microarrays are thus powerful tools that enhance our understanding of how lipids affect drug abuse, and neurological and psychiatric disorders. Lehrmann et al. (Chapter 9), present an extensive review of this budding subject matter.

EXPLORING THE ROLES OF PPARs, HNFs, AND SREBPs USING MICROARRAYS

In Chapters 2 and 3, and earlier work,[1] Berger and Roberts describe how dietary LC-PUFA affect a multitude of transcripts via common peroxisome proliferator activated receptor (PPAR) and SREBP signaling cascades. Leaders in the field, each having a different focus, expand this theme in Chapters 10–13. Yahagi and Shimano focus on transcripts altered in SREBP-1a overexpressing transgenic mice (Chapter 10). Kersten et al. (Chapter 11) focus on the transcripts altered in livers of fasted PPARα null mice when compared with wild-type mice. Yamazaki et al. (Chapter 12) focus on the effects of PPARα drug agonists on the broad transcriptome. Table 2.3 in Chapter 2 (Berger and Roberts) shows the similar transcripts affected by LC-PUFA feeding in liver, and PPARα drug agonists, adding credence to the notion that LC-PUFA signal through PPARs.

EXPLORING LIPIDS AND CARCINOGENESIS USING MICROARRAYS

Lipids have been known to affect cell growth both in vivo and in vitro for many years. Microarrays and beginning to rapidly and dramatically increase our mechanistic understanding of how cell growth is affected (Chapters 13–15). Anderle et al. (Chapter 13) focus on the mechanisms and signaling cascades that lead to LC-PUFA-induced changes in the transcriptome and in cell growth. Narayanan et al. (Chapter 14), focus on how LC-PUFA, particularly DHA, affect colon cancer and colonocyte cell growth. Narayanan et al. (Chapter 15) focus on how ellagic acid, affects prostate cancer and prostate cell growth. Ellagic acid is a hydrolyzable ellagitannin present in fruits, berries, and nuts. Understanding the effects of ellagic acid on prostate cell growth may elucidate how the aforementioned foods affected prostate cancer in scientific and epidemiologic studies, paving the way for more powerful neutraceutical and pharmaceutical approaches for controlling aberrant cell growth.

EXPLORING HOW LIPIDS AFFECT INSULIN SECRETION USING MICROARRAYS

Diabetes and metabolic syndrome, along with obesity, are major health concerns. Microarrays can be utilized to mechanistically understand how lipids may contribute, prevent, or reduce the incidence of these diseases (Chapters 16 and 17).

Type 2 diabetes is associated with elevated level of triglycerides (TAG) as well as non-esterified fatty acids (NEFA). In this context, Xiao et al. (Chapter 16) present the effects of the fatty acid palmitate (C16:0) on gene expression in insulin-secreting cells (INS-1 cells) and describe signaling pathways linking fatty acid and glucose.

Rondinone and Waring (Chapter 17) present the effects of protein tyrosine phosphatase IB (PTP1B) on diabetes and obesity. PTP1B is a protein tyrosine phosphatase and an important regulator of insulin signal transduction. PTP1B dephosphorylates tyrosine residues in the insulin receptor and its substrates. The effects of PTP1B antisense treatment on insulin resistance and regulation of fat metabolism is described in *ob/ob* diabetic mice. PTP1B treatment affected numerous signaling cascades, decreased SREBP-1-regulated transcripts involved in fatty acid synthesis in adipose and liver (Figs. 17.2 and 17.4), and decreased adipose fat content. Thus, PTP1B antisense treatment and dietary LC-PUFA (Chapter 2) share a common SREBP-1 signaling cascade which diminishes fat synthesis.

OTHER OMIC TECHNIQUES FOR UNRAVELING LIPID METABOLISM

Microarrays permit us to detect changes to thousands of transcripts simultaneously and to predict changes in signaling cascades and metabolism. The predictive power of microarrays is strengthened when numerous biochemical steps in the same pathway are affected in a consistent manner, and when common transcriptional signatures are identified.[1] Since proteins are post-translationally modified and affect biology by forming specific dimeric and multimeric complexes amongst themselves, and also form proteolipid complexes, changes to the proteome only predicts phenotypic changes to an organism. In small molecule omics (metabolomics or metabonomics), changes to amino acids, peptides, lipids, and carbohydrates are examined to provide for a more accurate assessment of phenotypic changes. Metabolomics, when combined with transcriptomics in a systems biology approach, is a particularly powerful tool.

In Chapters 18 and 19, the subject of metabolomics is introduced. Grainger et al. (Chapter 18) provide a summary of experiments performed using nuclear magnetic resonance (NMR) to examine metabolomic changes, and focus on changes to lipids, such as lipoproteins and triacylglycerol. NMR has the advantage of being able to detect isomeric changes in compounds, a weakness of mass spectroscopy (MS)-based approaches.

Animal models that permit researchers to examine metabolism in a whole body global approach are particularly powerful for understanding tissue–tissue cross-metabolism, and for understanding transport processes using specific markers. The transparent zebra fish has emerged as a powerful model to study whole body metabolism, and metabolite, nutrient, and drug transport and digestion. Whole body reporter assay approaches are also possible. Rubinstein et al. (Chapter 19) present an overview on how the zebra fish model can be used to advance our understanding of whole body lipid metabolism. The approach becomes omic when numerous transcripts and metabolites are examined throughout the organism.

Omic approaches, particularly microarray-based transcriptomics, have dramatically increased our understanding of lipid metabolism in a short time. Increasingly these omic datasets are publicly available[1] and provide common standards to compare lipid research at both the data and interpretive level. Our main hope is that more lipid researchers, particularly the readers of this book, will begin to apply these powerful new omic techniques to advance their own fields of scientific interest. The complexity inherent to lipid metabolism will thereafter become increasingly unraveled, revealing novel scientific insights.

Alvin Berger and Matthew A. Roberts

REFERENCES

1. Berger, A.; Mutch, D.M.; German, B.J.; Roberts, M.A. Dietary effects of arachidonate-rich fungal oil and fish oil on murine hepatic and hippocampal gene expression. Lipids Health Dis. **2002**, *1*, 2.
2. Fisher-Wilson, J. Long-suffering lipids gain respect: technical advances and enhanced understanding of lipid biology fuel a trend toward lipidomics. The Scientist **2003**, *17*, 5.
3. Berger, A. How lipidomic approaches will benefit the pharmaceutical industry. In *Metabolic Profiling and Metabolome Analyses*, Harrigan, G.G., Goodacre, R. Eds; Kluwer Academic Publishers: Boston, MA. 2004.
4. Berger, A.; Mutch, D.M.; German, J.B.; Roberts, M.A. Unraveling lipid metabolism with microarrays: effects of arachidonate and docosahexaenoate acid on murine hepatic and hippocampal gene expression. Genome Biol. **2002**, *3*.
5. Anderle, P.; Farmer, P.; Berger, A.; Roberts, M.-A. A nutrigenomic approach to understanding the mechanisms by which dietary fatty acids induce gene signals and control mechanisms involved in carcinogenesis. Nutrition **2004**, *20*, 103–108.
6. German, J.B.; Roberts, M.A.; Watkins, S.M. Genomics and metabolomics as markers for the interaction of diet and health: lessons from lipids. J. Nutr. **2003**, *133*, 2078S–2083S.
7. Miyazaki, M.; Dobrzyn, A.; Sampath, H.; Lee, S.H.; Man, W.C.; Chu, K.; Peters, J.M.; Gonzalez, F.J.; Ntambi, J.M. Reduced adiposity and liver steatosis by stearoyl-CoA desaturase deficiency are independent of peroxisome proliferator-activated receptor-α. J. Biol. Chem. **2004**, June 4 [E pub].
8. Di Marzo, V.; Goparaju, S.K.; Wang, L.; Liu, J.; Batkai, S.; Jarai, Z.; Fezza, F.; Miura, G.I.; Palmiter, R.D.; Sugiura, T.; Kunos, G. Leptin-regulated endocannabinoids are involved in maintaining food intake. Nature **2001**, *410*, 822–825.
9. Palinski, W.; Napoli, C. The fetal origins of atherosclerosis: maternal hypercholesterolemia, and cholesterol-lowering or antioxidant treatment during pregnancy influence in utero programming and postnatal susceptibility to atherogenesis. FASEB J. **2002**, *16*, 1348–1360.

---

[1]Chapters 2–3: www.ncbi.nlm.nih.gov/geo/ then type GEO accession # GSE91
Chapter 4: www.lsbm.org/database/index.html#

List of Contributors

Pascale Anderle *ISREC, Swiss Institute for Experimental Cancer Research, Epalinges, Lausanne, Switzerland*

Gwendolyn Barceló-Coblijn *Institute of Biochemistry, Biological Research Center, Hungarian Academy of Sciences, Hungary*

K. G. Becker *DNA Array Unit, Research Resources Branch, National Institute on Aging, NIH/DHHS, Baltimore, MD, USA*

Alvin Berger *Paradigm Genetics, Inc., Research Triangle Park, NC, USA*

Nancy E. J. Berman *University of Kansas Medical Center, Kansas City, KS, USA*

Susan E. Carlson *University of Kansas Medical Center, Kansas City, KS, USA*

J. Chen *Cellular Neurobiology Research Branch, National Institute on Drug Abuse, NIH/DHHS, Baltimore, MD, USA*

Mauro Delorenzi *Bioinformatics Core Facility, Institut Suisse de Recherche Expérimentale sur le Cancer, NCCR Molecular Oncology Epalinges, Switzerland*

Pascal Escher *Pfizer Global Research & Development, Ann Arbor Laboratories, Ann Arbor, MI, USA*

Steven A. Farber *Department of Microbiology and Immunology, Kimmel Cancer Center, Thomas Jefferson University, Philadelphia, PA, USA*

Tibor Farkas[*] *Institute of Biochemistry, Biological Research Center, Hungarian Academy of Sciences, Hungary*

---

[*]Deceased.

ix

Pierre Farmer *ISREC, Bioinformatic Core Facility (BCF), Epalinges, Lausanne, Switzerland*

W. J. Freed *Cellular Neurobiology Research Branch, National Institute on Drug Abuse, NIH/DHHS, Baltimore, MD, USA*

Yoko Fujiwara *Department of Nutrition and Food Science, Ochanomizu University, Japan*

Judy B. Gaines *University of Kansas Medical Center, Kansas City, KS, USA*

Otto Geoffroy *Institute for Cancer Prevention, Valhalla, NY, USA*

Darlene R. Goldstein[†] *Bioinformatics Core Facility, Institut Suisse de Recherche Expérimentale sur le Cancer, NCCR Molecular Oncology, Epalinges, Switzerland*

David J. Grainger *Department of Medicine, Cambridge University, Addenbrooke's Hospital, Cambridge, UK*

Kjeld Hermansen *Molecular Diagnostic Laboratory, Department of Clinical Biochemistry, Skejby Hospital, Aarhus University Hospital, Aarhus, Denmark*

Paul Hessler *Genomics and Molecular Biology, Global Pharmaceutical Research Division, Abbott Laboratories, Abbott Park, IL, USA*

Taro Hihara *Tsukuba Research Laboratories, Eisai Co., Ltd., Tokodai, Tsukuba, Ibaraki, Japan*

Shiu-Ying Ho *Department of Microbiology and Immunology, Kimmel Cancer Center, Thomas Jefferson University, Philadelphia, PA, USA*

Elaine Holmes *Division of Biomedical Sciences, Imperial College of Science, Technology, and Medicine, London, UK*

Per Bendix Jeppesen *Department of Endocrinology and Metabolism C, Aarhus Sygehus THG, Aarhus University Hospital, Aarhus, Denmark*

Wiweka Kaszubska[§] *Metabolic Disease Research, Abbott Laboratories, Abbott Park, IL, USA*

Sander Kersten *Nutrition, Metabolism and Genomics group, Division of Human Nutrition, Wageningen University, Wageningen, The Netherlands*

Klára Kitajka *Institute of Biochemistry, Biological Research Center, Hungarian Academy of Sciences, Hungary*

[†]*Current affiliation*: Institute of Mathematics, École Polytechnique Fédérale de Lausanne, Lausanne, Switzerland.

[§]*Current affiliation*: Global Product Development, Serono International S.A.; Genève, Switzerland.

Paul E. Kroeger *Genomics and Molecular Biology, Global Pharmaceutical Research Division, Abbott Laboratories, Abbott Park, IL, USA*

Mogens Kruhøffer *Molecular Diagnostic Laboratory, Department of Clinical Biochemistry, Skejby Hospital, Aarhus University Hospital, Aarhus, Denmark*

Junro Kuromitsu *Tsukuba Research Laboratories, Eisai Co., Ltd., Tokodai, Tsukuba, Ibaraki, Japan*

C.-T. Lee *Cellular Neurobiology Research Branch, National Institute on Drug Abuse, NIH/DHHS, Baltimore, MD, USA*

E. Lehrmann *Cellular Neurobiology Research Branch, National Institute on Drug Abuse, NIH/DHHS, Baltimore, MD, USA*

Akiyo Masumoto *Department of Nutrition and Food Science, Ochanomizu University, Japan*

David E. Mosedale *Department of Medicine, Cambridge University, Addenbrooke's Hospital, Cambridge, UK*

Bhagavathi A. Narayanan *Institute for Cancer Prevention, Valhalla, NY, USA*

Narayanan K. Narayanan *Institute for Cancer Prevention, Valhalla, NY, USA*

Jeremy K. Nicholson *Division of Biomedical Sciences, Imperial College of Science, Technology, and Medicine, London, UK*

Daniel W. Nixon *Institute for Cancer Prevention, Valhalla, NY, USA*

Torben Ørntoft *Molecular Diagnostic Laboratory, Department of Clinical Biochemistry, Skejby Hospital, Aarhus University Hospital, Aarhus, Denmark*

Jack Phan *Department of Medicine and Department of Human Genetics, University of California, Los Angeles, CA, USA*

László G. Puskás *Laboratory of Functional Genomics, Biological Research Center, Hungarian Academy of Sciences, Szeged, Hungary*

Bandaru S. Reddy *Institute for Cancer Prevention, Valhalla, NY, USA*

Karen Reue *Department of Medicine and Department of Human Genetics, University of California, Los Angeles, CA, USA*

Abudula Reziwanggu *Department of Endocrinology and Metabolism C, Aarhus Sygehus THG, Aarhus University Hospital, Aarhus, Denmark*

Matthew-Alan Roberts *Nestle Global Strategic Business Unit and R&D, St. Louis, MO, USA*

Cristina M. Rondinone *Metabolic Diseases Research and Department of Cellular and Molecular Toxicology, Global Pharmaceutical Research and Development, Abbott Laboratories, Abbott Park, IL, USA*

Amy L. Rubinstein *Zygogen, LLC, Atlanta, GA, USA*

Hitoshi Shimano *Department of Internal Medicine, Institute of Clinical Medicine, University of Tsukuba, Ibaraki, Japan*

Sherrie Tafuri *Pfizer Global Research & Development, Ann Arbor Laboratories, Ann Arbor, MI, USA*

Isao Tanaka *Tsukuba Research Laboratories, Eisai Co., Ltd., Tokodai, Tsukuba, Ibaraki, Japan*

Laurent Vergnes *Department of Medicine and Department of Human Genetics, University of California, Los Angeles, CA, USA*

Walter Wahli *Center for Integrative Genomics, University of Lausanne, Lausanne, Switzerland*

Jeffrey F. Waring *Metabolic Diseases Research and Department of Cellular and Molecular Toxicology, Global Pharmaceutical Research and Development, Abbott Laboratories, Abbott Park, IL, USA*

Jianzhong Xiao *Department of Endocrinology and Metabolism C, Aarhus Sygehus THG, Aarhus University Hospital, Aarhus, Denmark*

Naoya Yahagi *Department of Internal Medicine, Graduate School of Medicine, University of Tokyo, Tokyo, Japan*

Kazuto Yamazaki *Tsukuba Research Laboratories, Eisai Co., Ltd., Tokodai, Tsukuba, Ibaraki, Japan*

Hideki Yoshitomi *Tsukuba Research Laboratories, Eisai Co., Ltd., Tokodai, Tsukuba, Ibaraki, Japan*

Contents

---

*Deceased.

EXPLORING HOW LIPIDS AFFECT INSULIN SECRETION USING MICROARRAYS

Statistical Design and Data Analysis for Microarray Experiments

Darlene R. Goldstein*

Bioinformatics Core Facility, Institut Suisse de Recherche Expérimentale sur le Cancer, NCCR Molecular Oncology, Epalinges, Switzerland

Mauro Delorenzi

Bioinformatics Core Facility, Institut Suisse de Recherche Expérimentale sur le Cancer, NCCR Molecular Oncology Epalinges, Switzerland

INTRODUCTION

DNA microarrays are increasingly being used to study the molecular basis of complex disease traits, such as diabetes and coronary artery disease, as well as fundamental biological processes, such as lipid metabolism.[1–4] For example, microarrays can be used to investigate dietary effects on mRNA levels in cells or tissues[5–7] or effects of gene knockout.[8,9] Knowledge gained from such experiments advances understanding of molecular mechanisms and pathways involved in both normal and pathologic physiology, which can lead to improvements in food functionality, health, and disease treatment and prevention.

Measuring the amounts of mRNA can provide information on which genes are being expressed in a cell; microarrays provide a means to measure gene expression. There are several types of microarray systems, the types most commonly used at present are cDNA microarrays[10,11] and high-density oligonucleotide chips (Affymetrix GeneChips[12]). Other technologies, including

---

Current affiliation: Institute of Mathematics, École Polytechnique Fédérale de Lausanne, Lausanne, Switzerland.

tissue arrays and protein arrays, are likely to become more prominent in the future.

cDNA microarrays consist of thousands of individual cDNA *probe* sequences printed in a high-density array on a glass microscope slide (adopting the terminology convention in "The Chipping Forecast"[13]). The relative abundance of these spotted DNA sequences in two samples may be assessed by monitoring the differential hybridization of the two samples to the sequences on the array. The ratio of red to green fluorescence intensity for each spot is indicative of the relative abundance of the corresponding DNA probe in the two *target* samples. A more detailed introduction to the biology and technology of cDNA microarrays was published earlier.[13]

Affymetrix GeneChip arrays use a photolithography approach to synthesize probes directly onto a silicon chip. In addition to 11–20 short sequences used to probe each gene, the perfect match (PM) probes, there is an equal number of negative controls, the mismatch (MM) probes. A single labeled sample is hybridized to the array, so that absolute rather than relative measures of gene expression are obtained. Further details are available in Lockhart et al.[12] and The Chipping forecast.[13]

Common areas under study with microarray experiments include differential gene expression; gene expression patterns (profiles) across treatments; subclass identification using gene expression profiles; classification of samples into known classes; and identification of genes associated with outcomes, such as response to treatment or survival time. Analysis of the large, complex data sets resulting from microarray experiments presents a number of statistical problems and has initiated new statistical research, including creation of theory, improvement to existing methods, development of new methodology, and formulation of efficient computational algorithms.

The aim of this chapter is to give a current overview of the major statistical issues in microarray experiment design and data analysis, with special attention to identifying differentially expressed genes. Complete coverage of microarray statistics is not possible, as the field is rapidly expanding and quickly evolving; most technical elements are omitted here as well. Areas addressed in this review include experiment design, data preprocessing, identification of differential expression, multiple hypothesis testing, and classification (cluster analysis and discrimination). Pointers to the literature are provided for readers interested in further details. Slides covering this material and additional graphics are on the web (http://ludwig-sun2.unil.ch/~darlene/).

EXPERIMENTAL DESIGN

A number of decisions must be made before any type of experiment is performed, and microarray experiments are no exception. Microarray studies, particularly cDNA or other two-color arrays, confront researchers with a number of additional aspects that require consideration. We outline a few of these here: array design, some principles of study design, choice of samples for cohybridization (two-color arrays), amount and type of replication, and mRNA pooling. We conclude with a few general comments on the importance of pilot studies and validation of findings.

Array Design

The first design aspect involves *array design* which explains which genes should be represented on the array and where they should be placed. For commercially available arrays there is little possibility for user input, so the following discussion applies mainly to labs producing arrays in-house, most often cDNA arrays. We might simplify the possibilities by considering two broad types of arrays: those providing substantial, or "whole genome", coverage and those with limited, specialized coverage such as might be provided by arrays restricted to a few hundred metabolic genes. Arrays are often designed without any specific input from statisticians, which is less of an issue for most whole genome arrays. Specialized arrays are becoming more common, but their use is not without problems. The biggest difficulty with restricted coverage arrays at present is with normalization (discussed in detail later). Although array design has not been widely studied at the statistical level (but see Maindonald et al.[14]), control spots spaced across the array along with sufficient replication would seem to be required. Further careful study is needed before the routine use of specialized arrays can be whole-heartedly endorsed.

Study Design Principles

The ability to answer important scientific questions of interest depends to a large degree on the *study design*. Consider the following (hypothetical) example illustrating this point where interest is in the effect of diet on weight in an organism, such as a mouse. A group of mice of similar weight is divided into two groups based on age and gender. Group 1 (young females) is fed a low-fat diet and group 2 (old males) is fed a high-fat diet for a period of time, after which all mice are weighed. It is desired to attribute the difference in weight gain between the two groups to the difference in treatment; unfortunately, in this case that is not possible. The difference in weight gain could be attributable to age differences, or to gender differences, or even to synergistic effects of age and gender. This example contains an instance of a *confounder*, that is, a variable that is associated with both treatment and outcome. In this case, the effects of diet, age, and gender cannot be separated because of the poor choice of treatment allocation. Here, the confounding is very obvious. Study flaws can also occur in subtle ways. As an example, consider dividing a group of mice sharing a cage into two treatment groups by removing them one at a time and assigning the first half into one group, the remaining into the other. This does not represent a statistically random grouping, and in fact there are likely to be systematic differences between the two groups (e.g., the earlier ones may be more friendly or docile). Attention to the details ahead of time is critical to the success of the experiment (see, e.g., Cobb,[15] Quinn and Keough,[16] and Mead[17]), so consultation or collaboration with a statistician beginning early in the process is highly advisable in order to avoid pitfalls.

Sample Pairing for Two-Color Arrays

The first, and still most widely employed, cohybridization setup allowing for comparisons between cDNA arrays pairs each sample of interest with a *common reference*, which is usually labeled with Cy3.[18,19] This type of design allows the comparison of any number of samples to each other. All comparisons of interest, such as differences in gene expression between treatments, are made *indirectly* through this reference. For example, to find the difference in expression in samples A and B, each paired with reference sample R, we look at $(A-R) - (B-R)$ to obtain $A - B$. Unfortunately, the use of a common reference introduces additional noise, so that the variability associated with the difference $A - B$ is higher for the common reference design than for the *direct design*, in which A and B are compared on the same slide.[14,20,21] In addition, it is often the case that the reference sample is itself not of interest, which can also represent an inefficient use of expensive materials.

Several recent studies have been carried out without the use of a common reference, so that comparisons can be made directly rather than indirectly.[22,23] For estimating treatment differences, we are not limited to only those slides where both treatments are cohybridized. In fact, information can be combined across slides taking into account both direct and indirect comparisons. For example, consider an experiment with three samples A, B, C, and three slides such that each pair of samples is hybridized together once—slide 1 pairs A with B, slide 2 pairs B with C, and slide 3 pairs C with A. For comparing A and B, there is direct information from slide 1. However, there is also indirect information from slides 2 and 3, through sample C. Standard linear statistical methods lead to combined estimates for the difference $A - B$ (see, e.g., Mead,[17] Yang and Speed,[21] and Smyth et al.[24]).

Although there is appeal in making all pairwise comparisons of interest by direct cohybridization, this method becomes infeasible once the number of samples is even moderately large. Other pairing schemes have been proposed on the basis of statistical theory of optimal experimental designs. Single factor, such as time course, and factorial designs for sample pairing have been explored;[25] (see also Yang and Speed[20,21]). *Loop designs* have been considered as an alternative to both the common reference and the *all-pairs design*.[26,27] With the loop design, samples are ordered 1 through n, with adjacent samples cohybridized: samples 1 and 2, samples 2 and 3, and so on, finishing with samples n and 1. This reduces the number of hybridizations required when compared with the all-pairs design and is also more efficient than the common reference design in terms of average variance. However, there are drawbacks to this approach as well: some comparisons will be much less precise than others, because precision depends on the number of connecting steps between the sample pair. Additionally, if for some reason a particular slide fails and cannot be repeated then the loop is broken, rendering the analysis difficult or even impossible. Loop designs should therefore be avoided.

A major advantage of the common reference design is that the number of samples that can be readily and simply accommodated is limited only by the availability of reference material. The common reference design is equally simple for 10 or 100 samples, whereas other pairing designs are necessarily more complicated as the number of samples increases. The common reference design is also easily extensible if additional samples become available, as is often the case in clinical studies. Connecting two separate experiments together for other types of pairing is likely to be substantially more complex, so that the gains in precision the other designs afford may be considered too small to justify the additional effort.

There is insufficient knowledge at present for design recommendation when the aim is classification (clustering or discrimination, discussed in detail later). The common reference design can be used here certainly, but perhaps there are better design possibilities. To date there does not appear to be much literature on this topic, but some research in this direction has been initiated.[28]

It is also relevant to mention here the assignment of fluorescent dyes to each of the paired samples, most commonly Cy3 (green) and Cy5 (red). Systematic differences have been observed in dye behavior, with the amount of bias varying across the range of fluorescence intensities.[29,30] An effective way to reduce this bias is with dye swaps, that is, reversing the dye assignments on replicate hybridizations involving the same two samples[20,29–31] or balancing dye assignments.[32] Here, local median centering plays the main role in reducing this bias, as the intensity-specific bias is not constant across arrays.

There is no single design that is best in all cases. The choice of pairing design depends on amount of material available in each sample, number of samples, experimental cost, and scientific aims of the experiment, among other factors. A general principle though is that, insofar as possible, the design should yield greatest precision for those comparisons of most interest, which suggests direct rather than indirect comparisons of the relevant samples.

Replication

There are several forms of replication relevant to microarray experiments: *replicate spots*, where the same cDNA is spotted multiple times on each slide; *technical replicates*, or slides containing target mRNA from the same extraction; and *biological replicates*, in which different hybridizations involve mRNA from different individuals.[14,20] In the microarray context, replication allows the researcher to identify sources of variability as well as outliers, reduce variation through averaging, permit generalization to a population of interest, and make the conclusions robust. Each kind of replication involves a different type and degree of variability. The type of replication to be used in a given experiment depends on the precision and generalizability of experimental results the investigator seeks to achieve.

Replicate spots are usually duplicates, sometimes triplicates or more, and are most commonly placed in adjacent positions. Replication of spots may be considered part of array design, and as such is generally not subject to manipulation by individual investigators. There has been limited statistical examination of replicate spots.[33] Replicate spots can be useful for quality assessment of the experiment, but should not be considered as providing independent information that may be generalized to the population of interest (as would additional samples obtained from different individuals). It should also be noted that where feasible, replicate spots should be spaced across the slide rather than adjacent so that variability across the slide may be better assessed.

Because technical replicates share a great number of features, they cannot be considered as independent samples for the purpose of generalizing conclusions to the larger population. For this, biological replicates are required.[14,20,21] Independence assumptions standard in many statistical analyses will typically be more closely satisfied with biological rather than technical replicates.

Given the importance of replication,[34] an obvious practical question is how many arrays to use. The question is difficult to answer precisely in the context of microarray experiments because the variance of expression levels across hybridizations is not only unknown but also varies between genes.[35,36] Another crucial element lacking is some idea of the magnitude of effects to be detected.

One approach is to use an approximate value of gene variance that is not too small (e.g., the median of variances across genes) to plug in for power calculations. Research has also been done using all-gene ANOVA models to determine sample size for a given level of power.[32,34,37,38] General recommendations based on differing assumptions often suggest three arrays; for example, Affymetrix has suggested this number as a guideline. Frequently, the issue of sample size seems to be determined by the experiment budget rather than statistical considerations. In summary, there is not a single exact answer, but guidelines may be obtained using the methods described earlier. Alternatively, examination of the tradeoff between power and false positive rate across sample size may address the issue more adequately.[20] In general, though, more samples are better.

Pooling

The issue of mRNA pooling arises because single individuals may be unable to provide sufficient material for an experiment; thus, as an alternative to amplification, pools of mRNA from sets of individuals (e.g., genetically identical mice) are created so that the experiment may be carried out. Pooling has also been suggested even when there is enough material for individual hybridizations: for cost savings (subjects are generally less expensive than arrays), as an improvement to precision, or as a means to produce more usable spots on an array (as there should be fewer blank spots with more mRNA varieties). However, one problem with pooling is that the ability to distinguish variation between individual samples is lost. Use of pooling might also increase effects of RNA self-hybridization,[39] although this may not be strong enough to distort measurements.

Little experimental work exists that specifically examines different pooling options, and the evidence appears to conflict over the value of pooling.[40,41] A very recent simulation study considered the cost/benefit tradeoff of pooling, concluding that pooling can be financially beneficial.[42] A statistical examination of pooling in the microarray context indicates that pooling with additional subjects can decrease the number of arrays required without loss of precision, and included a small quantitative real-time PCR assessment.[43] This work appears to be very promising; however, open questions remain regarding validity of the assumptions of these studies, requiring further experimental evidence before pooling can be soundly endorsed.

Pilot Studies and Result Validation

Because, in many cases little is known about the sources, types, and amount of variability, it is advisable to carry out a small-scale pilot study. The sample size may be as low as only a few arrays for a relatively small size final experiment, or perhaps 10–20 arrays for a much larger clinical study. A pilot study aids in identifying unanticipated problems in biological or technical reproducibility in time to rectify them before substantial effort or cost has been expended. It also provides a first impression on the appropriateness of the data analytic methods and of the magnitude of the observable effects. On this basis, the project plan can be either confirmed or improved; in the face of insurmountable obstacles, it may be decided that the project should instead be completely abandoned rather than receiving any additional investment of time or money.

It is also good practice to give thought at the design stage to the amount and type of validation to be carried out on interesting findings, as these can influence decisions about sample size as well as statistical methods of data analysis. If result verification is relatively inexpensive and easy to carry out, then smaller sample sizes or less stringent analyses may be acceptable. Common methods include Northern or Western blots and quantitative real-time PCR.

In closing this section, it should also be stated that design issues are best addressed with a statistician before any steps of the experiment have been carried out. Once the experiment has started it may be too late to reverse serious problems. Care taken in advance saves time and money, and allows researchers to achieve the greatest benefits from their efforts.

DATA PREPROCESSING

The raw data resulting from a microarray experiment consist of image files produced by the array scanner. The preprocessing steps of *image analysis* and *normalization* are required to reliably quantify the fluorescence intensities.

These steps must be carried out prior to any analysis comparing expression levels within or between slides. Preprocessing can have a substantial impact on subsequent data analyses such as clustering and variable selection, so it is valuable to have some understanding of what goes on inside the computer at this stage.

Image Analysis

Image analysis combines a set of fluorescence intensities measured at a number of pixels for each spot into a single number. There are three major aspects to image analysis: *addressing*, also called *gridding*, or location of spots; *segmentation* or classification of pixels as either foreground or background; and *signal extraction*, the step creating a single measure of fluorescence intensity from the pixel level foreground and background intensity measurements. Image analysis can also be used to assess hybridization quality.

Most image analysis programs provide some form of automated addressing. A major difference between the many available programs is in the segmentation step. With *fixed circle segmentation*, all spots are assumed to be circles of the same size,[44] *adaptive circle segmentation* allows the size of the spots to vary,[45,46] *adaptive shape segmentation* allows spots of arbitrary shape and size,[47–51] *histogram segmentation* estimates foreground based on the distribution of the intensities within a region surrounding each spot.[52,53] When spots are fairly circular and uniform in size, fixed or adaptive circle segmentation works reasonably well. When the spots vary considerably in shape or size, though, adaptive shape segmentation performs better.

The signal in a spot is generally computed as the mean or median of pixel intensities. Usually, the overall (foreground) intensity is corrected by subtracting an estimate of background intensity. Background adjustment removes from the measured signal in a spot the components due to nonspecific hybridization, as well as fluorescence due to slide coating. For Affymetrix GeneChips, the MAS 5.0 software adjusts the PM measurements for nonspecific hybridization by subtracting a quantity based on the MM values; definition of an appropriate expression measure based on the signal is still an active area of research.[54–58] For two-color arrays, background intensity is determined from a region surrounding each spot; variations exist between programs in what region is used. It is also possible to use an estimate that includes the signal from negative spots. Background regions too close to the edge of the spot can result in an over-adjustment. In scatterplots of (log) red vs. green intensity, excessive background adjustment appears as a fantail or chevron shape, with spots at lower intensities spread quite far from the line of equality. Method comparisons have found that the background correction employed can greatly affect the resulting expression measures, even more than the segmentation method.[59,60] Better background estimation should therefore improve the quality of measurements obtained from microarrays. The resulting expression measure for two-color systems is the log base 2 ratio of background-subtracted red and green intensities: $\log_2(R/G)$.

The quality of fluorescence intensity measurements can vary substantially across spots within and between arrays. Quality assessment of microarray data occurs not only at the spot level but also at the array level. The distribution of pixel intensities across the spots on the array provides a simple example. Intensities encompassing only a small part of the possible range of values (0–16 on the log base 2 scale) suggest poor slide quality, possibly due to a failed label incorporation or unsuitable scanner settings. Saturation may also be apparent, suggesting a scanner PMT setting that is too high.

Most image analysis programs include procedures for "flagging" spots on the basis of one or more of the quality statistics, with flagged spots typically removed from subsequent analyses. From a statistical standpoint, it is more natural to include all spots in the analysis, but with weights indicating how much each spot should count. Spots of good quality receive full weight, whereas lower quality spots have smaller weights corresponding to the degree of quality reduction, with complete removal (weight 0) only for the very poorest quality spots. A number of papers address quality assessment,[24,61–63] but a broadly accepted scheme for weighting spots is not yet in place.

Exploratory Data Analysis

Once image processing is complete, the data should be displayed graphically in a number of ways as a prelude to more formal analysis. This type of exploratory analysis is useful for identifying specific problems, as well as giving an overall sense of experimental results.

Viewing a representation of the array images (.DAT file for Affymetrix GeneChips, pseudo-color overlay for two-color arrays) gives a quick impression of the quality of the hybridization. Strong artifacts such as dust, air pockets, or streaks are easily discovered in this way.

For two-color arrays, scatterplots provide a convenient visualization. The data for a single slide are most usefully displayed with the Tukey sum–difference (or mean–difference) plot,[64] most commonly referred to in the microarray context as the *MA* plot, where $M = \log_2(R/G)$ and $A = 1/2(\log_2 R + \log_2 G)$.[29,30,35] This same type of plot has also been referred to in the microarray literature by the terms SI (signal–intensity) or RI (ratio–intensity) plot. The *MA* plot is a rotated and rescaled version of the standard scatterplot of $\log_2 R$ vs. $\log_2 G$. The advantage is that the data represented in this way are more revealing of interesting and important features, which are more difficult to discern in the traditional $\log_2 R$ vs. $\log_2 G$ plot, because the high correlation between the two intensities affects our visual perception of distance.[65,66] Figures 1.1(a) and (b) shows examples of *MA* plots for all log-ratios from a single slide (before and after normalization, discussed later).

Histograms or boxplots of intensity measurements for individual arrays are also useful with all types of arrays. These plots provide different types of graphical summaries of the distribution of values in a data set. A *histogram* is

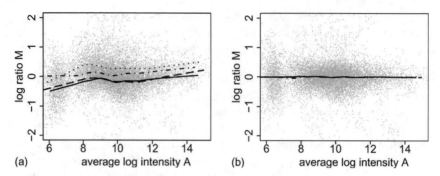

Figure 1.1 *MA* plots for a single cDNA slide. Curves are four pin-specific loess fits (a) before and (b) after normalization.

a type of bar chart, with area of the bars representing the proportion of the data falling into the corresponding interval of values. The bars may also be smoothed, in which case the plot may be referred to as a *density plot.* A boxplot provides a graphical summary of the 5-number summary: the minimum value, first quartile (25th percentile), median, third quartile (75th percentile), and maximum value. Boxplots in particular help to suggest appropriate normalization procedures. Differences in location or scale for different arrays or different regions within an array (e.g., print tip) are examples of artifacts that may be readily visualized and assessed with boxplots. Figure 1.2(a), for example, shows separate boxplots of log-ratios spotted by four different pins. It is easily seen that the log-ratios tend

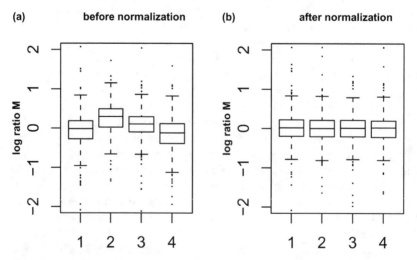

Figure 1.2 Boxplots stratified by print-tip group for the slide depicted in Fig. 1.1: (a) before and (b) after normalization.

to vary for different pins; pin 2 log-ratios tend to be higher and are also a bit more variable, as evidenced by the higher and slightly longer box. For two-color arrays, spatial plots of intensities or *M* values also aid in identifying spatially uneven foreground or background or other artifacts. All of these types of plots are described in greater detail elsewhere.[29,30,67,68] Software implementing the methods is available from the open source BioConductor project http://www.BioConductor.org/,[69,70] built on the R statistical software.[71]

Normalization

Detection of differential gene expression, or real biologically meaningful variation, is a major aim of many microarray experiments. However, nuisance systematic variation occurs within single arrays as well as across different arrays, making comparisons of uncorrected expression measures unreliable. The purpose of normalization is to identify and control for sources of nuisance systematic variation in the measured fluorescence intensities. For example, in cDNA microarrays there are different labeling efficiencies and scanning properties of fluors; the laser scanning parameters are often manually adjusted for different arrays in order to avoid optical saturation; spatial effects, which can originate from pins, plates, or technicians, are also usually apparent. None of these is of any biological importance, and must therefore be removed as far as possible.

Normalization for cDNA Arrays

The need for normalization is very readily seen in self–self hybridizations, in which identical mRNA samples are labeled with different dyes and cohybridized. In this case, there is no differential expression, but the red and green intensities are not equal. They also tend to differ by different amounts across the intensity range as well as by location on the slide.[29,30,72]

The simplest type of normalization is *global normalization*, which gives the same adjustment to each spot on the array without regard to individual spot features or location. Common choices are global mean, global median, or total intensity normalization. For global mean (median) normalization, the mean (median) of log-ratios across all spots on a slide is subtracted from the log-ratio for each spot. The result is that the mean (median) of the spots on each array becomes 0. Total intensity normalization scales the measurements by dividing by the total intensity on the slide so that total slide intensity becomes 1. Global ANOVA methods have also been proposed.[73,74] Global normalization, though widely used, is typically a very poor choice as it ignores the intensity- and spatial-dependent dye biases apparent in a large majority of experiments.

More flexible methods take into account intensity-dependent dye bias, as well as spatial variation in intensities across the slide, generally using some form of *local regression*, usually *lowess* or *loess* (two different types of local regression methods whose names derive from "locally weighted scatterplot smoothing"[29,30,75–77]). Classical (global) regression draws a single line through all points of a data set; every point, no matter how far from the A value of interest, has some influence on the predicted M value for the given A. The idea behind local regression methods is that the points in a region near each particular A value should have a stronger say (more weight) in the predicted M, whereas points farther away receive lower weight. The predictions are made in a moving window of points around each A, then the curve connecting the individual predictions is smoothed to yield a final prediction curve. The normalized M values are obtained by subtracting the loess prediction for each A from the corresponding observed M.

The loess curve can be constructed using all points on the array, under the assumption that either most genes are not differentially expressed or differential expression is equally likely to be up or down. For whole genome arrays, it is quite likely that only a small proportion of genes will be differentially expressed. For arrays with restricted coverage, neither assumption seems very realistic, as genes are generally selected for these arrays because they are expected to change; and changes may well be biased in one direction. An alternative to using all genes is to normalize on the basis of a subset of genes, for example, control sequences spotted for the purpose of normalization or a set of constantly expressed so-called "housekeeping" genes. One problem is that there is less information in a smaller number of genes, so the normalizing transformation will not be as stable. A different problem is with the concept of housekeeping genes, as there does not appear to be a set of constantly expressed genes. These are further reasons why restricted coverage arrays remain problematic.

Fitting separate loess curves within print-tip groups is very effective in correcting for both spatial- and intensity-dependent dye bias. The loess curves in the *MA* plots of Fig. 1.1(a) and (b) correspond to four different print tips. The curves are not all the same [Fig. 1.1(a)], indicating the need for print-tip normalization. After normalization, the curves are all roughly equal [Fig. 1.1(b)]. Figure 1.2(b) shows boxplots after print-tip (and scale, see later) normalization of the array from Fig. 1.2(a). For print-tip normalization, though, the assumptions are stronger as they need to hold not just across the entire array, but also *within* each print-tip group. Care must therefore be taken in applying this type of normalization; for example, if related genes are spotted together by the same print tip, the assumption is unlikely to hold. Fortunately, this type of purposeful spotting does not seem to be the norm, at least not for many whole genome arrays.

The procedures described earlier are forms of *location normalization*, which adjust observed log-ratios effectively by recentering them (subtraction). *Scale normalization* procedures adjust by rescaling observed log-ratios (division). Scale normalization is appropriate when the spread of log-ratios varies greatly between slides which are to be normalized together; it is straightforward to detect this type of problem with a boxplot.[29,30]

It is best to avoid excessive or unnecessary normalization, as this can make finding true biological differences more difficult. Insufficient normalization, on the other hand, makes it easier to find differences only due to artifacts. So how much normalization is the right amount? Specifying a standard operating procedure that should apply universally is not very realistic because different labs may experience different types of problems with their arrays. A good practice is to base normalization decisions on a thorough exploratory analysis, along with experience.

Normalization and Measuring Expression for Affymetrix GeneChips

An early expression measure for Affymetrix GeneChips was the average difference intensity (ADI), the average of PM − MM probes within a probe set (excluding outliers[78]). This expression measure is seriously defective, though, because the MM intensity is larger than the PM for around one-third of probes, yielding an illogical negative "signal". At present, the most commonly used measure of expression is that of MAS 5.0,[79] which addresses the problem by adjusting MM values exceeding the corresponding PM. Values are then scale normalized to a user-specified average value, generally set to be between 100 and 500.

Defining an appropriate measure of expression and normalization procedure for Affymetrix GeneChips is at present an area of intense research. Among the methods that have been proposed, the two most commonly used are the Li−Wong model-based expression index MBEI,[54,55,80] implemented in the software dChip; and a robust multichip (also called multiarray) average RMA,[56,57,81] implemented in several types of software: the affy package of Bio-Conductor; RMA Express, a freely available standalone GUI;[82] and also as part of some commercial software. The Li−Wong MBEI is based on a model of PM−MM that takes into account probe effects. RMA consists of three steps: a background adjustment on the original measurement scale followed by a log base 2 transformation, quantile normalization, and a summary statistic of the probe level data. In the present version of RMA only the PM probes are used, although incorporation of the information in the MM probes, along with other improvements, has been recently implemented as GC-RMA.[82a]

Quantile normalization is based on the assumption that the distributions of (background-adjusted) intensities should be the same on each chip; that is, that the boxplots of intensities should "line up"—be centered at the same place and of the same size, as in Fig. 1.2(b). It might be thought of as an extension of matching the median on every chip (see Bolstad et al.[83] for technical details). There has been discussion over whether it is preferable to normalize all chips in an experiment together or instead to separately normalize groups of chips corresponding to different treatments. There is a case to be made either way, but there is some consensus that all chip normalization is safer than separate group normalization because separate group normalization does not so easily reduce nonbiological variation and may in fact introduce artifactual differences between groups. Thus, although true differences between treatments may be more difficult to discover with all chip quantile normalization, one has more confidence in differences that do appear.

Recent studies have examined the performance of the MAS 5.0 signal, the Li–Wong MBEI, and RMA using data from specially designed test experiments; RMA was found to outperform the other methods.[56,83] Another study currently in progress also gives preliminary indications that RMA outperforms these other two in matching results of quantitative real-time PCR validation of selected findings.[84] Though not as well known now as the MAS 5.0 signal measure, the number of presentations at a recent Affymetrix User Group meeting (Amsterdam, June 2003) using RMA indicates that it is being applied by an increasing number of groups worldwide. Its superior performance and ready availability make it the method of choice for quantifying gene expression for Affymetrix GeneChips.

IDENTIFYING DIFFERENTIALLY EXPRESSED GENES

A very common goal of microarray experiments is to identify genes that are differentially expressed in two or more conditions. For example, which genes are expressed differently in lean and obese individuals or between diabetic and nondiabetic individuals? Although the question seems simple, there is not a unique statistical way to address it. Often it is desired to rank genes based on some statistic, and then to set a threshold for differential expression. Practical considerations play a role here: typically only a small number of genes, generally 10–50, will be considered for follow up or further study, which effectively translates into a threshold.

Statistics to Measure Differential Expression

Perhaps the most readily interpretable measure of differential expression is given by the fold change (ratio) in expression between two types of samples. It is more convenient to look at the average log base 2 fold change M. In this way, the lower end of the intensity scale becomes less compressed, and the measure treats overexpression or underexpression of a gene between two samples symmetrically around 0. For example, a gene which is expressed $4\times$ as much in sample A as in sample B is at the same time expressed one-fourth as much in sample B as in sample A. By taking logs, overexpression is positive, underexpression is negative, and the distance is the same in either direction. In this example, M between A and B is $+2$, whereas M between B and A is -2. It is simple to obtain fold change from M: fold change $= 2^M$.

The measure M has the shortcoming of not taking into account differing variability of different genes. The variability of M is not the same across the intensity range, and in particular, genes with larger variance across arrays are likely to produce large values of M even when they are not truly differentially expressed. Thus, setting a threshold based only on a fold change of 2 (corresponding to $M = 1$), say, is inadequate for identifying differential expression.

This deficiency of M has long been recognized (see, e.g., Berger et al.[5] and Callow et al.[8]). A way to deal with differing variability is by standardization, so that the difference in average expression is quantified with a t-statistic.[8,85] With this measure, differentially expressed genes are identified as those having a "sufficiently large" value in magnitude; that is, a t of $+3$ is as large as a t of -3. The familiar ANOVA F-statistic can be thought of as an extension to the t-statistic that deals with three or more groups.

However, a problem here is that the t-statistic performs very poorly with the small numbers of replicates found in typical microarray studies. Standard deviations (SD) in the denominator can therefore be unrealistic. In the case where the SD is extremely large, a truly differentially expressed gene is missed because of a t (or F) that is too small. On the other hand, an extremely small sample SD gives rise to large values of t (or F), even though the associated fold change may be small. In most circumstances, very small fold changes are unlikely to be considered as biologically meaningful, making the t-statistic on its own an unsatisfactory measure.

These two measures, M and t, provide one illustration of the difference between practical or biological importance and statistical significance. Large values of M may be important from a biological point of view, whereas large values of t are associated with statistical significance. This tradeoff can be visualized with a so-called "volcano plot", which plots the value of a test statistic, such as t (or its nominal p-value), against the log fold change M. Some combination of M and t could better advance biological knowledge while using resources efficiently.

There are several ways to deal with the problem of too small a denominator for t. One method is to add a "penalty" term to the individual gene SD or variance in the denominator, such as the 90th percentile of the individual gene SDs.[86] Other methods aim to pool information across genes, either empirically or using a statistical model, thereby stabilizing resulting estimates. For example, one simple but effective empirical approach computes a local mean and SD in a sliding window around each data point in the MA plot, pooling points with similar average intensity. An intensity-dependent Z-score for each point is obtained by subtracting the corresponding local mean and dividing by the local SD.[87]

Bayesian or empirical Bayes methods use data from *all* genes to improve estimation of differential expression for *single* genes.[36,84,88,88a] One example is to replace the single gene SDs in the denominator of t with a mix of the single gene SD and the overall SD (or alternatively a local SD). These mixed SDs are biased estimates of the single gene SDs, but with careful choice of the mixing factors the gain in precision is large enough to compensate for the small amount of bias typically incurred. Although the details involved with these approaches can be complicated, the general idea behind them (as described earlier) is straightforward. Software implementing many of the procedures is available from the BioConductor project, making these methods widely accessible.

With such a large number of possibilities, choosing one to implement seems a daunting task. There are a few comparative studies;[36,89] it is only a slight simplification to conclude that it may not make a very big difference exactly which type of modification to t is made, but that doing some modification clearly appears to be necessary.

The t-statistic assumes an underlying normal distribution for the data; the validity of this assumption will be hard to assess. One way around the assumption of normality is to use a nonparametric approach; see Troyanskaya et al.[90] for a comparison study. One type of nonparametric approach is to use a test based on rank, for example, a Wilcoxon (Mann–Whitney) test; however, such tests typically have low power. Another type of nonparametric alternative is a permutation test.[91] Permutation tests provide a means to assess significance of results. The test statistic is compared with a null distribution generated by first principles from the data, rather than by the conventional method of assuming a particular distributional form of the null such as a normal or t distribution. When distributional assumptions are likely to be far wrong, permutation-based p-values should be used (see "Multiple hypothesis testing and adjusted p-values" for more details on p-values).

An informal, graphical way of identifying differentially expressed genes is with a quantile–quantile plot (QQ plot). A QQ plot is a very useful visual tool for assessing whether the distribution for sample data follows a particular known distribution.[65] The most common example is checking whether data can be assumed to follow a normal distribution, although the method applies equally well to any distribution of interest.

To understand how to interpret a QQ plot, it is instructive to consider how one would go about making one by hand. Here is the idea for a normal QQ plot— associated with each data point is a probability, or percentage of data points smaller than it. The (horizontal) x-axis plots the standard normal values (quantiles) corresponding to the same probabilities as those observed in the data; on the (vertical) y-axis are the original data values for those same probability points. For example, for the median of the data, or 50th percentile, the corresponding normal quantile is the value 0, because 50% of the probability under the normal curve is to the left of 0. If the observed distribution of the data exactly matched the normal distribution, then the points would fall exactly on a line.

In most applications, interest is not so much in testing whether the test statistics follows a particular distribution, but in using the QQ plot as a visual aid for identifying genes with "unusual" test statistics. Thus, QQ plots informally "correct" for the large number of comparisons (see also "Multiple hypothesis testing and adjusted p-values"), so that the points that deviate markedly from an otherwise linear relationship are those most likely to correspond to genes whose expression levels differ between groups. The QQ plot shown in Fig. 1.3 contains three such points (plotted with a "*"). We can see that the log-ratios are not normally distributed (there are systematic deviations from a straight line relationship), but are rather heavy-tailed. There are more log-ratios farther away from the mean than there would be in a normal distribution. Interestingly,

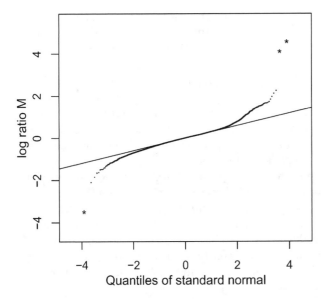

Figure 1.3 QQ normal plot of normalized log-ratios *M* for a self–self hybridization. Genes depicted as "*" correspond to outliers.

the plot is for log-ratios from a slide containing a self–self hybridization. That is, no gene is actually differentially expressed. Were we to declare the "*" points as differentially expressed, we would have three false positive results. This realization gives some perspective, and also gives a sense of the caution that must be exercised in interpretation.

Association of Expression Levels with Outcomes

The problem of identifying differentially expressed genes might also be thought of as a special case of the more general problem of detecting genes associated with an outcome of interest, such as change in weight or disease survival. The detection of gene subsets associated with outcomes can be viewed statistically as a *variable selection* problem. A standard approach to variable selection is to formulate a model connecting the variables to the response, then to search through subsets of variables until some criterion is optimized.[92] Here, however, the problem is that the number of available samples (arrays) is typically quite small, whereas the number of variables (genes) is huge. The large number of genes precludes consideration of all possible subsets of variables and restricts the class of models than can be considered. Techniques for dimensionality reduction such as principal component analysis do not improve the situation, as components consist of combinations of genes rather than individual genes; not being single genes, components are difficult to interpret in a biologically meaningful way.

In the microarray literature, this problem has tended to be addressed in an indirect and inefficient fashion (see, e.g., Bittner et al.[93]). A typical indirect approach involves first clustering the data (discussed later), deciding how many groups the data appear to contain, then to identify genes which appear to discriminate between these groups. The groups are then tested for association with covariates or responses; responses for which associations are found to be significant are then indirectly associated with the genes discriminating between the groups. In order to avoid well-known problems that can occur with clustering (discussed later), as well as to make more efficient use of limited and expensive microarray data, a direct approach is to be preferred.[94]

A number of methods have been proposed in the statistical and machine learning literature to deal this issue. Tree harvesting[95] and gene shaving[96] have been specifically proposed to deal with microarray data. The idea behind these is to identify interesting gene clusters by first applying a clustering algorithm, then averaging within cluster profiles, and finally choosing from these averages those to be included in a response. These methods still possess the drawbacks of clustering and, in addition, use forward variable selection, which tends to be unstable.[97]

Two new approaches appear to be very promising. *Least angle regression* (LARS),[98] related to the lasso selection algorithm,[99–101] has been applied to at least one microarray study.[102] The other approach is called *supervised clustering*, where genes are grouped into sets using knowledge of the outcome.[103] This method avoids problems with standard, "unsupervised" clustering, which avoids use of outcome information in forming data groups. Comparison studies of available methods, though, are few and to date it does not appear as though a clear winner has emerged.

MULTIPLE HYPOTHESIS TESTING AND ADJUSTED p-VALUES

The biological question of differential expression, or association of gene expression with outcome, can be restated statistically as a problem in *multiple hypothesis testing*—the simultaneous test for each gene (or many groups of genes) of the null hypothesis of no association between the expression level and the outcome, for example, no differential expression. As a typical microarray experiment measures expression levels for several thousands of genes simultaneously, we are faced with an extreme form of multiple testing when assessing the statistical significance of the results. For example, suppose that, on an array with several thousands of spots giving whole genome coverage, a particular gene shows a nominal p-value of 0.01. This finding cannot be characterized as "significant", as it is inevitable that such small p-values will occur by chance when considering a large set of genes, most of which will not be differentially expressed. Special problems arising from the multiplicity aspect include defining an appropriate false positive (type I error) rate for the *overall* experiment and devising powerful multiple testing procedures that control this error rate and account for the *joint* distribution of the gene expression levels. Recent developments in controlling the false discovery rate (FDR), rather than the more classically-oriented family-wise error rate (FWER), appear to provide a promising way to come up with meaningful significance measures among thousands of genes and groups of genes.[104,105]

When going from single to multiple hypothesis testing, several definitions of the false positive rate are possible. Two such definitions are the *family-wise error rate*, which is the probability of at least one false positive result in *all* tested hypotheses; and the *false discovery rate*, or expected proportion of false positive findings among the *rejected* hypotheses. In general, procedures controlling the FDR are typically less conservative than those controlling the FWER. The FDR may be viewed as an estimate of what proportion of effort will be wasted following up results that are false. One might be prepared to tolerate some false positives, as long as their number compared with the total number of positive results is not too high. This error definition seems highly suited to many microarray studies.

An important distinction in multiple testing is that between strong and weak control of the false positive rate. *Strong control* refers to control of the error rate under *any combination* of true and false hypotheses, that is, for any combination of differentially and constantly expressed genes. In contrast, *weak control* refers to control of the error rate only when *none* of the genes is differentially expressed (the "complete null hypothesis" that all individual null hypotheses are true). In the microarray setting, where it is very unlikely that none of the genes is differentially expressed, weak control is completely inadequate. Two early multiple testing methods proposed in the context of microarray experiments, significance analysis of microarrays (SAM)[106] and neighborhood analysis,[107] provide only weak false positive rate control,[108,109] and are therefore not preferred methods.

Rather than simply reporting rejection or not of the null hypothesis of no differential expression for a given gene, it is preferable to report a *p*-value. The interpretation of a *p*-value requires a little care. In particular, the *p*-value of a test statistic does not tell us the probability that the null hypothesis is true. Instead, the *p*-value represents the chance that we would see a difference as far from the null or farther, *if* there were really nothing happening other than chance variability. To put it another way, the *p*-value conveys an idea of how surprised we should be at the observed result if we really believe the null hypothesis. In the context of multiple testing, *adjusted p-values* provide useful and flexible summaries of the strength of the evidence of differential expression. The adjusted *p*-value for a particular gene reflects the overall false positive rate for the entire experiment when genes with smaller *p*-values are declared differentially expressed. This is in contrast to unadjusted *p*-values, which reflect significance only for each gene considered in isolation.

There has been some apparent hesitancy to correct for multiple testing, probably because of the fact that the best known such correction, the Bonferroni procedure, which provides strong FWER control, is overly conservative. It is an example of a *single-step* procedure; these provide the same adjustment to all *p*-values, regardless of the outcomes of the tests for other genes. There are more powerful, *sequential* adjustment procedures that also provide the same control for the same type of error, for example, max T and min P.[110] Sequential procedures give each p-value an individual adjustment, taking into account the outcomes of other tests, thereby achieving greater power (are less conservative) than single-step procedures. Software implementing adjustments for multiple hypothesis tests is widely available,[69,71,111] thus there should be no impediment to employing these procedures.

The performance of different multiple testing procedures has been examined using publicly available microarray and survival data,[112] as well as in other microarray contexts.[108,109] One general conclusion is that the FWER controlling procedures are probably too stringent in many microarray applications, so that FDR controlling procedures are generally preferred.

It is generally more difficult for an adjusted p-value to attain a low value than it is for an unadjusted one. To get around this, some have recommended *pre-screening* expression levels, for example, only testing those showing sufficient variation. This is an example of *data snooping*—looking at the data before deciding what to test. Unless the screening statistic is independent of the test statistic when the null is true, the nominal false positive rate will not be correct. In addition, any p-value for the test will be very difficult to interpret, as the decision to perform the test is based in part on the expectation of a finding—we have excluded from consideration genes whose expression is observed not to change across samples (i.e., are not differentially expressed). In practice, however, some type of gene filtering seems difficult to avoid.

There exists some controversy over whether multiple testing adjustments should be applied at all, and which tests to consider as part of the same experiment or "family" to which the adjustments will be applied. There are guidelines regarding when researchers should be concerned about inflated error rates.[110] Those relevant to microarray experiments include the case when a serious claim will be made whenever any (unadjusted) p-value is sufficiently small, much data manipulation may be performed to find a "significant" result, the analysis is planned to be *exploratory* but investigators wish to claim "significant" results are real, or the experiment is unlikely to be followed up before serious actions are taken. The safest course of action is to use an adjustment procedure to get a more realistic assessment of significance than that provided by the nominal, unadjusted p-value, and to carefully follow up results, for example, with an additional confirmatory assay or study, before taking actions that assume the truth of "significant" findings (e.g., before publishing a paper).

CLASSIFICATION PROBLEMS

Microarray experiments have revived interest in classification problems, by raising new methodological and computational challenges. There are two broad types of classification problems, with the term "classification" being applied by different authors to different problems, creating some confusion.

Here, we distinguish between the two problems by using different terminology for the two cases. In *discrimination* problems, often referred to in the machine learning literature as *supervised learning*, observations are known to belong to pre-specified classes, and the task is to build predictors for allocating new observations to these classes. In contrast, in *cluster analysis*, also called *unsupervised learning*, the classes are unknown *a priori* and the task is to determine these classes from the data. We give only brief coverage to these techniques; very good references in this area are available (see, e.g., Hastie et al.,[101] Everitt et al.,[113] and Hand[114]). Owing to its prominent use in the literature on microarray data analysis, cluster analysis is considered first.

Cluster Analysis

A seminal paper in microarray data analysis proposes hierarchical clustering of genes to identify patterns.[115] Clustering of samples may also be performed, as well as two-way clustering of genes and samples. Clustering methods have been used extensively and successfully in microarray experiments to organize and display very large genes-by-arrays data matrices.[115] Cluster analysis is commonly used to search for groups in data; it is most effective when the groups are not already known. It should be noted that there is difficulty even defining the term "cluster",[116] so it is not surprising that there are so many methods devised to search for them.

Cluster analysis seems much over-used in analyzing microarray data; it is one of the most common types of data analysis carried out, regardless of the specific questions under investigation. As demonstrated earlier (see also Goldstein et al.[94]), clustering does not address every interesting scientific question. It is thus better practice to allow the questions of interest to guide the choice of analysis tools and techniques. Unfortunately, in microarray studies the aims are often diverse, so that scientific questions may not be well formulated. This aspect should be addressed during the design phase of the experiment, where not only the design but also tentative analyses can be planned.

We provide here only a very brief introduction to a few of the widely used clustering techniques. Throughout, we assume that there are n *objects* to be clustered, with p *variables*, or features, measured for each. Depending on the aims of the analysis, an object could be either a sample, in which case the variables are gene expression measurements; or it could instead be a gene, then the variables are the samples.

Hierarchical Clustering

Hierarchical clustering is based on a dissimilarity measure for each pair of objects, however, there is no single best measure so that choice of dissimilarity requires some care. Often, a distance measure is used, one example being the Euclidean distance or the distance "as the crow flies". Some kind of scaling may be carried out so that larger measurements are treated similarly to smaller ones. Alternatively, the dissimilarity may be taken to be one minus some measure of association. For microarray studies, the dissimilarity used is typically one minus the correlation coefficient ρ, however, this choice is not without problems.

A long recognized deficiency of ρ is that objects that are more linearly related will be more "similar", although they may not seem to be "closer". As an example, consider three objects, with four measurements on each. All measurements for object 2 are double those of object 1. The measurements for object 3 are the same as those of object 1 on the first three variables, whereas the fourth variable for object 3 is the fourth value for object 1 plus a small amount say, 0.1. Then the correlation between objects 1 and 2 is perfect, whereas the correlation between objects 1 and 3 is not, although it seems intuitive that objects 1 and 3 are more similar.

Hierarchical clustering methods fall into two classes: agglomerative nesting methods and divisive analysis methods. *Agglomerative methods* proceed in the same general manner:

1. At the start, there are n clusters, with each single object belonging to a separate cluster.
2. Find the closest two objects and merge them into one cluster, so that there are now $n - 1$ clusters, one of which contains two objects.
3. Update the dissimilarity matrix (see later) to take into account the merging that has occurred.
4. Find the two closest distinct clusters and merge them into one cluster.
5. Repeat steps 3 and 4 until there remains one cluster consisting of all n objects.

There are many ways to update the dissimilarity matrix, the main issue is how to define a distance between two *clusters* of objects (as opposed to dissimilarity between *individual* objects, as defined earlier). A method commonly used in microarray data analysis is *average linkage clustering*, in which the distance between two clusters is the average of all pairwise distances between two objects, one from the first cluster and the other from the second. In *complete linkage clustering*, the distance between clusters is the maximum of all possible pairwise object distances; at the other extreme, for *single linkage clustering* the distance between clusters is the minimum of the pairwise object distances. A statistically appealing method is Ward's hierarchical clustering, which associates a measure of information with each cluster and merges the two for which doing so entails the least information loss (see Hartigan;[117] for an example application with microarrays see Diaz et al.[23]).

The opposite of agglomerative nesting is a *divisive analysis* approach. Heuristically, these algorithms proceed as follows:

1. All n objects are part of a single cluster.
2. The object that has the greatest dissimilarity to the other objects is then separated to form a "splinter group".

3. The remaining elements in the original cluster are examined to determine whether additional elements should be added to the splinter group. There are now two clusters.

4. For each cluster, the diameter, or largest distance between objects within the cluster, is computed. Steps 2 and 3 are repeated with the cluster having the greatest diameter until there are n clusters, each containing only a single object.

Clustering can be quite useful for data visualization. Hierarchical methods generate a sequence of cluster assignments for each object that can be visually represented as a *dendrogram* or tree diagram. Another tool is the *heat map*.[115] This is a false color image, typically with dendrograms added to the side and top; the colors represent gene expression measures. A heat map allows visualization of patterns, or profiles, of gene expression.

Hierarchical methods, particularly agglomerative ones, can be computationally quite fast. A major drawback, however, is that group assignments of objects cannot change once the object has been placed in (or removed from) a cluster. These methods cannot undo what has been done in previous steps. In contrast, methods based on partitioning can reconsider previous assignments so that improvements can be made.

Partition Clustering

Partition methods are nonhierarchical in nature. The general problem is to find a set of K cluster centers and an allocation into the K groups that optimizes some criterion measuring the "goodness" of the clustering. Several criteria have been proposed, the most commonly used ones derive from considering three (matrix) quantities: the total sum of squared deviations from the overall mean, the total within group sum of square deviations around group means, and the total of squared deviations between group means and the overall mean. An example of a criterion with only a single variable measured on each object is (minimum) within group sum of squares. Many possibilities exist for multivariate extensions, each with different properties.[113]

Examples of partitioning algorithms include the familiar K-means and its many variants.[101] Starting from a set of K cluster centers, the K-means procedure iteratively moves the centers until the total within cluster variance is minimized. Initial centers are usually randomly chosen objects, but resulting centers are combinations of objects rather than objects. Self-organizing maps (SOM) can be viewed as a version of K-means.[118] Another algorithm, partitioning around medoids (PAM) uses objects as centers at each stage.[119]

One difficulty with the use of partitioning methods is the large number of possible groupings of the data. The number of possible groupings very quickly becomes too large to "brute force" evaluate each possible grouping. For example, the number of possible partitions of 20 objects into four groups is over 45 billion. Thus, the development of optimization algorithms has been an important area.

Use of partitioning methods requires initial specification of the number of groups K, yet in practice this number will not be known. The usual way to deal with this problem is to try out several values of K, and then to somehow test or estimate its value. The task is not trivial, and very many ways have been proposed to determine K, each looking at different aspects of the problem.

There are still open problems in clustering. The estimation of the number of clusters, especially as it pertains to microarray data, is at present an area of intense research. One study examining several procedures shows the promise of a resampling-based approach.[120] Methods of assessing cluster strength and reproducibility are also beginning to be developed.[121–123] Methodological research is continuing in this area.

Other Clustering Approaches

Although most clustering algorithms fall into the categories discussed earlier, there are other approaches not included. Those which have been used in microarray applications include fuzzy clustering, which attributes cluster membership values to objects rather than giving an unequivocal assignment to a single group,[124] model-based clustering, where each group has an assumed probability distribution,[125,126] and hybrid methods attempting to blend the speed of hierarchical clustering with the reassignment possibilities of partitioning.[127]

In summary, no clustering algorithm performs best under all conditions. The performance of different algorithms depends on many factors, most of which will be typically unknown by the investigator. This aspect makes it impossible to make a single recommendation. Thus, if clustering is desired, specific implementation decisions must be based on the particular purpose at hand. Unlike a theory, a clustering result is neither true nor false—it cannot be proved or disproved; the results should therefore be judged according to their usefulness.[113] It is also crucial to bear in mind that different algorithms will give varying results, so there is no reason that the groups resulting from application of a clustering algorithm will necessarily correspond to something biologically meaningful. The biological knowledge and insight that a researcher brings to the problem therefore plays a major role in evaluating analysis results.

Discrimination

The ability to successfully distinguish between classes using gene expression data is an important aspect of genomic approaches to biology. In discrimination problems, the groups are assumed to be known for an initial *learning set* (also called *training set*) and the aim is to design a rule to allocate new *test samples* to these groups. Because the learning samples are assumed to be classified with certainty, one might wonder why not classify test samples using the same procedure. In practice, it may not be feasible or cost effective; for example, if the classification procedure is destructive. In medical settings, some conditions can only be determined by post-mortem examination; this is not much help to a living patient awaiting diagnosis and appropriate treatment. Thus, other rules for class discrimination are required. There are several discrimination methods available; here we highlight only those with highest relevance in the microarray arena.

Class Prediction Rules

An approach to the discrimination problem for two groups is to find the *linear combination* of the gene expression profiles, that is, a sum of constant numbers multiplied by gene expression values, which separates the two groups as much as possible. The predicted class of an observation X is the class whose mean vector is closest to X in terms of this *linear discriminant function*. This rule has good properties for groups whose probability distributions are (multivariate) normal, where the covariance relationships between expressions of different genes are the same for the two groups. When the covariance relationships are different, the prediction rule is in general quadratic rather than linear; that is, the rule involves squared values of gene expression. These approaches can also be generalized to handle more than two groups.

Nearest neighbor rules are based on a measure of distance between observations. The K-nearest neighbor rule (KNN) predicts the class of an object X to be the most common class among the K observations in the learning set closest to X. The number of neighbors is often chosen in an *ad hoc* fashion, but can be more carefully chosen by cross-validation (defined later).

Binary tree structured predictors, such as classification and regression trees (CART[128]), are constructed by repeated splits of subsets (nodes) of the space of gene expression profiles. Each terminal subset is assigned a predicted value. There are three main aspects to tree construction: the selection of the splits, the decision to declare a node terminal (sometimes called "pruning"), and assigning a predicted value to each terminal node. Trees can be used for prediction of continuous, categorical, or censored survival outcomes. Trees are appealing because they can give insight into the structure of gene expression data, including interactions between genes. They also tolerate the problem of missing data. The other classifiers mentioned earlier do not have these capabilities.

Aggregating Predictors

It is well known that trees are unstable predictors, that is, small changes in the learning set often result in large changes in the predictor. Substantial gains in accuracy can be obtained by creating multiple trees from the learning set, then allowing them to vote for the class assignment, assigning to an observation the most popular class among the resulting predictors.

There are many ways to generate multiple trees, possibilities include bagging and boosting. In the simplest form of *bagging* (a contraction of bootstrap aggregating), new trees are created from observations drawn at random with replacement from the original observations of the learning set.[97] In *boosting*, new predictors are created from adaptively resampled data, with increased weight for those cases most often misclassified.[129] Weighted voting, rather than equally likely voting, aggregates the predictors to decide on a class assignment for observations in the test set. *Logit boosting* gives a more robust rule in multicategory (more than two groups) classification.[130]

Performance Assessment

Once a discrimination, or class assignment, rule is in place, its performance must be evaluated. One obvious approach to this problem is to compute the resubstitution error rate—the misclassification rate on the original learning data set from which the discrimination rule was determined. This turns out to be a very poor way to assess performance, as it gives an overly optimistic view of how well the rule will perform on new observations.

An alternative is to estimate the error in a test set. If there is not a test set available, one could still use this approach by dividing the original data set into separate learning and test sets and using only the learning set to create a rule. A problem here, though, is that the effective sample size is reduced, as there is only the learning set available for determining the rule.

To make efficient use of scarce data while at the same time obtaining realistic misclassification error rate estimates, data resampling methods are very useful. Two important types of resampling methods in this context are cross-validation and out-of-bag error rate estimation. In (V-fold) *cross-validation*, the data are randomly divided into V subsets of (nearly) equal size (e.g., 5 or 10 subsets). A classifier is created by leaving one set out as a test set and using the remaining data for the learning set. The test set error rate is computed on left out set; these are then averaged to get an overall estimate of error. For example, in the case of five-fold cross-validation, one would first use subsets 1–4 to create a rule, and then test on subset 5 to get one estimate of the error rate. Next, subset 4 would be left out of the learning set, so that the rule on this turn would be determined from subsets 1–3 and 5 and then tested on subset 4. The entire procedure is carried out like this, each time leaving a different subset out of the learning set until each subset has had a turn being left out. A common version of this procedure is *leave one out cross-validation*, where each observation serves as a subset. *Out-of-bag error rate estimation* uses a similar idea but with a bagged classifier (see bagging discussed earlier), but here the test set consists of those observations not randomly selected to form the classifier.

Caution must be observed in applying these procedures; in order for them to work properly (e.g., give realistic error rates), the details are extremely important. For example, consider the common situation of gene prefiltering to reduce the number of possible genes for building the classifier. It has been common, but completely wrong, to use the entire data set to select candidate genes, followed by cross-validation only for model building and classification. However, usually which genes are relevant to the classification are unknown and the intended inference includes this gene selection. The genes should be selected only from the learning set used to build the classifier, and not the entire original set, in order to avoid bias.[131]

A recent comparison of discrimination methods for three cancer gene expression studies found that, for the data sets considered, simple classifiers such as diagonal linear discriminant analysis and nearest neighbors showed good performance.[132] The sample sizes of the data sets, around 60–70 arrays, are not sufficiently large for tree-based methods to work well. However, as larger clinical studies are currently underway, tree classifiers are proving to be of more widespread applicability.

SUMMARY AND PROSPECTS FOR THE FUTURE

Microarray technologies hold exceptional scientific promise, but experience is showing that gene expression experiments pose new challenges to the traditional practice of science. There are many steps to the completion of a microarray study, with ultimate success depending on avoiding serious flaws in any of them. As these steps require expertise from vastly divergent fields, microarray research requires a substantial team effort that brings together the eminently qualitative culture of biochemical experimentation with the quantitatively oriented culture of statisticians.

When different cultures meet, there is the need to develop a common language and to define common goals. It is with the purpose of contributing to these aims that we have tried to give an overview on the broadly relevant aspects for which interactions between biologists and statisticians are paramount. We have not mentioned methods that aim at very specific goals, for example, the identification of new biochemical pathways or the prediction of regulative interactions.

Where possible, we have provided recommendations or guidelines. However, appropriate data analysis is ideally performed as a collaborative effort between biological investigators, whose expertise extends to the substantive scientific questions and the characteristics of the experimental procedures, and statisticians with specialized knowledge of methods that could be applied. Even in a collaborative environment, it is still essential that microarray researchers grasp the logic underlying different methods in order to participate in rational data analysis decisions, as well as to evaluate analyses carried out by others.

In conclusion, microarray analysis is a very active, highly specialized area of research that is still in its infancy and undergoing brisk technological development. As additional experimental data and methodological comparisons become available, more definitive and well-founded guidelines are likely to emerge.

ACKNOWLEDGMENT

This work was supported by funds from the NCCR Molecular Oncology of the Swiss National Science Foundation.

REFERENCES

1. Nadler, S.T.; Attie, A.D. Please pass the chips: genomic insights into obesity and diabetes. J. Nutr. **2001**, *131*, 2078–2081.
2. Wells, J.M. Genes expressed in the developing endocrine pancreas and their importance for stem cell and diabetes research. Diabetes Metab. Res. Rev. **2003**, *19*, 191–201.
3. Cheek, D.J.; Cesan, A. Genetic predictors of cardiovascular disease: the use of chip technology. J. Cardiovasc. Nurs. **2003**, *18*, 50–56.
4. Eaves, I.A.; Wicker, L.S.; Ghandour, G.; Lyons, P.A.; Peterson, L.B.; Todd, J.A.; Glynne, R.J. Combining mouse congenic strains and microarray gene expression analyses to study a complex trait: the NOD model of type 1 diabetes. Genome Res. **2002**, *12*, 232–243.
5. Berger, A.; Mutch, D.M.; German, J.; Roberts, M.A. Dietary effects of arachidonate-rich fungal oil and fish oil on murine hepatic and hippocampal gene expression. Lipids Health Dis. **2002**, *1*, 2.
6. Narayanan, B.A.; Narayanan, N.K.; Simi, B.; Reddy, B.S. Modulation of inducible nitric oxide synthase and related proinflammatory genes by the omega-3 fatty acid docosahexaenoic acid in human colon cancer cells. Cancer Res. **2003**, *63*, 972–979.
7. Yang, X.; Pratley, R.E.; Tokraks, S.; Bogardus, C.; Permana, P.A. Microarray profiling of skeletal muscle tissues from equally obese, non-diabetic insulin-sensitive and insulin-resistant Pima Indians. Diabetologia **2002**, *45*, 1584–1593.
8. Callow, M.J.; Dudoit, S.; Gong, E.L.; Speed, T.P.; Rubin, E.M. Microarray expression profiling identifies genes with altered expression in HDL-deficient mice. Genome Res. **2000**, *10*, 2022–2029.
9. Kim, S.; Urs, S.; Massiera, F.; Wortmann, P.; Joshi, R.; Heo, Y.R.; Andersen, B.; Kobayashi, H.; Teboul, M.; Ailhaud, G.; Quignard-Boulange, A.; Fukamizu, A.; Jones, B.H.; Kim, J.H.; Moustaid-Moussa, N. Targeted expression to adipose tissue on lipid metabolism and renal gene expression. Horm. Metab. Res. **2002**, *3*, 721–725.
10. DeRisi, J.L.; Iyer, V.R.; Brown, P.O. Exploring the metabolic and genetic control of gene expression on a genomic scale. Science **1997**, *278*, 680–685.
11. Brown, P.O.; Botstein, D. Exploring the new world of the genome with {DNA} microarrays. Nat. Genet. **1999**, *21* (suppl), 33–37 (in The Chipping Forecast).
12. Lockhart, D.J.; Dong, H.; Byrne, M.C.; Follettie, M.T.; Gallo, M.V.; Chee, M.S.; Mittmann, M.; Wang, C.; Kobayashi, M.; Horton, H.; Brown, E.L. Expression monitoring by hybridization to high-density oligonucleotide arrays. Nat. Biotechnol. **1996**, *14*, 1675–1680.
13. The Chipping Forecast. Nat. Genet. **1999**, *21* (suppl).
14. Maindonald, J.H.; Pittelkow, Y.E.; Wilson, S.R. Some considerations for the design of microarray experiments. In *Science and Statistics: A Festschrift for Terry Speed*; Goldstein, D.R., Ed.; Institute for Mathematical Statistics: Beachwood, OH, 2003; 367–390.
15. Cobb, G.W. *Introduction to Design and Analysis of Experiments*; Springer: New York, 1998.
16. Quinn, G.P.; Keough, M.J. *Experimental Design and Data Analysis for Biologists*; Cambridge University Press: Cambridge, UK, 2002.

17. Mead, R. *The Design of Experiments: Statistical Principles for Practical Applications*, 1st paperback Ed.; Cambridge University Press: Cambridge, UK, 1990.
18. DeRisi, J.; Penland, L.; Brown, P.O.; Bittner, M.L.; Meltzer, P.S.; Ray, M.; Chen, Y.; Su, Y.A.; Trent, J.M. Use of a cDNA microarray to analyse gene expression patterns in human cancer. Nat. Genet. **1996**, *14*, 457–460.
19. Spellman, P.T.; Sherlock, G.; Zhang, M.Q.; Iyer, V.R.; Anders, K.; Eisen, M.B.; Brown, P.O.; Botstein, D.; Futche, B. Comprehensive identification of cell cycle-regulated genes of the yeast *Saccharomyces cerevisiae* by microarray hybridization. Mol. Biol. Cell **1998**, *9*, 3273–3297.
20. Yang, Y.H.; Speed, T. Design issues for cDNA microarray experiments. Nat. Rev. Genet. **2002**, *3*, 579–588.
21. Yang, Y.H.; Speed, T.P. Design and analysis of comparative microarray experiments. In *Statistical Analysis of Gene Expression Microarray Data*; Speed, T., Ed.; Chapman-Hall/CRC: Boca Raton, FL, 2003; 1–34.
22. Jin, W.; Riley, R.M.; Wolfinger, R.D.; White, K.P.; Passador-Gurgel, G.; Gibson, G. The contributions of sex, genotype and age to transcriptional variance in *Drosophila melanogaster*. Nat. Genet. **2001**, *29*, 389–395.
23. Diaz, E.; Yang, Y.H.; Ferreira, T.; Loh, K.C.; Okazaki, Y.; Hayashizaki, Y.; Tessier-Lavigne, M.; Speed, T.P.; Ngai, J. Analysis of gene expression in the developing mouse retina. Proc. Natl. Acad. Sci. USA **2003**, *100*, 5491–5496.
24. Smyth, G.K.; Yang, Y.H.; Speed, T. Statistical issues in cDNA microarray data analysis. Methods Mol. Biol. **2003**, *224*, 111–136.
25. Glonek, G.F.V.; Solomon, P.J. *Factorial Designs for Microarray Experiments*, Technical Report; Department of Applied Mathematics, University of Adelaide: Australia, 2002.
26. Kerr, M.K.; Churchill, G.A. Experimental design for gene expression microarrays. Biostatistics **2001**, *2*, 183–201.
27. Kerr, M.K.; Churchill, C.A. Statistical design and the analysis of gene expression microarray data. Genet. Res. **2001**, *77*, 123–128.
28. Dobbin, K.; Simon, R. Comparison of microarray designs for class comparison and class discovery. Bioinformatics **2002**, *18*, 1438–1445.
29. Yang, Y.H.; Dudoit, S.; Luu, P.; Speed, T.P. Normalization for cDNA microarray data. In *Microarrays: Optical Technologies and Informatics*; Bittner, M.L., Chen, Y., Dorsel, A.N., Dougherty, E.R., Eds.; Proceedings of SPIE, v. 4266, 2001.
30. Yang, Y.H.; Dudoit, S.; Luu, P.; Lin, D.M.; Peng, V.; Ngai, J.; Speed, T.P. Normalization for cDNA microarray data: a robust composite method addressing single and multiple slide systematic variation. Nucl. Acids Res. **2002**, *30* (4), e15.
31. Tseng, G.C.; Oh, M.K.; Rohlin, L.; Liao, J.; Wong, W.H. Issues in cDNA microarray analysis: quality filtering, channel normalization, models of variations and assessment of gene effects. Nucl. Acids Res. **2001**, *29*, 2549–2557.
32. Dobbin, K.; Shih, J.H.; Simon, R. Statistical design of reverse dye microarrays. Bioinformatics **2003**, *19*, 803–810.
33. Black, M.A.; Doerge, R.W. Calculation of the minimum number of replicate spots required for detection of significant gene expression fold change in microarray experiments. Bioinformatics **2002**, *18*, 1609–1616.

34. Lee, M.L.; Kuo, F.C.; Whitmore, G.A.; Sklar, J. Importance of replication in micro-array gene expression studies: statistical methods and evidence from repetitive cDNA hybridizations. Proc. Natl. Acad. Sci. USA **2000**, *97*, 9834–9839.

35. Dudoit, S.; Yang, Y.H.; Speed, T.P.; Callow, M.J. Statistical methods for identifying differentially expressed genes in replicated cDNA microarray experiments. Stat. Sinica **2002**, *12*, 111–139.

36. Lönnstedt, I.; Speed, T.P. Replicated microarray data. Stat. Sinica **2002**, *12*, 31–46.

37. Lee, M.L.; Whitmore, G.A. Power and sample size for DNA microarray studies. Stat. Med. **2002**, *21*, 3543–3570.

38. Hwang, D.; Schmitt, W.A.; Stephanopoulos, G. Determination of minimum sample size and discriminatory expression patterns in microarray data. Bioinformatics **2002**, *18*, 1184–1193.

39. Zuker, M. Calculating nucleic acid secondary structure. Curr. Opin. Struct. Biol. **2000**, *10*, 303–310.

40. Novak, J.P.; Sladek, R.; Hudson, T.J. Characterization of variability in large-scale gene expression data: implications for study design. Genomics **2002**, *79*, 104–113.

41. Bakay, M.; Chen, Y.W.; Borup, R.; Zhao, P.; Nagaraju, K.; Hoffman, E.P. Sources of variability and effect of experimental approach on expression profiling data interpretation. BMC Bioinformatics **2002**, *3*, 4.

42. Peng, X.; Wood, C.L.; Blalock, E.M.; Chen, K.; Landfield, P.W.; Stromberg, A.J. Statistical implications of pooling RNA samples for microarray experiments. BMC Bioinformatics **2003**, *4*, 26.

43. Kendziorski, C.M.; Zhang, Y.; Lan, H.; Attie, A.D. The efficiency of pooling mRNA in microarray experiments. Biostatistics **2003**, *4*, 465–477.

44. Eisen, M.B. ScanAlyze User Manual, 1999. http://rana.lbl.gov/software/

45. Axon Instruments Inc. *GenePix 4000A User's Guide*; Axon Instruments: Union City, CA, 1999.

46. Buhler, J.; Ideker, T.; Haynor, D. *Dapple: Improved Techniques for Finding Spots on DNA Microarrays*, Technical Report; Department of Computer Science and Engineering, University of Washington, 2000.

47. Beucher, S.; Meyer, F. The morphological approach to segmentation: the watershed transformation. In *Mathematical Morphology in Image Processing*; Dougherty, E., Ed.; Marcel Dekker: New York, 1993; 433–481.

48. Vincent, L.; Soille, P. Watersheds in digital spaces: an efficient algorithm based on immersion simulations. IEEE Trans. Pat. Anal. Machine Intel. **1991**, *13*, 583–598.

49. Soille, P. *Morphological Image Analysis: Principles and Applications*; Springer: New York, 1999.

50. Adams, R.; Bischof, L. Seeded region growing. IEEE Trans. Pat. Anal. Machine Intel. **1994**, *16*, 641–647.

51. Buckley, M.J. *The Spot User's Guide*; CSIRO Mathematical and Information Sciences: North Ryde, NSW 1670, Australia, 2000.

52. Chen, Y.; Dougherty, E.R.; Bittner, M.L. Ratio-based decisions and the quantitative analysis of cDNA microarray images. J. Biomed. Optics **1997**, *2*, 364–374.

53. *QuantArray Analysis Software, Operator's Manual*; GSI Lumonics, 1999.

54. Li, C.; Wong, W.H. Model-based analysis of oligonucleotide arrays: expression index computation and outlier detection. Proc. Natl Acad. Sci. USA **2001**, *98*, 31–36.

55. Li, C.; Wong, W.H. Model-based analysis of oligonucleotide arrays: model validation, design issues and standard error application. Genome Biol. **2001**, *2* (8), research0032.1–0032.11.

56. Irizarry, R.A.; Bolstad, B.M.; Collin, F.; Cope, L.M.; Hobbs, B.; Speed, T.P. Summaries of Affymetrix GeneChip probe level data. Nucl. Acids Res. **2003**, *31* (4), e15.

57. Irizarry, R.A. Measures of expression for Affymetrix high density oligonucleotide arrays. In *Science and Statistics: A Festschrift for Terry Speed*; Goldstein, D.R., Ed.; Institute for Mathematical Statistics: Beachwood, OH, 2003; 391–402.

58. Naef, F.; Socci, N.D.; Magnasco, M. A study of accuracy and precision in oligonucleotide arrays: extracting more signal at large concentrations. Bioinformatics **2003**, *19*, 178–184.

59. Glasbey, C.A.; Ghazal, P. Combinatorial image analysis of DNA microarray features. Bioinformatics **2003**, *19*, 194–203.

60. Yang, Y.H.; Buckley, M.J.; Dudoit, S.; Speed, T.P. Comparison of methods for image analysis on cDNA microarray data. J. Comp. Graph Stat. **2002**, *11*, 108–136.

61. Nadon, R.; Shoemaker, J. Statistical issues with microarrays: processing and analysis. Trends Genet. **2002**, *18*, 265–271.

62. Yang, M.C.; Ruan, Q.G.; Yang, J.J.; Eckenrode, S.; Wu, S.; McIndoe, R.A.; She, J.X. A statistical method for flagging weak spots improves normalization and ratio estimates in microarrays. Physiol. Genomics **2001**, *7*, 45–53.

63. Wang, X.; Ghosh, S.; Guo, S.W. Quantitative quality control in microarray image processing and data acquisition. Nucl. Acids Res. **2001**, *29*, E75–75.

64. Tukey, J.W. *Exploratory Data Analysis*; Addison-Wesley: Reading, MA, 1977.

65. Cleveland, W.S. *The Elements of Graphing Data*; Wadsworth: Monterey, CA, 1985.

66. Cleveland, W.S. *Visualizing Data*; Hobart: Summit, NJ, 1993.

67. Dudoit, S.; Yang, Y.H. Bioconductor R packages for exploratory analysis and normalization of cDNA microarray data. In The *Analysis of Gene Expression Data: Methods and Software*; Parmigiani, G., Garrett, E.S., Irizarry, R.A., Zeger, S.L., Eds.; Springer: New York, NY, 2003; 73–101.

68. Dudoit, S.; Yang, Y.H.; Bolstad, B. Using R for the analysis of DNA microarray data. R News **2002**, *2* (1), 24–32 (http://CRAN.R-project.org/doc/Rnews/)

69. Gentleman, R.; Carey, V. Bioconductor. R News **2002**, *2* (1), 11–16 (http://CRAN.R-project.org/doc/Rnews/)

70. Dudoit, S.; Gentleman, R.C.; Quackenbush, J. Open source tools for microarray analysis. *Biotechniques Supplements, Microarrays and Cancer: Research and Applications*; 2003; 45–51.

71. Ihaka, R.; Gentleman, R. R: a language for data analysis and graphics. J. Comput. Graph Statist. **1996**, *5*, 299–314.

72. Yang, I.V.; Chen, E.; Hasseman, J.P.; Liang, W.; Frank, B.C.; Wang, S.; Sharov, V.; Saeed, A.I.; White, J.; Li, J.; Lee, N.H.; Yeatman, T.J.; Quackenbush, J. Within the fold: assessing differential expression measures and reproducibility in microarray assays. Genome Biol. **2002**, *3*(11), research0062.1–0062.12.

73. Kerr, M.K.; Martin, M.; Churchill, G.A. Analysis of variance for gene expression microarray data. J. Comput. Biol. **2000**, *7*, 819–837.

74. Wolfinger, R.D.; Gibson, G.; Wolfinger, E.D.; Bennett, L.; Hamadeh, H.; Bushel, P.; Afshari, C.; Paules, R.S. Assessing gene significance from cDNA microarray expression data via mixed models. J. Comput. Biol. **2001**, *8*, 625–637.

75. Cleveland, W.S.; Devlin, S.J. Locally-weighted regression: an approach to regression analysis by local fitting. J. Am. Stat. Assoc. **1988**, *83*, 596–610.

76. Cleveland, W.S.; Loader, C.L. Smoothing by local regression: principles and methods. In *Statistical Theory and Computational Aspects of Smoothing*; Haerdle, W., Schimek, M.G., Eds.; Springer: New York, 1996; 10–49.

77. Kepler, T.B.; Crosby, L.; Morgan, K.T. Normalization and analysis of DNA microarray data by self-consistency and local regression. Genome Biol. **2002**, *3* (7), research0037.

78. Affymetrix Microarray Suite User Guide. Affymetrix, Santa Clara, CA, version 4 edition, 1999.

79. Affymetrix Microarray Suite User Guide. Affymetrix, Santa Clara, CA, version 5 edition, 2001.

80. Li, C.; Tseng, G.C.; Wong, W.H. Model-based analysis of oligonucleotide arrays and issues in cDNA microarray analysis. In *Statistical Analysis of Gene Expression Microarray Data*; Speed, T., Ed.; Chapman-Hall/CRC: Boca Raton, FL, 2003; 1–34.

81. Irizarry, R.A.; Hobbs, B.; Collin, F.; Beazer-Barclay, Y.D.; Antonellis, K.J.; Scherf, U.; Speed, T.P. Exploration, normalization, and summaries of high density oligonucleotide array probe level data. Biostatistics **2003**, *4*, 249–264.

82. http://www.stat.berkeley.edu/~bolstad/RMAExpress/RMAExpress.html.

82a. Wu, Z.; Irizarry, R.A.; Gentleman, R.; Martinez-Murillo, F.; Spencer, F. A model based background adjustment for oligonucleotide expression arrays. J. Am. Statist. Assoc. **2004**, *submitted*.

83. Bolstad, B.M.; Irizarry, R.A.; Åstrand, M.; Speed, T.P. A comparison of normalization methods for high density oligonucleotide array data based on variance and bias. Bioinformatics **2003**, *19*, 185–193.

84. Anderle, P.; Rumbo, M.; Goldstein, D.R. Comparison of expression measures and quantitative RT–PCR Results for Affymetrix GeneChips. Affymetrix User Group Meeting, Poster Abstract, 2003.

85. Baldi, P.; Long, A.D. A Bayesian framework for the analysis of microarray expression data: regularized *t*-test and statistical inferences of gene changes. Bioinformatics **2001**, *17*, 509–519.

86. Efron, B.; Tibshirani, R.; Goss, V.; Chu, G. *Microarrays and Their Use in a Comparative Experiment*, Technical Report; Department of Statistics, Stanford University, 2000.

87. Quackenbush, J. Microarray data normalization and transformation. Nat. Genet. **2002**, *32* (suppl), 496–501.

88. Newton, M.A.; Kendziorski, C.M.; Richmond, C.S.; Blattner, F.R.; Tsui, K.W. On differential variability of expression ratios: improving statistical inference about gene expression changes from microarray data. J. Comput. Biol. **2001**, *8*, 37–52.

88a. Smyth, G.K. Linear models and empirical Bayes methods for assessing differential expression in microarray experiments. Statist. Appl. Gen. Mol. Biol. **2004**, *3* (1).

89. Bhowmick, D.; Davison, A.C.; Goldstein, D.R. Comparison of microarray differential expression identification methods. **2004**, *submitted*.

90. Troyanskaya, O.G.; Garber, M.E.; Brown, P.O.; Botstein, D.; Altman, R.B. Nonparametric methods for identifying differentially expressed genes in microarray data. Bioinformatics **2002**, *18*, 1454–1461.

91. Good, P.I. *Permutation Tests: A Practical Guide to Resampling Methods for Testing Hypotheses*, 2nd Ed.; Springer: New York, 2000.

92. Miller, A.J. *Subset Selection in Regression*; Chapman-Hall/CRC: Boca Raton, FL, 2002.

93. Bittner, M.; Meltzer, P.; Chen, Y.; Jiang, Y.; Seftor, E.; Hendrix, M.; Radmacher, M.; Simon, R.; Yakhini, Z.; Ben-Dor, A.; Sampas, N.; Dougherty, E.; Wang, E.; Marincola, F.; Gooden, C.; Lueders, J.; Glatfelter, A.; Pollock, P.; Carpten, J.; Gillanders, E.; Leja, D.; Dietrich, K.; Beaudry, C.; Berens, M.; Alberts, D.; Sondak, V.; Hayward, N.; Trent, J. Molecular classification of cutaneous malignant melanoma by gene expression profiling. Nature **2000**, *406*, 536–540.

94. Goldstein, D.R.; Ghosh, D.; Conlon, E.M. Statistical issues in the clustering of gene expression data. Stat. Sinica **2002**, *12*, 219–240.

95. Hastie, T.; Tibshirani, R.; Botstein, D.; Brown, P. Supervised harvesting of expression trees. Genome Biol. **2001**, *2* (1), research0003.

96. Hastie, T.; Tibshirani, R.; Eisen, M.B.; Alizadeh, A.; Levy, R.; Staudt, L.; Chan, W.C.; Botstein, D.; Brown, P.O. 'Gene shaving' as a method for identifying distinct sets of genes with similar expression patterns. Genome Biol. **2000**, *1* (2), 1–21.

97. Breiman, L. Bagging predictors. Mach. Learn. **1996**, *24*, 123–140.

98. Efron, B.; Hastie, T.; Johnstone, I.; Tibshirani, R. *Least Angle Regression*, Technical Report; Department of Statistics, Stanford University, 2003.

99. Tibshirani, R. Regression shrinkage and selection via the lasso. J. Royal Statist. Soc. B **1996**, *58*, 267–288.

100. Tibshirani, R. The lasso method for variable selection in the Cox model. Stat. Med. **1997**, *16*, 385–395.

101. Hastie, T.; Tibshirani, R.; Friedman, J. *The Elements of Statistical Learning: Data Mining, Inference, and Prediction*; Springer: New York, 2001.

102. Segal, M.R.; Dahlquist, K.D.; Conklin, B.R. Regression approaches for microarray data analysis. J. Comput. Biol. **2003**, *10*, 961–980.

103. Dettling, M.; Bühlmann, P. Supervised clustering of genes. Genome Biol. **2002**, *3* (12), research0069.1–0069.15.

104. Benjamini, Y.; Hochberg, Y. Controlling the false discovery rate: a practical and powerful approach to multiple testing. J. Royal Statist. Soc. B **1995**, *57*, 289–300.

105. Reiner, A.; Yekutieli, D.; Benjamini, Y. Identifying differentially expressed genes using false discovery rate controlling procedures. Bioinformatics **2003**, *19*, 368–375.

106. Tusher, V.; Tibshirani, R.; Chu, G. Significance analysis of microarrays applied to transcriptional responses to ionizing radiation. Proc. Natl Acad. Sci. USA **2001**, *98*, 5116–5121.

107. Golub, T.R.; Slonim, D.K.; Tamayo, P.; Huard, C.; Gaasenbeek, M.; Mesirov, J.P.; Coller, H.; Loh, M.L.; Downing, J.R.; Caligiuri, M.A.; Bloomfield, C.D.; Lander, E.S. Molecular classification of cancer: class discovery and class prediction by gene expression monitoring. Science **1999**, *286*, 531–537.

108. Dudoit, S.; Shaffer, J.P.; Boldrick, J.C. Multiple hypothesis testing in microarray experiments. Statist. Sci. **2003**, *18*, 71–103.

109. Ge, Y.; Dudoit, S.; Speed, T.P. Resampling-based multiple testing for microarray data analysis. TEST **2003**, *12*, 1–78.

110. Westfall, P.H.; Young, S.S. *Resampling-based Multiple Testing: Examples and Methods for P-value Adjustment*; Wiley: New York, 1993.
111. *SAS Language and Procedures, Version 6*; SAS Institute: Cary, NC, 1990.
112. Correa, J.A.; Dudoit, S.; Goldstein, D.R. Multiple testing in the survival analysis of microarray data. **2004**, *submitted*.
113. Everitt, B.S.; Landau, S.; Leese, M. *Cluster Analysis*, 4th Ed.; Oxford University Press: New York, 2001.
114. Hand, D.J. *Construction and Assessment of Classification Rules*; Wiley: New York, 1997.
115. Eisen, M.B.; Spellman, P.T.; Brown, P.O.; Botstein, D. Cluster analysis and display of genome-wide expression patterns. Proc. Natl Acad. Sci. USA **1998**, *95*, 14863–14868.
116. Gordon, A.D. *Classification*, 2nd Ed.; Chapman-Hall/CRC: Boca Raton, FL, 1999.
117. Hartigan, J.A. *Clustering Algorithms*; Wiley: New York, 1975.
118. Kohonen, T. The self-organizing map. Proc IEEE *78*, 1464–1479.
119. Kaufman, L.; Rousseeuw, P.J. *Finding Groups in Data: An Introduction to Cluster Analysis*; Wiley: New York, 1990.
120. Dudoit, S.; Fridlyand, J. A prediction-based resampling method for estimating the number of clusters in a dataset. Genome Biol. **2002**, *3* (7), research0036.
121. Dudoit, S.; Fridlyand, J. Bagging to improve the accuracy of a clustering procedure. Bioinformatics **2003**, *19*, 1090–1099.
122. Kerr, M.K.; Churchill, G.A. Bootstrapping cluster analysis: assessing the reliability of conclusions from microarray experiments. Proc. Natl Acad. Sci. USA **2001**, *98*, 8961–8965.
123. McShane, L.M.; Radmacher, M.D.; Freidlin, B.; Yu, R.; Li, M.C.; Simon, R. Methods for assessing reproducibility of clustering patterns observed in analyses of microarray data. Bioinformatics **2002**, *18*, 1462–1469.
124. Dembele, D.; Kastner, P. Fuzzy C-means method for clustering microarray data. Bioinformatics **2003**, *19*, 973–980.
125. Yeung, K.Y.; Fraley, C.; Murua, A.; Raftery, A.E.; Ruzzo, W.L. Model-based clustering and data transformations for gene expression data. Bioinformatics **2001**, *17*, 977–987.
126. McLachlan, G.J.; Bean, R.W.; Peel, D. A mixture model-based approach to the clustering of microarray expression data. Bioinformatics **2002**, *18*, 413–422.
127. van der Laan, M.J.; Pollard, K.S. *Hybrid Clustering of Gene Expression data with Visualization and the Bootstrap*, Technical Report; Department of Biostatistics, University of California: Berkeley, 2001.
128. Breiman, L.; Friedman, J.H.; Olshen, R.; Stone, C.J. *Classification and Regression Trees*; Wadsworth International Group: Belmont, CA, 1984.
129. Freund, Y.; Schapire, R.E. A decision-theoretic generalization of on-line learning and an application to boosting. J. Comput. Sys. Sci. **1997**, *55*, 119–139.
130. Dettling, M.; Bühlmann, P. Boosting for tumor classification with gene expression data. Bioinformatics **2003**, *19*, 1061–1069.
131. Ambroise, C.; McLachlan, G.J. Selection bias in gene extraction on the basis of microarray gene-expression data. Proc. Natl Acad. Sci. USA **2002**, *99*, 6562–6566.
132. Dudoit, S.; Fridlyand, J.; Speed, T.P. Comparison of discrimination methods for the classification of tumors using gene expression data. J. Am. Statist. Assoc. **2002**, *97*, 77–87.

2

Dietary Effects of Arachidonate-Rich Fungal Oil and Fish Oil on Murine Hepatic Gene Expression: Focus on Newly Described LC-PUFA-Regulated Genes

Alvin Berger

Paradigm Genetics, Inc., Research Triangle Park, USA

Matthew A. Roberts

Nestle Global Strategic Business Unit and R&D, St. Louis, USA

INTRODUCTION

Dietary long-chain polyunsaturated fatty acids (LC-PUFA) affect processes including growth, neurological development, lean and fat mass accretion, reproduction, innate and acquired immunity, and infectious pathologies of viruses, bacteria, and parasites; and the incidence and severity of virtually all chronic and degenerative diseases including cancer, atherosclerosis, stroke, arthritis, diabetes, osteoporosis, and neurodegenerative, inflammatory, and skin diseases.[1–6]

An LC-PUFA-rich oil of considerable importance is fish oil. It contains several bioactive components including docosahexaenoic acid (DHA; 22:6n3) and eicosapentaenoic acid (EPA; 20:5n3). The EPA component of fish oil is linked to anti-inflammatory, antithrombotic, and generally, eicosanoid-mediated effects by antagonizing arachidonic acid (AA) metabolism. AA is present in AA-rich fungal oils, various meat sources, and breast milk. AA is generally considered to be a proinflammatory fatty acid.

The most often cited action of n3 LC-PUFA is that they induce alterations in eicosanoids by antagonizing the primary substrate of eicosanoid metabolism: AA.[7] In this context, new eicosanoid-like molecules are still being discovered, including endocannabinoid, N-acylated amino acids, N-acylated dopamine, and oxygenated and epoxylated bioactive derivatives of AA and DHA.[8−17]

Until the advent of microarrays and other "omic" technologies, the mechanisms by which LC-PUFA from fish oil and AA affected so many diverse processes has been a black box. The n3 LC-PUFA (or their eicosanoid derivatives described earlier) are known to alter the transcription of genes encoding proteins in lipid biosynthesis, lipid desaturation, lipid transport, and lipid oxidation.[18−22] LC-PUFA affect gene transcription by acting on peroxisome proliferator activated receptors (PPARs),[20,23] as well as other transcription factors, described herein.

In the present work, mice were fed control diets adequate in 18:2n6 and 18:3n3 but lacking LC-PUFA; or a diet enriched in fungal oil (FUNG) enriched in AA, the n6 elongation and desaturation product of 18:2n6; or a diet containing fish oil (FISH) enriched in 22:6n3 and 20:5n3, the major n3 LC-PUFA elongation and desaturation products of 18:3n3; or a diet containing both fungal oil and fish oil (FUNG + FISH). Thereafter, gene expression profiling was performed in parallel with quantitative metabolic profiling of a broad spectrum of liver fatty acids (FAs). Liver was of interest because it is the major lipid metabolizing tissue. Diets were fed for 57 days to study chronic and sustained alterations in control of gene expression rather than acute effects. Herein, we summarized and updated our previous work which focused on transcription factors such as PPARs, sterol regulatory element binding protein (SREBP), and hepatic nuclear factors (HNFs) which were previously identified as mediators of LC-PUFA actions.[24] Thereafter, we discuss changes to genes with important roles in diverse areas of metabolism, opening up new avenues of research for intrepid lipid researchers. Our statistical model does not use a fold change cut off rule to identify significantly changed genes.

MATERIALS AND METHODS

All experimental procedures utilized have been previously published.[24,25] This includes a description of the experimental diets, feeding and dissection conditions, nucleic acid preparation, gene expression analysis using the murine 11k GeneChip, and lipid analysis. The selection of genes was modified for the current chapter. The 5% LFC method[25] indicated that 489 genes in the liver were differentially regulated by LC-PUFA from the original 13,179 genes represented on the Mu11K Affymetrix GeneChip. After removing those genes that were identified as absent across all dietary conditions, 329 differentially expressed probe sets remained. Of these, 214 genes represented annotated sequences. Of these, 73 hepatic genes were selected because they fell into

Gene Ontology (GO) Consortium classifications of interest: nucleic acid binding, fatty acid synthesis, fatty acid oxidation, carbohydrate metabolism, cytochrome P450 metabolism, transport, sterol metabolism, and lipoprotein metabolism.[24] In our previous work,[24] within each GO category, we focused on genes with specific transcriptional signatures including genes with PPRE, SRE, HNF, or CRE response elements in their promoters. Herein, to expand our knowledge of how dietary LC-PUFA broadly affect metabolism, independent of GO category, another 50 gene transcripts were selected on the basis of having the largest interdietary group fold changes. Thus, 123 gene transcripts are included in Table 2.1.

RESULTS AND DISCUSSION

Body Weight, Feed Intake, and Incorporation of Dietary LC-PUFA into Tissue Membranes

There were no changes in body weight, liver weight, or feed intake with the various diets as previously reported.[24] Thus, differences in gene expression and metabolic profiles described herein were judged to be linked to the FAs present in FUNG and FISH oils in healthy, normal weight animals.

The incorporation of 20:4n6 and 22:4n6 following feeding of the AA-rich FUNG diet, and the incorporation of 20:5n3, 22:5n3, and 22:6n3 following feeding of the EPA/DHA-rich FISH diet, into hepatic phospholipid (PL) pools was further evidence the diets were consumed, resulting in significant changes to PL acyl composition and subsequent gene expression. Fish oil feeding is well established to increase n3 LC-PUFA in hepatic PL.[21] Our lipid profiling data revealed that FUNG and FISH FAs were not identically incorporated into individual hepatic PL,[24] which could affect downstream gene signaling.

Gene Expression Analysis

The statistical and modeling procedures for selecting our 329 differentially expressed genes in the liver using a limit fold change (LFC) model have been previously described.[24,25] Genes that are discussed are followed by parentheses indicating fold changes according to the following convention: FUNG vs. control diet, FISH vs. control diet, and FUNG + FISH vs. control diet [e.g., *G6pc* (1.1, 1.6*, −1.9*)]. Those fold changes that are significant by the 5% LFC model are further identified with an asterisk (*). It is important to note that the nature of the LFC model allows differentially expressed genes to be selected on the basis of differences between the treatments, and therefore explains why some genes discussed in the text have no fold changes marked with an asterisk (as significance was observed between treatments and not vs. the control diet). The high degree of concordance between real time-polymerase chain reaction (RT-PCR) and microarray data (86%—reported in detail elsewhere[25]) establishes confidence in the results discussed throughout this chapter.

Effects of LC-PUFA on Transcription Factors and Growth Factors

It is difficult to interpret transcriptional profiling data alone, since dimerization, posttranslational modifications, relative quantities of coactivators and corepressors, and the formation of multiple transcriptional complexes binding to common promoter regions ultimately determine the binding activation potential of transcription factors. Nevertheless, clustering analysis of differentially regulated genes revealed numerous genes known to be regulated by specific transcription factors, thereby enabling several transcription factor signatures to be established. These signatures then provide clues concerning the various pathways through which FA may effect gene transcription.

PPARs, SREBPs, HNF, and cAMP Signaling (General Update)

In our previous work, we described how mRNA for several key transcription factors were upregulated and downregulated in response to dietary LC-PUFA, leading to changes in genes that regulate FA oxidation and synthesis and lipid metabolism. This work will be summarized pictorially, and new information about these transcripton factors will be updated from the original publication[24] describing these changes (Fig. 2.1). See the section "Effects of LC-PUFA on Lipid and Energy Metabolism" for further discussion.

LC-PUFA can affect not only *SREBP1* mRNA (-1.1, -2.1^*, -1.8^*), but also prevent maturation of SREBP1 protein.[26] LC-PUFA could speculatively affect SREBP processing proteins such as Insig and SREBP cleavage-activating protein (SCAP).[27] We may thus have underestimated the effects of our n3 diets on SREBP signaling, as we did not evaluate SREBP processing in our experiments.

HNF foxhead transcription factors such as HNF3β (FOXA2) have emerged as key genes involved in diabetes type II disease etiology.[28] Fish oil feeding is now known to influence diabetes etiology through genes such as stearoyl CoA desaturase (*SCD*), which was downregulated with combination diet; and PPARγ.[29] Foxheads such as FoxO3a are known to be activated in response to oxidative stress.[30] FoxO3a in turn activates sterol carrier protein 2 (SCP2) which protects FA from further oxidation. PUFA also activate PPARα which may activate sterol carrier protein X (SCPx), another factor that protects FA from oxidation.[30] We previously described the -2.0^*-fold downregulation of *HNF3γ* (winged helix protein, forkhead, and Foxa3) with FUNG + FISH, but not the individual LC-PUFA; and lack of effect on the more well known Hnf4α factor. HNF3γ protein may be involved in etiology of obesity, hyperlipidemia, and diabetes, by regulating glucagon transcription, insulin resistance, and pancreatic γ-cell function.[31,32] To our knowledge, the regulation of hepatic *HNF3γ* expression by LC-PUFA had not been previously reported. Another group found all three members of the *HNF3* family (*HNF3α*, *HNF3β*, and *HNF3γ*) to be coordinately regulated in response to angiotensin II exposure.[33]

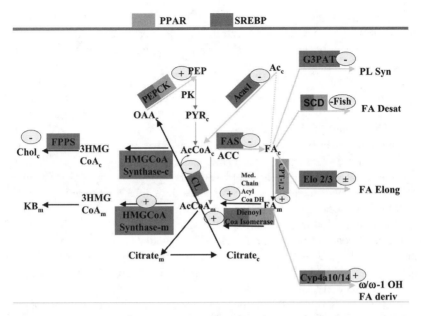

Figure 2.1 Overview of gene transcripts regulating hepatic lipid metabolism. The predominant pattern is that LC-PUFA downregulated FA synthetic pathways; upregulated FA degradative (β-oxidative) pathways; and increased gluconeogenesis. In essence, LC-PUFA emulated the fasted metabolic state. The signaling cascades mediated by PPAR and SREBP are shown. Note that PEPCK is also regulated by HNF and cAMP (not shown in the figure). c, cytoplasmic; m, mitochondrial; +, −, up- or downregulation with LC-PUFA; ±, differentially regulated with FISH and FUNG diets. See tables for other abbreviations.

Initiation of Transcription/bZIP Transcription Factors

Lipids have been reported to affect basal transcription. Lipids such as phosphoinositides may affect DNA polymerase via direct physical binding of DNA polymerase to a phosphoinositide binding site.[34] Phosphoinositides may bind to histones, preventing the histones from binding to RNA polymerase II and initiating transcription. Feeding of our diets did affect the acyl composition of phosphatidylinositol, which could affect the subsequent synthesis of phosphoinositides. These are a few recent examples of how LC-PUFA could affect transcription initiation.

Narayanan et al. (Chapters 14 and 15), found that DHA downregulated transcription of basal transcription factors in human colon cancer cells in a consistent manner. We did not observe a consistent transcriptional downregulation of basal factors with FISH feeding. This could be related to *in vivo* vs. *in vitro* differences; the fact that we used a fish oil mixture rather than purified DHA; and our use of mice, not humans.

Activating transcription factor (ATF) family comprises basic region-leucine zipper (bZIP) proteins.[35] ATF4 is homologous to ApCREB-2 (CREB-2), and related to C/EBP transcription factors.[36] ATF4 interacts with RPB3, the core subunit of RNA polymerase II (pol II), affecting promoter recognition.[37] ATF4 is also a component of intracellular stress pathways (e.g., activating transcription of ER chaperone proteins).[38] ATF1 and CREB1 act as nuclear cofactors, complexing with endogenous ATF4.[39] In hepatic HepG2 cells, *atf4* was downregulated with fish oil relative to safflower oil (Table 4.6). In our *in vivo* experiments, *atf4* was downregulated with the combination diet (-1.2, -1.1, -1.7^*). The related family member ATF5 is 58% identical to mouse ATF4 in the carboxy-terminal bZIP region[35] and may have overlapping roles with ATF4.[40] ATF5 can repress human cAMP-induced gene transcription.[41] *Atf5* was downregulated with all three LC-PUFA diets (-1.7^*, -1.5^*, -3.1^*). Thus, there is possible evidence that LC-PUFA signals through at least two AT.

RNA polymerase II activators recruit cofactors to unpack DNA from chromatin for transcription to commence.[42] Rpo2tc1 encodes RNA polymerase II transcriptional coactivator. *Rpo2tc1* was downregulated with combination diet relative to FUNG (1.2, -1.3, -1.6). Tceb1l auxiliary factor assists RNA polymerase II in mRNA chain elongation.[43] *Tceb1l* was upregulated with FISH and combination diet (1.3, 1.6^*, 1.8^*).

Albumin transcription is controlled by binding of factors to albumin promoter D site and other sites.[44] The bZIP protein D site albumin promoter binding protein (DBP) dimerizes with DPDCCAAT/enhancer binding protein-α (C/EBPα) (from the C/EBP family of bZIP proteins). In rats, dbp mRNA is affected by starvation and circadian rhythms.[44] *Dbp* was upregulated by FUNG and the combination diet (3.4^*, 1.0, 4.4^*), but intriguingly was not affected by FISH alone. A next step is to examine if levels of albumin are changed in response to various LC-PUFA. Lipophilic LC-PUFA are transported in plasma bound to albumin, so there could be adaptive advantages of increasing transcription of albumin.

Other Transcription Factors Linked to Growth and Development/G Protein Pathways

The five most 5' *HoxD* genes pattern axial and appendicular skeletal elements and the nervous system of developing vertebrate embryos. Hoxd12 binds Maf family bZip oncoproteins.[45] Its role in the liver, however, is unclear. Hoxd12 was downregulated with combination diet relative to FUNG in liver (1.4, -1.5, -2.9), as well as in hippocampus (see Chapter 3).

Numb and Notch control many aspects of neuronal cell differentiation, as Numb and Numb-like are Notch receptor inhibitors.[46] The role of Numb in the liver is not clear. *Numb* was highly upregulated with FISH alone (0.6, 5.5^*, 2.4). In hippocampus, there were indications Notch signaling was affected via NOV (see Chapter 3).

Neuronal guanine nucleotide exchange factor (NGEF) is a *Dbl* gene family member, which functions as a guanine nucleotide exchange factor for Rho-type GTPases. NGEF is predominantly expressed in brain. NGEF has transforming potential in cell culture and nude mice.[47] *Ngef* was downregulated with FISH relative to FUNG and combination diet (1.7, −6.0, 1.0). It should be investigated whether fish oil may affect cell transformation via NGEF signaling.

G protein γ subunits determine specificity of receptor–G protein interactions. The γ-4 subunits are brain specific.[48] *Gng4* was highly upregulated with FISH (1.0, 9.2*, 1.0). FUNG is not only inactive, but represses FISH induction of *gng4*. This is potentially yet another pathway differentiating FAs in FUNG from those in FISH.

Hmga1 and -2 encode high mobility group A (HMGA) nonhistone chromosomal proteins. HMGA are architectural transcription factors acting on numerous genes. *Hmg* genes are dysregulated in human cancer, highlighting a growth role.[49] *Hmga1* was upregulated with the combination diet (−1.8, 1.2, 6.2*). This may highlight a unique interaction of FUNG and FISH FAs in regulating cell growth in some systems.

Pbx1 encodes pre-B-cell-leukemia transcription factor 1 (*PBX1*), a TALE homeodomain transcription factor of the PBC family, regulating developmental gene expression.[50] PBC and HOX proteins cooperatively bind DNA.[51] Cyclin-dependent kinase inhibitors (CDKI) downregulated *Pbx1* expression.[52] *Pbx1* was downregulated with combination diet (−1.5, −1.2, −1.7*).

Herpud1 gene is overexpressed in Wnt-1-transformed cells. Wnt signal transduction pathway regulates morphogenesis and mitogenesis of cells.[53] No other meaningful information was obtained on this gene, which was upregulated with combination diet (1.6, 1.4, 1.8*).

Kai1 is a tumor metastasis suppressor gene in cancer encoding KAI1/CD82 protein, which associates with p53 tumor suppressor.[54] *Kai1* was downregulated with FISH relative to FUNG (1.3, −1.7, −1.1).

Transcription Factors Linked To Heat, Stress, and Inflammation (Immunity)

TCF2 is a T-cell-specific transcription factor class member. *Tcf2* was largely downregulated with combination diet, but not with FISH (−1.2, 1.6, −7.6*). TCF7 similarly activates genes involved in immune regulation, and TCF7 polymorphism may be a risk factor for type 1 diabetes.[55] In contrast to *tcf2*, *tcf7* was upregulated with the combination diet only (1.1, 1.0, 2.7*). LC-PUFA are established to affect various aspects of T-cell signaling, and TCF signaling may explain some immune-related actions of LC-PUFA.

DMBT1 homologs may regulate hepatic tissue repair and fate decision and differentiation paths of specific cell populations in the hepatic lineage.[56] *Dmbt1* was downregulated with FISH and combination diets relative to FUNG (1.6, −5.7, −5.7).

Heat shock proteins (HSP) enable cell survival and recovery from stressful conditions. Hspb1 is equivalent to HSP25 and HSP27.[57] Hspb1 overexpression induced the antioxidant defense gene manganese superoxide dismutase

(*SOD2*) and its enzyme activity, and increased gene expression of the inflammatory factors tumor necrosis factor-α (TNFα), interleukin 1-β (Il1b), and Nfkb.[58]
Nfkb1 was activated in hippocampus in our experiments (Chapter 3). Hsps also
have roles in cytoskeletal actin dynamics in response to stress, and are phosphorylated via p38 MAPK and PKCδ.[59] The large decrease in *hspb1* expression
with FISH relative to FUNG and combination diet is intriguing (1.4, −4.0, 1.7).
Farkas et al. (Chapter 5) found fish oil feeding to decrease the related *hsp86-1*
transcript in 3-month-old rat forebrains. Effects of LC-PUFA on hsps are an
important area of future investigation for understanding how LC-PUFA affect
stress-related immune reactions.

Macrophage migration inhibitory factor (MIF) delays hypersensitivity.
LC-PUFA, particularly from fish oil, affect neutrophil and macrophage functioning.[60] It is intriguing that only FISH downregulated *mif*, which may increase
hypersensitivity (the opposite direction one might expect from fish oil which is
not known to increase hypersensitivity) (1.3, −2.3*, −1.1).

Rgs16 encodes GTPase-activating protein for Gα subunits, which regulates
G protein signaling (RGS16). RGS16 attenuates chemokine receptor signaling in
T lymphocytes, regulating lymphocyte trafficking.[61] *Rgs16* was upregulated only
with the combination diet (1.1, −1.3, 2.2*). Rgs16 is likely SREBP1a-regulated.[62]

SERPINC1 (antithrombin III; ATIII) is an inhibitor of the blood coagulation
cascade.[63] ATIII activity is increased after fish oil consumption in patients with
angina pectoris;[64] but ATIII activity was not changed in numerous other fish oil
studies. Consistent with fish oil's known antithrombotic properties, only FISH increased *serpinc1* (−1.2, 1.6*, −1.4). FUNG feeding did not increase ATIII, which
is consistent with the thrombotic properties of the n6 LC-PUFA in the FUNG diet.

MHC class II determinants present antigenic peptides to T cells. Pre-
assembled regulatory factor X (RFX) and nuclear factor Y (NFY) bind DNA
and recruit class II transactivator (CIITA) and other activators including
NFYB, NFYC, RFX5, RFXAP, and RFXANK/B.[65] *Nfyb* was downregulated
with FISH relative to FUNG and combination diet (1.8, −1.7, 1.6).

Miz1 (Msx-interacting-zinc finger) is a member of PIAS (protein inhibitor
of activated STAT; signal transducer and activator of transcription) protein family
implicated in cytokine signaling inhibition.[66] PIAS proteins interact with glucocorticoid receptor (GR)-interacting protein 1 (GRIP1) to activate GR.[66] The
oncogene myc may repress *miz1*.[67] Miz1/mPIASxβ proteins interact with
TFII-I.[68] *Miz1* was downregulated with combination diet relative to FUNG
and FISH (1.6, 1.5, −1.9).

Effects of LC-PUFA on Amino Acid and Protein Metabolism

L23 is part of the L23 mitochondrial ribosomal protein family. No specific
functional information was found. *Mrpl23* was hugely downregulated with
combination diet (−1.2, 1.1, −12.3*). L28 is required for ribosome assembly
in *Escherichia coli*.[69] *RpL28* mRNA expression was altered in colorectal

cancer.[70] *Rpl28* was similarly downregulated with combination diet (− 1.3, 1.0, − 1.7*). LC-PUFA regulated several ribosomal proteins in the hippocampus (see Chapter 3 for a detailed discussion).

Lysosomes are not solely involved in degradation. Lysosome-associated membrane protein LAMP-2 is involved in lysosomal enzyme targeting, autophagy, and lysosomal biogenesis. In humans, LAMP-2 deficiency leads to Danon disease, a fatal cardiomyopathy and myopathy.[71] *Lamp2* was upregulated with combination diet (− 3.5, 1.2, 3.7*). We are unaware of previous links between LC-PUFA and lysosomal protein degradation.

Plod3 (lysyl hydroxylase; LH3) is involved in collagen synthesis. LH3 possesses lysyl hydroxylase- and galactosylhydroxylysyl glucosyltransferase activity.[72] There are reports that fish oil may increase collagen synthesis. Our data showing *plod3* was downregulated with FUNG relative to FISH and combination diet (− 4.9, 1.1, 1.0) may suggest more collagen is synthesized with FISH than FUNG feeding. Various groups report that fish oils can be beneficial in osteoporosis.[73] Thus, the effects of LC-PUFA on lysyl hydroxylase activity is an important area for future investigation.

Dopachrome tautomerase (DCT) is a tyrosinase member involved in coat color determination.[74] *Dct* was downregulated with FUNG and combination diet (− 5.2*, − 1.1, − 2.2*). Interestingly, *dct* was also altered in PTP1B antisense-treated ob/ob mice with decreased SREBP-regulated lipogenic gene expression (Chapter 17). A common signaling pathway may be responsible for changes in *dct* expression in both biological systems. The physiological relevance of the finding is uncertain.

Eukaryotic elongation factor 2 (eEF2) is involved in translational elongation. The unphosphorylated form promotes translation; whereas the phosphorylated form reduced translational elongation of cyclin D3 in Jurkat cells.[75] *Eef2* was downregulated with FISH (− 1.3, − 1.6*, − 1.3).

Eukaryotic elongation factor 1A (eEF1A; formerly EF1α) promotes transfer of aminoacylated tRNAs to ribosomal A site during protein synthesis. eEF1A1 and -2 isoforms are found in mammals.[76] *Eef1a1* was downregulated with FUNG (− 1.8*, 1.2, 1.1). EIF4A1 is a eukaryotic initiation factor (eIF) member.[77] EIF5 factor complexes with eIF1, eIF2, and eIF3 to stimulate Met-tRNA(i)(Met) binding to 40S ribosomes, promoting scanning or AUG recognition.[77] *Eif4a1* (1.4, − 1.8, 1.7) and *Eif5* (1.6, − 1.5, 1.3) were consistently downregulated with FISH relative to FUNG and combination diet.

Effects of LC-PUFA on Lipid and Energy Metabolism

This section updates changes to genes involved in lipid metabolism that we previously discussed[24] on the basis of the new literature. We previously described how LC-PUFA activated a number of PPARα-regulated genes coding for enzymes involved in FA β-oxidation; and decreased a number of SREBP-regulated transcripts involved in FA synthesis (Tables 2.1 and 2.2; Fig. 2.1).

Table 2.1 Differentially Expressed Genes in Response to Dietary LC-PUFAs in Liver

Gene symbol	Description	Unigene	Fold change (FUNG/C, FISH/C, FUNG + FISH/C)	FUNG	FISH	COMBO
Structural role						
Pxmp2	Peroxisomal membrane protein 2	Mm.21853	(1.1, −1.3, −1.5)			
Tuba1	Tubulin, alpha 1	Mm.231463	(−1.3, 1.3, −1.3)			
Nucleic acid binding						
Atf4	Activating transcription factor 4	Mm.641	(−1.2, −1.1, −1.7*)			
Atf5	Activating transcription factor 5	Mm.1566	(−1.7*, −1.5*, −3.1*)			×
Dbp	D site albumin promoter binding protein	Mm.3459	(3.4*, 1.0, 4.4*)			
Eef1a1	Eukaryotic translation elongation factor 1 alpha 1	Mm.196614	(−1.8*, 1.2, 1.1)			
Eif4a1	Eukaryotic translation initiation factor 4A1	Mm.12858	(1.4, −1.8, 1.7)			
Ercc2	Excision repair cross-complementing rodent repair deficiency, complementation group 2	Mm.56990	(1.0, 2.2*, 2.0*)			
Gtf2i	General transcription factor II I	Mm.22593	(−1.9, 1.7, −4.5)			×
HNF3γ	Forkhead box A3 (Foxa3)	Mm.42260	(1.2, 1.1, −2.0*)			
Hoxd12	Homeo box D12	Mm.57124	(1.4, −1.5, −2.9)			×

Hzf-pending	Hematopoietic zinc finger	Mm.14099	(1.4, 2.2*, −1.4)
Klf13	Kruppel-like factor 13	Mm.41170	(1.8, 1.5, 2.0*)
Miz1	Msx-interacting-zinc finger	Mm.6370	(1.6, 1.5, −1.9)
Nfyb	Nuclear transcription factor-Y beta (NF-Y)	Mm.3259	(1.8, −1.7, 1.6)
Pbx1	Pre B-cell leukemia transcription factor 1	Mm.61526	(−1.5, −1.2, −1.7*)
Rpo2tc1	RNA polymerase II transcriptional coactivator	Mm.41746	(1.2, −1.3, −1.6)
Srebp1	Ethanol induced 6 (sterol regulatory element binding protein-1); Etohi6	Mm.30133	(−1.1, −2.1*, −1.8*)
Tceb1l	Transcription elongation factor B (SIII), polypeptide 1 (15 kDa),-like	Mm.42944	(1.3, 1.6*, 1.8*)
Tcf2	Transcription factor 2	Mm.7226	(−1.2, 1.6, −7.6*)
Tcf7	Transcription factor 7, T-cell specific	Mm.31630	(1.1, 1.0, 2.7*)
Tcfe3	EST highly similar to transcription factor E3	Mm.25762	(1.2, 1.8, −5.7)
Thrsp; Spot14	Thyroid hormone responsive SPOT14 homolog (Rattus)	Mm.28585	(1.2, −1.1, −1.6*)
Usf2	Upstream transcription factor 2	Mm.15781	(1.2, −5.6*, 1.1)

(*continued*)

Table 2.1   Continued

Gene symbol	Description	Unigene	(FUNG/C, FISH/C, FUNG+FISH/C)	Fold change		
				FUNG	FISH	COMBO
Cell growth factors, angiogenic factors, oncogenes, protein kinases, TFs						
Ang	angiogenin	Mm.202665	(1.2, −1.5, 1.3)			
Anp32a	Acidic (leucine-rich) nuclear phosphoprotein 32 family, member A	Mm.613	(−4.2*, −5.1*, −4.1*)			
Cdk4	Cyclin-dependent kinase 4	Mm.6839	(−1.7*, 1.0, 1.0)			
Col18a1	Procollagen, type XVIII, alpha 1	Mm.4352	(−1.3, −1.6*, −1.2)			
Dmbt1	Deleted in malignant brain tumors 1	Mm.4138	(1.6, −5.7, −5.7)			
Gng4	Guanine nucleotide binding protein (G protein), gamma 4 subunit	Mm.215394	(1.0, 9.2*, 1.0)			
Hmga1	High mobility group AT-hook 1	Mm.4438	(−1.8, 1.2, 6.2*)			✕
Kai1	Kangai 1 (suppression of tumorigenicity 6, prostate)	Mm.4261	(1.3, −1.7, −1.1)			
LOC223775	Similar to Serine/threonine-protein kinase pim-3	Mm.30018	(14.6*, 6.8*, 8.6*)			
Ngef	Neuronal guanine nucleotide exchange factor	Mm.143717	(1.7, −6.0, 1.0)			

Gene	Description	Accession	Values
Numbl	Numb-like	Mm.255487	(−1.7, 5.5*, 2.4)
Rac2	RAS-related C3 botulinum substrate 2	Mm.1972	(13.1*, 1.0, 5.0)
Rgs16	Regulator of G-protein signaling 16	Mm.181709	(1.1, −1.3, 2.2*)
Spi1-3	Serine protease inhibitor 1-3	Mm.196591	(1.0, 1.4, −1.1)

Energy metabolism, electron transport

Gene	Description	Accession	Values
Atp5g1	ATP synthase, H+ transporting, mitochondrial F0 complex, subunit c (subunit 9), isoform 1	Mm.258	(1.1, −6.5*, 2.4*)
Atp6v1e1	ATPase, H+ transporting, V1 subunit E isoform 1	Mm.29045	(1.7, 1.0, 2.0*)
Ndufa10	NADH dehydrogenase (ubiquinone) 1 alpha subcomplex 10	Mm.28293	(7.2*, 6.3, 8.6*)

Amino acid and protein metabolism

Gene	Description	Accession	Values
Eef1a1	Eukaryotic translation elongation factor 1 alpha 1	Mm.196614	(−1.8*, 1.2, 1.1)
Eef2	Eukaryotic translation elongation factor 2	Mm.27818	(−1.3, −1.6*, −1.3)
Eif5	Eukaryotic translation initiation factor 5	Mm.260384	(1.6, −1.5, 1.3)
Lamp2	Lysosomal membrane glycoprotein 2	Mm.486	(−3.5, 1.2, 3.7*)
Mrpl23	Mitochondrial ribosomal protein L23	Mm.12144	(−1.2, 1.1, −12.3*)

(continued)

Table 2.1 Continued

Gene symbol	Description	Unigene	Fold change (FUNG/C, FISH/C, FUNG + FISH/C)	FUNG	FISH	COMBO
Rpl28	Ribosomal protein L28	Mm.3111	(−1.3, 1.0, −1.7*)	gray		gray
Plod3	Procollagen-lysine, 2-oxoglutarate 5-dioxygenase 3	Mm.251003	(−4.9, 1.1, 1.0)			
Fatty acid and lipid synthesis						
Acas1	Acetyl-Coenzyme A synthetase 1 (AMP forming)	Mm.22719	(−2.5*, −1.7*, −4.4*)	gray	gray	gray
Acly	ATP citrate lyase (ACL)	N/A	(−2.7*, −2.2*, −4.7*)	gray	gray	gray
Acly	ATP citrate lyase (ACL)	N/A	(−2.0*, −1.8*, −3.6*)	gray	gray	gray
Acly	ATP citrate lyase (ACL)	N/A	(−4.0*, −7.1*, −8.9*)	gray	gray	gray
Acly	ATP citrate lyase (ACL)	N/A	(−2.2*, −2.4*, −3.7*)	gray	gray	gray
Elovl2; Ssc2	Elongation of very long chain fatty acids (FEN1/Elo2, SUR4/Elo3, yeast)-like 2	Mm.2567	(−1.3, −1.7*, −1.8*)	gray	gray	gray
Elovl3; Cig30	Elongation of very long chain fatty acids (FEN1/Elo2, SUR4/Elo3, yeast)-like 3	Mm.21806	(−1.2, 1.9*, −1.6)	gray	black	gray
Facl2	Fatty acid coenzyme A ligase, long chain 2	Mm.28962	(1.4, 1.3, 1.8*)	black	black	black

Gene	Description	Accession	Values
Fasn	Fatty acid synthase (FAS)	Mm.3760	(−1.9*, −1.3, −5.4*)
Gpam	Glycerol-3-phosphate acyltransferase, mitochondrial (GPAT)	Mm.87773	(−1.7*, −1.2, −2.3*)
Scd1	Stearoyl-Coenzyme A desaturase 1	Mm.140785	(−1.1, 1.0, −1.8*)
Fatty acid oxidation			
Acadm	Acetyl-Coenzyme A dehydrogenase, medium chain (MCAD)	Mm.10530	(1.5, 1.3, 1.7*)
Aldh1a1	Aldehyde dehydrogenase family 1, subfamily A1	Mm.4514	(1.6*, 1.8*, 1.7*)
Aldh1a7	aldehyde dehydrogenase family 1, subfamily A7	Mm.14609	(1.2, 1.9*, 1.3)
Aldh2	Aldehyde dehydrogenase 2, mitochondrial	Mm.2621	(1.0, −1.6*, −1.1)
Crat	Carnitine acetyltransferase	Mm.20396	(1.4, 3.3*, 1.2)
Cpt1a	Carnitine palmitoyltransferase 1, liver	Mm.18522	(1.8*, 1.4, 1.7*)
Cpt2	Carnitine palmitoyltransferase 2	Mm.29499	(1.6, 1.9*, 1.7*)
Ech1	Enoyl coenzyme A hydratase 1 (delta 3,5-delta 2,4-dienoyl-CoA isomerase)	Mm.2112	(1.1, 1.6*, 1.4)

(continued)

Table 2.1 Continued

Gene symbol	Description	Unigene	Fold change (FUNG/C, FISH/C, FUNG+FISH/C)	FUNG	FISH	COMBO
Carbohydrate metabolism						
Camk2b	Calcium/calmodulin-dependent protein kinase II, β	Mm.4857	(−1.1, −7.6*, −1.8)			
Cam-Pde 1c	Calcium/calmodulin-dependent 3′,5′-cyclic nucleotide phosphodiesterase 1c	N/A	(−1.6, −1.3, 6.0*)			×
G6pc	Glucose-6-phosphatase, catalytic (G6Pase)	Mm.18064	(1.1, 1.6*, −1.9*)			×
Got1	Glutamate oxaloacetate transaminase 1, soluble	Mm.19039	(−1.2, −1.5, −2.1*)			
Hspa5	Heat shock 70 kDa protein 5 (glucose-regulated protein, 78kDa)	Mm.918	(−1.4, 1.1, −1.5*)			
Pck1	Phosphoenolpyruvate carboxykinase 1, cytosolic (PEPCK)	Mm.42246	(2.1*, 1.6*, 1.7*)			
Pck1	Phosphoenolpyruvate carboxykinase 1, cytosolic (PEPCK)	Mm.42246	(1.6, 1.4, 1.7*)			
Pck1	Phosphoenolpyruvate carboxykinase 1, cytosolic (PEPCK)	Mm.42246	(2.7*, 1.9*, 2.6*)			

Pck1	Phosphoenolpyruvate carboxykinase 1, cytosolic (PEPCK)	Mm.42246	(2.1*, 1.8*, 2.3*)
Cytochrome P450 metabolism			
Cyp1a2	Cytochrome P450, 1a2, aromatic compound inducible	Mm.15537	(1.3, −1.4*, 1.1)
Cyp2a4	Cytochrome P450, 2a4	Mm.14781	(2.9*, 1.2, 2.6*)
Cyp2b9	Cytochrome P450, 2b9, phenobarbitol inducible type a	Mm.876	(2.8, 4.2*, 2.6)
Cyp3a11	Cytochrome P450, steroid inducible 3a11	Mm.21193	(1.5, 1.8*, 1.4)
Cyp3a16	Cytochrome P450, 3a16	Mm.30303	(2.0*, 2.0*, 1.6*)
Cyp4a10	Cytochrome P450, 4a10	Mm.10742	(2.7*, 4.7*, 3.0*)
Cyp4a14	Cytochrome P450, 4a14	Mm.7459	(8.0*, 19.3*, 2.6)
Cypf13	Cytochrome P450 CYP4F13	Mm.22045	(1.4, −1.1, −1.4)
Por	P450 (cytochrome) oxidoreductase	Mm.3863	(3.9*, 2.2, 4.3*)
Sterol metabolism			
Fpps	Farnesyl pyrophosphate synthase	Mm.39472	(−2.4*, −1.6*, −1.9*)
Hmgcs2	3-Hydroxy-3-methylglutaryl-Coenzyme A synthase 2, mitochondrial (HMG-CoA2)	Mm.10633	(1.3, 1.3, 1.6*)

(continued)

Table 2.1 Continued

Gene symbol	Description	Unigene	Fold change (FUNG/C, FISH/C, FUNG + FISH/C)	FUNG	FISH	COMBO
Lipoprotein metabolism						
Apoa4	Apolipoprotein A-IV	Mm.4533	$(-2.6^*, -3.2^*, -17.2^*)$			
Apoa4	Apolipoprotein A-IV precursor	Mm.4533	$(-3.2^*, -4.2^*, -9.7^*)$			×
ApoB-100	EST highly similar to apolipoprotein B-100 precursor	Mm.29123	$(2.1^*, 1.9^*, 2.0^*)$			
Apoc1	Apolipoprotein CI	Mm.182440	$(-1.7^*, 1.1, 1.0)$			
Apoc2	Apolipoprotein CII	Mm.28394	$(-1.3, -1.5^*, -1.6^*)$			
Apoe	Apolipoprotein E	Mm.138866	$(1.0, 1.5^*, -1.2)$			
Transport						
Abca1	ATP-binding cassette, sub-family A (ABC1) member 1	Mm.369	$(-4.5^*, -2.5^*, -2.0^*)$			
Abca7	ATP-binding cassette, sub-family A (ABC1) member 7	Mm.103351	$(1.0, 1.0, 7.2^*)$			
Abce1	ATP-binding cassette, sub-family E (OABP)member 1	Mm.5831	$(1.0, 2.6^*, 1.1)$			
Aqp8	Aquaporin 8	Mm.9970	$(-2.2^*, -2.1^*, -3.6^*)$			

Gene	Description	Unigene ID	Values
Cd36	CD36 antigen	Mm.18628	(1.1, 2.4*, −1.5)
Dbi	Diazepam binding inhibitor	Mm.2785	(1.0, 1.3, −1.3)
Fabp5	Fatty acid binding protein 5, epidermal (E-FABP)	Mm.741	(−5.5*, −8.4*, −13.6*)
Slc10a1	Solute carrier family 10 (sodium/bile acid cotransporter family), member 1 (Ntcp)	Mm.104295	(1.0, −1.6*, 1.2)
Synpr	Synaptoporin	Mm.41729	(1.0, 1.0, 21.4*)
Oxygen transport, heme biosynthesis			
Alad	Aminolevulinate, delta-, dehydratase	Mm.6988	(1.3, 1.6*, 1.3)
Alas1	Aminolevulinic acid synthase 1	Mm.19143	(3.7*, 1.8*, 2.7*)
Hba-a1	Hemoglobin alpha, adult chain 1	Mm.196110	(−1.0, 1.3, −1.4)
Hbb-b1	Hemoglobin, beta adult major chain	Mm.233825	(−1.1, 1.4, −1.6*)
Response to heat, stress, inflammation			
Herpud1	Homocysteine-inducible, endoplasmic reticulum stress-inducible, ubiquitin-like domain member 1	Mm.29151	(1.6, 1.4, 1.8*)
Hpxn	Hemopexin	Mm.3485	(1.2, 1.0, −1.5)
Hspb1	Heat shock protein 1	Mm.13849	(1.4, −4.0, 1.7)
Mif	Macrophage migration inhibitory factor	Mm.2326	(1.3, −2.3*, −1.1)

(continued)

Table 2.1 Continued

Gene symbol	Description	Unigene	Fold change (FUNG/C, FISH/C, FUNG+FISH/C)	FUNG	FISH	COMBO
Mpeg1	Macrophage expressed gene 1	Mm.3999	(−1.1, −3.6*, −1.4)			
Mup1	Major urinary protein 1	Mm.27508	(1.1, 1.6*, −1.1)			
Other						
Dct	Dopachrome tautomerase	Mm.19987	(−5.2*, −1.1, −2.2*)			
Gtpat12	Gene trap PAT 12	Mm.27668	(3.3, −2.7, 1.2)			
Mpdu1	Mannose-P-dolichol utilization defect 1	Mm.89579	(4.9*, 4.5*, 3.4*)			
Serpinc1	Serine (or cysteine) proteinase inhibitor, clade C (antithrombin), member 1	Mm.30025	(−1.2, 1.6*, −1.4)			
Sdh1	Sorbitol dehydrogenase 1	Mm.104920	(1.6*, 1.2, 1.3)			
Ubc	Ubiquitin C	Mm.331	(−1.6*, −1.2, 1.0)			

Note: The gene selection process is described in the "Materials and Methods" section. Hepatic genes differentially regulated by dietary LC-PUFA. Genes are grouped by functional category, sorted alphabetically, and annotated with Netaffx and NCBI LocusLink. Average difference intensity was used to calculate fold changes. *Fold changes significantly different from the control group, as determined using the 5% LFC gene selection model equation: LFC = 1.52 + (100/min ADI). Genes without an asterisk are included in the table because there is at least one significant group difference between FUNG, FISH, and FUNG+FISH. Heat map on the right—clear box: a fold change of less than 1.3; black box: an upregulation of at least 1.3-fold; gray box: a downregulation of at least 1.3; a × in the FUNG+FISH (Combo) box indicates an unexpected from the fold changes with FUNG and FISH alone (e.g., a synergistic positive or negative effect). A fold change of 1.3 was chosen arbitrarily as a means to maximize the chances of detecting global trends in the data, such as a pattern of upregulation with all three diets (three green bars), and so on. The genes are classified according to Gene Ontology (GO) Consortium classifications.

Table 2.2 Liver Signaling Pathways

Pathway/physiological function	Signaling pathway	Specific metabolism	Specific genes
Lipid metabolism	PPARs	FA β oxidation	*Ech1, Cpt1a, Cpt2, Crat*
		Lipid transport	*Abca1*
		OH-lipid metabolism	*Cyps* (numerous)
		Dolichol metabolism	*Mpdu1*
		PL metabolism	*Gpam*
		Cell growth	*Cd36*
	SREBP	Lipid transport	*Dbi*
		FA synthesis	*Acas1, Fasn, Acly, Thrsp*
		PL metabolism	*Gpam*
		FA desaturation	*Scd*
		Electron transport	*Atp5g1, Ndufa10, Atp6v1e1*
		Oxygen transport and heme	*Alas1, Alad, Hbb-a1*
		Lipid oxidation	*Aldh2, Hpxn, Hspb1*
		Bile secretion?	*Aqp8*
Obesity, hyperlipidemia, diabetes	HNF	Broad	*HNF3γ*
Gluconeogenesis	cAMP		*Pck1*
Transcription initiation/ RNA polymerase	bZIP		*ATF4/5, Dbp*
Hereditary disease	Zn finger		*Hzf*-pending
DNA repair			*Ercc2*
Protein translation			*L23/L28, Rpl28, Lamp2, Dct*
Collagen formation			*Plod3*
Growth and development			*Hoxd12, Pbx1, Numbl, Ngef, Gng4, Rac2, Cd36*
Structural role			*Synpr*
Inflammation/immunity/ antioxidant defense/ coagulation			*TCFs, Dmbt1, Hsps, Serpinc1, Mif*

Note: Summary of Table 2.1, highlighting the key functions of lipids in the liver. A blank cell indicates information is either not available or not relevant. FA, fatty acid; OH-lipid metabolism, referring to metabolism of hydroxylated lipids by Cyps; PL, phospholipids.

PPARα-Regulated Lipid Metabolic Genes

Yamazaki et al. (Chapter 13) have examined the effects of two PPARα agonists on gene transcription in mouse liver (summarized in Table 2.3). Dietary DHA and the potent PPARα agonist Wy-14,643 both activated transcription of genes

Table 2.3   Effects of FISH Feeding and a PPARα Agonist on
Transcription of Selected Genes

Gene symbol	Fold change	
	(FISH/C)	Wy-14,643, 30 mg/kg, Day 3
Fatty acid oxidation		
Gpam	−1.2	1.5
Acadm	1.3	2.0
Aldh1a1/Aldh1a17	1.8−1.9	1.8, aldh3a2
Crat	3.3	3.4
Ech1	1.6	1.7
Cytochrome P450 metabolism		
Cyp2b9	4.2	5.3
Cyp4a10	4.7	2.0
Cyp4a14	19.3	1.8
Lipoprotein metabolism		
Apoa4	−3.2	−2.1
Fabp5	−8.4*	2.4
Slc10a1	−1.6*	4.0 Slc27a1

Note: Fold changes for FISH are extracted from Table 2.1. Fold changes
for Wy-14,643 are from Chapter 12.

involved in FA oxidation; both increased transcription of several *cyp* genes
involved in P450 metabolism; and both decreased the lipoprotein transport
gene *Apoa4* similarly. This data reassures us that DHA is likely signaling
through PPARα. Interestingly, the FA binding protein fabp5 (with a purported
role in psoriasis) was downregulated with oral DHA feeding, but upregulated
with Wy-14,643, suggesting DHA likely affects pathways other than PPARα
signaling.

Pank1 codes for pantothenate kinase, which synthesizes CoA for FACoA
production and over 100 reactions. This includes production of carnitinoyl
CoA for β-oxidation, as well as acyl transferases involved in synthetic processes.
PPARα ligands are now known to activate Pank1.[78] Although we did not find
that LC-PUFA affected pank1 expression, this should be closely examined in
future LC-PUFA experiments because of the central metabolic role of Pank1.

SREBP-Regulated Lipid Metabolic Genes

Maxwell et al.[79] used a combination of cholesterol feeding and SREBP
transgenic mice to identify SREBP targets. Genes upregulated in transgenic
mice overexpressing SREBP1a included genes involved in cholesterol
metabolism: *HMGCoA* reductase, *HMGCoA* synthase, *Aqp8* (−2.2*, −2.1*,
−3.6*; fold changes shown refer to the changes observed in Table 2.1 of our

experiments), *fpps* (-2.4^*, -1.6^*, -1.9^*;); fatty acid synthesis [*Acas1* (-2.5^*, -1.7^*, -4.4^*), *Acly* (-2.7^*, -2.2^*, -4.7^*)]; lipid transport [diazepam binding inhibitor (*dbi*; acyl-CoA binding protein) (1.0, 1.3, -1.3), epidermal *FABP5* (-5.5^*, -8.4^*, -13.6^*)]; G protein signaling linked to inflammation [*Rgs16* (1.1, -1.3, 2.2^*)]; and cAMP signaling (*Camk1D*). Acas1 and *acly* are two cytoplasmic enzymes responsible for synthesizing acetyl CoA for FA synthesis and other purposes. The coordinated downregulation of both acetyl CoA producing transcripts would be consistent with the observation that *acly* heterogeneous knockout mice do not show a compensatory upregulation in *Acas1*.[80] *Acas1* and *dbi* were also upregulated by an LXR agonist, whereas *Rgs16* was downregulated.[79] We observed a downregulation in the related calcium/calmodulin-dependent protein kinase II, *β-Camk2b* (-1.1, -7.6^*, -1.8), and a slight upregulation in the related *Hmgcs2* involved in cholesterol metabolism (1.3, 1.3, 1.6^*). In contrast to Maxwell et al.[79] where cholesterol feeding increased *Aqp8*, Vergnes et al. (Chapter 8) found cholesterol feeding in mice decreased *Aqp8*.

Remarkably similar overall results on SREBP regulated genes were reported by Brown and coworkers.[62] Overexpression of SREBP1a, increased genes involved in cholesterol metabolism (*fpps*, *Aqp8*); FA synthesis (*acas1*, *acly*, *ACC-1*, *FAS*), FA elongation (long chain fatty acyl elongase, *ELOVL2*, *ELOVL5*); Δ5, Δ6, and Δ9 (SCD) desaturation (some metabolic and transcript evidence of downregulation in our experiments); PL metabolism (*gpam*; -1.7^*, -1.2, -2.3^*); FA β-oxidation (*Ech1*; 1.1, 1.6^*, 1.4; *Ech1* is also PPARα-regulated); and lipid peroxidation (*SCALD*, short chain aldehyde reductase[81]). *Aqp8* was increased, but unchanged in Scap$^{-/-}$ mice having no SREBPs. *Spot14* (1.2, -1.1, -1.6^*) was also upregulated. ELOVL2 elongates 20:4n6 to 22:4n6, 20:5n3 to 22:5n, and 22:5n3 to 24:5n3; whereas ELOVL5 elongates 18:3n6 to 20:3n6 and 18:4n3 to 20:4n3.[62] Studies by Maxwell et al.[79] and Horton et al.[62] provide further evidence that LC-PUFA in FUNG and FISH downregulated a number of transcripts via SREBP-mediated signaling. These studies conservatively expand the number of known SREBP-regulated targets altered in our LC-PUFA experiements to include *dbi* involved in lipid transport, *Aqp8* likely involved in bile secretion, and *Rgs16* involved in G protein signaling linked to inflammation. *DBI* and *AQP8* are further described subsequently.

Cholesterol Metabolism

The water channel aquaporin-8 (AQP8) is expressed in hepatocyte intracellular vesicles. The cAMP-induced translocation of AQP8 to the plasma membrane facilitates osmotic water movement during canalicular bile secretion, facilitating hepatic bile secretion.[82] As described earlier, *Aqp8* is an SREBP1a-regulated gene.[62,79] SREBP transgenic mice reportedly have increased bilary cholesterol secretion which may be coupled to this increased water secretion into bile.[62] The efficiency of bilary cholesterol enterohepatic recirculation can affect circulating cholesterol. All three LC-PUFA diets downregulated *Aqp8* (-2.2^*,

−2.1*, −3.6*), consistent with LC-PUFA-downregulated SREBP1a-mediated transcription of *Aqp8*.

We previously showed that FUNG and FISH decreased hepatic *Abca1*[24] (Table 2.1). ABCA1 is a member of the ATP-binding cassette (ABC) superfamily and is induced by PPARγ and LXRα.[83] It has a role in HDL reverse cholesterol pathways. In macrophages, 18:2n6 and AA increased ABCA1 protein degradation without affecting transcriptional levels.[84] It is reassuring to note that Fujiwara and Masumoto (Chapter 4) similarly found AA and DHA as well as EPA to decrease *Abca1* transcripts in hepatic HepG2 cells, relative to olive oil.

Dolichol metabolism

Dolichol is an isoprenoid lipid functioning in protein glycosylation. Mannose-P-dolichol utilization defect 1 (MPDU1) is involved in dolichylphosphomannose and dolichylphosphoglucose metabolism. Mpdu1 mutations exist in congenital disorders of glycosylation.[85] All three LC-PUFA diets increased *mpdu1* (4.9*, 4.5*, 3.4*), perhaps via PPARα signaling since dolichol syntheis from mevalonate is induced by the PPARα agonist clofibrate.[86] Verification at the metabolite level would be an important next step.

Electron Transport

Following glycolysis, FA oxidation, and the citric acid cycle, ATP is formed as NADH and FADH$_2$ transfer electrons to molecular oxygen. In the presence of a proton gradient, the F$_1$ unit of F$_1$F$_0$ ATP synthase [H(+)-ATPase] catalyzes ATP synthesis. The other major unit of ATP synthase is F$_0$, forming a proton channel and containing four types of polypeptide chains. Carriers and translocases move ADP into the inner mitochondrial matrix as ATP exits.

Hepatic ATP synthase H transporting mitochondrial F$_0$ complex subunit c subunit 9 isoform 1 (*Atp5g1*), was downregulated −6.5*-fold with FISH alone, and, curiously, upregulated 2.4*-fold with the combination diet (1.1, −6.5*, 2.4*). The NADH:ubiquinone oxidoreductase (complex I) is the first enzyme complex in the electron transport chain of mitochondria. NDUFA10 is one of eight complex I subunits. NDUFA10 is found within the hydrophobic protein fractions of complex I, and likely participates in proton transfer from mitochondrial matrix to intermembrane space, and may reside in the lipid bilayer.[87] *Ndufa10* was largely upregulated with all three diets (7.2*, 6.3, 8.6*). Our group (Chapter 3) and Farkas et al. (Chapter 5) reported that various electron transport genes are altered by LC-PUFA in rodent brain (see chapters for discussion).

Effects of LC-PUFA on Oxygen Transport and Heme Biosynthesis

There may have been increased hepatic heme synthesis, evidenced by the increase in transcripts for the rate-limiting heme synthetic enzyme *ALAS1* (3.7*, 1.8*, 2.7*), and the second enzyme in the pathway, *ALAD* (1.3, 1.6*,

1.3). Interestingly, there was a decrease or no change in hemoglobulins *Hbb-a1* (-1.0, 1.3, -1.4) and *Hbb-b1* (-1.1, 1.4, -1.6^*). Thus, more heme is seemingly available, but less could be incorporated into hemoglobulins for oxygen transport with the combination diet. The excess heme could be used for many other purposes. For example, it could serve as the heme component of lipidic enzymes needed to handle the increased flux of fatty acids, of eicosanoid enzymes with a heme core such as cyclooxygenase (COX), and be used for heme-catalyzed lipid peroxidation, involving transport proteins such as hemopexin,[88] which was differentially regulated in our experiments. Hemopexin is an acute phase protein which binds heme, inhibiting heme-catalyzed lipid peroxidation.[88] *Hpxn* was downregulated with combination diet relative to FUNG and FISH (1.2, 1.0, -1.5).

To our knowledge, this is the first evidence that dietary LC-PUFA may affect heme synthesis and transport. Next steps are to verify changes to these transcripts by RT-PCR, to examine levels of the actual metabolites involved, and to perform specific functional tests related to hemes.

Lipid Oxidative Damage

Reactive aldehydes are common with alcohol usage, Alzheimer's disease, and following excessive consumption of highly oxidizable LC-PUFA such as DHA. 4-Hydroxy-2-nonenal, for example, is generated by the reaction of superoxide with LC-PUFA. Aldehyde dehydrogenases prevent lipid damage by destroying such reactive aldehydes. Aldehyde dehydrogenase 2 (ALDH2) is a mitochondrial enzyme; *aldh2* transcripts were surprisingly decreased with FISH feeding (1.0, -1.6^*, -1.1). It is important to assess effects of FISH on direct ALDH2 activity. LC-PUFA feeding did increase transcripts of *Aldh1a1* (1.6^*, 1.8^*, 1.7^*) and *Aldh1a7*, consistent with a need to control LC-PUFA-induced reactive aldehyde generation. As described earlier, the related short chain aldehyde dehydrogenase known as SCALD is SREBP1a-regulated.[62,81] It is thus conceivable that other aldehyde dehydrogenases may be SREBP-regulated.

As described earlier, there is the speculative possibility that LC-PUFA may also limit oxidative damage via activation of foxhead transcription factors, and via PPARα-activation of *SCPx*.[30]

Lipid Transport

Diazepam binding inhibitor (DBI) mediates neurotransmission via the γ-aminobutyric acid (GABA) receptor. It is actually identical to acyl-CoA binding protein (ACBP), binding acyl-CoA in nonneural and neural tissues.[89] It has an essential role in transporting acyl-CoA intracellularly. *Dbi* was downregulated with combination diet relative to FUNG and FISH (1.0, 1.3, -1.3). This could be one mechanism of how dietary LC-PUFA feeding ultimately affects acyl CoA transport and acyl distribution in PL.[90] As described earlier, *dbi* is regulated by SREBP1a.[79] Dbi was also altered in PTP1B antisense-treated ob/ob mice having a metabolism that resembles a fish oil fed mice: decreased SREBP-regulated lipogenic genes and altered PPARγ expression (Chapter 17). Taken together, LC-PUFA may affect *dbi* via SREBP1a and possibly PPAR γ-mediated signal transduction.

CONCLUSIONS AND KEY FINDINGS

Figure 2.1 and Table 2.2 represent a summary of the effects of LC-PUFA on hepatic gene expression, categorized by common pathways/physiological function, signaling pathways, and specific metabolism. The n6- and n3-derived LC-PUFA influenced the usual lipid metabolic enzymes, but also affected a plethora of metabolic systems. Separate studies have previously shown that LC-PUFA can affect most of these pathways, but microarrays yield the power to detect global changes to metabolism in a single experiment.

It is notable that LC-PUFA affected genes involved in cell structure, oxygen transport and heme synthesis, and osmotic water movement, as these processes are less commonly known to be LC-PUFA-mediated. It is particularly interesting that LC-PUFA affected other unusual pathways, such as dopachrome tautomerase, a tyrosinase member involved in coat color determination (see the "Other" category of Table 2.1)!

In a striking number of examples, interdiet (FUNG vs. FISH vs. FUNG + FISH) differences in the transcriptional profile were observed, indicating that the bioactive FAs in fish oil (20:5n3, 22:5n3, and 22:6n3) and AA-enriched fungal oil (20:4n6) do not affect gene transcription identically.

Within each category, there are few obvious patterns of complete upregulation or complete downregulation with the three LC-PUFA diets. An exception is that FA β-oxidation was upregulated with all three diets, and FA synthesis was downregulated with all three diets. This is a consistent finding in many recent microarray studies of LC-PUFA and occurs because both AA and DHA can signal through the same PPAR and SREBP pathways, and are not antagonistic to one another.

The combined diet was expected to produce an effect intermediate to that observed with the two individual diets. This was clearly not the case, as there were surprising examples of synergistic positive and negative effects (marked with a × in Table 2.1). Examples include *Hmga1* (− 1.8, 1.2, 6.2*) and *Synpr* (1.0, 1.0, 21.4*). These results should be confirmed with RT-PCR as a next step.

Genes that were notable for their large upregulation were: *Synpr* (21-fold upregulation with combination; encoding structural proteins of synaptic vesicles); *LOC223775*, *pim-3* (7–15, LC-PUFA); *Rac2* (13, FUNG); *Gng4* (9, FISH; specificity for receptor–G protein interactions); *Ndufa10* (6–9, LC-PUFA; role in electron transport); *Numbl* (6, FISH); *Hmga1* (6, combination; nonhistone chromosomal proteins with growth roles); *Mpdu1* (5, FUNG and FISH; dolichol glycosylation); *Dbp* (3–4, FUNG and combination; albumin transcription); and *Alas1*

(4, FUNG; heme synthesis). Genes that were notable for their large downregulation were: *Atp5g1* (-7 fold downregulation with FISH; electron transport); *Dmbt1* (-6, FISH and combination; hepatic tissue repair); *Ngef* (-6, FISH; guanine nucleotide exchange factors for Rho-type GTPases); *Plod3* (-5, FUNG; collagen synthesis); *Anp32a* (-4 to 5, LC-PUFA); *Dct* (-5, FUNG; tyrosinase, coat color determination); *Lamp2* (-4, FUNG; lysosomal enzyme targeting); *Aqp8* (-4, combination; water channel, role in bile secretion); *Mpeg1* (-4, FISH); and *Hspb1* (-4, FISH; related to antioxidant defense). These genes should be the focuses of future inquiry.

Other notable genes were those changed in both the liver (present chapter) and hippocampus, (Chapter 3) including *Dbp* (transcription of albumin); *Hoxd12* (patterning axial and appendicular skeletal elements and the nervous system); *Rpl28* (ribosome assembly); *Hbb-a1*, *Hbb-b1* (hemoglobin classes); *mif* (macrophage motility in immune and inflammation reactions), *ndufa10* (electron transport), and other similar electron transport genes. See Chapter 3 for interorgan comparisons.

Currently, nutritional FA recommendations are made largely by examining levels of key LC-PUFA in plasma; and less often, red blood cells, platelets, and biopsied tissue. Transcriptional gene profiles have not been considered in making such recommendations. The present work convincingly shows the potential importance of examining transcriptional profiles in addition to FA profiles; and the importance of examining more than one tissue, in making such LC-PUFA nutritional and clinical recommendations. The current results provide a first glimpse of potentially significant molecular "players" in a long, but increasingly unraveled, continuing theater play.

REFERENCES

1. Calder, P.C. n-3 Polyunsaturated fatty acids and cytokine production in health and disease. Ann. Nutr. Metab. **1997**, *41*, 203–234.
2. Volker, D.; Fitzgerald, P.; Major, G.; Garg, M. Efficacy of fish oil concentrate in the treatment of rheumatoid arthritis. J. Rheumatol. **2000**, *27*, 2343–2346.
3. Ziboh, V.A.; Miller, C.C.; Cho, Y. Metabolism of polyunsaturated fatty acids by skin epidermal enzymes: generation of antiinflammatory and antiproliferative metabolites. Am. J. Clin. Nutr. **2000**, *71*, 361S–366S.
4. Martinez, M.; Vazquez, E.; Garcia-Silva, M.T.; Manzanares, J.; Bertran, J.M.; Castello, F.; Mougan, I. Therapeutic effects of docosahexaenoic acid ethyl ester in patients with generalized peroxisomal disorders. Am. J. Clin. Nutr. **2000**, *71*, 376S–385S.
5. Bougnoux, P. n-3 Polyunsaturated fatty acids and cancer. Curr. Opin. Clin. Nutr. Metab. Care **1999**, *2*, 121–126.
6. Hu, F.B.; van Dam, R.M.; Liu, S. Diet and risk of type II diabetes: the role of types of fat and carbohydrate. Diabetologia **2001**, *44*, 805–817.
7. Kelley, D.S. Modulation of human immune and inflammatory responses by dietary fatty acids. Nutrition **2001**, *17*, 669–673.

8. Berger, A.; Crozier, G.; Bisogno, T.; Cavaliere, P.; Innis, S.; Di-Marzo, V. Anandamide and diet: inclusion of dietary arachidonate and docosahexaenoate leads to increased brain levels of the corresponding *N*-acylethanolamines in piglets. Proc. Natl. Acad. Sci. USA **2001**, *98*, 6402–6406.

9. Burstein, S.H.; Huang, S.M.; Petros, T.J.; Rossetti, R.G.; Walker, J.M.; Zurier, R.B. Regulation of anandamide tissue levels by *N*-arachidonylglycine. Biochem. Pharmacol. **2002**, *64*, 1147–1150.

10. Cowart, L.A.; Wei, S.; Hsu, M.H.; Johnson, E.F.; Krishna, M.U.; Falck, J.R.; Capdevila, J.H. The CYP4A isoforms hydroxylate epoxyeicosatrienoic acids to form high affinity peroxisome proliferator-activated receptor ligands. J. Biol. Chem. **2002**, *277*, 35105–35112.

11. Huang, S.M.; Bisogno, T.; Trevisani, M.; Al-Hayani, A.; De Petrocellis, L.; Fezza, F.; Tognetto, M.; Petros, T.J.; Krey, J.F.; Chu, C.J.; Miller, J.D.; Davies, S.N.; Geppetti, P.; Walker, J.M.; Di Marzo, V. An endogenous capsaicin-like substance with high potency at recombinant and native vanilloid VR1 receptors. Proc. Natl. Acad. Sci. USA **2002**, *99*, 8400–8405.

12. Serhan, C.N.; Hong, S.; Gronert, K.; Colgan, S.P.; Devchand, P.R.; Mirick, G.; Moussignac, R.L. Resolvins: a family of bioactive products of omega-3 fatty acid transformation circuits initiated by aspirin treatment that counter proinflammation signals. J. Exp. Med. **2002**, *196*, 1025–1037.

13. Walker, J.M.; Huang, S.M. Endocannabinoids in pain modulation. Prostaglandins Leukot Essent. Fatty Acids **2002**, *66*, 235–242.

14. Amer, R.K.; Pace-Asciak, C.R.; Mills, L.R. A lipoxygenase product, hepoxilin A(3), enhances nerve growth factor-dependent neurite regeneration post-axotomy in rat superior cervical ganglion neurons *in vitro*. Neuroscience **2003**, *116*, 935–946.

15. Capdevila, J.H.; Nakagawa, K.; Holla, V. The CYP P450 arachidonate monooxygenases: enzymatic relays for the control of kidney function and blood pressure. Adv. Exp. Med. Biol. **2003**, *525*, 39–46.

16. Chu, C.J.; Huang, S.M.; De Petrocellis, L.; Bisogno, T.; Ewing, S.A.; Miller, J.D.; Zipkin, R.E.; Daddario, N.; Appendino, G.; Di Marzo, V.; Walker, J.M. *N*-Oleoyldopamine, a novel endogenous capsaicin-like lipid that produces hyperalgesia. J. Biol. Chem. **2003**, *278*, 13633–13639.

17. Hong, S.; Gronert, K.; Devchand, P.R.; Moussignac, R.L.; Serhan, C.N. Novel docosatrienes and 17S-resolvins generated from docosahexaenoic acid in murine Brain, human blood, and glial cells. J. Biol. Chem. **2003**, *278*, 14677–14687.

18. Abeywardena, M.Y.; Head, R.J. Longchain n-3 polyunsaturated fatty acids and blood vessel function. Cardiovasc. Res. **2001**, *52*, 361–371.

19. Escudero, A.; Montilla, J.C.; Garcia, J.M.; Sanchez-Quevedo, M.C.; Periago, J.L.; Hortelano, P.; Suarez, M.D. Effect of dietary (n-9), (n-6) and (n-3) fatty acids on membrane lipid composition and morphology of rat erythrocytes. Biochim. Biophys. Acta **1998**, *1394*, 65–73.

20. Jump, D.B. Dietary polyunsaturated fatty acids and regulation of gene transcription. Curr. Opin. Lipidol. **2002**, *13*, 155–164.

21. Kitajka, K.; Puskas, L.G.; Zvara, A.; Hackler, L., Jr.; Barcelo-Coblijn, G.; Yeo, Y.K.; Farkas, T. The role of n-3 polyunsaturated fatty acids in brain: modulation of rat brain gene expression by dietary n-3 fatty acids. Proc. Natl. Acad. Sci. USA **2002**, *99*, 2619–2624.

22. Nakamura, M.T.; Cho, H.P.; Xu, J.; Tang, Z.; Clarke, S.D. Metabolism and functions of highly unsaturated fatty acids: an update. Lipids **2001**, *36*, 961–964.

23. Joseph, S.B.; Castrillo, A.; Laffitte, B.A.; Mangelsdorf, D.J.; Tontonoz, P. Reciprocal regulation of inflammation and lipid metabolism by liver X receptors. Nat. Med. **2003**, *9*, 213–219.

24. Berger, A.; Mutch, D.M.; German, B.J.; Roberts, M.A. Dietary effects of arachidonate-rich fungal oil and fish oil on murine hepatic and hippocampal gene expression. Lipids Health Dis. **2002**, *1*, 2.

25. Mutch, D.M.; Berger, A.; Mansourian, R.; Rytz, A.; Roberts, M.A. The limit fold change model: a practical approach for selecting differentially expressed genes from microarray data. BMC Bioinformatics **2002**, *3*, 17.

26. Nakatani, T.; Kim, H.J.; Kaburagi, Y.; Yasuda, K.; Ezaki, O. A low fish oil inhibits SREBP-1 proteolytic cascade, while a high-fish-oil feeding decreases SREBP-1 mRNA in mice liver: relationship to anti-obesity. J. Lipid Res. **2003**, *44*, 369–379.

27. Yabe, D.; Komuro, R.; Liang, G.; Goldstein, J.L.; Brown, M.S. Liver-specific mRNA for Insig-2 down-regulated by insulin: implications for fatty acid synthesis. Proc. Natl. Acad. Sci. USA **2003**, *100*, 3155–3160.

28. Braun, A. Genome-wide SNP scan for diabetes type 2 susceptibility genes. *Genomics of Diabetes and Associated Diseases in the Post Genome Era Conference*, Lille, France, August 22–24, 2003.

29. Rahman, S.M.; Dobrzyn, A.; Dobrzyn, P.; Lee, S.H.; Miyazaki, M.; Ntambi, J.M. Stearoyl-CoA desaturase 1 deficiency elevates insulin-signaling components and down-regulates protein-tyrosine phosphatase 1B in muscle. Proc. Natl. Acad. Sci. USA **2003**, *100*, 11110–11115.

30. Dansen, T.B.; Kops, G.J.; Denis, S.; Jelluma, N.; Wanders, R.J.; Bos, J.L.; Burgering, B.M.; Wirtz, K.W. Regulation of sterol carrier protein gene expression by the Forkhead transcription factor FOXO3a. J. Lipid Res. **2003**, *45*.

31. Navas, M.A.; Vaisse, C.; Boger, S.; Heimesaat, M.; Kollee, L.A.; Stoffel, M. The human HNF-3 genes: cloning, partial sequence and mutation screening in patients with impaired glucose homeostasis. Hum. Hered. **2000**, *50*, 370–381.

32. Cederberg, A.; Gronning, L.M.; Ahren, B.; Tasken, K.; Carlsson, P.; Enerback, S. *FOXC2* is a winged helix gene that counteracts obesity, hypertriglyceridemia, and diet-induced insulin resistance. Cell **2001**, *106*, 563–573.

33. Braam, B.; Allen, P.; Benes, E.; Koomans, H.A.; Navar, L.G.; Hammond, T. Human proximal tubular cell responses to angiotensin II analyzed using DNA microarray. Eur. J. Pharmacol. **2003**, *464*, 87–94.

34. Ledeen, R.W.; Wu, G. Nuclear lipids: key signaling effectors in the nervous system and other tissues. J. Lipid Res. **2003**, *45*.

35. Hansen, M.B.; Mitchelmore, C.; Kjaerulff, K.M.; Rasmussen, T.E.; Pedersen, K.M.; Jensen, N.A. Mouse Atf5: molecular cloning of two novel mRNAs, genomic organization, and odorant sensory neuron localization. Genomics **2002**, *80*, 344–350.

36. Chen, A.; Muzzio, I.A.; Malleret, G.; Bartsch, D.; Verbitsky, M.; Pavlidis, P.; Yonan, A.L.; Vronskaya, S.; Grody, M.B.; Cepeda, I.; Gilliam, T.C.; Kandel, E.R. Inducible enhancement of memory storage and synaptic plasticity in transgenic mice expressing an inhibitor of ATF4 (CREB-2) and C/EBP proteins. Neuron **2003**, *39*, 655–669.

37. De Angelis, R.; Iezzi, S.; Bruno, T.; Corbi, N.; Di Padova, M.; Floridi, A.; Fanciulli, M.; Passananti, C. Functional interaction of the subunit 3 of RNA polymerase II (RPB3) with transcription factor-4 (ATF4). FEBS Lett. **2003**, *547*, 15–19.

38. Rutkowski, D.T.; Kaufman, R.J. All roads lead to ATF4. Dev. Cell. **2003**, *4*, 442–444.

39. Luo, S.; Baumeister, P.; Yang, S.; Abcouwer, S.F.; Lee, A.S. Induction of grp78/BiP by translational block: activation of the Grp78 promoter by ATF4 through an upstream ATF/CRE site independent of the endoplasmic reticulum stress elements. J. Biol. Chem. **2003**, *278*, 37375–37385.

40. Morris, J.A.; Kandpal, G.; Ma, L.; Austin, C.P. DISC1 (Disrupted-In-Schizophrenia 1) is a centrosome-associated protein that interacts with MAP1A, MIPT3, ATF4/5 and NUDEL: regulation and loss of interaction with mutation. Hum. Mol. Genet. **2003**, *12*, 1591–1608.

41. Pati, D.; Meistrich, M.L.; Plon, S.E. Human Cdc34 and Rad6B ubiquitin-conjugating enzymes target repressors of cyclic AMP-induced transcription for proteolysis. Mol. Cell. Biol. **1999**, *19*, 5001–5013.

42. Lewis, B.A.; Reinberg, D. The mediator coactivator complex: functional and physical roles in transcriptional regulation. J. Cell. Sci. **2003**, *116*, 3667–3675.

43. Sowden, J.; Morrison, K.; Schofield, J.; Putt, W.; Edwards, Y. A novel cDNA with homology to an RNA polymerase II elongation factor maps to human chromosome 5q31 (TCEB1L) and to mouse chromosome 11 (Tceb11). Genomics **1995**, *29*, 145–151.

44. Ogawa, A.; Tsujinaka, T.; Yano, M.; Morita, S.; Taniguchi, M.; Kaneko, K.; Doki, Y.; Shiozaki, H.; Monden, M. Changes of liver-enriched nuclear transcription factors for albumin gene in starvation in rats. Nutrition **1999**, *15*, 213–216.

45. Kataoka, K.; Yoshitomo-Nakagawa, K.; Shioda, S.; Nishizawa, M. A set of Hox proteins interact with the Maf oncoprotein to inhibit its DNA binding, transactivation, and transforming activities. J. Biol. Chem. **2001**, *276*, 819–826.

46. Roncarati, R.; Sestan, N.; Scheinfeld, M.H.; Berechid, B.E.; Lopez, P.A.; Meucci, O.; McGlade, J.C.; Rakic, P.; D'Adamio, L. The gamma-secretase-generated intracellular domain of beta-amyloid precursor protein binds Numb and inhibits Notch signaling. Proc. Natl. Acad. Sci. USA **2002**, *99*, 7102–7107.

47. Rodrigues, N.R.; Theodosiou, A.M.; Nesbit, M.A.; Campbell, L.; Tandle, A.T.; Saranath, D.; Davies, K.E. Characterization of Ngef, a novel member of the Dbl family of genes expressed predominantly in the caudate nucleus. Genomics **2000**, *65*, 53–61.

48. Kalyanaraman, S.; Copeland, N.G.; Gilbert, D.G.; Jenkins, N.A.; Gautam, N. Structure and chromosomal localization of mouse G protein subunit gamma 4 gene. Genomics **1998**, *49*, 147–151.

49. Beitzel, B.; Bushman, F. Construction and analysis of cells lacking the HMGA gene family. Nucl. Acids Res. **2003**, *31*, 5025–5032.

50. Schnabel, C.A.; Godin, R.E.; Cleary, M.L. Pbx1 regulates nephrogenesis and ureteric branching in the developing kidney. Dev. Biol. **2003**, *254*, 262–276.

51. LaRonde-LeBlanc, N.A.; Wolberger, C. Structure of HoxA9 and Pbx1 bound to DNA: Hox hexapeptide and DNA recognition anterior to posterior. Genes Dev. **2003**, *17*, 2060–2072.

52. Chao, S.H.; Walker, J.R.; Chanda, S.K.; Gray, N.S.; Caldwell, J.S. Identification of homeodomain proteins, PBX1 and PREP1, involved in the transcription of murine leukemia virus. Mol. Cell. Biol. **2003**, *23*, 831–841.

53. Chtarbova, S.; Nimmrich, I.; Erdmann, S.; Herter, P.; Renner, M.; Kitajewski, J.; Muller, O. Murine Nr4a1 and Herpud1 are up-regulated by Wnt-1, but the homologous human genes are independent from beta-catenin activation. Biochem. J. **2002**, *367*, 723–728.

54. Wu, Q.; Ji, Y.; Zhang, M.Q.; Chen, Y.Q.; Chen, F.; Shi, D.L.; Zheng, Z.H.; Huang, Y.J.; Su, W.J. Role of tumor metastasis suppressor gene KAI1 in digestive tract carcinomas and cancer cells. Cell. Tissue Res. **2003**.

55. Noble, J.A.; White, A.M.; Lazzeroni, L.C.; Valdes, A.M.; Mirel, D.B.; Reynolds, R.; Grupe, A.; Aud, D.; Peltz, G.; Erlich, H.A. A polymorphism in the TCF7 gene, *C883A*, is associated with type 1 diabetes. Diabetes **2003**, *52*, 1579–1582.

56. Bisgaard, H.C.; Holmskov, U.; Santoni-Rugiu, E.; Nagy, P.; Nielsen, O.; Ott, P.; Hage, E.; Dalhoff, K.; Rasmussen, L.J.; Tygstrup, N. Heterogeneity of ductular reactions in adult rat and human liver revealed by novel expression of deleted in malignant brain tumor 1. Am. J. Pathol. **2002**, *161*, 1187–1198.

57. Fontaine, J.M.; Rest, J.S.; Welsh, M.J.; Benndorf, R. The sperm outer dense fiber protein is the 10th member of the superfamily of mammalian small stress proteins. Cell Stress Chaperones **2003**, *8*, 62–69.

58. Yi, M.J.; Park, S.H.; Cho, H.N.; Yong Chung, H.; Kim, J.I.; Cho, C.K.; Lee, S.J.; Lee, Y.S. Heat-shock protein 25 (Hspb1) regulates manganese superoxide dismutase through activation of Nfkb (NF-kappaB). Radiat. Res. **2002**, *158*, 641–649.

59. Maizels, E.T.; Peters, C.A.; Kline, M.; Cutler, R.E.; Shanmugam, M.; Hunzicker-Dunn, M. Heat-shock protein-25/27 phosphorylation by the delta isoform of protein kinase C. Biochem. J. **1998**, *332* (Pt 3): 703–712.

60. Adam, O. Dietary fatty acids and immune reactions in synovial tissue. Eur. J. Med. Res. **2003**, *8*, 381–387.

61. Lippert, E.; Yowe, D.L.; Gonzalo, J.A.; Justice, J.P.; Webster, J.M.; Fedyk, E.R.; Hodge, M.; Miller, C.; Gutierrez-Ramos, J.C.; Borrego, F.; Keane-Myers, A.; Druey, K.M. Role of regulator of g protein signaling 16 in inflammation- induced T lymphocyte migration and activation. J. Immunol. **2003**, *171*, 1542–1555.

62. Horton, J.D.; Shah, N.A.; Warrington, J.A.; Anderson, N.N.; Park, S.W.; Brown, M.S.; Goldstein, J.L. Combined analysis of oligonucleotide microarray data from transgenic and knockout mice identifies direct SREBP target genes. Proc. Natl. Acad. Sci. USA **2003**, *100*, 12027–12032.

63. Terp, B.N.; Cooper, D.N.; Christensen, I.T.; Jorgensen, F.S.; Bross, P.; Gregersen, N.; Krawczak, M. Assessing the relative importance of the biophysical properties of amino acid substitutions associated with human genetic disease. Hum. Mutat. **2002**, *20*, 98–109.

64. Schmidt, E.B.; Kristensen, S.D.; Sorensen, P.J.; Dyerberg, J. Antithrombin III and protein C in stable angina pectoris–influence of dietary supplementation with polyunsaturated fatty acids. Scand. J. Clin. Lab. Invest. **1988**, *48*, 469–473.

65. Jabrane-Ferrat, N.; Nekrep, N.; Tosi, G.; Esserman, L.; Peterlin, B.M. MHC class II enhanceosome: how is the class II transactivator recruited to DNA-bound activators? Int. Immunol. **2003**, *15*, 467–475.

66. Kotaja, N.; Vihinen, M.; Palvimo, J.J.; Janne, O.A. Androgen receptor-interacting protein 3 and other PIAS proteins cooperate with glucocorticoid receptor-interacting protein 1 in steroid receptor-dependent signaling. J. Biol. Chem. **2002**, *277*, 17781–17788.

67. Schneider, A.; Peukert, K.; Eilers, M.; Hanel, F. Association of Myc with the zinc-finger protein Miz-1 defines a novel pathway for gene regulation by Myc. Curr. Top. Microbiol. Immunol. **1997**, *224*, 137–146.

68. Tussie-Luna, M.I.; Michel, B.; Hakre, S.; Roy, A.L. The SUMO ubiquitin-protein iso-peptide ligase family member Miz1/PIASxbeta/Siz2 is a transcriptional cofactor for TFII-I. J. Biol. Chem. **2002**, *277*, 43185–43193.

69. Maguire, B.A.; Wild, D.G. The roles of proteins L28 and L33 in the assembly and function of *Escherichia coli* ribosomes *in vivo*. Mol. Microbiol. **1997**, *23*, 237–245.

70. Frigerio, J.M.; Dagorn, J.C.; Iovanna, J.L. Cloning, sequencing and expression of the L5, L21, L27a, L28, S5, S9, S10 and S29 human ribosomal protein mRNAs. Biochim. Biophys. Acta **1995**, *1262*, 64–68.

71. Eskelinen, E.L.; Tanaka, Y.; Saftig, P. At the acidic edge: emerging functions for lysosomal membrane proteins. Trends Cell. Biol. **2003**, *13*, 137–145.

72. Ruotsalainen, H.; Vanhatupa, S.; Tampio, M.; Sipila, L.; Valtavaara, M.; Myllyla, R. Complete genomic structure of mouse lysyl hydroxylase 2 and lysyl hydroxylase 3/collagen glucosyltransferase. Matrix Biol. **2001**, *20*, 137–146.

73. Hankenson, K.D.; Watkins, B.A.; Schoenlein, I.A.; Allen, K.G.; Turek, J.J. Omega-3 fatty acids enhance ligament fibroblast collagen formation in association with changes in interleukin-6 production. Proc. Soc. Exp. Biol. Med. **2000**, *223*, 88–95.

74. Beermann, F.; Rossier, A.; Guyonneau, L. Inactivation of the mouse Dct gene. Pigment Cell. Res. **2003**, *16*, 577–578.

75. Gutzkow, K.B.; Lahne, H.U.; Naderi, S.; Torgersen, K.M.; Skalhegg, B.; Koketsu, M.; Uehara, Y.; Blomhoff, H.K. Cyclic AMP inhibits translation of cyclin D3 in T lymphocytes at the level of elongation by inducing eEF2-phosphorylation. Cell Signal **2003**, *15*, 871–881.

76. Bischoff, C.; Kahns, S.; Lund, A.; Jorgensen, H.F.; Praestegaard, M.; Clark, B.F.; Leffers, H. The human elongation factor 1 A-2 gene (*EEF1A2*): complete sequence and characterization of gene structure and promoter activity. Genomics **2000**, *68*, 63–70.

77. Valasek, L.; Mathew, A.A.; Shin, B.S.; Nielsen, K.H.; Szamecz, B.; Hinnebusch, A.G. The yeast eIF3 subunits TIF32/a, NIP1/c, and eIF5 make critical connections with the 40S ribosome in vivo. Genes Dev. **2003**, *17*, 786–799.

78. Ramaswamy, G.; Karim, M.A.; Murti, K.G.; Jackowski, S. PPARα controls the intracellular coenzyme A concentration via regulation of PANK1α gene expression. J. Lipid Res. **2003**, *45*.

79. Maxwell, K.N.; Soccio, R.E.; Duncan, E.M.; Sehayek, E.; Breslow, J.L. Novel putative SREBP and LXR target genes identified by microarray analysis in liver of cholesterol-fed mice. J. Lipid Res. **2003**, *44*, 2109–2119.

80. Beigneux, A.P.; Kosinski, C.; Gavino, B.; Horton, J.D.; Young, S.G.; Shah, N.A.; Warrington, J.A.; Anderson, N.N.; Park, S.W.; Brown, M.S.; Goldstein, J.L. ATP-citrate lyase deficiency in the mouse. Combined analysis of oligonucleotide microarray data from transgenic and knockout mice identifies direct SREBP target genes. J. Biol. Chem. **2003**, *100*, 12027–12032.

81. Kasus-Jacobi, A.; Ou, J.; Bashmakov, Y.K.; Shelton, J.M.; Richardson, J.A.; Goldstein, J.L.; Brown, M.S. Characterization of mouse short-chain aldehyde reductase (SCALD), an enzyme regulated by sterol regulatory element-binding proteins. J. Biol. Chem. **2003**, *278*, 32380–32389.

82. Carreras, F.I.; Gradilone, S.A.; Mazzone, A.; Garcia, F.; Huang, B.Q.; Ochoa, J.E.; Tietz, P.S.; Larusso, N.F.; Calamita, G.; Marinelli, R.A. Rat hepatocyte aquaporin-8 water channels are down-regulated in extrahepatic cholestasis. Hepatology **2003**, *37*, 1026–1033.
83. Chinetti, G.; Lestavel, S.; Bocher, V.; Remaley, A.T.; Neve, B.; Torra, I.P.; Teissier, E.; Minnich, A.; Jaye, M.; Duverger, N.; Brewer, H.B.; Fruchart, J.C.; Clavey, V.; Staels, B. PPARα and PPARγ activators induce cholesterol removal from human macrophage foam cells through stimulation of the ABCA1 pathway. Nat. Med. **2001**, *7*, 53–58.
84. Wang, Y.; Oram, J.F. Unsaturated fatty acids inhibit cholesterol efflux from macrophages by increasing degradation of ATP-binding cassette transporter A1. J. Biol. Chem. **2002**, *277*, 5692–5697.
85. Schenk, B.; Imbach, T.; Frank, C.G.; Grubenmann, C.E.; Raymond, G.V.; Hurvitz, H.; Korn-Lubetzki, I.; Revel-Vik, S.; Raas-Rotschild, A.; Luder, A.S.; Jaeken, J.; Berger, E.G.; Matthijs, G.; Hennet, T.; Aebi, M. MPDU1 mutations underlie a novel human congenital disorder of glycosylation, designated type If. J. Clin. Invest. **2001**, *108*, 1687–1695.
86. Shiota, Y.; Ikeda, M.; Hashimoto, F.; Hayashi, H. Effects of peroxisome proliferators gemfibrozil and clofibrate on syntheses of dolichol and cholesterol in rat liver. J. Biochem. (Tokyo) **2003**, *134*, 197–202.
87. Loeffen, J.L.; Triepels, R.H.; van den Heuvel, L.P.; Schuelke, M.; Buskens, C.A.; Smeets, R.J.; Trijbels, J.M.; Smeitink, J.A. cDNA of eight nuclear encoded subunits of NADH:ubiquinone oxidoreductase: human complex I cDNA characterization completed. Biochem. Biophys. Res. Commun. **1998**, *253*, 415–422.
88. Vincent, S.H.; Grady, R.W.; Shaklai, N.; Snider, J.M.; Muller-Eberhard, U. The influence of heme-binding proteins in heme-catalyzed oxidations. Arch. Biochem. Biophys. **1988**, *265*, 539–550.
89. Yanase, H.; Shimizu, H.; Yamada, K.; Iwanaga, T. Cellular localization of the diazepam binding inhibitor in glial cells with special reference to its coexistence with brain-type fatty acid binding protein. Arch. Histol. Cytol. **2002**, *65*, 27–36.
90. Kannan, L.; Knudsen, J.; Jolly, C.A. Aging and acyl-CoA binding protein alter mitochondrial glycerol-3-phosphate acyltransferase activity. Biochim. Biophys. Acta **2003**, *1631*, 12–16.

3

Dietary Effects of Arachidonate-Rich Fungal Oil and Fish Oil on Murine Hippocampal Gene Expression

Alvin Berger

Paradigm Genetics, Inc., Research Triangle Park, USA

Matthew A. Roberts

Nestle Global Strategic Business Unit and R&D, St. Louis, USA

INTRODUCTION

The incorporation of dietary long-chain polyunsaturated fatty acids (LC-PUFAs) into various brain regions, and the subsequent effects on behavior have been studied for many years. In humans, LC-PUFA are reported to affect diverse processes including sleep,[1] psychiatric disorders (schizophrenia, autism, bipolar disorder, depression, violent and suicidal behavior, dyslexia, attention-deficit/ hyperactivity disorder, and apraxis of speech),[2-9] and alcohol intake.[10] Recently, the importance of lipid storage in nerves for proper functioning has also been studied.[11] As arachidonic acid (AA) and docosahexaenoic acid (DHA) are both components of breast milk, there is also keen interest on the effects of these LC-PUFA on cognition of the developing offspring.[12,13]

Rodents represent ideal animals to study behavioral influences of LC-PUFA because of the ease of controlling experimental parameters and similarities in LC-PUFA metabolism. In many rodent studies, overt deficiencies of n3 LC-PUFA were used to exaggerate any observed effects on behavior and neural transmission.[14-19] In rodents, LC-PUFA are reported to affect auditory

brain stem conduction times,[20] Morris water maze ability,[21] habituation, exploratory activity, active avoidance,[22,23] and aggressive behavior.[24] The precise
mechanisms of action of LC-PUFA on behavior are not known. Model, simple
organisms are now being used to understand better the mechanistic effects of
LC-PUFA on behavior. For example, LC-PUFA may affect neurotransmitter
release independent of conversion to eicosanoids in *Caenorhabditis elegans*,
which does not produce eicosanoids.[25]

To date, most researchers have administered LC-PUFA and then, acting on
a priori information or hunches, tediously measured changes to particular metabolites and hormones (e.g., the serotonin and dopaminergic systems), hoping to see
an effect and then make a soft, mechanism of action connection. Microarrays and
metabolomic approaches, combined with receptor-inhibition approaches, will
allow us to more efficiently and expeditiously understand how LC-PUFA
affects diverse behaviors. A combined microarray, and lipid-metabolomic
approach was thus utilized in the present experiment. Our behavioral outcomes
will be published in the near future.

Mice were fed control diets adequate in 18:2n6 and 18:3n3 but lacking
LC-PUFA; or a diet enriched in fungal oil (FUNG) enriched in AA, the n6
elongation and desaturation product of 18:2n6; or a diet containing fish oil
(FISH) enriched in 22:6n3 and 20:5n3, the major n3 LC-PUFA elongation and
desaturation products of 18:3n3; or a diet containing both fungal oil and fish
oil (FUNG + FISH). Thereafter, gene expression profiling was performed in
the hippocampus in parallel with quantitative metabolic profiling of a broad spectrum of hippocampal fatty acids (FAs).

The brain is the major neurological tissue and principle site for accumulation and functionality of DHA (and other fish oil components). The hippocampus brain region was specifically chosen because of its importance in
memory and learning.[26] Diets were fed for 57 days to study chronic and sustained alterations in control of gene expression rather than acute effects.

MATERIALS AND METHODS

All of the experimental procedures utilized have been previously published.[27,32]
This includes a description of the experimental diets, feeding and dissection conditions, nucleic acid preparation, gene expression analysis using the murine 11k
GeneChip, and lipid analysis. The selection of genes was modified for the current
chapter. The 5% LFC method[32] indicated that 513 genes in the hippocampus
were differentially regulated by LC-PUFA from the original 13,179 genes represented on the Mu11K Affymetrix GeneChip. After removing those genes that
were identified as absent across all dietary conditions, 356 differentially
expressed probe sets remained. Of these, 221 genes represented annotated
sequences. Of these, 22 genes were selected because they fell into Gene Ontology
(GO) Consortium classifications of interest: nucleic acid binding, lipoprotein
metabolism, and miscellaneous genes.[27] Herein, to expand our knowledge of
how dietary LC-PUFA broadly affect metabolism, independent of GO category,

another 42 genes were selected on the basis of having the largest interdietary group fold changes. Thus, 64 genes are included in Table 3.1. A summary is presented in Table 3.2.

In addition to the lipid analyses reported previously, brain samples were also separately extracted, and NAEs and monoacylglycerols (MAGs) purified and characterized (Table 3.3).[31]

RESULTS AND DISCUSSION

Incorporation of LC-PUFA into Phospholipid Pools

The incorporation of 20:4n6 and 22:4n6 following feeding of the AA-rich FUNG diet, and the incorporation of 20:5n3, 22:5n3, and 22:6n3 following feeding of the eicosapentaenoic acid (EPA)/DHA-rich FISH diet, into whole brain, hippocampal, and hepatic phospholipid (PL) pools.[27] provided evidence the diets were consumed, resulting in significant changes to PL acyl composition. Fish oil feeding is well established to increase n3 LC-PUFA in hepatic PL as well as whole brain PL,[28] whereas deficiency of n3 LC-PUFA can decrease 22:6n3 in hippocampal PL.[29] DHA present in the brain can be formed from precursors in the liver or astrocytes, and intact DHA passes the blood–brain barrier via selective transport mechanisms.[30]

The feeding of FUNG also led to a 5.8*-fold increase in 20:4n6 *N*-acylethanolamine (NAE) in whole mouse brain (Table 3.3), similarly to that observed in piglets.[31] NAEs could not be detected in individual mouse brain regions because of insufficient mass of the brain regions. In contrast to findings in piglets,[31] FISH feeding did not increase 22:6n3 NAE in the mouse, for reasons that are not evident.

Gene Expression Analysis Overview

The statistical and modeling procedures for selecting our 356 differentially expressed genes in the hippocampus using a limit fold change (LFC) model have been previously described.[27,32] The interpretation of all 356 hippocampal transcripts is beyond the scope of the current chapter, but all raw data is deposited at the NCBI gene expression Omnibus.[33]

In the present chapter, genes that are discussed are followed by parentheses indicating fold changes according to the following convention: FUNG vs. control diet, FISH vs. control diet, and FUNG + FISH vs. control diet [e.g., *G6pc* (1.1, 1.6*, − 1.9*)]. Those fold changes that are significant by the 5% LFC model are further identified with an asterisk (∗). It is important to note that the nature of the LFC model allows differentially expressed genes to be selected on the basis of differences between the treatments, and therefore explains why some genes discussed in the text have no fold changes marked with an asterisk (as significance was observed between treatments and not vs. the control diet). The high degree of concordance between real time-polymerase chain reaction (RT-PCR)

Table 3.1 Differentially Expressed Genes in Response to Dietary LC-PUFAs in Hippocampus

Gene symbol	Description	Unigene	Fold change (FUNG/C, FISH/C, FUNG + FISH/C)	FUNG	FISH	COMBO
Structural role						
Acta2	Actin, alpha 2, smooth muscle, aorta	Mm.16537	(5.3*, 2.1, 2.7*)			
Actb	Actin, beta, cytoplasmic	Mm.297	(1.4*, 1.5*, 1.6*)			
Lama3	Laminin, alpha 3	Mm.42012	(−3.3, 3.1*, −3.3)			×
Rpa1	Replication protein A1	Mm.180734	(1.9, −4.3, 1.3)			
Smarca4	SWI/SNF related, matrix associated, actin-dependent regulator of chromatin, subfamily a, member 4	Mm.200406	(1.0, 1.1, −3.3*)			×
Nucleic acid binding						
Ar	Androgen receptor	Mm.4470	(2.5*, 1.6, 1.2)			
Atoh4	Atonal homolog 4 (Drosophila)	Mm.42017	(4.6, 3.5, 5.7*)			
Btf3	Basic transcription factor 3	Mm.1538	(−1.9*, −2.1*, −1.1)			
Dbp	D-site albumin promoter binding protein	Mm.3459	(1.9, −1.3, 1.4)			
Ddb1	Damage specific DNA binding protein 1	Mm.29623	(1.3, 1.0, −1.7)			
Dscr1	Down syndrome critical region homolog 1 (human)	Mm.56	(−1.2, −2.2*, −2.3*)			

Gene	Description	UniGene	Values
E2f3	E2F transcription factor 3	Mm.6333	(−2.0, 1.9, −1.4)
Elk1	ELK1, member of ETS oncogene family	Mm.3064	(−1.4, 1.1, −1.5)
Hdac1 (Rpd3)	High homology to histone deacetylase 1	Mm.2602	(−1.7, 1.0, −1.1)
Hoxd12	Homeo box D12	Mm.57124	(−1.2, −1.1, −1.9*)
Idb1	Inhibitor of DNA binding 1	Mm.444	(1.1, 2.3*, 1.6)
Idb3	Inhibitor of DNA binding 3	Mm.110	(2.9, 3.8*, 3.4)
Nhlh1	Nescient helix loop helix 1	Mm.2474	(1.8, −1.4, 1.2)
Pax9	Paired box gene 9	Mm.5035	(−5.2, 1.6, −1.4)
Pcbp3	Poly(rC) binding protein 3	Mm.143816	(1.0, −3.6*, 1.0)
Purb	Purine-rich element binding protein B	Mm.154651	(−1.2, −2.3*, −1.6)
Cell growth factors, angiogenic factors, oncogenes, protein kinases, TFs			
Basp1	Brain abundant, membrane-attached signal protein 1	Mm.29586	(−1.1, 1.1, 1.1)
Enpp2	Ectonucleotide pyrophosphatase/phosphodiesterase 2	Mm.28107	(1.3, −1.2, 1.7*)
Fcer1g	Fc receptor, IgE, high affinity I, gamma polypeptide	Mm.22673	(1.9*, −1.3, 2.6*)
Gnb2	Guanine nucleotide binding protein, beta 2	Mm.30141	(−1.1, −1.1, −2.0*)
Gria2	Glutamate receptor, ionotropic, AMPA2 (alpha 2)	Mm.220224	(1.2, 1.4, 1.5*)

(continued)

Table 3.1 Continued

Gene symbol	Description	Unigene	Fold change (FUNG/C, FISH/C, FUNG+FISH/C)	FUNG	FISH	COMBO
Htr4	5 hydroxytryptamine (serotonin) receptor 4 (5-HT)	Mm.20440	(6.5*, 3.8, 4.5)			
Igf2	Insulin-like growth factor 2	Mm.3862	(2.3*, 1.2, 2.1*)			
Il6st	Interleukin 6 signal transducer	Mm.4364	(1.8, −5.3, −1.1)			
Nov	Nephroblastoma overexpressed gene	Mm.5167	(2.8*, 2.1*, 3.5*)			
Polr2i	Polymerase (RNA) II (DNA directed) polypeptide I	Mm.29917	(7.7*, 6.3*, 8.5*)			
Pomc1	Pro-opiomelanocortin-alpha	Mm.21878	(6.0*, 1.6, 1.4)			
Prkcd	Protein kinase C, delta (PKC)	Mm.2314	(−1.6, −1.3, −1.8*)			
Ramp2	Receptor (calcitonin) activity modifying protein 2	Mm.218611	(1.4, 1.4*, 1.4*)			
Sst	Somatostatin	Mm.2453	(1.8*, 1.8*, 2.1*)			
Energy metabolism, electron transport						
Ant1/Slc25a4	Solute carrier family 25 (mitochondrial carrier; adenine nucleotide translocator), member 4	Mm.16228	(1.3, 1.4, 1.5)			

	Gene	Description	Accession	Values
	ATP5D	Homolgous to ATP synthase, H+ transporting, mitochondrial F1 complex, delta subunit	Mm.22514	(−1.5, −1.6*, −1.1)
	Atp5c1	ATP synthase, H+ transporting, mitochondrial F1 complex, gamma polypeptide 1	Mm.12677	(−1.3, 1.0, 1.2)
	Atp6m	ATPase, H+ transporting lysosomal (vacuolar proton pump)	Mm.30206	(−1.1, 1.0, 1.4)
	Cox8a	Cytochrome c oxidase, subunit VIIIa	Mm.14022	(−1.1, 1.1, 1.0)
	Ndufa10	NADH dehydrogenase (ubiquinone) 1 alpha subcomplex 10	Mm.28293	(7.7*, 1.0, 6.2*)
Amino acid and protein metabolism				
	Cpe	Carboxypeptidase E	Mm.31395	(1.1, 1.1, −1.2)
	Mrps11	Mitochondrial ribosomal protein S11	Mm.28378	(−2.0, −11.2*, −1.1)
	Rpl27a	Ribosomal protein L27a	Mm.142380	(−1.3, 1.1, −1.1)
	Rpl28	Ribosomal protein L28	Mm.3111	(−1.2, −1.2, −1.5*)
	Ubb	Ubiquitin B	Mm.235	(1.2, 1.2, 1.0)
	Vars2	Valyl-tRNA synthetase 2	Mm.28420	(2.3*, −10.2*, 2.1*)
Lipid and lipoprotein metabolism				
	Apod	Apolipoprotein D	Mm.2082	(1.3, 1.8*, 1.4)
	Ptgds	Prostaglandin D2 synthase (brain)	Mm.1008	(5.0*, 3.0*, 5.3*)

(continued)

Table 3.1 Continued

Gene symbol	Description	Unigene	Fold change (FUNG/C, FISH/C, FUNG + FISH/C)	FUNG	FISH	COMBO
Transport						
Ap2a2	Adaptor protein complex AP-2, alpha 2 subunit	Mm.180577	(−2.0*, −1.1, −1.6*)	gray		gray
Oxygen transport, heme biosynthesis						
Hba-a1	Hemoglobin alpha, adult chain 1	Mm.196110	(1.8*, 1.4, 1.3)	black	black	black
Hbb-b1	Hemoglobin, beta adult major chain	Mm.233825	(1.4, 1.4, 1.5*)	black	gray	black
Response to heat, stress, inflammation						
Mif	Macrophage migration inhibitory factor	Mm.2326	(−3.3, −4.9, 1.6)	gray	gray	gray
Other						
Cd81	CD 81 antigen	Mm.806	(1.0, −1.1, −1.7*)			gray
Chgb	Chromogranin B	Mm.1339	(−1.4, −1.1, −1.4*)	gray	black	gray
Dcn	Decorin	Mm.56769	(3.7, 2.5, 4.6)	black	black	black
Nnat	Neuronatin	Mm.140956	(1.4, 1.2, 1.6*)		black	black
Ryr3	Ryanodine receptor 3	Mm.3441	(1.0, 1.0, 8.0*)			black

Scn1b	Sodium channel, voltage-gated, type I, beta polypeptide	Mm.1418	(−1.2, −1.1, −1.3)
Siat8e	Sialyltransferase 8 (alpha 2, 8 sialytransferase) E	Mm.5173	(1.6, −2.6, −6.4*)
Timp2	Tissue inhibitor of metalloproteinase 2	Mm.206505	(1.4, −3.1*, 1.3)
Ttr	Transthyretin	Mm.2108	(1.6, −1.9*, 2.3*)
Vamp3	Vesicle-associated membrane protein 3	Mm.168744	(9.1*, 9.6*, 8.2*)
Zfp385	Zinc finger protein 385	Mm.14099	(1.4, −1.8*, −1.3)

Note: The gene selection process is described in the "Mateials and Methods" section. Hippocampal genes differentially regulated by dietary LC-PUFA. Genes are grouped by functional category, sorted alphabetically, and annotated with Netaffx and NCBI LocusLink. Average difference intensity was used to calculate fold changes. An asterisk indicates fold changes significantly different from the control group, as determined using the 5% LFC gene selection model equation: LFC = 1.52 + (100/min ADI). Genes without an asterisk are included because there is at least one significant group difference between FUNG, FISH, and FUNG + FISH. Heat map on the right—clear box, a fold change of less than 1.3; black box, an upregulation of at least 1.3-fold; gray box, a downregulation of at least 1.3; an × in the FUNG + FISH (Combo) box, an unexpected from the fold changes with FUNG and FISH alone (e.g., a synergistic positive or negative effect). A fold change of 1.3 was chosen arbitrarily as a means to maximize the chances of detecting global trends in the data, such as a pattern of upregulation with all three diets (three black bars), and so on. The genes are classified according to Gene Ontology (GO) Consortium classifications.

Table 3.2 Hippocampus Signaling Pathways

Pathway/physiological function	Signaling pathway	Specific metabolism	Specific genes
Lipid metabolism	PG	PG metabolism, sleep	*Ptgds*
		Electron transport	*ATP5g1*, *Ndufa10*, *Atp6v1e1*
		Oxygen transport and heme	*Hbb-a1*, *Hbb-b1*
Albumin synthesis/fat transport	bZIP		*Dbp*
Growth, differentiation, development	Ar	Differentiation	*Ar*
	bHLH	Proliferation	*Atoh4*
	Rb	Development	*E2f3*
	Hox-MAF	Proliferation, development	*Hoxd12*
			Pax9
	Nov–Notch	Proliferation, differentiation, adhesion	*Nov*
	TGFβ-r, PDGF-r	Growth, FFA release, amyloid deposition	*Dcn*
	TIMP-MMP	Proliferation, membrane integrity, tumoral invasion	*Timp-2*
Structural role	AP-1, USF-1	Assessibility nucleosomal DNA	*Smarca4*
			Lama3
	Inositol	Intracellular transport, clathrin coat assembly, cargo selection	*Ap2a2*
		Inhibit platelet secretion	*Vamp3*
			Acta2

Category	Molecule / pathway	Role (RNA pol II)	Gene(s) (BTF3)
Transcription initiation/RNA polymerase	Sp1	RNA pol II	BTF3
			Polr2i
Protein translation/AA metabolism		Ribosome assembly	Rpl27a, Rpl28
		Val transfer	Vars2
	G protein	Broad	Gnb2
Inflammation/immunity/antioxidant defense/coagulation/Ca metabolism	Cytokine	IL-6 and -11, leukemia inhibitory factor, oncostatin M, neurotrophic factor	Il6st
	IgE-r	Allergic Rxns	Fcer1g
	O_2 stress, Ca inducible	Oxidative stress	Dscr1
	Ryanodine-r	Ca release	Ryr3
	Macrophage signaling	Inflammation, immune	Mif
Memory consolidation?	PKCδ?	Polysialyltransferase activity; neural cell adhesion	Siat8e, Prkcd
Normal learning/memory, energy homeostasis, growth	Leptin cross-talk?		Sst
Neuronal growth, survival, cell motility/differentiation, learning ability?	gp78-r		Gpi1
Normal synaptic functions	Glutamate-r		gria2
Appetite suppression	POMC → Adrenocorticotrophin, β-endorphin melanocyte-stimulating hormone → neuronal melanocortin receptors. Leptin cross-talk?		Pomc

Note: Summary of Table 1, highlighting the key functions of lipids in the hippocampus. AA, amino acid; Ca, calcium; PG, prostaglandin; r, receptor.

Table 3.3 Levels of NAE in Whole Brain

	NAE		
Diet	16:0	20:4	22:6
Control	558.9 ± 75.0	21.8 ± 8.7	13.4 ± 2.8
AA	488.9 ± 36.1	125.8 ± 18.9**	14.8 ± 4.3
DHA	496.6 ± 188.4	16.6 ± 4.5	3.4 ± 0.7*
AA + DHA	330.5 ± 20.4*	15.7 ± 7.0	2.9 ± 0.9*

Note: There was evidence that FUNG feeding increased levels of the corresponding NAE product. NAE, *N*-acylethanolamine; refer to previous tables for other abbreviations. Values represent mean NAE concentration in pmol/g wet weight brain tissue, following five determinations ± standard error.
Source: Data adapted from Berger et al. 1998.
*$p < 0.05$; **$p < 0.01$ vs. control; one-sided Student's *t*-test.

and microarray data (86%—reported in detail elsewhere[32]) establishes a level of confidence in the results discussed throughout this manuscript.

There have been few studies to date examining the effects of LC-PUFA on hippocampal and brain gene transcription using microarrays, with most coming from two groups besides our own.[28,34–36] In a recent work, genetic mutations in an enzyme involved in fat oxidation on hippocampal gene expression were studied.[37] Others have used microarrays to examine the effects of nutrients (*Gingko biloba*),[38] ethanol,[39] acute and chronic exercise,[40] disease such as Alzheimer's,[41] aging[42] and learning tasks[43–45] on hippocampal genes. It should be appreciated that the effects of LC-PUFA on hippocampal gene expression are best studied by examining specific hippocampal neurons rather than the entire hippocampus.[46]

Effects of LC-PUFA on Transcription Factors and Growth Factors (Liver vs. Hippocampus)

A notable observation was that in the liver dietary LC-PUFA signaled through the well-known peroxisome proliferator activated receptor (PPAR), sterol regulatory element binding protein (SREBP), and hepatic nuclear factor (HNF) transcription factors,[27] whereas in the hippocampus, a completely different set of transcription factors were activated as described subsequently. Nuclear factor of κ light chain gene enhancer in B cells 1 p105 (*Nfκb1*), is biologically important, and had a rank of 139 (thus, not included in the tables).

It should not be surprising that FA metabolism is differentially regulated in liver and hippocampus, since lipids serve a myriad of different purposes in the two organs. As one example, LC-PUFA downregulated both acyl and acas1 in the liver. These cytosolic enzymes are needed to produce acetyl CoA for FA synthesis and other purposes. But in the brain, acyl is needed to produce acetyl-CoA in cholinergic neurons for synthesis of the key neurotransmitter acetylcholine synthesis;[47] thus a downregulation in hippocampal acyl could be detrimental to the organism.

Androgen Receptor

Ar encodes androgen receptor (AR), mediating androgen-induced cell growth. The LC-PUFA AA and EPA can inhibit binding of androgen to AR in cell fractions[48] and whole prostate cells,[49] thus affecting androgen signaling.

Further, the LC-PUFA DPA and EPA have been shown to signal through androgens in decreasing prostate cancer cell growth;[50] and DHA can inhibit androgen-dependent gene transcription (*PSA, ornithine decarboxylase, NKX 3.1, immunophilin fkbp 51,* and *Drg-1*). Androgens may signal through SREBP in stimulating lipogenic genes (*fatty acid synthase* and *hydroxymethyl-glutaryl-coenzyme A synthase*).[51] FUNG significantly upregulated *Ar* (2.5*, 1.6, 1.2). How this increase in *Ar* ultimately affects AR signaling is unclear; and we did not observe differential transcription of AR-dependent genes. The possibility that LC-PUFA inhibits SREBP-mediated transcription through effects on androgen signaling is an important area of future signal transduction research.

Basic Helix–Loop–Helix Transcription Factors

In liver, but not in hippocampus, *SREBP-1*, a member of the basic helix–loop–helix (bHLH) family, was downregulated with FISH and FUNG + FISH feeding, which is expected to downregulate FA synthetic pathways. Thus, LC-PUFA signal through different pathways in hippocampus than in liver.

Atonal homolog proteins are bHLH proteins inducing neuronal differentiation.[52,53] All three LC-PUFA diets upregulated *Atoh4*, encoding atonal homolog protein 4 (4.6, 3.5, 5.7*).

NH1H1 is a member of the bHLH family of transcription factors, whose expression is restricted to the nervous system; it may play a role in neuronal differentiation[54] and maintaining parasympathetic tone.[55] *Nhlh1* was upregulated with FUNG relative to FISH (1.8, −1.4, 1.2).

Initiation of Transcription/bZIP Transcription Factors

DHA was reported to downregulate transcription of basal transcription factors in human colon cancer cells in a consistent manner (Narayanan et al., Chapters 14 and 15). We did not observe a consistent transcriptional downregulation of basal factors with FISH feeding. This could be related to *in vivo* vs. *in vitro* differences, the fact that we used a fish oil mixture rather than purified DHA, and our use of mice, not humans.

Btf3 encodes basic transcription factor 3 (active α-isoform), required for transcription initiation of RNA polymerase II.[56] FUNG and FISH downregulated *btf3* (−1.9*, −2.1*, −1.1), which could decrease basal transcription.

Polr2i [RNA polymerase II (DNA directed) polypeptide I] was strongly upregulated with all three LC-PUFA diets (7.7*, 6.3*, 8.5*); no specific information on this gene could be found.

D-site albumin promoter binding protein (DBP) is a member of the proline- and acidic amino acid-rich (PAR) basic-leucine zipper (bZIP) transcription subfamily.[57] Transcription of albumin is controlled by binding of factors to the D-site and other sites.[58] *Dbp* was upregulated with FUNG relative to FISH feeding (1.9, −1.3, 1.4), and the same pattern was observed in the liver (Chapter 2). A next step is to examine if levels of albumin are changed in response to various LC-PUFA. Lipophilic LC-PUFA are transported in plasma bound to albumin, so there could be adaptive advantages of increasing transcription of albumin.

Retinoblastoma Pathway

Retinoblastoma (Rb)/E2F pathways control neuronal cellular proliferation. Rb activates E2F1, E2F2, and E2F3, leading to G1/S progression. E2F1, but not E2F3 can trigger apoptosis.[59] *E2f3* was downregulated with FUNG and FUNG + FISH relative to FISH (−2.0, 1.9, −1.4), which could have the effect of decreasing G1/S progression, consistent with LC-PUFA-known inhibition of cell proliferation. The lack of effect with FISH alone is surprising given that DHA has strong antiproliferative effects (see Chapters 13–15).

HDAC1 (Rpd3, trichostatin A, TSA) is an inhibitor of histone deacetylase, which inhibits cell proliferation via several pathways.[60] The tumor suppressor ARF represses the transcriptional activation domain of the NFκB family member RelA by inducing its association with HDAC1.[61] *Hdac1* (*Rpd3*) was downregulated with FUNG relative to FISH (−1.7, 1.0, −1.1).

Ras–Raf–MEK–ERK Pathway

The Ras–Raf–MEK–ERK pathway activates ERK and in turn Elk1. This enhances transcription of serum response element (SRE)-dependent growth promoting genes, such as *c-fos*.[62] Elk1 was downregulated with FUNG and FUNG + FISH relative to FISH (−1.4, 1.1, −1.5).

G Protein Pathways

Eight different G protein-binding α-subunit proteins have been described. GNB2 is a β-subunit member.[63] *Gnb2* was downregulated with combination diet relative to FUNG and FISH diets alone (−1.1, −1.1, −2.0*).

Receptor activity modifying proteins (RAMPs) are accessory proteins for G protein-coupled receptors (GPCRs), functioning as receptor modulators determining ligand specificity of receptors for calcitonin gene-related peptide (CGRP), amylin, and adrenomedullin (ADM). *Ramp2* is abundantly expressed in hippocampus, cerebellum, pia mater, and blood vessels.[64] *Ramp2* was significantly upregulated with FISH and combination diet (1.4, 1.4*, 1.4*).

Effects on G protein signaling could explain how LC-PUFA affects numerous diverse, G protein mediated pathways, and is an important area for future investigation.

Other Transcription Factors Linked to Growth and Development of Nervous System

HoxD genes pattern axial and appendicular skeletal elements and the nervous system of developing vertebrate embryos. Hoxd12 acts by binding Maf family bZip oncoproteins.[65] *Hoxd12* was downregulated with combination diet (−1.2, −1.1, −1.9*); a similar pattern was observed in liver (Chapter 2) (1.4, −1.5, −2.9). The physiological sequelae of changes in *Hoxd12* merits further investigation.

Pax1 and Pax9 control cell proliferation during early sclerotome development,[66] and affect vertebral column morphogenesis.[67] The severe downregulation of *Pax9* with FUNG feeding relative to FISH feeding is notable (−5.2, 1.6, −1.4).

Nephroblastoma overexpressed (NOV) is a member of the connective tissue growth factor *CCN* gene family, and is found in brain, lung, and skeletal muscle.[68] NOV–Notch1 dimers regulate proliferation, differentiation, and adhesion in a variety of cell types.[69] *Nov* was upregulated with all three LC-PUFA diets (2.8*, 2.1*, 3.5*). Thus, LC-PUFA may affect neuronal proliferation, differentiation, and adhesion via NOV–Notch signaling.

Decorin is a small leucine rich proteoglycan (SLRP), with roles in tissue development and assembly, and collagen fibrillogenesis.[70] It inhibits transforming growth factor-β (TGF-β) activity and PDGF receptor phosphorylation.[71] Decorin may enhance Aβ2m amyloid fibril deposition.[72] Decorin also binds secreted phospholipases A$_2$ (sPLA$_2$), which hydrolyzes glycerophospholipids at sn-2 position to release LC-PUFA such as AA and DHA.[73] *Dcn* was upregulated with all three diets (3.7*, 2.5*, 4.6*). An LC-PUFA-induced upregulation in *Dcn* could affect TGF-β and PDGF growth-mediated processes, release of free FAs, and even amyloid deposition. Ultimately, this could link LC-PUFA to normal and cancerous growth, and etiology of Alzheimer's disease. There is already evidence to suggest LC-PUFA may be linked to these diseases and signaling through Dcn is an important new area of inquiry.

Tissue inhibitors of metalloproteinases (TIMPs) suppress extracellular matrix turnover associated with tissue remodeling, by affecting matrix metalloproteinase (MMP) activity. Independent of MMP, TIMP-2 also inhibits endothelial cell proliferation *in vitro* and angiogenesis *in vivo*.[74] In rat corpus luteum, prolactin increased timp-2 transcription, whereas the arachidonate-derived eicosanoid PGF2α inhibited the related transcript, *timp-3*.[75] Only FISH decreased *timp-2* (1.4, −3.1*, 1.3). Signaling through TIMPs could explain how fish oils (and their eicosanoid derivatives) affect processes such as cell growth, tumor invasion, angiogenesis, and tissue remodeling.[76]

BASP1 (also known as CAP-23 and NAP-22) is a novel myristoylated, acid-soluble, calmodulin-binding protein, abundant in nerve terminals. It participates in neurite outgrowth and synaptic plasticity,[77] and is present in brain, and in kidney, testis, and lymphoid tissues. *Basp1* was downregulated slightly with FUNG (−1.1, 1.1, 1.1).

Neuronatin (Nnat) (α- and β-spliced forms) is present in central and peripheral nervous system, in postmitotic and differentiating neuroepithelial cells.[78] It participates in maintenance of segment identity I hindbrain (whereas hox genes pattern the hindbrain) and pituitary development and maturation, or maintenance of the overall nervous system structure.[79] During development, neuronatin could protect developing cells from toxic insults.[80] Nnat 5′-flanking region contains SP-1, AP-2 (two sites), δ-subunit, SRE-2, NF-A1, and ETS transcription factor binding sites; and intron 1 SP-1 and AP-3 sites.[81] *Nnat* was upregulated with combination diet (1.4, 1.2, 1.6*).

CD81 (TAPA) is a member of the tetraspanin family of proteins, which induces cell migration and mitotic activity. CD81 regulates astrocyte and microglial number, perhaps by regulating cell proliferation by a contact inhibition-dependent mechanism.[82] *Cd81* was downregulated with combination diet (1.0, −1.1, −1.7*).

Chromogranins are acidic, soluble polypeptides found in the central nervous system in dense core vesicles of nerve terminals.[83] CHGB is expressed in hippocampus and other brain regions.[84] *Chgb* was downregulated with combination diet (−1.4, −1.1, −1.4*).

Transcription Factors Linked to Heat, Stress, and Inflammation (Immunity)

Human gp130 (IL6ST) is cytokine receptor chain component involved in signal transduction of interleukins-6 and −11, leukemia inhibitory factor, oncostatin M, and ciliary neurotrophic factor.[85] FISH decreased *il6st* relative to FUNG (1.8, −5.3, −1.1). The n6 and n3 LC-PUFA are known to have different effects on the immune system, the latter tending to downregulate the immune system; LC-PUFA of the n6 and n3 series could differentially signal through gp10 to affect immune functioning.

Fcer1g encodes the γ-subunit of Fc epsilon RI, an immunoglobulin E receptor, and a key molecule triggering allergic reactions. The receptor is present on mast cells and basophils.[86] Its role in neuronal cells is not known. *Fcer1g* was increased with FUNG and combination diet (1.9*, −1.3, 2.6*).

Macrophage migration inhibitory factor (MIF) is a cytokine involved in local and systemic inflammatory and immune responses, and implicated in pathogenesis of sepsis and inflammatory and autoimmune diseases.[87] Fish oil feeding is known to affect macrophage migration and macrophage functioning. FUNG and FISH downregulated *mif* relative to combination diet (−3.3, −4.9, 1.6);

the role of MIF in the hippocampus is not clear. In liver, mif was downregulated with FISH relative to FUNG (1.3, −2.3*, −1.1).

Novel and Undefined Transcription Factors

PURb binds a purine-rich negative regulatory (PNR) element of α-myosin heavy chain, the principle molecule of sarcomere thick filament. PUR also binds α-MHC mRNA in cardiac myocytes and attenuates its translational efficiency.[88] There was a large decrease in *Purb* with FUNG (−1.2, −2.3*, −1.6), the effects of which are unclear in the hippocampus.

Ddb1 encodes DNA damage-binding protein complex (DDB), involved in DNA repair, and nucleotide excision repair in chromatin. The 127 kDa DDB1 subunit also interacts with proteins involved in ubiquitin-mediated proteolysis.[89] It was upregulated with FUNG relative to FUNG + FISH (1.3, 1.0, −1.7).

Effects of LC-PUFA on Gene Products Having Structural Roles

The SWI/SNF family consists of ATP-dependent remodeling proteins making nucleosomal DNA accessible to regulatory factors.[90] No specific information on smarca4 was found. Only combination diet downregulated *smarca4* (1.0, 1.1, −3.3*).

Lama3 encodes laminin α3A chain of specialized epithelia. *Lama3* expression was elevated in hepatocellular carcinoma.[91] AP-1 and USF-1 transcription factors are involved in *lama3* activation.[92] Herein, it was downregulated by FUNG and upregulated with FISH (−3.3, 3.1*, −3.3).

Adaptor proteins (AP) function in clathrin coat assembly and cargo selection at trans-Golgi network (TGN).[93] 3-Phospho-inositide lipids interact with peptides with tyrosine YXXphi sorting motifs and AP-2 mu 2 subunit.[94] *Ap2a2* was downregulated with FISH and combination diet (−2.0*, −1.1, −1.6*). LC-PUFA is known to affect inositide signaling and inositide acyl composition, which could ultimately have an effect on Ap2a2 and intracellular transport processes. *Acta2* (actin, α2, smooth muscle) was significantly upregulated with FUNG and combination diet (5.3*, 2.1, 2.7*); no specific function information on this gene was obtained.

Platelet granular secretion activates platelets during clotting. The vesicle-associated membrane protein VAMP3 inhibits platelet secretion of α-granules and dense granules, preventing clotting.[95] We observed a large fold increase in *vamp3* with all three LC-PUFA diets (9.1*, 9.6*, 8.2*), which could lead to less platelet degranulation and clotting. This is an intriguing observation since n3 LC-PUFA-derived eicosanoids are already known to reduce platelet clotting via effects on prostacyclin acting on smooth muscle cells and thromboxanes acting on platelets.

Actb encodes for ACTB which is a common house keeping gene, but many classical house keeping genes with purported stable ubiquitous expression can actually be variable.[96] Thus, it is not surprising that *actb* was upregulated by all three diets in our experiments (1.4*, 1.5*, 1.6*).

Effects of LC-PUFA on Amino Acid and Protein Metabolism

In *Escherichia coli*, ribosomal protein L28 is required for ribosome assembly.[97] *RpL28* mRNA expression was altered in colorectal cancer.[98] *Rpl28* was down-regulated with combination diet in both hippocampus (-1.2, -1.2, -1.5*) and liver (-1.3, 1.0, -1.7*).

Rpl27a encodes mouse r-protein 27A of the r-protein family (*RpL27a* gene in humans). The promoter contains Sp1 binding sites and transcriptional regu-latory elements, Box-A and GABP.[99] *Rpl27a* was downregulated with FUNG relative to FISH (-1.3, 1.1, -1.1). In the posterior cerebrum of rats, soybean oil feeding (rich in 18:3n3, which increased DHA in membranes) similarly decreased mRNA coding for the related ribosomal L13 protein relative to sun-flower oil feeding (Gaines et al. chapter in this book). Farkas et al. (Chapter 5) found mRNA encoding ribosomal protein L13 was decreased in 3-month-old forebrains of rats fed fish oil. Overall, it appears ribosomal *rpl* genes are a target of dietary LC-PUFA.

Valyl-tRNA synthetase 2 transfers valine to tRNA during translation.[100] *Vars2* was strikingly downregulated with FISH (2.3*, -10.2*, 2.1*). It is notable that only the tRNA synthetase for valine and not other amino acids was LC-PUFA sensitive.

Ubiquitin is a small, highly conserved protein covalently attaching to proteins forming a unique branched protein structure marking the modified protein for proteasomal degradation. PPARs are known to be ubiquitinated.[101] Impairment of this system may play a role in neurodegenerative disorders such as Alzheimer's and Parkinson's diseases.[102] *Ubb* was only slighty affected by our interventions (1.2, 1.2, 1.0).

Carboxypeptidase E (CPE) is a transmembrane, lipid raft-associated intra-cellular prohormone sorting receptor, found in the TGN and secretory granules of (neuro)endocrine cells.[103] *Cpe* was slighty affected by our diets (1.1, 1.1, -1.2).

Effects of LC-PUFA on Lipid and Energy Metabolism

Prostaglandins

Prostaglandin D_2 (PGD_2) is a major prostaglandin (PG) in the brain of rodents and humans, and is present in the hippocampus.[104,105] AA is converted to PGG_2 via cyclooxygenase, and then to PGH_2, which in turn is converted to PGD_2 via PGD_2 synthase in leptomeninges and choroid plexus.[106] Infusion or overexpression of PGD_2 can promote sleep induction in monkeys and rats.[107,108] In addition to being a known somnogen acting in the preoptic area, PGD_2 also modulates pain responses, odor responses, and temperature regu-lation.[104,105]

PGD_2 synthase (*ptgds* transcript; 21 kDa brain; RN 1) gene expression in the hippocampus was upregulated by all treatments (5.0*, 3.0*, 5.3*). Principle substrates for cyclooxygenase are 20:4n6 and 20:5n3 (the latter producing

PGD$_3$; also catalyzed by PGD$_2$ synthase). We did not quantify levels of the more direct substrates of PGD$_2$ synthase, namely free 20:4n6 and 20:5n3, and PGH$_2$, for evidence of a substrate push mechanism, leading to upregulation of PGD$_2$ synthase. Combined levels (μg/g hippocampus) of 20:4n6 and 20:5n3 in total hippocampal PL pools did not correlate with PGD$_2$ synthase expression (data not shown). The PUFA-induced upregulation of *ptgds* may have led to higher hippocampal levels of PGD$_2$, and this could have resulted in physiological seque-lae, perhaps related to calming, that should be investigated in future studies. In contrast to our results, in intestinal CaCo-2 cells, DHA treatment decreased *ptgds* (Chapter 15, Fig. 15.6).

Electron Transport

In hippocampus (as well as in the liver), LC-PUFA affected genes involved in the electron transport machinery involved in ATP production following FA oxidation. LC-PUFA also affected *ant* genes involved in ATP/ADP porting.[109]

Following glycolysis, FA oxidation, and the citric acid cycle, ATP is formed as NADH and FADH$_2$ transfer electrons to molecular oxygen. In the presence of a proton gradient, the F$_1$ unit of F$_1$F$_0$ ATP synthase [H(+)-ATPase] catalyzes the synthesis of ATP. The other major unit of ATP synthase is F$_0$, forming the proton channel of the complex, and containing four kinds of polypeptide chains. Carriers and translocases move ADP into the inner mitochondrial matrix as ATP exits. Vacuolar (V)-ATPases are an important class of proton pumps in endomembrane systems, structurally similar to mitochondrial F$_1$F$_0$ ATP synthase, consisting of a catalytic head, a stalk, and a membrane domain. They are involved/implicated in pH regulation, intracellular trafficking, neo-intimal cell growth, tumor metastasis, and multidrug resistance.[110] The multi-subunit NADH : ubiquinone oxidoreductase (complex I) is the first enzyme complex in the electron transport chain of mitochondria. NDUFA10 is one of eight complex I subunits. NDUFA10 is found within the hydrophobic protein fractions of complex I. The hydrophobic fraction likely participates in proton transfer from mitochondrial matrix to intermembrane space, and may reside in the lipid bilayer.[111]

Hippocampal differentially expressed transcripts were: ATP synthase δ chain (ATP5D; -1.5, -1.6^*, -1.1); ATP synthase H transporting mito-chondrial F$_1$ complex γ polypeptide 1 (*Atp5c1*; -1.3, -1.0, 1.2; significantly different between FUNG and the combination diet); V-ATP synthase subunit D (*Atp6m*; -1.1, -1.0, 1.4; significantly different between FUNG and the combi-nation diet); and *Ndufa10*, which was strikingly upregulated with FUNG and the combination diet (7.7*, 1.0, 6.2*). Interestingly, *Ndufa10* was largely upregulated with all three diets in the liver as well (7.2*, 6.3, 8.6*). The interorgan pattern was somewhat similar considering that the hippocampal increase with FISH was not significant. In rat brain, Farkas et al. (Chapter 5) found that fish oil feeding gen-erally increased various cytochrome *c* oxidase and ATP synthase transcripts in rat brain, but these authors did not report that LC-PUFA feeding affected *Ndufa10* transcripts.

We did not quantify mitochondrial electron transport enzymatic activities, but other authors have found that fish oil feeding can affect, for example, cytochrome *c* oxidase activity,[112] possibly because of changes in the inner mitochondrial membrane CL. We observed only a very slight effect of our diets on cytochrome *c* oxidase, subunit VIIIa mRNA (-1.1, 1.1, 1.0).

Effects of LC-PUFA on Oxygen Transport and Heme Biosynthesis

In liver, hemoglobins *Hbb-a1* (-1.0, 1.3, -1.4) and *Hbb-b1* (-1.1, 1.4, -1.6^*) were not increased, despite increased hemoglobin synthetic enzymes alas1 and alad (1.3, 1.6^*, 1.3). Conversely, in hippocampus, Hbb-a1 (1.8^*, 1.4, 1.3) and *Hbb-b1* (1.4, 1.4, 1.5^*) were somewhat increased with our LC-PUFA rich diets. See Chapter 2 for a complete discussion.

Lipid Oxidative Damage

DSCR1 (Adapt78) is a novel, early stage, calcium-inducible oxidative stress-activated cytoprotectant.[113] *Dscr1* was downregulated with FISH and FISH + FUNG diets (-1.2, -2.2^*, -2.3^*). LC-PUFA may activate *Dscr1* by inducing cellular oxidant stress.

Effects of LC-PUFA on Miscellaneous Genes (Not Classified as Transcription Factors)

Calcium and Sodium Channels

Ryanodine receptors are involved with calcium release. Palmitoyl carnitine produced a direct stimulation of rabbit and pig skeletal muscle ryanodine receptor. Fatty acyl-CoA esters in combination fatty acyl-CoA-binding protein (ACBP) also activated ryanodine receptor/Ca^{2+} release channels in rabbit skeletal muscle.[114] The interaction of esterified fatty acids and ryanodine receptors in skeletal muscle thus influences intracellular Ca^{2+}. Effects of esterified or carnitinoylated LC-PUFA were not examined in our studies. Only combination diet hugely increased *Ryr3* transcription (1.0, 1.0, 8.0^*), which might affect hippocampal calcium homeostasis and calcium-dependent processes.

Transcriptomic Signaling Pathways Implicated in Behavior Mediation

Transthyretin Signaling

Thyroid hormones, including transthyretin, are generally involved in neuronal proliferation and differentiation, and are required for normal cytoskeletal assembly. Transthyretin is specifically involved in transporting thyroxine (T4) and retinol-binding protein in cerebrospinal fluid and brain serum. AA has been found to bind transthyretin and inhibit thyroxin associations.[115] Transthyretin may also sequester β amyloid, thereby having neuroprotective properties. Dietary FISH had the effect of decreasing *Ttr*; whereas, FUNG and particularly the combination diet, increased *Ttr* (1.6, -1.9^*, 2.2^*). Similar to our findings in mice, feeding young rats fish oil significantly decreased *Ttr* expression in whole brain 2.9-fold.[28] In a later experiment, older rats (2-year-old), however, showed an increase in hippocampal *Ttr* expression after 1 month of feeding.[36] Interestingly, feeding of *G. biloba* extract (publicized to affect mental function) in mice increased hippocampal *Ttr*,[38,116] but not cortical *Ttr*.[116] Levels of transthyretin and associations of transthyretin with thyroxine and retinol-binding protein should be assessed in future LC-PUFA feeding experiments.

5-Hydroxytryptamine Signaling

Dietary LC-PUFA may affect 5-hydroxytryptamine (serotonin, 5-HT) signaling. Specifically, 20:4 NAE can displace binding of ligands to HT receptors,[117] and displace 5-HT_3 receptor currents,[118] and we report herein and previously[31] that dietary AA feeding increased whole brain 20:4 NAE. In the present work, dietary FUNG and FISH, alone or together, increased *Htr4* receptor levels (6.5*, 3.8, 4.4). 5-HT_4 receptor (*Htr4* transcript) increases in expression have been shown to augment hippocampal acetylcholine outflow, thereby affecting cognitive processes.[119]

α-2,8-Sialyltransferase

Hippocampal α-2,8-sialyltransferase (*Siat8e*) was downregulated with FISH and the combination diet, but not with FUNG; difference relative to control was only significant for the combination diet (1.6, -2.6, -6.4^*). Hippocampal protein kinase Cδ (Prkcd) was slightly downregulated with the three LC-PUFA diets (-1.6, -1.3, -1.8^*). PUFA feeding can influence PKC activity[120] and one mechanism of action may be via generation of unique diacylglycerol molecular species.[121] Residues of α-2,8-linked disialic acid are found in glycoproteins.[122] During learning, there is a transient increase in neuronal polysialylation in the dentate gyrus of the hippocampus. This has been associated with selective retention and/or elimination of synapses that are transiently overproduced during memory consolidation.[123] In rat hippocampus, during development, PKCδ can negatively regulate polysialyltransferase activity and neural cell adhesion molecule (NCAM) polysialylation state.[123] NCAM is implicated in neural differentiation and cellular plasticity.[124] Our observed decrease in *Prkcd* expression may have affected hippocampal polysialyltransferase activity. It is not known if *Prkcd* regulates expression of Siat8e.

Proopiomelanocortin

Proopiomelanocortin (POMC) is a hypothalamic neuropeptide, and a target for leptin. POMC is cleaved posttranslationally producing bioactive peptides (adrenocorticotrophin, β-endorphin and α-, β-, and γ-melanocyte stimulating hormones) that interact with neurons expressing melanocortin receptors, leading to appetite suppression.[125] High fat feeding has been found to increase POMC. Induction of POMC may be a second defense against obesity.[125,126] We found that FUNG, rich in arachidonic acid, strikingly increased *pomc* (6.0*, 1.6, 1.4). This would not be consistent with the observation that AA feeding can possibly induce obesity.[127] As a next step, levels of POMC protein and downstream POMC peptides should be quantified.

Somatostatin

Somatostatin (SST) in the hippocampus reverses learning and memory-related behavior impairment induced by olfactory bulbectomy.[128] Fish oil feeding has similarly been found to increase *sst* 1.8-fold in rat brain (Chapter 5). DHA feeding to pregnant rats has been to elevate hippocampal SST in the offspring, along with 5-HT, dopamine (DA), and norepinephrine (NE).[129] Consistent with this observation, all three LC-PUFA diets increased *sst* (1.8*, 1.8*, 2.1*).

Glucose-6-Phosphate Isomerase

In the liver, glucose-6-phosphate isomerase (GPI) reversibly converts glucose-6-phosphate to fructose-6-phosphate. However, in the brain, it is involved in neuronal growth and survival, cell motility, and differentiation, and is known as neuroleukin (NLK), autocrine motility factor, and maturation factor.[130] It is intriguing to note that Stone T maze and Morris water maze behavioral testing showed increased hippocampal mRNA and protein levels for GPI/NLK and its gp78 receptor in mice.[130] In the current experiment, hippocampal *Gpi1* was significantly upregulated by all dietary treatments (1.7*, 1.9*, 1.7*). Whether this change in *Gpi1* could affect learning ability was not assessed, and should be examined using receptor inhibition approaches in the future (see following behavioral section). Previous work has shown DHA feeding to have positive effects on various aspects of learning. For example, reference memory-related learning ability in the baited eight-arm radial maze test was increased following DHA feeding.[131]

Ionotropic Glutamate Receptor 2

Gria2 encodes the brain ionotropic glutamate receptor 2 (AMPA 2, GRIA2, GluR2). *Gria2* was decreased with schizophrenia in one study, consistent with a decrease in glutamate synaptic function.[132] During seizures, deacetylation of histone H4 in rat hippocampal CA3 neurons at the *gria2* promoter down-regulated gria2.[133] Only combination diet increased *gria2* (1.2, 1.4, 1.5*).

CONCLUSIONS AND KEY FINDINGS

A notable observation was that in the liver dietary LC-PUFA signaled through the well-known PPAR, SREBP, and HNF transcription factors,[27] whereas in the hippocampus, a completely different set of transcription factors were activated. It may be that these transcription factors bind to specific consensus sequences accounting for some of the activation of other genes described subsequently.

Genes that were notable for their large upregulation were: *Vamp3* (8– 10-fold upregulation with LC-PUFA diets), *Polr2i* (6–9, LC-PUFA diets; RNA synthesis), *Ndufa10* (6–8, FUNG and combination diet; electron transport), *pomc1* (6, FUNG; see later), *htr4* (7, FUNG; see later), *acta2* (5, FUNG; actin metabolism), and *Ptgds* (3–5, LC-PUFA diets; prostaglandin D synthesis). Genes that were notable for their large downregulation were: *vars2* (−10, FISH; introduction of valine during translation) and *pax9* (−5, FUNG; cell proliferation during sclerotome development and roles in vertebral column morphogenesis). These genes should be the focuses of future inquiry.

Other notable genes were those changed in both the liver (Chapter 2) and the hippocampus, including: *Dbp* (transcription of albumin), *Hoxd12* (patterning axial and appendicular skeletal elements and the nervous system), *Rpl28* (ribosome assembly); *Hbb-a1*, *Hbb-b1* (hemoglobin classes), *mif* (macrophage motility in immune and inflammation reactions), *ndufa10* (electron transport), and other similar electron transport genes.

The link between LC-PUFA and hippocampal *Ptgds* merits further study, first to assess if levels of PGD_2 were increased, and second to determine the physiological effects (e.g., calming) using specific inhibitors and behavioral assays.

Of the genes involved in signal transduction linked to behavior, the following are noteworthy: *Ttr*, *htr4* (see earlier), *Siat8e*, *Prkcδ*, *Pomc*, and *Gpi1*. It is of course very difficult to link changes in these transcripts to changes in behaviors. We examined changes in behaviors linked to CB-1 receptors since we had previously shown that LC-PUFA may signal through this cascade. We did observe that LC-PUFA may in fact affect specific behaviors via signaling through CB-1 receptors, but we cannot relate this back to changes in the specific transcripts earlier described without using specific inhibitors to further dissect the pathways.

During learning tasks in rodents, microarrays are now being used to determine which hippocampal genes and neurons are activated during specific aspects of learning.[43–45] If common genes are changed during LC-PUFA supplementation or deficiency, and during learning, it may be possible to extrapolate that LC-PUFA are involved in the signal transduction cascade leading to the particular behavior, paving the way for focused future behavioral experimentation to test the hypothesis. A similar logic could be employed for searching for genes that are altered in a specific disease, and with LC-PUFA supplementation or deficiency.

REFERENCES

1. Cheruku, S.R.; Montgomery-Downs, H.E.; Farkas, S.L.; Thoman, E.B.; Lammi-Keefe, C.J. Higher maternal plasma docosahexaenoic acid during pregnancy is associated with more mature neonatal sleep-state patterning. Am. J. Clin. Nutr. **2002**, *76*, 608–613.

2. Chalon, S.; Vancassel, S.; Zimmer, L.; Guilloteau, D.; Durand, G. Polyunsaturated fatty acids and cerebral function: focus on monoaminergic neurotransmission. Lipids **2001**, *36*, 937–944.

3. Fenton, W.S.; Dickerson, F.; Boronow, J.; Hibbeln, J.R.; Knable, M. A placebo-controlled trial of omega-3 fatty acid (ethyl eicosapentaenoic acid) supplementation for residual symptoms and cognitive impairment in schizophrenia. Am. J. Psychiatr. **2001**, *158*, 2071–2074.

4. Hibbeln, J.R. Seafood consumption and homicide mortality. A cross-national ecological analysis. World Rev. Nutr. Diet **2001**, *88*, 41–46.

5. Locke, C.A.; Stoll, A.L. Omega-3 fatty acids in major depression. World Rev. Nutr. Diet. **2001**, *89*, 173–185.

6. Hibbeln, J.R. Seafood consumption, the DHA content of mothers' milk and prevalence rates of postpartum depression: a cross-national, ecological analysis. J. Affect. Disord. **2002**, *69*, 15–29.

7. Tanskanen, A.; Hibbeln, J.R.; Hintikka, J.; Haatainen, K.; Honkalampi, K.; Viinamaki, H. Fish consumption, depression, and suicidality in a general population. Arch. Gen. Psychiatr. **2001**, *58*, 512–513.

8. Vancassel, S.; Durand, G.; Barthelemy, C.; Lejeune, B.; Martineau, J.; Guilloteau, D.; Andres, C.; Chalon, S. Plasma fatty acid levels in autistic children. Prostaglandins Leukot. Essent. Fatty Acids **2001**, *65*, 1–7.

9. Peet, M.; Horrobin, D.F. A dose-ranging study of the effects of ethyl-eicosapentaenoate in patients with ongoing depression despite apparently adequate treatment with standard drugs. Arch. Gen. Psychiatr. **2002**, *59*, 913–919.

10. Cabanes, A.; de Assis, S.; Gustafsson, J.A.; Hilakivi-Clarke, L. Maternal high n-6 polyunsaturated fatty acid intake during pregnancy increases voluntary alcohol intake and hypothalamic estrogen receptor alpha and beta levels among female offspring. Dev. Neurosci. **2000**, *22*, 488–493.

11. Verheijen, M.H.; Chrast, R.; Burrola, P.; Lemke, G. Local regulation of fat metabolism in peripheral nerves. Genes Dev. **2003**, *17*, 2450–2464.

12. Anderson, J.W.; Johnstone, B.M.; Remley, D.T. Breast-feeding and cognitive development: a meta-analysis. Am. J. Clin. Nutr. **1999**, *70*, 525–535.

13. Wainwright, P.E. Dietary essential fatty acids and brain function: a developmental perspective on mechanisms. Proc. Nutr. Soc. **2002**, *61*, 61–69.

14. Ikemoto, A.; Nitta, A.; Furukawa, S.; Ohishi, M.; Nakamura, A.; Fujii, Y.; Okuyama, H. Dietary n-3 fatty acid deficiency decreases nerve growth factor content in rat hippocampus. Neurosci. Lett. **2000**, *285*, 99–102.

15. Ikemoto, A.; Ohishi, M.; Sato, Y.; Hata, N.; Misawa, Y.; Fujii, Y.; Okuyama, H. Reversibility of n-3 fatty acid deficiency-induced alterations of learning behavior in the rat: level of n-6 fatty acids as another critical factor. J. Lipid Res. **2001**, *42*, 1655–1663.

16. Okuyama, H.; Ohishi, M.; Fukuma, A.; Sato, Y.; Ikemoto, A.; Fujii, Y. Alpha-linolenate-deficiency-induced alterations in brightness discrimination learning behavior and retinal function in rats. World Rev. Nutr. Diet. **2001**, *88*, 35–40.

17. Kodas, E.; Vancassel, S.; Lejeune, B.; Guilloteau, D.; Chalon, S. Reversibility of n-3 fatty acid deficiency-induced changes in dopaminergic neurotransmission in rats: critical role of developmental stage. J. Lipid Res. **2002**, *43*, 1209–1219.

18. Zimmer, L.; Vancassel, S.; Cantagrel, S.; Breton, P.; Delamanche, S.; Guilloteau, D.; Durand, G.; Chalon, S. The dopamine mesocorticolimbic pathway is affected by deficiency in n-3 polyunsaturated fatty acids. Am. J. Clin. Nutr. **2002**, *75*, 662–667.
19. Aid, S.; Vancassel, S.; Poumes-Ballihaut, C.; Chalon, S.; Guesnet, P.; Lavialle, M. Effect of a diet-induced (n-3) polyunsaturated fatty acid depletion on cholinergic parameters in the rat hippocampus. J. Lipid Res. **2003**.
20. Stockard, J.E.; Saste, M.D.; Benford, V.J.; Barness, L.; Auestad, N.; Carver, J.D. Effect of docosahexaenoic acid content of maternal diet on auditory brainstem conduction times in rat pups. Dev. Neurosci. **2000**, *22*, 494–499.
21. Wauben, I.P.; Xing, H.C.; McCutcheon, D.; Wainwright, P.E. Dietary trans fatty acids combined with a marginal essential fatty acid status during the pre- and postnatal periods do not affect growth or brain fatty acids but may alter behavioral development in B6D2F(2) mice. J. Nutr. **2001**, *131*, 1568–1573.
22. Chalon, S.; Delion-Vancassel, S.; Belzung, C.; Guilloteau, D.; Leguisquet, A.M.; Besnard, J.C.; Durand, G. Dietary fish oil affects monoaminergic neurotransmission and behavior in rats. J. Nutr. **1998**, *128*, 2512–2519.
23. Carrie, I.; Smirnova, M.; Clement, M.; De, J.D.; Frances, H.; Bourre, J.M. Docosahexaenoic acid-rich phospholipid supplementation: effect on behavior, learning ability, and retinal function in control and n-3 polyunsaturated fatty acid deficient old mice. Nutr. Neurosci. **2002**, *5*, 43–52.
24. Raygada, M.; Cho, E.; Hilakivi-Clarke, L. High maternal intake of polyunsaturated fatty acids during pregnancy in mice alters offsprings' aggressive behavior, immobility in the swim test, locomotor activity and brain protein kinase C activity. J. Nutr. **1998**, *128*, 2505–2511.
25. Lesa, G.M.; Palfreyman, M.; Hall, D.H.; Clandinin, M.T.; Rudolph, C.; Jorgensen, E.M.; Schiavo, G. Long chain polyunsaturated fatty acids are required for efficient neurotransmission in *C. elegans*. J. Cell Sci. **2003**, *116*, 4965–4975.
26. Abel, T.; Lattal, K.M. Molecular mechanisms of memory acquisition, consolidation and retrieval. Curr. Opin. Neurobiol. **2001**, *11*, 180–187.
27. Berger, A.; Mutch, D.M.; German, B.J.; Roberts, M.A. Dietary effects of arachidonate-rich fungal oil and fish oil on murine hepatic and hippocampal gene expression. Lipids Health Dis. **2002**, *1*, 2.
28. Kitajka, K.; Puskas, L.G.; Zvara, A.; Hackler, L., Jr.; Barcelo-Coblijn, G.; Yeo, Y.K.; Farkas, T. The role of n-3 polyunsaturated fatty acids in brain: modulation of rat brain gene expression by dietary n-3 fatty acids. Proc. Natl. Acad. Sci. USA **2002**, *99*, 2619–2624.
29. Murthy, M.; Hamilton, J.; Greiner, R.S.; Moriguchi, T.; Salem, N., Jr.; Kim, H.Y. Differential effects of n-3 fatty acid deficiency on phospholipid molecular species composition in the rat hippocampus. J. Lipid Res. **2002**, *43*, 611–617.
30. Qi, K.; Hall, M.; Deckelbaum, R.J. Long-chain polyunsaturated fatty acid accretion in brain. Curr. Opin. Clin. Nutr. Metab. Care **2002**, *5*, 133–138.
31. Berger, A.; Crozier, G.; Bisogno, T.; Cavaliere, P.; Innis, S.; Di-Marzo, V. Anandamide and diet: inclusion of dietary arachidonate and docosahexaenoate leads to increased brain levels of the corresponding *N*-acylethanolamines in piglets. Proc. Natl. Acad. Sci. USA **2001**, *98*, 6402–6406.
32. Mutch, D.M.; Berger, A.; Mansourian, R.; Rytz, A.; Roberts, M.A. The limit fold change model: a practical approach for selecting differentially expressed genes from microarray data. BMC Bioinformatics **2002**, *3*, 17.

33. Full Gene set-Gen accession No: GSE91: GSE91. http://wwwncbinlmnihgov/geo/query/acccgi?acc=GSE91.

34. Barcelo-Coblijn, G.; Kitajka, K.; Puskas, L.G.; Hogyes, E.; Zvara, A.; Hackler, L., Jr.; Farkas, T. Gene expression and molecular composition of phospholipids in rat brain in relation to dietary n-6 to n-3 fatty acid ratio. Biochim. Biophys. Acta **2003**, *1632*, 72–79.

35. Gaines, J.; Levant, B.; Radel, J.; Berman, N.; Carlson, S. Cerebral gene expression in a rat model of "physiology relevant" DHA reduction. *AOCS Maternal and Infant LCPUFA Workshop*, May 3–4, 2003, Kansas City, MO 2003.

36. Puskas, L.G.; Kitajka, K.; Nyakas, C.; Barcelo-Coblijn, G.; Farkas, T. Short-term administration of omega 3 fatty acids from fish oil results in increased transthyretin transcription in old rat hippocampus. Proc. Natl. Acad. Sci. USA **2003**, *100*, 1580–1585.

37. Tafti, M.; Petit, B.; Chollet, D.; Neidhart, E.; De Bilbao, F.; Kiss, J.Z.; Wood, P.A.; Franken, P. Deficiency in short-chain fatty acid beta-oxidation affects theta oscillations during sleep. Nat. Genet. **2003**.

38. Watanabe, C.M.; Wolffram, S.; Ader, P.; Rimbach, G.; Packer, L.; Maguire, J.J.; Schultz, P.G.; Gohil, K. The *in vivo* neuromodulatory effects of the herbal medicine *Ginkgo biloba*. Proc. Natl. Acad. Sci. USA **2001**, *98*, 6577–6580.

39. Saito, M.; Smiley, J.; Toth, R.; Vadasz, C. Microarray analysis of gene expression in rat hippocampus after chronic ethanol treatment. Neurochem. Res. **2002**, *27*, 1221–1229.

40. Molteni, R.; Ying, Z.; Gomez-Pinilla, F. Differential effects of acute and chronic exercise on plasticity-related genes in the rat hippocampus revealed by microarray. Eur. J. Neurosci. **2002**, *16*, 1107–1116.

41. Colangelo, V.; Schurr, J.; Ball, M.J.; Pelaez, R.P.; Bazan, N.G.; Lukiw, W.J. Gene expression profiling of 12633 genes in Alzheimer hippocampal CA1: transcription and neurotrophic factor down-regulation and up-regulation of apoptotic and pro-inflammatory signaling. J. Neurosci. Res. **2002**, *70*, 462–473.

42. Blalock, E.M.; Chen, K.C.; Sharrow, K.; Herman, J.P.; Porter, N.M.; Foster, T.C.; Landfield, P.W. Gene microarrays in hippocampal aging: statistical profiling identifies novel processes correlated with cognitive impairment. J. Neurosci. **2003**, *23*, 3807–3819.

43. D'Agata, V.; Cavallaro, S. Gene expression profiles—a new dynamic and functional dimension to the exploration of learning and memory. Rev. Neurosci. **2002**, *13*, 209–219.

44. Leil, T.A.; Ossadtchi, A.; Nichols, T.E.; Leahy, R.M.; Smith, D.J. Genes regulated by learning in the hippocampus. J. Neurosci. Res. **2003**, *71*, 763–768.

45. Robles, Y.; Vivas-Mejia, P.E.; Ortiz-Zuazaga, H.G.; Felix, J.; Ramos, X.; Pena de Ortiz, S. Hippocampal gene expression profiling in spatial discrimination learning. Neurobiol. Learn Mem. **2003**, *80*, 80–95.

46. Kamme, F.; Salunga, R.; Yu, J.; Tran, D.T.; Zhu, J.; Luo, L.; Bittner, A.; Guo, H.Q.; Miller, N.; Wan, J.; Erlander, M. Single-cell microarray analysis in hippocampus CA1: demonstration and validation of cellular heterogeneity. J. Neurosci. **2003**, *23*, 3607–3615.

47. Beigneux, A.P.; Kosinski, C.; Gavino, B.; Horton, J.D.; Young, S.G.; Shah, N.A.; Warrington, J.A.; Anderson, N.N.; Park, S.W.; Brown, M.S.; Goldstein, J.L.

ATP-citrate lyase deficiency in the mouse. Combined analysis of oligonucleotide microarray data from transgenic and knockout mice identifies direct SREBP target genes. J. Biol. Chem. **2003**, *100*, 12027–12032.

48. Kato, J. Arachidonic acid as a possible modulator of estrogen, progestin, androgen, and glucocorticoid receptors in the central and peripheral tissues. J. Steroid Biochem. **1989**, *34*, 219–227.

49. Prinsloo, S.E.; van Aswegen, C.H. Effect of fatty acids on estradiol and testosterone binding to whole DU-145 prostate cells. Prostaglandins Leukot. Essent. Fatty Acids **2002**, *66*, 419–425.

50. Chung, B.H.; Mitchell, S.H.; Zhang, J.S.; Young, C.Y. Effects of docosahexaenoic acid and eicosapentaenoic acid on androgen-mediated cell growth and gene expression in LNCaP prostate cancer cells. Carcinogenesis **2001**, *22*, 1201–1206.

51. Heemers, H.; Maes, B.; Foufelle, F.; Heyns, W.; Verhoeven, G.; Swinnen, J.V. Androgens stimulate lipogenic gene expression in prostate cancer cells by activation of the sterol regulatory element-binding protein cleavage activating protein/ sterol regulatory element-binding protein pathway. Mol. Endocrinol. **2001**, *15*, 1817–1828.

52. Sommer, L.; Ma, Q.; Anderson, D.J. Neurogenins, a novel family of atonal-related bHLH transcription factors, are putative mammalian neuronal determination genes that reveal progenitor cell heterogeneity in the developing CNS and PNS. Mol. Cell. Neurosci. **1996**, *8*, 221–241.

53. Schwab, M.H.; Druffel-Augustin, S.; Gass, P.; Jung, M.; Klugmann, M.; Bartholomae, A.; Rossner, M.J.; Nave, K.A. Neuronal basic helix-loop-helix proteins (NEX, neuroD, NDRF): spatiotemporal expression and targeted disruption of the NEX gene in transgenic mice. J. Neurosci. **1998**, *18*, 1408–1418.

54. Mullick, A.; Groulx, N.; Trasler, D.; Gros, P. Nhlh1, a basic helix-loop-helix transcription factor, is very tightly linked to the mouse looptail (Lp) mutation. Mamm. Genome **1995**, *6*, 700–704.

55. Cogliati, T.; Good, D.J.; Haigney, M.; Delgado-Romero, P.; Eckhaus, M.A.; Koch, W.J.; Kirsch, I.R. Predisposition to arrhythmia and autonomic dysfunction in Nhlh1-deficient mice. Mol. Cell. Biol. **2002**, *22*, 4977–4983.

56. Grein, S.; Pyerin, W. BTF3 is a potential new substrate of protein kinase CK2. Mol. Cell. Biochem. **1999**, *191*, 121–128.

57. Inukai, T.; Inaba, T.; Yoshihara, T.; Look, A.T. Cell transformation mediated by homodimeric E2A-HLF transcription factors. Mol. Cell. Biol. **1997**, *17*, 1417–1424.

58. Ogawa, A.; Tsujinaka, T.; Yano, M.; Morita, S.; Taniguchi, M.; Kaneko, K.; Doki, Y.; Shiozaki, H.; Monden, M. Changes of liver-enriched nuclear transcription factors for albumin gene in starvation in rats. Nutrition **1999**, *15*, 213–216.

59. Hallstrom, T.C.; Nevins, J.R. Specificity in the activation and control of transcription factor E2F-dependent apoptosis. Proc. Natl. Acad. Sci. USA **2003**.

60. Di Padova, M.; Bruno, T.; De Nicola, F.; Iezzi, S.; D'Angelo, C.; Gallo, R.; Nicosia, D.; Corbi, N.; Biroccio, A.; Floridi, A.; Passananti, C.; Fanciulli, M. CHE-1 arrests human colon carcinoma cell proliferation by displacing HDAC1 from theP21WAF1/CIP1 promoter. J. Biol. Chem. **2003**.

61. Rocha, S.; Campbell, K.J.; Perkins, N.D. p53- and Mdm2-independent repression of NF-kappa B transactivation by the ARF tumor suppressor. Mol. Cell **2003**, *12*, 15–25.

62. Galetic, I.; Maira, S.M.; Andjelkovic, M.; Hemmings, B.A. Negative regulation of ERK and Elk by protein kinase B modulates c-Fos transcription. J. Biol. Chem. **2003**, *278*, 4416–4423.

63. Blatt, C.; Eversole-Cire, P.; Cohn, V.H.; Zollman, S.; Fournier, R.E.; Mohandas, L.T.; Nesbitt, M.; Lugo, T.; Jones, D.T.; Reed, R.R.; Weiner, L.P.; Sparkes, R.S.; Simon, M.I. Chromosomal localization of genes encoding guanine nucleotide-binding protein subunits in mouse and human. Proc. Natl. Acad. Sci. USA **1988**, *85*, 7642–7646.

64. Ueda, T.; Ugawa, S.; Saishin, Y.; Shimada, S. Expression of receptor-activity modifying protein (RAMP) mRNAs in the mouse brain. Brain Res. Mol. Brain Res. **2001**, *93*, 36–45.

65. Kataoka, K.; Yoshitomo-Nakagawa, K.; Shioda, S.; Nishizawa, M. A set of Hox proteins interact with the Maf oncoprotein to inhibit its DNA binding, transactivation, and transforming activities. J. Biol. Chem. **2001**, *276*, 819–826.

66. Peters, H.; Wilm, B.; Sakai, N.; Imai, K.; Maas, R.; Balling, R. Pax1 and Pax9 synergistically regulate vertebral column development. Development **1999**, *126*, 5399–5408.

67. Neubuser, A.; Koseki, H.; Balling, R. Characterization and developmental expression of Pax9, a paired-box-containing gene related to Pax1. Dev. Biol. **1995**, *170*, 701–716.

68. Liu, C.; Liu, X.J.; Crowe, P.D.; Kelner, G.S.; Fan, J.; Barry, G.; Manu, F.; Ling, N.; De Souza, E.B.; Maki, R.A. Nephroblastoma overexpressed gene (*NOV*) codes for a growth factor that induces protein tyrosine phosphorylation. Gene **1999**, *238*, 471–478.

69. Perbal, B. *NOV* (nephroblastoma overexpressed) and the CCN family of genes: structural and functional issues. Mol. Pathol. **2001**, *54*, 57–79.

70. Reed, C.C.; Iozzo, R.V. The role of decorin in collagen fibrillogenesis and skin homeostasis. Glycoconj. J. **2002**, *19*, 249–255.

71. Nili, N.; Cheema, A.N.; Giordano, F.J.; Barolet, A.W.; Babaei, S.; Hickey, R.; Eskandarian, M.R.; Smeets, M.; Butany, J.; Pasterkamp, G.; Strauss, B.H. Decorin inhibition of PDGF-stimulated vascular smooth muscle cell function: potential mechanism for inhibition of intimal hyperplasia after balloon angioplasty. Am. J. Pathol. **2003**, *163*, 869–878.

72. Yamaguchi, I.; Suda, H.; Tsuzuike, N.; Seto, K.; Seki, M.; Yamaguchi, Y.; Hasegawa, K.; Takahashi, N.; Yamamoto, S.; Gejyo, F.; Naiki, H. Glycosaminoglycan and proteoglycan inhibit the depolymerization of beta2-microglobulin amyloid fibrils in vitro. Kidney Int. **2003**, *64*, 1080–1088.

73. Valentin, E.; Lambeau, G. Increasing molecular diversity of secreted phospholipases A(2) and their receptors and binding proteins. Biochim. Biophys. Acta **2000**, *1488*, 59–70.

74. Seo, D.W.; Li, H.; Guedez, L.; Wingfield, P.T.; Diaz, T.; Salloum, R.; Wei, B.Y.; Stetler-Stevenson, W.G. TIMP-2 mediated inhibition of angiogenesis: an MMP-independent mechanism. Cell **2003**, *114*, 171–180.

75. Stocco, C.; Callegari, E.; Gibori, G. Opposite effect of prolactin and prostaglandin F(2 alpha) on the expression of luteal genes as revealed by rat cDNA expression array. Endocrinology **2001**, *142*, 4158–4161.

76. Anderle, P.; Farmer, P.; Berger, A.; Roberts, M.-A. Nutrigenomic approach to understanding the mechanisms by which dietary long-chain fatty acids induce gene signals

and control mechanisms involved in carcinogenesis. Nutr.: Int. J. Appl. Basic Nutr. Sci. **2004**, *20*, 103–108.

77. Zakharov, V.V.; Capony, J.P.; Derancourt, J.; Kropolova, E.S.; Novitskaya, V.A.; Bogdanova, M.N.; Mosevitsky, M.I. Natural N-terminal fragments of brain abundant myristoylated protein BASP1. Biochim. Biophys. Acta **2003**, *1622*, 14–19.
78. Kikyo, N.; Williamson, C.M.; John, R.M.; Barton, S.C.; Beechey, C.V.; Ball, S.T.; Cattanach, B.M.; Surani, M.A.; Peters, J. Genetic and functional analysis of neuronatin in mice with maternal or paternal duplication of distal Chr 2. Dev. Biol. **1997**, *190*, 66–77.
79. Wijnholds, J.; Chowdhury, K.; Wehr, R.; Gruss, P. Segment-specific expression of the neuronatin gene during early hindbrain development. Dev. Biol. **1995**, *171*, 73–84.
80. Zheng, S.; Chou, A.H.; Jimenez, A.L.; Khodadadi, O.; Son, S.; Melega, W.P.; Howard, B.D. The fetal and neonatal brain protein neuronatin protects PC12 cells against certain types of toxic insult. Brain Res. Dev. Brain Res. **2002**, *136*, 101–110.
81. Dou, D.; Joseph, R. Structure and organization of the human neuronatin gene. Genomics **1996**, *33*, 292–297.
82. Geisert, E.E., Jr.; Williams, R.W.; Geisert, G.R.; Fan, L.; Asbury, A.M.; Maecker, H.T.; Deng, J.; Levy, S. Increased brain size and glial cell number in CD81-null mice. J. Comp. Neurol. **2002**, *453*, 22–32.
83. Pirker, S.; Czech, T.; Baumgartner, C.; Maier, H.; Novak, K.; Furtinger, S.; Fischer-Colbrie, R.; Sperk, G. Chromogranins as markers of altered hippocampal circuitry in temporal lobe epilepsy. Ann. Neurol. **2001**, *50*, 216–226.
84. Kandlhofer, S.; Hoertnagl, B.; Czech, T.; Baumgartner, C.; Maier, H.; Novak, K.; Sperk, G. Chromogranins in temporal lobe epilepsy. Epilepsia **2000**, *41* (suppl 6), S111–114.
85. Rodriguez, C.; Grosgeorge, J.; Nguyen, V.C.; Gaudray, P.; Theillet, C. Human gp130 transducer chain gene (IL6ST) is localized to chromosome band 5q11 and possesses a pseudogene on chromosome band 17p11. Cytogenet. Cell. Genet. **1995**, *70*, 64–67.
86. Le Coniat, M.; Kinet, J.P.; Berger, R. The human genes for the alpha and gamma subunits of the mast cell receptor for immunoglobulin E are located on human chromosome band 1q23. Immunogenetics **1990**, *32*, 183–186.
87. Calandra, T.; Roger, T. Macrophage migration inhibitory factor: a regulator of innate immunity. Nat. Rev. Immunol. **2003**, *3*, 791–800.
88. Gupta, M.; Sueblinvong, V.; Raman, J.; Jeevanandam, V.; Gupta, M.P. Single-stranded DNA-binding proteins, PURa and PURb, bind to a purine-rich negative regulatory element of the a-myosin heavy chain gene and control transcriptional and translational regulation of the gene expression: implications in the repression of a-MHC during heart failure. J. Biol. Chem. **2003**, *278*, 44935–44948.
89. Wittschieben, B.B.; Wood, R.D. DDB complexities. DNA Repair (Amsterdam) **2003**, *2*, 1065–1069.
90. Fan, H.Y.; He, X.; Kingston, R.E.; Narlikar, G.J. Distinct strategies to make nucleosomal DNA accessible. Mol. Cell **2003**, *11*, 1311–1322.
91. Midorikawa, Y.; Tsutsumi, S.; Taniguchi, H.; Ishii, M.; Kobune, Y.; Kodama, T.; Makuuchi, M.; Aburatani, H. Identification of genes associated with

dedifferentiation of hepatocellular carcinoma with expression profiling analysis. Jpn. J. Cancer Res. **2002**, *93*, 636–643.

92. Virolle, T.; Coraux, C.; Ferrigno, O.; Cailleteau, L.; Ortonne, J.P.; Pognonec, P.; Aberdam, D. Binding of USF to a non-canonical E-box following stress results in a cell-specific derepression of the *lama3* gene. Nucl. Acids Res. **2002**, *30*, 1789–1798.

93. Yeung, B.G.; Phan, H.L.; Payne, G.S. Adaptor complex-independent clathrin function in yeast. Mol. Biol. Cell. **1999**, *10*, 3643–3659.

94. Rapoport, I.; Chen, Y.C.; Cupers, P.; Shoelson, S.E.; Kirchhausen, T. Dileucine-based sorting signals bind to the beta chain of AP-1 at a site distinct and regulated differently from the tyrosine-based motif-binding site. EMBO J. **1998**, *17*, 2148–2155.

95. Polgar, J.; Chung, S.H.; Reed, G.L. Vesicle-associated membrane protein 3 (VAMP-3) and VAMP-8 are present in human platelets and are required for granule secretion. Blood **2002**, *100*, 1081–1083.

96. Lee, P.D.; Sladek, R.; Greenwood, C.M.; Hudson, T.J. Control genes and variability: absence of ubiquitous reference transcripts in diverse mammalian expression studies. Genome Res. **2002**, *12*, 292–297.

97. Maguire, B.A.; Wild, D.G. The roles of proteins L28 and L33 in the assembly and function of *Escherichia coli* ribosomes *in vivo*. Mol. Microbiol. **1997**, *23*, 237–245.

98. Frigerio, J.M.; Dagorn, J.C.; Iovanna, J.L. Cloning, sequencing and expression of the L5, L21, L27a, L28, S5, S9, S10 and S29 human ribosomal protein mRNAs. Biochim. Biophys. Acta **1995**, *1262*, 64–68.

99. Kusuda, J.; Hirai, M.; Tanuma, R.; Hirata, M.; Hashimoto, K. Genomic structure and chromosome location of *RPL27A/Rpl27a*, the genes encoding human and mouse ribosomal protein L27A. Cytogenet. Cell. Genet. **1999**, *85*, 248–251.

100. Doring, V.; Mootz, H.D.; Nangle, L.A.; Hendrickson, T.L.; de-Crecy-Lagard, V.; Schimmel, P.; Marliere, P. Enlarging the amino acid set of *Escherichia coli* by infiltration of the valine coding pathway. Science **2001**, *292*, 501–504.

101. Blanquart, C.; Barbier, O.; Fruchart, J.C.; Staels, B.; Glineur, C. Peroxisome proliferator-activated receptors: regulation of transcriptional activities and roles in inflammation. J. Steroid Biochem. Mol. Biol. **2003**, *85*, 267–273.

102. Lindsten, K.; Menendez-Benito, V.; Masucci, M.G.; Dantuma, N.P. A transgenic mouse model of the ubiquitin/proteasome system. Nat. Biotechnol. **2003**, *21*, 897–902.

103. Arnaoutova, I.; Jackson, C.L.; Al-Awar, O.S.; Donaldson, J.G.; Loh, Y.P. Recycling of Raft-associated prohormone sorting receptor, carboxypeptidase E requires interaction with ARF6. Mol. Biol. Cell **2003**.

104. Mong, J.A.; Devidze, N.; Frail, D.E.; O'Connor, L.T.; Samuel, M.; Choleris, E.; Ogawa, S.; Pfaff, D.W. Estradiol differentially regulates lipocalin-type prostaglandin D synthase transcript levels in the rodent brain: evidence from high-density oligonucleotide arrays and *in situ* hybridization. Proc. Natl. Acad. Sci. USA **2003**, *100*, 318–323.

105. Mong, J.A.; Devidze, N.; Goodwillie, A.; Pfaff, D.W. Reduction of lipocalin-type prostaglandin D synthase in the preoptic area of female mice mimics estradiol effects on arousal and sex behavior. Proc. Natl. Acad. Sci. USA **2003**, *100*, 15206–15211.

106. Beuckmann, C.T.; Lazarus, M.; Gerashchenko, D.; Mizoguchi, A.; Nomura, S.; Mohri, I.; Uesugi, A.; Kaneko, T.; Mizuno, N.; Hayaishi, O.; Urade, Y. Cellular localization of lipocalin-type prostaglandin D synthase (beta-trace) in the central nervous system of the adult rat. J. Comp. Neurol. **2000**, *428*, 62–78.

107. Hayaishi, O. Molecular mechanisms of sleep-wake regulation: a role of prostaglandin D_2. Philos. Trans. R Soc. Lond. B Biol. Sci. **2000**, *355*, 275–280.

108. Mizoguchi, A.; Eguchi, N.; Kimura, K.; Kiyohara, Y.; Qu, W.M.; Huang, Z.L.; Mochizuki, T.; Lazarus, M.; Kobayashi, T.; Kaneko, T.; Narumiya, S.; Urade, Y.; Hayaishi, O. Dominant localization of prostaglandin D receptors on arachnoid trabecular cells in mouse basal forebrain and their involvement in the regulation of nonrapid eye movement sleep. Proc. Natl. Acad. Sci. USA **2001**, *98*, 11674–11679.

109. Vyssokikh, M.Y.; Katz, A.; Rueck, A.; Wuensch, C.; Dorner, A.; Zorov, D.B.; Brdiczka, D. Adenine nucleotide translocator isoforms 1 and 2 are differently distributed in the mitochondrial inner membrane and have distinct affinities to cyclophilin D. Biochem. J. **2001**, *358*, 349–358.

110. Wilkens, S. Structure of the vacuolar adenosine triphosphatases. Cell. Biochem. Biophys. **2001**, *34*, 191–208.

111. Loeffen, J.L.; Triepels, R.H.; van den Heuvel, L.P.; Schuelke, M.; Buskens, C.A.; Smeets, R.J.; Trijbels, J.M.; Smeitink, J.A. cDNA of eight nuclear encoded subunits of NADH:ubiquinone oxidoreductase: human complex I cDNA characterization completed. Biochem. Biophys. Res. Commun. **1998**, *253*, 415–422.

112. Yamaoka-Koseki, S.; Urade, R.; Kito, M. Cardiolipins from rats fed different dietary lipids affect bovine heart cytochrome *c* oxidase activity. J. Nutr. **1991** *121*, 956–958.

113. Lin, H.Y.; Michtalik, H.J.; Zhang, S.; Andersen, T.T.; Van Riper, D.A.; Davies, K.K.; Ermak, G.; Petti, L.M.; Nachod, S.; Narayan, A.V.; Bhatt, N.; Crawford, D.R. Oxidative and calcium stress regulate DSCR1 (Adapt78/MCIP1) protein. Free Radic. Biol. Med. **2003**, *35*, 528–539.

114. Fulceri, R.; Knudsen, J.; Giunti, R.; Volpe, P.; Nori, A.; Benedetti, A. Fatty acyl-CoA-acyl-CoA-binding protein complexes activate the Ca2+ release channel of skeletal muscle sarcoplasmic reticulum. Biochem. J. **1997**, *325* (Pt 2), 423–428.

115. Lim, C.F.; Munro, S.L.; Wynne, K.N.; Topliss, D.J.; Stockigt, J.R. Influence of nonesterified fatty acids and lysolecithins on thyroxine binding to thyroxine-binding globulin and transthyretin. Thyroid **1995**, *5*, 319–324.

116. Gohil, K. Genomic responses to herbal extracts: lessons from *in vitro* and *in vivo* studies with an extract of *Ginkgo biloba*. Biochem. Pharmacol. **2002**, *64*, 913–917.

117. Kimura, T.; Ohta, T.; Watanabe, K.; Yoshimura, H.; Yamamoto, I. Anandamide, an endogenous cannabinoid receptor ligand, also interacts with 5-hydroxytryptamine (5-HT) receptor. Biol. Pharm. Bull. **1998**, *21*, 224–226.

118. Fan, P. Cannabinoid agonists inhibit the activation of 5-HT3 receptors in rat nodose ganglion neurons. J. Neurophysiol. **1995**, *73*, 907–910.

119. Matsumoto, M.; Togashi, H.; Mori, K.; Ueno-, K.; Ohashi, S.; Kojima, T.; Yoshioka, M. Evidence for involvement of central $5-HT_4$ receptors in cholinergic function associated with cognitive processes: behavioral, electrophysiological, and neurochemical studies. J. Pharmacol. Exp. Ther. **2001**, *296*, 676–682.

120. Raygada, M.; Cho, E.; Hilakivi-Clarke, L. High maternal intake of polyunsaturated fatty acids during pregnancy in mice alters offsprings' aggressive behavior,

immobility in the swim test, locomotor activity and brain protein kinase C activity. J. Nutr. **1998**, *128*, 2505–2511.

121. Madani, S.; Hichami, A.; Legrand, A.; Belleville, J.; Khan, N.A. Implication of acyl chain of diacylglycerols in activation of different isoforms of protein kinase C. FASEB J. **2001**, *15*, 2595–2601.

122. Sato, C.; Fukuoka, H.; Ohta, K.; Matsuda, T.; Koshino, R.; Kobayashi, K.; Troy, F.A.; Kitajima, K. Frequent occurrence of pre-existing α2 → 8-linked disialic and oligosialic acids with chain lengths up to 7 Sia residues in mammalian brain glycoproteins. Prevalence revealed by highly sensitive chemical methods and anti-, di-, oligo-, and poly-Sia antibodies specific for defined chain lengths. J. Biol. Chem. **2000**, *275*, 15422–15431.

123. Gallagher, H.C.; Murphy, K.J.; Foley, A.G.; Regan, C.M. Protein kinase C delta regulates neural cell adhesion molecule polysialylation state in the rat brain. J. Neurochem. **2001**, *77*, 425–434.

124. Seidenfaden, R.; Hildebrandt, H. Retinoic acid-induced changes in polysialyltransferase mRNA expression and NCAM polysialylation in human neuroblastoma cells. J. Neurobiol. **2001**, *46*, 11–28.

125. Pritchard, L.E.; Turnbull, A.V.; White, A. Pro-opiomelanocortin processing in the hypothalamus: impact on melanocortin signalling and obesity. J. Endocrinol. **2002**, *172*, 411–421.

126. Ziotopoulou, M.; Mantzoros, C.S.; Hileman, S.M.; Flier, J.S. Differential expression of hypothalamic neuropeptides in the early phase of diet-induced obesity in mice. Am. J. Physiol. Endocrinol. Metab. **2000**, *279*, E838–E845.

127. Ahren, B.; Magrum, L.J.; Havel, P.J.; Greene, S.F.; Phinney, S.D.; Johnson, P.R.; Stern, J.S. Augmented insulinotropic action of arachidonic acid through the lipoxygenase pathway in the obese Zucker rat. Obes. Res. **2000**, *8*, 475–480.

128. Nakagawasai, O.; Hozumi, S.; Tan-No, K.; Niijima, F.; Arai, Y.; Yasuhara, H.; Tadano, T. Immunohistochemical fluorescence intensity reduction of brain somatostatin in the impairment of learning and memory-related behaviour induced by olfactory bulbectomy. Behav. Brain Res. **2003**, *142*, 63–67.

129. Li, H.; Liu, D.; Zhang, E. Effect of fish oil supplementation on fatty acid composition and neurotransmitters of growing rats. Wei Sheng Yan Jiu **2000**, *29*, 47–49.

130. Luo, Y.; Long, J.M.; Lu, C.; Chan, S.L.; Spangler, E.L.; Mascarucci, P.; Raz, A.; Longo, D.L.; Mattson, M.P.; Ingram, D.K.; Weng, N.P. A link between maze learning and hippocampal expression of neuroleukin and its receptor gp78. J. Neurochem. **2002**, *80*, 354–361.

131. Gamoh, S.; Hashimoto, M.; Sugioka, K.; Shahdat Hossain, M.; Hata, N.; Misawa, Y.; Masumura, S. Chronic administration of docosahexaenoic acid improves reference memory-related learning ability in young rats. Neuroscience **1999**, *93*, 237–241.

132. Vawter, M.P.; Crook, J.M.; Hyde, T.M.; Kleinman, J.E.; Weinberger, D.R.; Becker, K.G.; Freed, W.J. Microarray analysis of gene expression in the prefrontal cortex in schizophrenia: a preliminary study. Schizophr. Res. **2002**, *58*, 11–20.

133. Huang, Y.; Doherty, J.J.; Dingledine, R. Altered histone acetylation at glutamate receptor 2 and brain-derived neurotrophic factor genes is an early event triggered by status epilepticus. J. Neurosci. **2002**, *22*, 8422–8428.

4

Exploring the Effects of Polyunsaturated Fatty Acids on Gene Expression in Human Hepatocyte (HepG2 Cells) Using DNA Chip

Yoko Fujiwara and Akiyo Masumoto

Department of Nutrition and Food Science, Ochanomizu University, Japan

Department of Clinical Dietetics and Human Nutrition, Josai University, Japan

INTRODUCTION

Fatty acids play an important role not only as an energy source but also as the components of cell membrane because their composition influences membrane fluidity and the function of receptors or channels.[1] Polyunsaturated fatty acids (PUFAs) are essential for mammalians and play an important role in maintaining their physiological conditions. Metabolites of PUFAs, eicosanoids including prostaglandins and leukotrienes, regulate platelet aggregation, blood vessels construction, immune reaction, and inflammation. In addition to these roles it is well known that PUFAs, such as linoleic acid, eicosapentaenoic acid (EPA), and docosahexaenoic acid (DHA), decrease plasma concentrations of cholesterol and triacylglycerol, and influence lipid metabolism.[2–5] Magnitude of these effects is different between the types of PUFA, for example, the numbers and the position of double bonds and their chain length. Furthermore, PUFAs, especially DHA, are related to the development of nervous system and retinal function.[6–8] Many studies within the past decade have shown that PUFAs function as mediators of gene transcription. One of the mechanisms is explained by the downregulation of sterol regulatory element binding protein (SREBP), which regulates the intracellular cholesterol metabolism.[9,10] Peroxisome proliferator activated receptor (PPAR), which binds fatty acids and their metabolites, regulates the gene expression concerned with the β-oxidation and biosynthesis of fatty acids.[11–14] Several other factors such as liver X receptor (LXR),[15] hepatocyte nuclear factor 4 (HNF4),[16] *c-fos*, and *nur-77*[17] are also thought to regulate the expression of genes that respond to fatty acids.

The mechanism of how PUFA controls gene expression is summarized by Duplus and Forest.[18] Fatty acids or their derivatives function as ligands for a transcription factor, which then binds DNA at the fatty acid response element and activates or represses transcription. Fatty acids or their derivatives thereby modify transcriptional potency and initiate a signal transduction cascade to induce covalent modification of the transcription factor. Fatty acids act indirectly via alterations in either transcription factor mRNA stability or gene transcription, resulting in variations of *de novo* synthesis of transcription factor with impact on the transcription rate of genes encoding proteins related to fatty acid transport and metabolism. Despite much recent progress, the mechanism(s) by which fatty acid modulate gene transcription remains largely unknown.

To explore the comprehensive effects of PUFA on hepatocytes, we analyzed mRNA expression profiles in PUFA-treated HepG2 cells using a DNA microarray. Here, we describe the effect of oleic acid (OA, 18:1n-9), arachidonic acid (AA, 20:4n-6), EPA (20:5n-3), and DHA (22:6n-3).

GENES EXPRESSED IN HepG2 CELLS

We investigated the effect of PUFA on HepG2 cells using DNA array (HuGene FL Array, Affymetrix Inc., Santa, CA, USA), which contained ~6000 human genes.[19] We chose HepG2 cells as a model of human liver cells, as they are widely used to investigate the lipoprotein and cholesterol metabolism.[20–23]

At first, we listed the genes highly expressed in HepG2 cells without PUFA treatment, even though a part of the high intensities of the array may be explained by the high affinity between probe sets and cDNA (Table 4.1). Although HepG2 cells expressed many kinds of ribosomal protein, the expressions of apolipoprotein were quite remarkable. Almost all the apolipoproteins, such as apolipoproteins AI, AII, E, B-100, CI, and CIII, ranked until 90th high expression levels. This evidence indicates that HepG2 cell is a good model for investigation of lipoprotein metabolism, and it also shows that HepG2 cells might produce not only the major apolipoprotein secreted by liver but also the minor one such as apolipoprotein H. Genes related to cholesterol metabolism are listed in Table 4.2. HepG2 cells highly expressed LDL receptor, SREBP, HDL binding protein, lecithin:cholesterol acyltransferase (LCAT), and HMG CoA reductase. However, microsomal triglyceride transfer protein (MTP), cholesterol 7-α-hydroxylase (Cyp7A1), and cholesterol ester transfer protein (CETP) were hardly detected by DNA chip.

Table 4.1 Genes Highly Expressed in HepG2 Cells

Order	Accession number	Average difference	Description
1	M11147	21,843	Ferritin L chain
2	L06499	21,593	Ribosomal protein L37a (RPL37A) mRNA, complete cds
3	X01038	19,326	Apolipoprotein AI precursor
4	M12529	18,908	Apolipoprotein E mRNA, complete cds
5	M17885	18,683	Acidic ribosomal phosphoprotein P0
6	J04617	17,770	Human elongation factor EF-1-alpha
7	K01396	17,371	Alpha-1-antitrypsin
8	M16961	17,347	Alpha-2-HS-glycoprotein alpha and beta chain
9	X17206	16,992	LLRep3
10	M24194	16,977	Human MHC protein homologous to chicken B complex protein mRNA, complete cds
11	M81757	16,469	S19 ribosomal protein mRNA, complete cds
12	X56932	16,290	mRNA for 23 kDa highly basic protein
13	L04483	15,822	Ribosomal protein S21 (RPS21)
14	X62691	15,804	Ribosomal protein (homologuous to yeast S24)
15	M60854	15,606	Ribosomal protein S16
16	L47125	14,926	*Homo sapiens* (chromosome X) glypican (GPC3) mRNA, complete cds
17	K02765	14,615	Human complement component C3 mRNA, alpha and beta subunits, complete cds
18	X04898	14,610	Human gene for apolipoprotein AII
19	X69150	14,536	*H. sapiens* mRNA for ribosomal protein S18.
20	M17886	14,447	Human acidic ribosomal phosphoprotein P1 mRNA, complete cds
21	X12447	14,394	Human aldolase A gene (EC 4.1.2.13)
22	X53595	14,287	Apolipoprotein H
23	X01677	13,853	Human liver mRNA for glyceraldehyde-3-phosphate dehydrogenase (G3PD, EC 1.2.1.12)
24	U22961	12,938	Human mRNA clone with similarity to L-glycerol-3-phosphate:NAD oxidoreductase and albumin gene sequences
25	S95936	12,765	Transferrin
26	X80822	12,730	ORF
27	V01514	12,158	Alpha-fetoprotein (AFP). AFP is a major serum protein (MG: 70000) synthesized during fetal life
28	X00351	11,944	Beta-actin
29	U14969	11,732	Ribosomal protein L28 mRNA, complete cds
30	M63379	11,589	TRPM-2 protein gene
31	X06617	11,513	Ribosomal protein S11

(*continued*)

Table 4.1   Continued

Order	Accession number	Average difference	Description
32	U14970	11,386	Ribosomal protein S5 mRNA, complete cds
33	M36072	11,384	Ribosomal protein L7a (surf 3) large subunit mRNA, complete cds
34		11,357	Metallopanstimulin 1
35		11,234	Ribosomal protein L37
36	X63527	11,205	Ribosomal protein L19
37	D23660	11,175	Ribosomal protein, complete cds
38	AB00058	10,809	TGF-beta superfamily protein, complete cds
39	U12465	10,740	Ribosomal protein L35 mRNA, complete cds
40		10,402	Albumin
41	M18000	10,370	Human ribosomal protein S17 gene, complete cds
42	M13934	10,273	5551–5557, *RPS14* gene (unknown protein) extracted from human ribosomal protein S14 gene
43	U49869	10,133	Human ubiquitin gene, complete cds
44	Z49148	10,069	Ribosomal protein L29
45	AB00253	9,749	Human mRNA for Qip1, complete cds
46	U14972	9,431	Human ribosomal protein S10 mRNA, complete cds
47	L19686	9,312	*H. sapiens* macrophage migration inhibitory factor (MIF) gene, complete cds
48	M14328	9,269	Human alpha enolase mRNA, complete cds
49	M20902	9,231	Human apolipoprotein C-I (VLDL) gene, complete cds
50	Z26876	9,197	Ribosomal protein L38

Note: HepG2 cells were incubated with 10% LPDS–DME for 24 h. Average differences express the intensities of mRNA levels by DNA microarray described in Chapter 1, p. 14.

HepG2 cells also highly expressed some metal-containing proteins, such as ferritin and transferin; various kinds of cytokines, tumor growth factor β (TGF-β), insulin growth factor 2 (IGF2), *v*-fos, and macrophage migration inhibitory factor (MIF); and the genes related to adhesion molecule, such as β-tublin, laminin receptor (data not shown).

EFFECT OF PUFA ON GENE EXPRESSION PROFILE IN HepG2 CELLS

HepG2 cells were incubated in Dulbecco's modified Eagle's medium (DMEM) containing 10% lipoprotein-deficient fetal calf serum (LPDS) with 0.5 mM PUFAs, OA, AA, EPA, and DHA, dissolved in essential fatty acid-free bovine

Table 4.2 Expression Levels of Genes Related to Cholesterol Metabolism in Control HepG2 Cells

Order	Average difference	Absolute call	Gene
3	19,326	P	Apolipoprotein AI
4	18,908	P	Apolipoprotein E
20	14,610	P	Apolipoprotein AII
24	14,287	P	Apolipoprotein H
53	9,231	P	Apolipoprotein C-I
84	6,653	P	Apolipoprotein B-100
348	1,358	P	LDL receptor
377	6,016	P	Apolipoprotein CII
466	1,105	P	SREBP-1
486	942	P	HDL binding protein (HBP)
475	957	P	Acid cholesteryl esterase
584	741	P	LCAT
592	731	P	CD36
683	614	P	HMG-CoA reductase
1256	297	A	Apolipoprotein D
1264	294	P	Apolipoprotein AIV
1456	245	A	Phospholipid transfer protein
1537	226	P	HMG-CoA syntase
1631	207	A	PPAR gamma
2603	89	A	Apolipoprotein (a)
2617	87	M	ALCAM (HB2)
2878	68	A	VLDL receptor
3149	52	A	Lipoprotein lipase
	≤ 0	A	MTP
	≤ 0	A	CLA-1 (SR-BI)
	≤ 0	A	ACAT
	≤ 0	A	Cholesterol 7-α-hydroxylase (Cyp7A1)
	≤ 0	A	APOBEC1
	≤ 0	A	CETP

Note: Hep G2 cells were incubated in 10% LPDS–DME for 24 h without addition of PUFA. Genes related to cholesterol metabolism were shown in order with high amounts of average difference.

serum albumin (BSA) solution for 24 h. After the treatment of PUFAs, the cells were harvested to isolate total RNA. Studies using the GeneChip proceeded according to the technical manual supplied with the Affymetrix GeneChip expression analysis system.[24,25] After hybridization of the GeneChips array, the intensity of each feature was captured by Affymetrix GeneChip Software (Affymetrix Inc.) according to standard Affymetrix procedures with a class AB mask file.

Of ~6000 genes investigated in this study, 44 genes showed greater than twofold change in expression levels after DHA treatment. Number of genes changed over two-fold were 19, 29, and 26 by OA, AA, and EPA treatments, respectively. DHA was the most effective in changing the mRNA expression. Whereas the effects of PUFAs, DHA, EPA, and AA, were seen in both induced and reduced genes, OA treatment did not reduce the expression level of mRNA significantly. This suggested that OA was different from PUFAs in the manner of regulating gene expression. In any case, the effects of PUFA on gene expression seemed to be moderate compared with those elicited by other drugs or chemicals.[26,27] This may be due to PUFAs being nutrients that are naturally catabolized to produce energy.

Table 4.5 shows the effect of PUFA on genes except for those related to cholesterol and lipoprotein metabolism. PUFA treatment affected several transcription factors. They suppressed the mRNA levels of SREBPs, and slightly changed the expression of LXR, NF-κB p65, nuclear factor I-X (NFI-X), PPARs, and Rad2. The expressions of HNF, *c-fos*, and *nur-77*, which have been previously reported to respond to fatty acids,[17] did not change. All fatty acids largely increased mRNA levels of metallothionein-IG, ventricular/slow twitch myosin alkali light chain (MLC-1V/Sb isoform), and deleted split hand/split foot 1(DDS1). PUFA also affected the expression of genes involved in cell differentiation and proliferation.

PUFA repressed the expression of lipogenic genes such as fatty acid synthase (*FAS*) and stearoyl-CoA desaturase (*SCD1*). However, gene expression of enzymes that catabolize fatty acids, namely, carnitine:palmitoyl-CoA acyltransferase 1 (CPT1), acyl-CoA oxidase (AOX), and acyl-CoA synthetase (ACS), which are induced by PUFA, did not change in our study. These enzymes are related to fatty acid oxidation and are generally believed to be regulated by PPAR.[28–32] PPAR-α and PPAR-γ are located in the liver and adipocytes, respectively.[14] The expression level of PPAR-α in human liver[33,34] is much lower than that in mouse liver, and overexpression of PPAR-α in HepG2 cells leads to the induction of mitochondrial HMG-CoA synthase, CPT, and ACS mRNA.[35] The present study detected only weak expression of PPAR-α and PPAR-γ in HepG2 cells (Table 4.3).

PPAR also regulates fatty acid transport, fatty acid binding protein (FABP), and desaturation. As shown in Table 4.3, *FAS, FABP,* and *SCD1* were suppressed by PUFA. However, fatty acid metabolism is regulated not only by PPAR but also by SREBP. Whereas SREBP-1 regulates lipogenic genes expression, SREBP-2 controls the cholesterol metabolism.[36,37] Therefore, reduced lipogenesis by PUFA in our study was mainly caused by SREBP-1, and SREBPs are suggested to be the key transcription factor to regulate gene expressions in HepG2 cells. Although mRNA expression related to fatty acid oxidation was not changed, PUFA upregulated 2-oxoglutarate dehydrogenase, isocitrate dehydrogenase, and succinyl-CoA synthetase, all of which are involved in the TCA cycle. In summary, PUFA suppressed lipogenesis, but induced ATP generation by activation of the TCA cycle in HepG2 cells.

Table 4.3 Fold Changes of mRNA Levels by Fatty Acid Treatments in HepG2 Cells

Gene	Accession number	Average difference[a]	Fold change				Function
			OA	AA	EPA	DHA	
Interferon-gamma receptor alpha chain	U19247	226	−1.1	−2.3	−2.3	−2.2	Antiviral activity
Mitochondrial NADH dehydrogenase	U65579	407	1.8	2.0	3.1	2.7	Aspiratory chain
Heparan sulfate proteoglycan (HSPG2)	M85289	146	−1.5	1.3	5.3	1.3	Cell adhesion
cdc25Hs	M34065	−26	2.7[b]	1.9[b]	1.4[b]	2.1[b]	Cell differentiation
Interleukin 1 alpha (IL 1)	M28983	−51	2.6[b]	2.0[b]	2.7[b]	1.8[b]	Cell differentiation
MAC30	L19183	1769	1.0	−2.8	−2.1	−1.8	Cell differentiation
Protein tyrosine phosphatase (PTP-PEST)	M93425	73	2.0[b]	1.9[b]	−1.0[b]	3.1[b]	Cell differentiation
Small proline-rich protein 2 (SPRR2B)	L05188	−110	2.9[b]	2.8[b]	3.0[b]	1.8[b]	Cell differentiation
SWI/SNF complex 155 kDa subunit (BAF155)	U66615	197	1.4	1.5[b]	2.3	2.2[b]	Cell differentiation
Drosophila female sterile homeotic (FSH)	X62083	4	1.5	1.2	13.3[b]	2.9[b]	Cell proliferation
Glial growth factor 2		394	−5.5[b]	−3.0	−5.3[b]	−5.6[b]	Cell proliferation
Membrane-associated protein (HEM-1)	M58285	193	1.8	2.5[b]	2.9	2.4[b]	Cell proliferation
Sec23A isoform	X97064	51	3.3[b]	1.1[b]	2.1[b]	2.6[b]	Cell proliferation
Sec23B isoform	X97065	230	2.3	1.8	2.4	2.1	Cell proliferation
S-lac lectin L-14-II (LGALS2)	M87860	−3	1.5[b]	2.0[b]	3.2[b]	4.5[b]	Cell proliferation
Microsomal glutathione S-transferase (GST-II)	U77604	2836	1.0	−1.1	1.2	−2.0	Detoxification
FDXR gene (adrenodoxin reductase)	M58509	287	1.2	1.5	1.6	2.2	Electron transport system
Uncoupling protein homolog (UCPH)	U94592	169	2.8	−2.7	2.2	−2.0	Energy consumption

(*continued*)

Table 4.3 Continued

Gene	Accession number	Average difference[a]	Fold change OA	AA	EPA	DHA	Function
Fatty acid synthase	S80437	4358	-1.0	-2.1	-2.1	-2.3	Fatty acid synthesis
Stearoyl-CoA desaturase	M10050	1416	1.1	-2.9	-2.9	-3.1	Fatty acid synthesis
Liver fatty acid binding protein (FABP)	M13699	6859	1.1	-2.0	-1.6	-1.5	Fatty acid transport
Ceruloplasmin (ferroxidase)		309	1.2	-1.9	-2.9	-3.1	Fe oxidation
Galactokinase (GALK1)	L76927	120	2.5	2.8	-1.9[b]	3.1	Glycogenesis/glycolysis
RASF-A PLA2	M22430	122	2.0	2.2	1.5	2.7	Inflamation
S-lac lectin L-14-II (LGALS2)	M87860	-3	1.5[b]	2.0[b]	3.2[b]	4.5[b]	Lectin
Deleted in split hand/split foot 1 (DSS1)	U41515	102	1.4	3.7	4.4	4.8	Limb development
Urokinase-type plasminogen activator receptor	U09937	-19	1.6[b]	1.7[b]	4.9[b]	2.4[b]	Platelet coagulation
Metallothionein-IG (MT1G)	J03910	195	1.8	3.6	2.3	5.6	Protection against heavy metal toxicity
Inter-alpha-trypsin inhibitor subunit 3	X16260	238	-1.1	-4.8[b]	-4.2[b]	-2.8	Proteinase inhibitor
Vacuolar proton pump, 116-kDa subunit	U45285	41	2.0[b]	4.4[b]	4.3[b]	7.1[b]	Proton pump
Prostasin	L41351	749	-1.2	-2.6	-8.3	-3.8	Serine proteinase
Extracellular-superoxide dismutase (SOD3)	J02947	124	1.7	1.3	3.6[b]	2.2	Superoxiside scavenger

Manganese superoxide dismutase (SOD2)	X65965	611	-1.2	-1.2	-2.0	-1.1	Superoxiside scavenger
2-Oxoglutarate dehydrogenase	D10523	143	1.4	-1.0	2.0	1.4	TCA cycle
Isocitrate dehydrogenase	Z68129	202	1.5	1.9	2.5	2.0	TCA cycle
Succinate dehydrogenase (SDH)	L21936	496	1.9	1.4	2.1	2.5	TCA cycle
Succinyl-CoA synthetase	Z68204	6	2.1[b]	1.2[b]	2.1[b]	2.1[b]	TCA cycle
LXR-alpha	U22662	67	1.4[b]	-1.3[b]	2.1[b]	1.3[b]	Transcription factor
NF-κ-B p65 subunit	L19067	201	2.3	1.7	1.4	1.8	Transcription factor
Nuclear factor I-X	L31881	94	1.4	-1.2	4.4	1.4	Transcription factor
PPAR alpha	L02932	4	-1.4[b]	1.2[b]	1.5[b]	1.1[b]	Transcription factor
PPAR gamma	L40904	99	2.0	-1.6	-1.1	1.0	Transcription factor
Rad2		40	2.6	2.1[b]	3.5[b]	2.6[b]	Transcription factor
SREBP-1	U00968	1105	1.0	-1.7	-1.2	-1.8	Transcription factor
SREBP-2	U02031	559	1.1	-1.5	-1.9	-1.7	Transcription factor
KIAA0030	D21063	45	2.5[b]	2.2[b]	8.1[b]	1.9[b]	Unknown
KIAA0092	D42054	325	-1.1	-1.3	-1.7	-3.8	Unknown
KIAA0219	D86973	96	4.2	1.9	2.7	3.2	Unknown
Inducible protein	L47738	-55	3.9[b]	3.5[b]	4.5[b]	3.5[b]	Unknown

Note: HepG2 cells were treated with 0.25 mM of oleic acid (OA), arachidonic acid (AA), eicosapentaenoic acid (EPA), or docosahexaenoic acid (DHA) for 24 h.

Source: Fujiwara et al.[19]

[a] Average difference were expressed the intensities of the mRNA levels in control HepG2 cells.

[b] The value of fold change was calculated using the backgroud value, as the average difference of the transcript in either the control or the FA-treated group was smaller than the background.

PUFA SUPPRESSES THE CHOLESTEROL METABOLISM

As PUFA is well known to reduce plasma cholesterol and affect the cholesterol metabolism, we showed the changes in genes related to cholesterol and lipo-protein metabolism by fatty acid treatments in Table 4.4. PUFA suppressed the mRNA levels of LDL receptor, HMG-CoA synthase, and HMG-CoA reductase, all of which are SREBP targets.[9,38,36] Moreover, the expression of mevalonate pyrophosphate decarboxylase (MPD) and squalene epoxidase, which function in the cholesterol synthetic pathway, were downregulated. Lysosomal acid lipase (LAL/CE) is one of the most PUFA-sensitive genes that has not been reported previously. Results of gene chip did not show significant changes in the expression of LXR-α. The levels of the mRNA expression of acyl-CoA:choles-terol acyltransferase (ACAT) and CETP, which are thought to be regulated by PUFA,[39,40] were undetectable. It might be due to their very low expression levels in HepG2 cells.

As the MPD mRNA level was decreased by -9.5 (11%) with DHA by gene chip analysis, we measured the mRNA levels of MPD using a quantitative RT–PCR in order to confirm the result of gene chip analysis. MPD is one of the enzymes involved in the cholesterol biosynthesis pathway and Sakakura et al.[41] reported that SREBP regulates the gene expression of all of the enzymes involved in cholesterol synthesis including MPD. The gene chip analysis also showed reduced mRNA levels by -1.7, -1.2, and -1.8 in SREBP-1, and -1.5, -1.9, and -1.7 in SREBP-2 with AA, EPA, and DHA, respectively. Therefore, we also measured the mRNA levels of SREBPs in HepG2 cells treated with fatty acid (Fig. 4.1). AA, EPA, and DHA reduced the mRNA levels of SREBP-1 and SREBP-2, whereas OA did not affect the mRNA levels. Figure 4.2 shows that PUFA also decreased the mRNA levels of MPD by 50%. We confirmed that PUFA suppressed MPD and SREBPs but the magnitude of the effect was not equivalent between the data obtained from the gene chip analysis and that from RT–PCR. However, the gene chip analysis reflects the ten-dencies of PUFA, and is a useful tool for investigating the comprehensive effect.

Using the quantitive RT–PCR, we tried further investigation about the effect of PUFA as follows. We also evaluated the effects on HDL metabolism and cholesterol reverse transport because the expression levels were too low to detect by DNA microarray.

As shown in Fig. 4.3, mRNA level of SREBP-1 was decreased by AA, EPA, and DHA with increasing fatty acid concentrations in the medium. However, OA did not reduce the SREBP-1 expression. Expression of SREBP-2 was similar. The expression levels of HMG-CoA reductase and LAL/CE showed quite similar response to PUFA as SREBP-1. These major enzymes that control cholesterol metabolism are suggested to be regulated via SREBP. Although LAL/CE has not been reported as the SREBP target, our result strongly suggested that PUFA downregulated LAL/CE by the mediation of SREBP.

Table 4.4 Changes of mRNA Levels in Genes Related to Cholesterol and Lipoprotein Metabolism by Fatty Acid Treatments

Gene	Accession number	Average difference[a]	Fold change			
			OA	AA	EPA	DHA
Repressed						
HMG-CoA reductase	M11058	614	−1.5	−2.9	−2.2	−3.1
HMG-CoA synthase	L25798	226	−1.5	−2.9	−2.4[b]	−2.0
Mevalonate kinase	M88468	276	−1.2	−1.2	−2.7	−1.1
Mevalonate pyrophosphate decarboxylase	U49260	1,638	−1.3	−3.4	−1.9	−9.5
Squalene epoxidase	D78129	1,782	−1.0	−2.0	−1.2	−2.2
2,3-Oxidosqualene-lanosterol cyclase	U22526	200	−1.0	−2.5	−2.8[b]	−4.8[b]
LDL receptor	L00352	1,358	−1.1	−2.6	−2.1	−2.3
Lysosomal acid lipase	U04285	957	−1.1	−1.6	−2.2	−1.6
Induced						
Hepatic triglyceride lipase	M29194	−1	1.7[b]	1.2[b]	1.8[b]	2.6[b]
Apolipoprotein(a)	X06290	89	2.2	1.3	2.4	−1.3[b]
ICAM-2	M32334	5	2.0[b]	1.3[b]	3.1[b]	−1.3[b]
No change						
Apolipoprotein AI regulatory protein (ARP-1)	M64497	40	1.2[b]	1.1[b]	−1.7[b]	1.1[b]
Apolipoprotein AI precurser	X01038	19,326	1.0	−1.0	−1.0	1.0
Apolipoprotein AII	X04898	14,610	1.0	1.0	−1.0	1.0
Ear-3		75	1.0	1.1[b]	−1.1[b]	−1.5[b]
Lectin-like oxidized LDL receptor	D89050	−21	1.1[b]	−2.0[b]	−1.1[b]	1.1[b]
Lipoprotein lipase	M15856	52	−1.4[b]	−1.0[b]	−1.3[b]	1.1[b]
Scavenger receptor type I	D13264	−13	−1.2[b]	−1.3[b]	1.0[b]	1.2[b]
CLA-1 (SR-BI)	Z22555	0	0.0[b]	0.0[b]	0.0[b]	0.0[b]

(continued)

Table 4.4 Continued

Gene	Accession number	Average difference[a]	Fold change			
			OA	AA	EPA	DHA
CD36	Z32765	731	1.1	-1.3	1.3	1.5
HDL binding protein	M64098	942	1.2	1.2	1.2	1.4
CD6 ligand (*ALCAM/HB2*)	L38608	87	1.0	-1.8[b]	-1.5[b]	1.2
Cdc42 GTPase-activating protein	U02570	310	1.3	1.2	1.2	1.4
LCAT	M12625	741	1.0	1.1	-1.2	1.2
ACAT	L21934	-12	1.4[b]	1.1[b]	1.2[b]	1.2[b]
CETP	M30185	-140	-2.9[b]	-1.2[b]	1.3[b]	-1.9[b]
Phospholipid transfer protein		245	1.4	-1.2	1.4	-1.0
MTP	X91148	0	0.0[b]	0.0[b]	0.0[b]	0.0[b]

Note: HepG2 cells were treated with 0.25 mM of oleic acid (OA), arachidonic acid (AA), eicosapentaenoic acid (EPA), or docosahexaenoic acid (DHA) for 24 h.
Source: Ref.[19].
[a] Average difference were expressed the intensities of the mRNA levels in control HepG2 cells.
[b] The value of fold change was calculated using the background value, as the average difference of the transcript in either the control or the FA-treated group was smaller than the background.

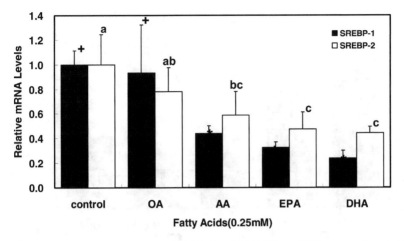

Figure 4.1 Effect of PUFA on SREBP-1 and SREBP-2 mRNA expressions in HepG2 cells. HepG2 cells were incubated with 0.25 mM OA, AA, EPA, and DHA for 24 h. Total RNA was extracted, then mRNA expression levels of SREBPs were measured using a real time RT–PCR (GeneAmp 5700, Applied Biosystems). Single-strand cDNA was synthesized from 1 μg of total RNA using random hexamer and TaqMan reverse transcription reagents (Applied Biosystems, Foster City, CA, USA). Primer sequences were 5′-GCAAGGCCATCGACTACATTC-3′; (forward) and 5′-TTGCTTTTGTGGA CAGCAGTG-3′ (backward) for SREBP-1; 5′-AGGCGGACAACCCATAATATCA-3′ (forward) and 5′-GACTTGTGCATCTTGGCGTCT-3′ (backward) for SREBP-2. Real-time quantitative RT–PCR proceeded in a reaction mixture containing 10 ng of first-strand cDNA, 300 nM of each primer set in a final volume of 25 μL and SYBR Green PCR core reagent (Applied Biosystems). Relative mRNA levels were normalized to those of GAPDH. Values are mean ± SD ($n = 3$). Significance of differences between PUFA was analyzed by multiple comparisons. Bars not sharing a common letter are significantly different ($p < 0.05$). (from Fujiwara et al.[19])

Results of gene chip analysis did not show significant changes in LXR expression by PUFA. DNA microarray was not able to evaluate mRNA levels of LXR because of its low intensity. As shown in Fig. 4.4, results obtained by RT–PCR show that all fatty acids increased LXR mRNA levels even though they had differences in their optimum concentration Similarly, ABCA1[42,43] and Cyp7A1,[44,45] known to be the LXR-α target genes were increased by OA, AA, EPA, and DHA. Therefore, these proteins that play an important role in cholesterol catabolism should be regulated mainly by LXR. And, results of dose–response studies suggest that we have to be careful to compare the effect of each PUFA at the same concentration, as the effective concentration of each PUFA should be different. Together with these results, PUFA have antiathero-genic effect to reduce the cholesterol synthesis and induce the cholesterol catabolism.

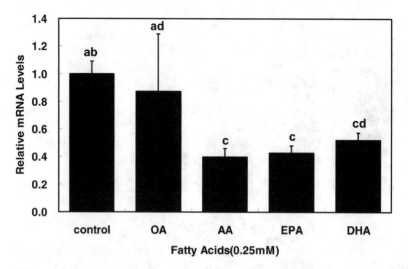

Figure 4.2 Effect of PUFA on MPD expression in HepG2 cells. HepG2 cells were incubated with 0.25 mM OA, AA, EPA, and DHA for 24 h. Total RNA was extracted, then mRNA expression levels of SREBPs were measured using a real time RT–PCR (GeneAmp 5700, Applied Biosystems). Single-strand cDNA was synthesized from 1 μg of total RNA using random hexamer and TaqMan reverse transcription reagents (Applied Biosystems, Foster City, CA, USA). Primer sequences for MPD were 5′-CGAGTCACACTGGCCTGAACT-3′ (forward) and 5′-CACGGTACTGCCTGTCA GCTT-3′ (backward). Real-time quantitative RT–PCR proceeded in a reaction mixture containing 10 ng of first-strand cDNA, 300 nM of each primer set in a final volume of 25 μL, and SYBR Green PCR core reagent (Applied Biosystems). Relative mRNA levels were normalized to those of GAPDH. Values are mean $\pm$ SD ($n = 3$). Significance of differences between PUFA was analyzed by multiple comparisons. Bars not sharing a common letter are significantly different. ($p < 0.05$) (from Fujiwara et al.[19])

PUFA AND THE CROSS-TALK OF TRANSCRIPTION FACTORS

As described earlier, PUFA treatment in HepG2 cells altered the expression of many transcription factors. The most remarkable downregulation was seen in SREBPs (Table 4.5). Our data showed that PUFA repressed the expression of hepatic lipogenic genes and almost all of the genes associated with the cholesterol synthetic pathway. In this study using HepG2 cells, it was suggested that one of the major mechanisms of downregulation of mRNA expression caused by PUFA was mediated through SREBPs. PUFAs reduce the mRNA expression of SREBPs,[46–50] which regulate lipogenic gene transcription (*SREBP-1*) and control the cholesterol metabolism (*SREBP-2*).[51,36,52,37] These results indicate that PUFA downregulates the entire cholesterol synthetic pathway. Downregulation of SREBP expression by PUFA is explained by its mRNA stability.[49]

LXR binds oxysterols and directly regulates the expressions of genes involved in hepatic bile acid synthesis. On the other hand, our data showed

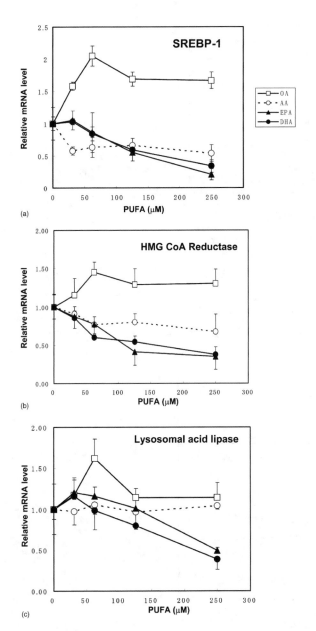

Figure 4.3 Dose effect of PUFA on SREBP-1, HMG-CoA reductase, and lysosomal acid lipase expressions in HepG2 cells. HepG2 cells were incubated with OA, AA, EPA, and DHA at 31.3, 62.5, 125, and 250 μM for 24 h. Total RNA was extracted, then mRNA expression levels were measured using real time RT–PCR (Geneamp 5700, Applied Biosystems). Relative mRNA levels were normalized to those of GAPDH. Values are mean ± SD ($n = 3$)

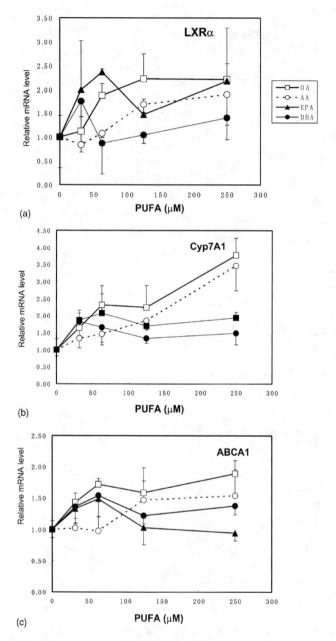

Figure 4.4 Dose effect of PUFA on LXRα, Cyp7A1 and ABCA1 in HepG2 cells. HepG2 cells were incubated with OA, AA, EPA, and DHA at 31.3, 62.5, 125, and 250 μM for 24 h. Total RNA was extracted, then mRNA expression level were measured using real time RT–PCR (Geneamp 5700, Applied Biosystems). Relative mRNA levels were normalized to those of GAPDH. Values are mean ± SD ($n = 3$).

Table 4.5 Fish oil Diet-Induced Increases in Gene Expression in Liver

Order	Accession number	Fold increase	Gene description	Average difference Carbohydrate	Sallower	Fish	Putative function	Group
1	V00802	37	Kappa-immunoglobulin (constant region)	27	14	1797↑	Inflammatory	1
2	X03690	24	Ig heavy-chain constant region mu(b) allele.	3	1	519↑	Inflammatory	1
3	J04953	10	Gelsolin	27	71	245↑	Phagocytosis	1
4	AA145127	8.9	Homologous to *H. sapiens* leukocyte elastase inhibitor	52	67	600↑	Protease inhibitor	1
5	J04695	6.6	Alpha-2 type IV collagen	32	12	170↑	Matrix	
6	W29430	5.7	Hepatocyte growth factor activator inhibitor type 2	97	22	158↑	Protease inhibitor	
7	U15976	5.6	Fatty acid transport protein	32	45	253↑	Fatty acid transport	2
8	M23998	4.3	Testosterone 16-alpha-hydroxylase	109	121	440↑	Steroid hormone degradation	2
9	U69135	4.1	Uncoupling protein 2	381	576	2415↑	Uncoupler	3
10	U48420	4.1	Theta class glutathione transferase type 2	68	206	840↑	Conjugation with glutathione	3
11	L06047	3.8	Glutathione transferase	68	59	222↑	Conjugation with glutathione	3
12	W29265	3.2	Glutathione transferase Ya subunit	221	169	616↑	Conjugation with glutathione	3
13	X69296	2.9	Cytochrome P-450, 4a10	412	1296↑	3813↑	Fatty acid omega oxidation	2
14	U15977	2.8	Long-chain fatty acyl-CoA synthetase	306	1124↑	3941↑	Fatty acid beta oxidation	2

(continued)

Table 4.5 Continued

Order	Accession number	Fold increase	Gene description	Average difference			Putative function	Group
				Carbohydrate	Sallower	Fish		
15	M32599	2.5	Glyceraldehyde-3-phosphate dehydrogenase	2873	783↓	2052↑	Glycolysis	
16	X70303	2.5	Proteosome subunit, alpha type 2	146	140	421↑	Protein turnover	
17	U15636	2.4	T-cell specific GTPase, GTP-binding protein	139	123	334↑	Unknown	
18	AA028398	2.4	Beta-tubulin	218	106	255↑	Cytoskeleton, chaperone	
19	L05439	2.3	Insulin-like growth factor binding protein 2	690	923	1602↑	Modify IGF function	
20	X62940	2.3	Transforming growth factor-beta-stimulated clone-22	163	133	308↑	Transcription factor	
21	U01163	2.2	Carnitine palmitoyltransferase 11	655	767	1653↑	Fatty acid beta oxidation	2
22	L20276	2.2	Biglycan	538	442	1090↑	Matrix	
23	L35528	2.1	Manganese superoxide dismutase	447	443	925↑	Antioxidant	3
24	Z38015	2.0	Myotonin protein kinase	368	289	574↑	Ca^{2+} homeostasis	

Note: Group 1, immunoreaction; group 2, fat oxidation; group 3, antioxidant. In order of fold change levels in gene expression in liver from fish-oil fed mice relative to safflower-oil fed mice, genes increased more than two-fold are presented. Fold change calculation was carried out as an indication of the relative change of each transcript represented on the probe array. The average difference value is a marker of abundance of each gene. Putative gene functions are based on literature review. Change (↑/↓) in fish-oil fed mice indicates that difference call is significantly increased/decreased, compared with difference call is significantly increased/decreased, compared with safflower oil-fed mice, whereas that in safflower oil-fed mice was compared with that in high-carbohydrate-fed mice.

Source: Takahashi et al.[68]

that fatty acid treatment seemed to upregulate the expression of LXR-α. This result agrees with the report by Tobin et al.[15] LXR-α also regulates the expression of SREBP that has LXR response element of its promoter region. Yoshikawa et al.[53] reported that the PUFA suppression of SREBP-1c expression is mediated through competition with LXR ligand during activation of the ligand binding domain of LXR.

PUFA and their metabolites act as a ligand of PPAR. HepG2 cells do not highly express PPAR and the genes involved in β-oxidation were not changed, suggesting that PPAR slightly contributed to the decrease of mRNA expression by PUFA in HepG2 cells. However, PPAR-α was reported to inhibit the binding of LXR-α/RXR-α to LXR response element.[54] They also showed that LXR-α reduced the binding of PPAR-α/RXR-α to PPAR-α response element and the inhibitory effect was attenuated by addition of RXR-α, suggesting that LXR-α/PPAR-α heterodimers lead to a reduction of the other types of RXR heterodimers formation.[55] So they proposed the presence of an intricate network of nutritional transcription factors with mutual interactions resulting in efficient reciprocal regulation of lipid degradation and lipogenesis. Especially, there are many reports about fatty acid desaturation and its transcriptional regulation with regard to the relationship between fatty acid and cholesterol metabolism.

The PUFA response region is located in the promoter of the stearoyl-CoA desaturase 1 (*SCD1*) gene.[56] SREBP may play an important role in regulating the SCD1 as its rate of downregulation is similar to those of the genes related to cholesterol metabolism. However, Kim et al.[57] demonstrated that cholesterol overrides the PUFA-mediated repression of the *SCD1* gene and regulates *SCD1* gene expression through a mechanism independent of SREBP-1 maturation *in vivo*. The detailed mechanism of the downregulation of *SCD1* caused by PUFA has not been resolved. Matsuzaka et al.[58] reported that Δ6-desaturase and Δ5-desaturase expression is dually regulated by SREBP-1c and PPAR-α. PUFAs are thought to also autoregulate their biosynthesis through SREBP. In addition, CETP might interact with SREBP-1[59] and is downregulated by PUFA.[60,39] As the expression level of CETP in HepG2 cells is very low, we could not evaluate the effect of PUFA on its expression using the oligonucleotide chip system (Table 4.2).

HNF-4α, one of the important transcription factors in liver, is reported to induce the genes encoding apolipoproteins CII, CIII, AII, AIV, transferrin, carbohydrate metabolism, cytochrome P450 monooxigenases, and bile acid synthesis.[61-64] It is reported that binding of saturated fatty acid CoA activated HNF-4α whereas binding of n-3 PUFA-CoA inhibited HNF-4α effect on gene transcription. In our data, mRNA levels of HNF-4α and the expression of genes regulated by HNF-4α did not change. While focusing on the cross-talk between HNF-4α, Torra et al.[65,66] identified the promoter of human PPAR-α and showed that HNF-4α enhanced human PPAR-α promoter activity via binding the HNF-4α-regulatory element. Furthermore, farnesol X receptor

(FXR) activated with bile acid also interacts with PPAR-α.[67] Molecular cross-talk between these transcription factors is advancing to clarify the mechanism of the regulation and interaction of fatty acid and cholesterol.

EFFECT OF FISH OIL FEEDING ON mRNA EXPRESSION PROFILE IN MICE LIVER

On the contrary to our *in vitro* study, Takahashi et al.[68] have examined the effect of dietary fish oil on the gene expression profile in mouse liver using high-density oligonucleotide arrays (Tables 4.5 and 4.6). They fed mice with high-fat diet (60% of total energy intake), either safflower oil or fish oil (tuna), for 6 months. They also found that fish oil feeding downregulated the cholesterol and fatty acid synthesis-related genes and upregulated the lipid catabolism-related genes. Although our findings were similar to theirs, they showed that immune reaction-related genes, antioxidant genes (several glutathione transferase, uncoupling protein 2, and Mn-superoxide dismutase), and genes involved in lipid catabolism were significantly upregulated, indicating that dietary fish oil downregulated the endogenous PPAR-α-activation system and increased the antioxidant gene expression that protects against excess reactive oxygen species (ROS). Our data also suggested that PUFA induced antioxidant genes, such as metallothionein-IG and extracellular-superoxide dismutase (SOD3). However, the overall response to oxidation was much less because the expressions of microsomal glutathione *S*-transferase and manganese superoxide dismutase (SOD2) were not significantly changed (Table 4.5). We suppose that little oxidative stress was induced by adding PUFA to HepG2 cells even though the PUFAs were extremely pure (99%). The induction of immunological and antioxidant genes in their study might have been caused by adaptation to excess ROS production, as they fed the diet containing a very high concentration of fish oil (60% of total energy intake) for 6 months.

This idea was supported by the study of Berger et al.,[69] which reported the dietary effects of arachidonate-rich oil and fish oil on murine hepatic and hippo-campal gene expression. They fed mice with the diet containing relatively low fat to avoid the confounding effect of PUFA. Their expression profile was very similar to our results (see Chapter 2). Both studies *in vivo* using mice showed upregulation of genes related to cytochrome P450 metabolism, which did not change in our study using HepG2 cells.

PUFA AND ITS NOVEL FUNCTION

Gene chip analysis found that PUFA affected the expression of some genes that have not been known previously. We confirmed the effect of PUFA on prostasin gene expression by quantitative RT–PCR (Fig. 4.5). Prostasin is a new serine protease that was purified from seminal fluid, and its cDNA has been sequenced.[70,71] It is expressed in the human prostate, kidney, and lung, as

Table 4.6 Fish Oil Diet-Induced Decreases in Gene Expression in Liver

				Average difference				
Order	Accession number	Fold change	Gene description	Carbohydrate	Safflower	Fish	Putative function	Group
1	M21285	−83	Stearoyl-CoA desaturase gene	13,098	7,644	81↓	PUFA synthesis	1
2	L27121	−34	Hydroxysteroid sulfotransferase	952	1,330↑	3↓	Phase II xenobiotic metabolizing enyzmes	2
3	M64863	−29	Cytochrome P-450 17-alpha hydroxylase/C17-20 lyase	1,143	952	−77↓	Extra glandular steroid genesis	2
4	W89667	−15	Homologous to rat sterol regulatory element binding protein-1c	266	181	14↓	Transcription factor	
5	L41631	−9.2	Glucokinase	375	338	4↓	Glucose uptake	
6	W17745	−7.2	Homologous to rat ATP citrate-lyase	2,815	1,647↓	229↓	Fatty acid synthesis	1
7	AA137659	−6.5	Cytochrome P-450 IIC40	1,505	1,652	252↓	Inflammatory	
8	AA139907	−5.8	Spot14	1,182	269↓	46↓	Unknown	
9	X13135	−5.7	Fatty acid synthase	2,781	1,505↓	275↓	Fatty acid synthesis	1
10	M64250	−5.4	Apolipoprotein A-IV	331	275	71↓	HDL cholesterol metabolism	
11	W48402	−4.3	Silent mating type information regulation 2	78	212↑	50↓	Silencing gene transcription in yeast	3
12	D42048	−4.1	Squalene epoxidase	364	539↑	114↓	Cholesterol synthesis	1
13	X05475	−3.8	Complement component C9	919	753	229↓	Cell lysis	
14	AA036251	−3	Homologous to rat farnesyl pyrophosphate synthetase	385	403	156↓	Cholesterol synthesis	1
15	W81960	−2.5	Phenol/aryl form sulfotransferase	764	722	249↓	Phase II xenobiotic metabolising enyzmes	2
16	M33212	−2.5	Nucleolar protein N038	113	304↑	121↓	Unknown	

(continued)

Table 4.6 Continued

Order	Accession number	Fold change	Gene description	Average difference			Putative function	Group
				Carbohydrate	Safflower	Fish		
17	M58588	−2.5	Plasma kallikrein	160	289	117↓	Blood coagulation	
18	U51805	−2.4	Arginase	2,206	2,641	1,124↓	Urea production	
19	M73329	−2.3	Phospholipase C-alpha	535	1,057↑	424↓	Chaperone	
20	M94087	−2.3	ATF-4 gene for activating transcription factor 4	568	800↑	351↓	Transcription factor	3
21	M19960	−2.3	cAMP-dependent protein kinase catalytic submit	192	341↑	146↓	Protein kinase A	
22	M16359	−2.2	Major urinary protein III	3,491	4,110	2,144↓	Unknown	
23	X64414	−2.2	Low-density lipoprotein receptor	436	602	271↓	Lipid intake	1
24	W47892	−2.2	Calcium-binding protein P22	226	446↑	182↓	Protein phosphatase	3
25	M69293	−2.1	Id2 protein	608	703	340↓	Transcription factor, inhibit bHLH transcription factors	3

Note: Group 1, cholesterol and fatty acid synthesis; group 2, reactive oxygen species (ROS) and peroxisomal proliferator-activated receptor α (PPARα) activator production; group 3, transcription HDL, high-density lipoprotein; PUFA, polyunsaturated fatty acids. In order of fold change levels in gene expression in liver from fish oil-fed mice relative to safflower oil-fed mice, genes change less than two-fold are presented. Fold change calculation was carried out as an indication of the relative change of each transcript represented on the probe array. The average difference value is a marker of abundance of each gene. Putative gene functions are based on literature review. Some of the genes are classified by their functions as "↑/↓" in fish oil-fed mice indicates that difference call is significantly increased/ decreased, compared with safflower oil-fed mice, whereas that in safflower oil-fed mice was compared with that in high-carbohydrate-fed mice.

Source: Takahashi et al.[68]

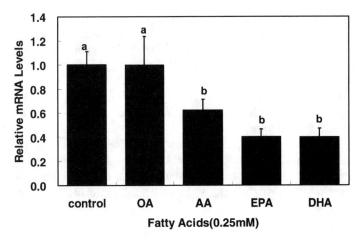

Figure 4.5 Effect of PUFA on prostasin mRNA levels in HepG2 cells. HepG2 cells were incubated with 0.25 mM OA, AA, EPA, and DHA for 24 h. Total RNA was extracted, then mRNA expression levels of SREBPs were measured using a real time RT–PCR (GeneAmp 5700, Applied Biosystems). Single-strand cDNA was synthesized from 1 μg of total RNA using random hexamer and TaqMan reverse transcription reagents (Applied Biosystems, Foster City, CA). Real-time quantitative RT–PCR proceeded in a reaction mixture containing 10 ng of first-strand cDNA, 300 nM of each primer set in a final volume of 25 μL, and SYBR Green PCR core reagent (Applied Biosystems). Relative mRNA levels were normalized to those of GAPDH. Values are mean ± SD ($n = 3$). Significance of differences between PUFA was analyzed by multiple comparisons. Bars not sharing a common letter are significantly different. ($p < 0.05$) (from Fujiwara et al.[19])

well as in body fluids, including seminal fluid and urine.[70] The relationship between prostasin and prostate cancer has been investigated.[72–75] Prostasin is also suggested to act as an extracellular regulator of epithelial sodium channels.[76,77] However, its physiological role in humans is not known. Prostasin was significantly suppressed by PUFA in this study and an SRE was located in its promoter region,[78] suggesting that prostasin plays an important role in processing some proteins in response to cellular cholesterol concentrations.

PUFA also affected the genes involved in cell proliferation and differentiation. Further analysis using the data obtained by this study is needed in order to clarify the mechanism.

CONCLUSION

To investigate the comprehensive effects of PUFA on gene expression, we analyzed changes of mRNA expression in PUFA-treated HepG2 cells. Of the 6000 genes we assessed using Affymetrix DNA microarray, the most remarkable changes shown by PUFA was downregulation of lipogenic genes and cholesterol metabolism. It was suggested that SREBP played the most important role when we used HepG2 cells as a model to investigate lipid metabolism. Results of dose–response study showed that effects of PUFA on SREBP seemed different from that of OA, monounsaturated fatty acid. PUFA upregulated the mRNA levels of genes involved in the cholesterol catabolism, such as Cyp7A1, ABCA1, and HTGL. PUFA also upregulated the genes related to TCA cycle. It seemed that PUFA suppressed lipogenesis and cholesterol synthesis but activated TCA cycle and cholesterol catabolism.

Although SREBP has a strong effect on HepG2 cells, several transcription factors, PPAR, LXR, HNF-4α, SREBP, RXR, and so on, cross-talk each other. Fatty acid must be the most important intracellular mediator in these molecular transcriptional cross-talks in order to control the metabolism of cholesterol, fat, and glucose.

REFERENCES

1. Spector, A.A.; York, M. Membrane lipid composition and cellular function. J. Lipid Res. **1985**, *26*, 1015–1035.
2. Khosla, P.; Hayes, K.C. Comparison between the effects of dietary saturated (16:0), monounsaturated (18:1), and polyunsaturated (18:2) fatty acids on plasma lipoprotein metabolism in cebus and rhesus monkeys fed cholesterol-free diets. Am. J. Clin. Nutr. **1992**, *55*, 51–62.
3. Khosla, P.; Hayes, K.C. Dietary palmitic acid raises plasma LDL cholesterol relative to oleic acid only at a high intake of cholesterol. Biochim. Biophys. Acta **1993**, *1210*, 13–22.
4. Spady, D.K. Regulatory effects of individual n-6 and n-3 polyunsaturated fatty acids on LDL transport in the rat. J. Lipid Res. **1993**, *34*.
5. Nestel, P. Fish oil and cardiovascular disease: lipids and arterial function. Am. J. Clin. Nutr. **2000**, *71*, 228S–231S.
6. Carlson, S.E.; Werkman, S.H. A randomized trial of visual attention of preterm infants fed docosahexaenoic acid until two months. Lipids **1996**, *31*, 85–90.
7. Hoffman, D.R.; Birch, E.E.; Birch, D.G.; Uauy, R. Fatty acid profile of buccal cheek cell phospholipids as an index for dietary intake of docosahexaenoic acid in preterm infants. Lipids **1999**, *34*, 337–342.
8. Birch, E.E.; Garfield, S.; Hoffman, D.R.; Uauy, R.; Birch, D.G. A randomized controlled trial of early dietary supply of long-chain polyunsaturated fatty acids and mental development in term infants. Dev. Med. Child. Neurol. **2000**, *42*, 174–181.
9. Wang, X.; Sato, R.; Brown, M.S.; Hua, X.; Goldstein, J.L. SREBP-1, a membrane-bound transcription factor released by sterol-regulated proteolysis. Cell **1994**, *77*, 53–62.
10. Brown, M.S.; Goldstein, J.L. A proteolytic pathway that controls the cholesterol content of membranes, cells, and blood. Proc. Natl. Acad. Sci. USA **1999**, *96*, 11041–11048.
11. Dreyer, C.; Krey, G.; Keller, H.; Givel, F.; Helftenbein, G.; Wahli, W. Control of the peroxisomal beta-oxidation pathway by a novel family of nuclear hormone receptors. Cell **1992**, *68*, 879–887.

12. Schmidt, A.; Endo, N.; Rutledge, S.J.; Vogel, R.; Shinar, D.; Rodan, G.A. Identification of a new member of the steroid hormone receptor superfamily that is activated by a peroxisome proliferator and fatty acids. Mol. Endocrinol. **1992**, *6*, 1634–1641.

13. Keller, H.; Mahfoudi, A.; Dreyer, C.; Hihi, A.K.; Medin, J.; Ozato, K.; Wahli, W. Peroxisome proliferator-activated receptors and lipid metabolism. Ann. NY Acad. Sci. **1993**, *684*, 157–173.

14. Desvergne, B.; Wahli, W. Peroxisome proliferator-activated receptors: nuclear control of metabolism. Endocr. Rev. **1999**, *20*, 649–688.

15. Tobin, K.A.; Steineger, H.H.; Alberti, S.; Spydevold, O.; Auwerx, J.; Gustafsson, J.A.; Nebb, H.I. Cross-talk between fatty acid and cholesterol metabolism mediated by liver X receptor-alpha. Mol. Endocrinol. **2000**, *14*, 741–752.

16. Hertz, R.; Sheena, V.; Kalderon, B.; Berman, I.; Bar-Tana, J. Suppression of hepatocyte nuclear factor-4alpha by acyl-CoA thioesters of hypolipidemic peroxisome proliferators. Biochem. Pharmacol. **2001**, *61*, 1057–1062.

17. Roche, E.; Buteau, J.; Aniento, I.; Reig, J.A.; Soria, B.; Prentki, M. Palmitate and oleate induce the immediate-early response genes *c-fos* and *nur-77* in the pancreatic beta-cell line INS-1. Diabetes **1999**, *48*, 2007–2014.

18. Duplus, E.; Forest, C. Is there a single mechanism for fatty acid regulation of gene transcription? Biochem. Pharmacol. **2002**, *64*, 893–901.

19. Fujiwara, Y.; Yokoyama, Sawada, Seyama, Y.; Ishii, M.; Tsutsumi, S.; Aburatani, H.; Hanaka, S.; Itakura, H.; Matsumoto, A. Investigating comprehensive effects of polyunsaturated fatty acid on the mRNA expression using a gene chip. J. Nutr. Sci. Vitaminol. (Tokyo) **2003**, *49*, 125–132.

20. Dashti, N.; Alaupovic, P.; Knight-Gibson, C.; Koren, E. Identification and partial characterization of discrete apolipoprotein B containing lipoprotein particles produced by human hepatoma cell line HepG2. Biochemistry **1987**, *26*, 4837–4846.

21. Dashti, N.; Wolfbauer, G. Secretion of lipids, apolipoproteins, and lipoproteins by human hepatoma cell line, HepG2: effects of oleic acid and insulin. J. Lipid Res. **1987**, *28*, 423–436.

22. Faust, R.A.; Cheung, M.C.; Albers, J.J. Secretion of cholesteryl ester transfer protein–lipoprotein complexes by human HepG2 hepatocytes. Atherosclerosis **1989**, *77*, 77–82.

23. Forte, T.M.; McCall, M.R.; Knowles, B.B.; Shore, V.G. Isolation and characterization of lipoproteins produced by human hepatoma-derived cell lines other than HepG2. J. Lipid Res. **1989**, *30*, 817–829.

24. Lockhart, D.J.; Dong, H.; Byrne, M.C.; Follettie, M.T.; Gallo, M.V.; Chee, M.S.; Mittmann, M.; Wang, C.; Kobayashi, M.; Horton, H.; Brown, E.L. Expression monitoring by hybridization to high-density oligonucleotide arrays. Nat. Biotechnol. **1996**, *14*, 1675–1680.

25. Lee, C.K.; Klopp, R.G.; Weindruch, R.; Prolla, T.A. Gene expression profile of aging and its retardation by caloric restriction. Science **1999**, *285*, 1390–1393.

26. Takabe, W.; Mataki, C.; Wada, Y.; Ishii, M.; Izumi, A.; Aburatani, H.; Kamakubo, T.; Niki, E.; Kodama, T.; Noguchi, N. Gene expression induced by BO-654, probucol and BHQ in human endotherial cell. J. Atheroscler. Thromb. **2000**, *7*, 223–230.

27. Akiyoshi, S.; Ishii, M.; Nemoto, N.; Kawabata, M.; Aburatani, H.; Miyazono, K. Targets of transcriptional regulation by transforming growth factor-beta: expression profile analysis using oligonucleotide arrays. Jpn. J. Cancer Res. **2001**, *92*, 258–268.

28. Clarke, S.D.; Jump, D. Polyunsaturated fatty acids regulate lipogenic and peroxisomal gene expression by independent mechanisms. Prostaglandins Leukot. Essent. Fatty Acids **1997**, *57*, 65–69.

29. Brandt, J.M.; Djouadi, F.; Kelly, D.P. Fatty acids activate transcription of the muscle carnitine palmitoyltransferase I gene in cardiac myocytes via the peroxisome proliferator-activated receptor alpha. J. Biol. Chem. **1998**, *273*, 23786–23792.

30. Bremer, J. The biochemistry of hypo- and hyperlipidemic fatty acid derivatives: metabolism and metabolic effects. Prog. Lipid Res. **2001**, *40*, 231–268.

31. Louet, J.F.; Chatelain, F.; Decaux, J.F.; Park, E.A.; Kohl, C.; Pineau, T.; Girard, J.; Pegorier, J.P. Long-chain fatty acids regulate liver carnitine palmitoyltransferase I gene (L-CPT I) expression through a peroxisome-proliferator-activated receptor alpha (PPARalpha)-independent pathway. Biochem. J. **2001**, *354*, 189–197.

32. Reddy, J.K.; Hashimoto, T. Peroxisomal beta-oxidation and peroxisome proliferator-activated receptor alpha: an adaptive metabolic system. Annu. Rev. Nutr. **2001**, *21*, 193–230.

33. Palmer, C.N.; Hsu, M.H.; Griffin, K.J.; Raucy, J.L.; Johnson, E.F. Peroxisome proliferator activated receptor-alpha expression in human liver. Mol. Pharmacol. **1998**, *53*, 14–22.

34. Gervois, P.; Torra, I.P.; Chinetti, G.; Grotzinger, T.; Dubois, G.; Fruchart, J.C.; Fruchart-Najib, J.; Leitersdorf, E.; Staels, B. A truncated human peroxisome proliferator-activated receptor alpha splice variant with dominant negative activity. Mol. Endocrinol. **1999**, *13*, 1535–1549.

35. Hsu, M.H.; Savas, U.; Griffin, K.J.; Johnson, E.F. Identification of peroxisome proliferators-responsive human genes by elevated expression of the peroxisome proliferators-activated receptor alpha in HepG2 cells. J. Biol. Chem. **2001**, *276*, 27950–27958.

36. Horton, J.D.; Shimomura, I.; Brown, M.S.; Hammer, R.E.; Goldstein, J.L.; Shimano, H. Activation of cholesterol synthesis in preference to fatty acid synthesis in liver and adipose tissue of transgenic mice overproducing sterol regulatory element-binding protein-2. J. Clin. Invest. **1998**, *101*, 2331–2339.

37. Shimomura, I.; Bashmakov, Y.; Horton, J.D. Increased levels of nuclear SREBP-1c associated with fatty livers in two mouse models of diabetes mellitus. J. Biol. Chem. **1999**, *274*, 30028–30032.

38. Shimano, H.; Horton, J.D.; Hammer, R.E.; Shimomura, I.; Brown, M.S.; Goldstein, J.L. Overproduction of cholesterol and fatty acids causes massive liver enlargement in transgenic mice expressing truncated SREBP-1a. J. Clin. Invest. **1996**, *98*, 1575–1584.

39. Hirano, R.; Igarashi, O.; Kondo, K.; Itakura, H.; Matsumoto, A. Regulation by long-chain fatty acids of the expression of cholesteryl ester transfer protein in HepG2 cells. Lipids **2001**, *36*, 401–406.

40. Seo, T.; Oelkers, P.M.; Giattina, M.R.; Worgall, T.S.; Sturley, S.L.; Deckelbaum, R.J. Differential modulation of ACAT1 and ACAT2 transcription and activity by long chain free fatty acids in cultured cells. Biochemistry **2001**, *40*, 4756–4762.

41. Sakakura, Y.; Shimano, H.; Sone, H.; Takahashi, A.; Inoue, N.; Toyoshima, H.; Suzuki, S.; Yamada, N.; Inoue, K. Sterol regulatory element-binding proteins induce an entire pathway of cholesterol synthesis. Biochem. Biophys. Res. Commun. **2001**, *286*, 176–183.

42. Venkateswaran, A.; Laffitte, B.A.; Joseph, S.B.; Mak, P.A.; Wilpitz, D.C.; Edwards, P.A.; Tontonoz, P. Control of cellular cholesterol efflux by the nuclear oxysterol receptor LXR alpha. Proc. Natl. Acad. Sci. USA **2000**, *97*, 12097–12102.

43. Laffitte, B.A.; Repa, J.J.; Joseph, S.B.; Wilpitz, D.C.; Kast, H.R.; Mangelsdorf, D.J.; Tontonoz, P. LXRs control lipid-inducible expression of the apolipoprotein E gene in macrophages and adipocytes. Proc. Natl. Acad. Sci. USA **2001**, *98*, 507–512.

44. Lehmann, J.M.; Kliewer, S.A.; Moore, L.B.; Smith-Oliver, T.A.; Oliver, B.B.; Su, J.L.; Sundseth, S.S.; Winegar, D.A.; Blanchard, D.E.; Spencer, T.A.; Willson, T.M. Activation of the nuclear receptor LXR by oxysterols defines a new hormone response pathway. J. Biol. Chem. **1997**, *272*, 3137–3140.

45. Peet, D.J.; Turley, S.D.; Ma, W.; Janowski, B.A.; Lobaccaro, J.M.; Hammer, R.E.; Mangelsdorf, D.J. Cholesterol and bile acid metabolism are impaired in mice lacking the nuclear oxysterol receptor LXR alpha. Cell **1998**, *93*, 693–704.

46. Worgall, T.S.; Sturley, S.L.; Seo, T.; Osborne, T.F.; Deckelbaum, R.J. Polyunsaturated fatty acids decrease expression of promoters with sterol regulatory elements by decreasing levels of mature sterol regulatory element-binding protein. J. Biol. Chem. **1998**, *273*, 25537–25540.

47. Kim, H.J.; Takahashi, M.; Ezaki, O. Fish oil feeding decreases mature sterol regulatory element-binding protein 1 (SREBP-1) by down-regulation of SREBP-1c mRNA in mouse liver. A possible mechanism for down-regulation of lipogenic enzyme mrnas. J. Biol. Chem. **1999**, *274*, 25892–25898.

48. Mater, M.K.; Thelen, A.P.; Pan, D.A.; Jump, D.B. Sterol response element-binding protein 1c (SREBP1c) is involved in the polyunsaturated fatty acid suppression of hepatic *S14* gene transcription. J. Biol. Chem. **1999**, *274*, 32725–32732.

49. Xu, J.; Nakamura, M.T.; Cho, H.P.; Clarke, S.D. Sterol regulatory element binding protein-1 expression is suppressed by dietary polyunsaturated fatty acids. A mechanism for the coordinate suppression of lipogenic genes by polyunsaturated fats. J. Biol. Chem. **1999**, *274*, 23577–23583.

50. Xu, J.; Teran-Garcia, M.; Park, J.H.; Nakamura, M.T.; Clarke, S.D. Polyunsaturated fatty acids suppress hepatic sterol regulatory element-binding protein-1 expression by accelerating transcript decay. J. Biol. Chem. **2001**, *276*, 9800–9807.

51. Shimomura, I.; Bashmakov, Y.; Shimano, H.; Horton, J.D.; Goldstein, J.L.; Brown, M.S. Cholesterol feeding reduces nuclear forms of sterol regulatory element binding proteins in hamster liver. Proc. Natl. Acad. Sci. USA **1997**, *94*, 12354–12359.

52. Horton, J.D.; Shimomura, I. Sterol regulatory element-binding proteins: activators of cholesterol and fatty acid biosynthesis. Curr. Opin. Lipidol. **1999**, *10*, 143–150.

53. Yoshikawa, T.; Shimano, H.; Yahagi, N.; Ide, T.; Amemiya-Kudo, M.; Matsuzaka, T.; Nakakuki, M.; Tomita, S.; Okazaki, H.; Tamura, Y.; Iizuka, Y.; Ohashi, K.; Takahashi, A.; Sone, H.; Osuga Ji, J.; Gotoda, T.; Ishibashi, S.; Yamada, N. Polyunsaturated fatty acids suppress sterol regulatory element-binding protein 1c promoter activity by inhibition of liver X receptor (LXR) binding to LXR response elements. J. Biol. Chem. **2002**, *277*, 1705–1711.

54. Yoshikawa, T.; Ide, T.; Shimano, H.; Yahagi, N.; Amemiya-Kudo, M.; Matsuzaka, T.; Yatoh, S.; Kitamine, T.; Okazaki, H.; Tamura, Y.; Sekiya, M.; Takahashi, A.; Hasty, A.H.; Sato, R.; Sone, H.; Osuga, J.I.; Ishibashi, S.; Yamada, N. Cross-talk between peroxisome proliferator activated receptor (PPAR) {alpha} and liver X receptor (LXR) in nutritional regulation of fatty acid metabolism. PPARs suppress sterol

regulatory element-binding protein-1c promoter through inhibition of LXR signaling. Mol. Endocrinol. **2003**.

55. Ide, T.; Shimano, H.; Yoshikawa, T.; Yahagi, N.; Amemiya-Kudo, M.; Matsuzaka, T.; Nakakuki, M.; Yatoh, S.; Iizuka, Y.; Tomita, S.; Ohashi, K.; Takahashi, A.; Sone, H.; Gotoda, T.; Osuga, J.I.; Ishibashi, S.; Yamada, N. Cross-talk between peroxisome proliferator activated receptor (PPAR) {alpha} and liver X receptor (LXR) in nutritional regulation of fatty acid metabolism. II. LXRs suppress lipid degradation gene promoters through inhibition of PPAR signaling. Mol. Endocrinol. **2003**.

56. Zhang, L.; Ge, L.; Tran, T.; Stenn, K.; Prouty, S.M. Isolation and characterization of the human stearoyl-CoA desaturase gene promoter: requirement of a conserved CCAAT cis-element. Biochem. J. **2001**, *357*, 183–193.

57. Kim, H.J.; Miyazaki, M.; Ntambi, J.M. Dietary cholesterol opposes PUFA-mediated repression of the stearoyl-CoA desaturase-1 gene by SREBP-1 independent mechanism. J. Lipid Res. **2002**, *43*, 1750–1757.

58. Matsuzaka, T.; Shimano, H.; Yahagi, N.; Amemiya-Kudo, M.; Yoshikawa, T.; Hasty, A.H.; Tamura, Y.; Osuga, J.; Okazaki, H.; Iizuka, Y.; Takahashi, A.; Sone, H.; Gotoda, T.; Ishibashi, S.; Yamada, N. Dual regulation of mouse Delta(5)- and Delta(6)-desaturase gene expression by SREBP-1 and PPARalpha. J. Lipid Res. **2002**, *43*, 107–114.

59. Chouinard, R.A., Jr.; Luo, Y.; Osborne, T.F.; Walsh, A.; Tall, A.R. Sterol regulatory element binding protein-1 activates the cholesteryl ester transfer protein gene in vivo but is not required for sterol up-regulation of gene expression. J. Biol. Chem. **1998**, *273*, 22409–22414.

60. Fusegawa, Y.; Kelley, K.L.; Sawyer, J.K.; Shah, R.N.; Rudel, L.L. Influence of dietary fatty acid composition on the relationship between CETP activity and plasma lipoproteins in monkeys. J. Lipid Res. **2001**, *42*, 1849–1857.

61. Hertz, R.; Seckbach, M.; Zakin, M.M.; Bar-Tana, J. Transcriptional suppression of the transferrin gene by hypolipidemic peroxisome proliferators. J. Biol. Chem. **1996**, *271*, 218–224.

62. Pan, D.A.; Mater, M.K.; Thelen, A.P.; Peters, J.M.; Gonzalez, F.J.; Jump, D.B. Evidence against the peroxisome proliferator-activated receptor alpha (PPARalpha) as the mediator for polyunsaturated fatty acid suppression of hepatic L-pyruvate kinase gene transcription. J. Lipid Res. **2000**, *41*, 742–751.

63. Hayhurst, G.P.; Lee, Y.H.; Lambert, G.; Ward, J.M.; Gonzalez, F.J. Hepatocyte nuclear factor 4alpha (nuclear receptor 2A1) is essential for maintenance of hepatic gene expression and lipid homeostasis. Mol. Cell. Biol. **2001**, *21*, 1393–1403.

64. Jump, D. Dietary polyunsaturated fatty acid and regulation of gene transcription. Curr. Opin. Lipidol. **2002**, *13*, 155–164.

65. Pineda Torra, I.; Jamshidi, Y.; Flavell, D.M.; Fruchart, J.C.; Staels, B. Characterization of the human PPARalpha promoter: identification of a functional nuclear receptor response element. Mol. Endocrinol. **2002**, *16*, 1013–1028.

66. Barbier, O.; Duran-Sandoval, D.; Pineda Torra, I.; Kosykh, V.; Fruchart, J.C.; Staels, B. PPARalpha induces hepatic expression of the human bile acid glucuronidating UGT2B4 enzyme. J. Biol. Chem. **2003**.

67. Pineda Torra, I.; Claudel, T.; Duval, C.; Kosykh, V.; Fruchart, J.C.; Staels, B. Bile acids induce the expression of the human peroxisome proliferator-activated receptor

alpha gene via activation of the farnesoid X receptor. Mol. Endocrinol. **2003**, *17*, 259–272.

68. Takahashi, M.; Tsuboyama-Kasaoka, N.; Nakatani, T.; Ishii, M.; Tsutsumi, S.; Aburatani, H.; Ezaki, O. Fish oil feeding alters liver gene expressions to defend against PPARalpha activation and ROS production. Am. J. Physiol. Gastrointest. Liver Physiol. **2002**, *282*, G338–348.

69. Berger, A.; Mutch, D.M.; Bruce German, J.; Roberts, M.A. Dietary effects of arachidonate-rich fungal oil and fish oil on murine hepatic and hippocampal gene expression. Lipids Health Dis. **2002**, *1*, 2.

70. Yu, J.X.; Chao, L.; Chao, J. Prostasin is a novel human serine proteinase from seminal fluid. Purification, tissue distribution, and localization in prostate gland. J. Biol. Chem. **1994**, *269*, 18843–18848.

71. Yu, J.X.; Chao, L.; Chao, J. Molecular cloning, tissue-specific expression, and cellular localization of human prostasin mRNA. J. Biol. Chem. **1995**, *270*, 13483–13489.

72. Chen, L.M.; Hodge, G.B.; Guarda, L.A.; Welch, J.L.; Greenberg, N.M.; Chai, K.X. Down-regulation of prostasin serine protease: a potential invasion suppressor in prostate cancer. Prostate **2001**, *48*, 93–103.

73. Laribi, A.; Berteau, P.; Gala, J.; Eschwege, P.; Benoit, G.; Tombal, B.; Schmitt, F.; Loric, S. Blood-borne RT–PCR assay for prostasin-specific transcripts to identify circulating prostate cells in cancer patients. Eur. Urol. **2001**, *39*, 65–71.

74. Mok, S.C.; Chao, J.; Skates, S.; Wong, K.; Yiu, G.K.; Muto, M.G.; Berkowitz, R.S.; Cramer, D.W. Prostasin, a potential serum marker for ovarian cancer: identification through microarray technology. J. Natl. Cancer Inst. **2001**, *93*, 1458–1464.

75. Narikiyo, T.; Kitamura, K.; Adachi, M.; Miyoshi, T.; Iwashita, K.; Shiraishi, N.; Nonoguchi, H.; Chen, L.M.; Chai, K.X.; Chao, J.; Tomita, K. Regulation of prostasin by aldosterone in the kidney. J. Clin. Invest. **2002**, *109*, 401–408.

76. Adachi, M.; Kitamura, K.; Miyoshi, T.; Narikiyo, T.; Iwashita, K.; Shiraishi, N.; Nonoguchi, H.; Tomita, K. Activation of epithelial sodium channels by prostasin in Xenopus oocytes. J. Am. Soc. Nephrol. **2001**, *12*, 1114–1121.

77. Donaldson, S.H.; Hirsh, A.; Li, D.C.; Holloway, G.; Chao, J.; Boucher, R.C.; Gabriel, S.E. Regulation of the epithelial sodium channel by serine proteases in human airways. J. Biol. Chem. **2002**, *277*, 8338–8345.

78. Yu, J.X.; Chao, L.; Ward, D.C.; Chao, J. Structure and chromosomal localization of the human prostasin (*PRSS8*) gene. Genomics **1996**, *32*, 334–340.

5

Neurological Effects of Dietary n-3 Fatty Acids in Rat Brain

Tibor Farkas,* Klára Kitajka, and Gwendolyn Barceló-Coblijn
*Institute of Biochemistry, Biological Research Center,
Hungarian Academy of Sciences, Hungary*

László G. Puskás
*Laboratory of Functional Genomics, Biological Research Center,
Hungarian Academy of Sciences, Szeged, Hungary*

INTRODUCTION

Long-chain n-3 polyunsaturated fatty acids (linolenic acid, 18:3n-3; LNA; eicosapentaenoic acid, 20:5n-3, EPA; and docosahexaenoic acid 22:6n-3, DHA) are essential for vertebrates, because, in contrast to saturated and monounsaturated fatty acids, they cannot be formed *de novo*, and at least their precursor, the 18:3n-3, has to be taken up from the diet. These fatty acids, and the complex lipids formed from them, are important constituents of cellular membranes and contribute to maintain the structural and functional integrity of these structures. Though these polyunsaturated fatty acids occur in various proportions in all organs, the nervous tissue is characterized by the absence of LNA and EPA and presence of high levels DHA. Level of DHA is strictly controlled in brain and retina: any deviation from the physiological level results in impairments in cognitive or visual functions.[1] It has been proposed that the nervous system requires specific molecular species for its functions.[2] The exact mode of action of DHA-containing phospholipids (first of all ethanolamine and serine phosphoglycerides)

---

*Deceased.

is not known but it might be possible that they exert their beneficial effect by regulating blood–brain barrier,[3] membrane fluidity,[4] activity of certain enzymes,[5] ionic channels,[6] and nerve growth factor.[7] However, the question arises whether DHA-containing phospholipid molecules control neural functions on some other levels than their effect on membrane molecular architecture. It has already been shown that polyunsaturated fatty acids of both structures control gene expression in a variety of tissues.[8–10] One of the major functions of neural tissue is to receive, store, process, and retrieve information. Extensive studies have been done at morphological, electrophysiological, and biochemical levels to understand the underlying mechanisms. Effect of DHA on learning and memory in animal and human models has been demonstrated,[11] but the question remains unanswered whether these highly complex and sophisticated processes can be explained by the sole effect of DHA in particular, or of long-chain poly-unsaturated fatty acids in general, on biophysical properties and molecular archi-tecture of neural membranes. To find a relation between the effects of DHA on molecular composition of membranes, on cognitive functions, and on genetic machinery of neurons is an exciting research task.

LNA and DHA alone or in combination with linoleic acid (LA) have been found to improve cognitive potential of rats and humans.[12,13] In studies on rats and mice, the experiments were done on essential fatty acid deficient animals, after refeeding them with LNA or DHA,[14,15] whereas in human studies either aged persons,[16] or persons with certain neurodegenerative disease, accompanied with loss of brain DHA content were investigated.[17] In these studies LNA or DHA re-established the "physiological level" of DHA accompanied with improvement of learning performance. However, these experiments do not give information about the "direct" role of DHA in cognitive functions. In the exper-iments described later, essential fatty acid adequate rats were supplied with either LNA (in the form of perilla oil) or DHA (in the form of tuna fish oil). In other experiments a mixture of oils was given, consisting of perilla oil (rich in LNA) with soybean oil (rich in LA) or fish oil (rich in DHA) with sunflower oil (rich in LA). The distribution of DHA-containing ethanolamine phosphoglycerides, as well as gene expression patterns, was assessed. LA/LNA and LA/DHA mix-tures were also assessed for gene expression and lipid biochemistry because LA is a constant component of the western diet and because a given mixture of LA/LNA proved very efficient in supporting mental activities.[11,12]

Fatty acid compositions of the ingested foods are given in Table 5.1. Incor-poration of DHA into ethanolamine phosphoglyceride molecular species was fol-lowed by an HPLC technique,[18] whereas a DNA microarray technique[19–21] was used to follow the effect of dietary fatty acids on gene expression.

EFFECTS OF n-3 FATTY ACIDS ON GENE EXPRESSION DURING ONTOGENY

The final fatty acid composition of brain is determined during embryogenesis, particularly during rapid brain growth, and in rats 12–15 days after delivery.[22]

Table 5.1 Major Fatty Acids in the Diet (Percentage of Total)

Fatty acid	Control	Fish oil	Perilla oil	Perilla/soybean	LA/fish oil
16:0	13.6	7.5	8.9	11.8	13.3
16:1	0.8	1.3	Tr	—	1.4
18:0	3.1	2.2	2.5	3.9	5.4
18:1n-9	17.0	16.2	19.5	24.5	21.9
18:2n-6	51.0	23.0	28.0	49.4	37.4
18:3n-3	6.2	2.5	39.4	10.4	1.5
20:5n-3	0.7	12.0	—	—	5.2
22:6n-3	1.2	26.9	—	—	11.3

It is important that during this time, for its functions, the brain is supplied with fatty acids of proper composition. In the experiments presented here rats were supplied with either fish oil or perilla oil or with mixture of sunflower and fish oil or with mixture of soybean oil and perilla oil from conception till adulthood (infant rats).

Table 5.2 shows that level of DHA-containing phosphatidylethanolamine molecular species, particularly of 18:0/22:6, becomes higher in rat brains after consuming any combination of the fatty acid mixtures, indicating that even essential fatty acid adequate rats can incorporate more DHA in their neural membranes. Whereas the role of the essential fatty acids in controlling learning and memory has well been studied, information about the effect of these fatty acids on genetic machinery of neurons is rather poor.

Table 5.3 compiles the list of genes affected by dietary DHA, LNA, LA/LNA, and LA/DHA. The expression levels of 102 cDNAs, representing 3. 4% of the total DNA elements on the array, were significantly altered in brains of rats fed with the experimental diets. Fifty-five genes were found overexpressed and forty-seven were repressed relative to controls by the four dietary regimens. The altered genes include those involved in synaptic plasticity, cytoskeleton, signal transduction, ion channel formation, energy household, and regulatory proteins. It was interesting to observe that 14 genes responded more intensively to the dietary fatty acid mixtures, LA/LNA and LA/DHA, than to either DHA or LNA. These genes include those which encode clathrin-associated adaptor

Table 5.2 DHA-Containing Phosphatidylethanolamine in Brains of Infant Rats

Fatty acid	Control	Fish oil	Perilla oil	Perilla/ soybean	LA/fish oil
16:0/22:6	6.55 ± 2.18	8.17 ± 1.73	6.12 ± 1.32	12.31 ± 2.10	8.28 ± 1.69
18:1/22:6	1.93 ± 0.34	2.10 ± 0.61	1.72 ± 0.49	2.62 ± 0.80	2.43 ± 0.50
18:0/22:6	30.12 ± 2.42	34.15 ± 0.38	32.37 ± 1.09	36.31 ± 4.09	32.66 ± 1.00

Table 5.3 Changes in Gene Expression in Rat Brains Fed from Conception Till Adulthood

Δ-Fold					
Fish oil	Fish oil + sunflower oil	Perilla oil	Perilla oil + soybean oil	Gene product	Accession number
0.28	0.51	b(0.32)	b(1.85)	L1cam (*M. musculus*)	AA409744
0.32	0.44	0.2	0.3	Putative E1–E2 ATPase	RSU78977
0.39	0.21	0.09	0.18	Transthyretin	NM_012681
0.39	b(0.21)	b(1.16)	b(2.6)	Membrane protein	L09260
0.41	0.26	0.32	0.26	Dihydrolipoamide transacylase E2	AA410009
0.43	0.48	b(0.27)	b(—)	Bromodomain protein CELTIX1	AF213969
0.46	0.43	0.35	0.55	Kruppel-type zinc finger (C2H2)	AW540949
0.47	0.39	0.42	b(1.4)	Methyl CpG binding protein 2	NM_022673
0.48	0.41	0.21	0.19	Unconventional myosin 3 (Myr5)	NM_012981
0.51	b(1.54)	b(1.3)	b(0.8)	Capping protein (actin filament)	AW541453
0.52	0.51	0.47	0.48	Syntaxin binding protein Munc18-2	U20283
0.57	0.7	0.34	0.5	STK-1 (serine_threonine kinase)	AA410169
0.57	0.93	0.46	0.49	Phospholipase D	D8862
0.58	0.73	0.42	0.53	Citb homolog	AA407729
0.63	0.43	0.52	b(0.33)	Nonreceptor tyrosine kinase 2	W07947
0.63	0.44	0.44	0.58	RNA-binding protein homolog	AA038452
0.66	0.39	0.8	0.59	Integrin alpha 5	AW544851
0.66	0.54	0.46	0.85	Transcriptional repressor deltaEF1	D76433S
0.74	0.53	0.52	0.49	RNA1 homolog (Fug1)	AW536168
0.76	0.53	0.5	b(0.8)	Microtubule-associated protein 4	AW142042
0.78	0.39	0.61	0.7	Ajuba (Jub)	AW536166
0.82	0.41	0.39	0.45	NIPSNAP2 protein	AA925098
0.84	0.48	0.51	0.62	Protein gamma-subtype	AW542425

				Protein	Accession
0.87	0.41	0.51	1.14	Goliath protein hom., Zn-finger protein	AA288977
0.87	0.46	0.54	0.59	Ras oncogene neuroblastoma, Nras	W15662
0.87	1.63	1.71	2.34	Microtubule-associated protein 1A_1B chain 3	NM_022867
0.89	0.45	b(0.47)	b(1.01)	Axoneural dynein heavy chain 8	U61747
1.17	2.65	2.13	2.7	Cytochrome c oxidase subunit VIc	X06146
1.54	b(10)	1.92	1.6	Tax1 binding protein TXBP151	W35945
1.55	1.81	1.54	2.42	Calmodulin (RCM3)	NM_017326
1.69	2.17	2.03	2.56	Calmodulin 1 (Calm 1)	NM_031969
1.77	1.27	1.97	1.87	B cell receptor associated protein 37	U75392
1.77	2.17	1.54	2.31	Ubiquinol–cytochrome c reductase	XM_009253
1.77	2.49	1.78	2.44	Pleiotrophin (Hbnf, HB/GAM)	NM_017066
1.8	2.03	1.75	2.28	Vasopressin V1b receptor	AF314527
1.81	1.87	1.62	b(3.86)	Cytochrome c oxidase subunit I	M27315
1.83	1.41	1.6	1.99	LIM-domain protein LMP-1	AF095585
1.89	1.55	2.37	b(4.4)	Profilin IIa (Pfn2)	AF228737
1.89	1.67	1.89	2.83	Alpha-tubulin (Tuba1)	NM_022298
1.91	1.5	1.88	2.82	Alpha-tubulin (Tuba1) (*M. musculus*)	BC008117
1.92	1.94	1.62	2.81	Protein phosphatase 2A	NM_022209
1.96	1.91	1.85	2.71	Synuclein, alpha	NM_019169
2.02	2.08	1.9	2.64	D-cadherin precursor	AF135156
2.07	1.52	b(1.7)	2.02	Protein phosphatase 2, alpha isoform	NM_017039
2.07	1.74	1.64	b(3.8)	Dipeptidyl peptidase III	D89340
2.08	2.71	2.29	2.18	Somatostatin	NM_012659
2.1	1.45	1.68	1.93	Calmodulin	AF178845
2.1	2.64	2.55	2.12	Synuclein, gamma	NM_031688
2.19	b(10.6)	2.09	b(4.52)	RAB6B small GTPase	W17800
2.26	4.04	2.76	1.94	Adaptor-related protein AP-3 mu2 subunit	L07074

(continued)

Table 5.3 Continued

Δ-Fold

Fish oil	Fish oil + sunflower oil	Perilla oil	Perilla oil + soybean oil	Gene product	Accession number
2.26	b(5.01)	b(2.02)	2.96	ATP synthase, H_transport, F0	NM_013795
2.33	6.4	4.47	6.41	Translin	NM_021762
2.42	1.18	1.7	1.86	Beta-globin	X16417
2.55	5.01	3.72	5.39	NADH dehydrog. hom._cytochrome c. I	X13220
2.58	b(12)	2.97	b(4.4)	Actin-related protein 2	XM_002674
2.68	6.25	3.56	5.04	Cytochrome *b*	AF295545
2.82	6.02	4.76	7.79	Elongation factor 1-alpha	X63561
2.88	5.4	5.6	8.4	Ribosomal protein L7a	X15013
3.07	2.91	3.35	3.92	SH3-containing protein SH3P4	AF009603
3.17	b(3.9)	1.94	b(4.5)	ATP synthase subunit d (Atp5jd)	NM_019383
3.26	b(3.5)	b(2.4)	b(3.91)	Parathyroid hormone reg. sequence	AA290355
3.65	6.14	5.57	5.54	NAD-isocitrate dehydrogenase gamma	X74125
3.99	8.87	5.52	6.55	Sec24 protein (Sec24A isoform)	AJ131244
4.15	4.85	5.39	3.47	Clathrin-associated adaptor chain mu 1A	AF139405
4.16	8.31	4.81	5.42	Cytochrome *c* oxidase subunit II	M64496
4.2	2.87	2.94	b(6.83)	Liver cytochrome *c* oxidase VIII	L48209
4.21	b(7.10)	3.34	4.92	Serine palmitoyl transferase	AF111168
4.4	8.74	5.56	9.83	Ubiquitin–protein ligase Nedd4-2	AF277232
4.5	6.2	3.94	6.51	Farnesyl pyrophosphate synthase	NM_031840
4.75	b(10.2)	6.57	7.71	Transcription factor-like protein MRGX	AF100620
5.55	6.7	3.84	7.01	Cytochrome *c* oxidase subunit II	AF378830

Note: "b" denotes nonsignificant data because of high background or high statistical deviation.

protein, farnesyl pyrophosphatase synthetase, Sec24 protein, NADH dehydrogenase/cytochrome *c* oxidase, cytochrome *b*, elongation factor 1-alpha, ribosomal protein L7a, cytochrome *c* oxidase subunit II, ubiquitin–protein ligase Nedd4-2, adult mice cerebellum cDNA, transcription factor-like protein, and RAB6B small GTPase. Several genes participating in signal transduction (the calmodulins for instance) were overexpressed to the same degree by all the four dietary fatty acids tested.

It is noteworthy that the genes encoding the synucleins were also overexpressed. To the best of our knowledge, this is the first report on regulation of synuclein gene expression by n-3 fatty acids, like DHA. These proteins play a role in neural plasticity and are related to learning in brains of songbirds in times of learning to sing.[23] In addition, these proteins may play a role in the development and maturation of some neurons.[24] In avian system sex steroids have been shown to upregulate synuclein transcription.[25]

The observation that some mitochondrial enzymes were overexpressed by the fatty acid/mixtures can be interpreted as brains being in an elevated metabolic state. Indeed, ATP is required for many biochemical processes. An energy requiring process, among others, is the formation of farnesyl pyrophosphate, which is an intermediary in cholesterol and also in CoQ biosynthesis. The latter is involved in electron transport system as electron carrier. On the other hand, cholesterol is required for synaptogenesis.[26] It has been shown that learning can cause synaptogenesis in cerebellum.[27]

It is interesting to point out that transthyretin was downregulated by all the fatty acids and their mixtures tested at our experimental conditions. It has to be noted that Berger and his coworkers found the same regulatory effect of fish oil in mice: the expression of transthyretin was decreased in the hippocampus, whereas diets rich in arachidonic acid (AA) induced transcription of transthyretin. This protein sequesters amyloid β-protein that is deeply involved in development of Alzheimer's disease (AD)[28]on one hand and binds thyroid hormones on the other. It has been shown that thyroid hormone deficiency during brain development impairs cognitive functions.[29]

EFFECT OF n-3 FATTY ACIDS ON GENE EXPRESSION IN AGING BRAIN

Advanced age is accompanied with reduced level of long-chain polyunsaturated fatty acids, AA and DHA, in the brain.[30,31] Memory impairment is also a common phenomenon in this age. We also found reduced level of DHA in brains of 2-year-old rats, which could be elevated to the level of 3-month-old (control, young) rats by feeding a fish-oil diet containing 11% DHA (data not shown) for 1 month. This diet resulted in an increase of diacyl 18:0/22:6 phosphatidylethanolamine species from $28.95 \pm 1.90\%$ to $32.98 \pm 1.25\%$ ($p < 0.05$) in young rat brains and from $25.13 \pm 1.38\%$ to $29.62 \pm 1.49\%$ ($p < 0.05$) in old rat brains.

Table 5.4 shows that in brains of young rats, six genes were upregulated during 1 month feeding period. The rats started to receive fish oil only when they were 3 months old. As seen in Table 5.4, the response to DHA was very poor. Of the upregulated genes, transthyretin was slightly upregulated by a factor of 1.9 and synuclein was dramatically upregulated by a factor of 8.6. This would indicate that DHA might have a beneficial effect on learning and memory of these rats. Indeed, we found that these rats performed better in Morris water maze test when compared with the rats on control diet (data not shown).

In 2-year-old rat brains there was no significant change at the gene expression level (data not shown). The DHA supplied old rats did not perform better in Morris water maze test (data not shown) which could be, at least partially, due to the absence of synuclein among the upregulated genes. Although these experiments prove that old brains respond to dietary n-3 fatty acids in an age-dependent manner, the possibility that longer feeding time might be also a factor affecting the gene expression profile cannot be ruled out.

SITE-SPECIFIC EFFECTS OF n-3 FATTY ACIDS ON GENE EXPRESSION IN BRAINS OF OLD RATS

It is seen that forebrains of old and young rats give different responses to dietary DHA. To extend this study, hippocampus and cerebellum of 2-year-old rats receiving fish oil for 1 month were investigated. Seven genes were found over-expressed and eleven genes were found repressed in the hippocampus of old

Table 5.4 Changes in Gene Expression in 3-Month-Old Rat Forebrains After 1 Month of Feeding with Fish Oil

Δ-Fold	Gene product	Accession number
0.53	CDC10	AW539811
0.55	Heat shock protein, 86 kDa 1	AW536206
0.57	Hypothetical	AW544599
0.57	DEK oncogene (DNA binding) (DEK)	AW539649
0.58	Hypothetical	AW544613
0.58	H3 histone, family 3B (H3f3b)	AW539780
0.59	Similar to golgi peripheral membrane protein p65	AA213185
0.60	Protein kinase C inhibitor (mPKCI)	NM_022192
1.93	Transthyretin	W17647
1.99	Mesoderm development candiate 2	AA087768
2.56	Similar to kinesin light chain	W15957
2.82	EST	W14581
3.78	Interferon regulatory factor 3	W41937
8.64	Alpha-synuclein	W41663

rats (Table 5.5). For the time being we do not have data on infant or young rats but it is interesting to mention that in young murine hippocampus twenty-two genes were affected, but only three of them were overexpressed.[32] There are a few genes responding in similar way in different brain regions of young rats receiving fish oil for 1 month (e.g., the 86 kDa heat shock protein gene). Whereas transthyretin was significantly downregulated in young murine hippocampus[32] and slightly overexpressed in brains of young rats fed on fish oil for 1 month (Table 5.4), in old rat hippocampus it was overexpressed by a factor of 11,[20] indicating that transthyretin fulfills different functions in different brain regions. Figure 5.1 shows the real-time quantitative RT–PCR analysis of cDNA vs. β-actin from control and fish oil treated rat brains. In old rat hippocampus, its major function might be the sequestering of amyloid β-protein, thus protecting the brain from developing AD. Deposition of amyloid-β polypeptide (Aβ) plays a causative role in the pathogenesis in AD. Amyloid aggregates are formed from the soluble form of Aβ, which is a normal metabolic product detectable in the ventricular cerebrospinal fluid (CSF) and plasma of healthy and AD subjects. Different extracellular proteins found in CSF are implicated in the transport and deposition of Aβ, such as α_1-antichymotripsin, ApoJ, ApoE, and transthyretin.[33] Purified transthyretin has been shown to bind Aβ *in vitro*[34] and the overexpressed protein prevented the formation of insoluble amyloid *in vivo*.[35] Transthyretin is known to transport thyroxine and retinol-binding protein in CSF.[36] In this context, it is interesting to note that an

Table 5.5 Changes in Gene Expression in Old Rat Hippocampus

Δ-Fold	Gene name	Accession number
0.47	RNA binding protein	
0.47	Makorin RING zinc-finger protein 2 (Mkrn2)	BG016106
0.51	Similar to lactate dehydrogenase	AW140623
0.55	Anaphylatoxin C3a receptor	U86379
0.55	Similar to autoantigen homolog	AA270674
0.57	EST	AA230444
0.58	Apolipoprotein CI	NM_012824
0.59	Toll-associated serine protease (Tasp)	AF057025
0.60	EST	AW544503
0.60	RecQ protein-like (Recql)	W14898
0.61	K-Cl cotransporter 3b	U55816
1.62	CD39 antigen-like 4 (Cd39l4)	AW544751
1.66	Similar to BAT2	BE329313
1.67	EST	AA245586
1.88	EST	AA221692
1.86	Mitochondrial creatine kinase	X59736
2.15	Similar to ER lumen protein retaining receptor 2	W46090
9.48	Transthyretin	NM_012681

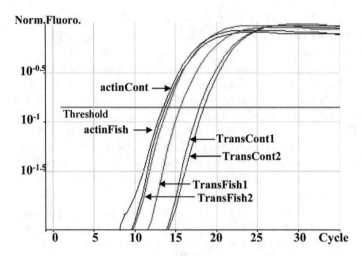

Figure 5.1   Relative quantitative real time RT–PCR analysis of cDNA samples (~40 ng in each PCR) in experiment prepared from hippocampus RNA (5 μg of total RNA was reverse transcribed in 20 μL reaction mixture) of fish-oil fed and control animal groups. For better visualization, only one result is shown for β-actin expression from the treated and control groups. Relative transcript ratios of different samples were determined by normalizing Ct values to β-actin expression.

inverse relationship has been found between transthyretin level in CSF and the severity of dementia in AD patients.[37] We suggest that natural transthyretin inducers such as fish oil or DHA could be of benefit in the prevention of AD.

However, gene expression pattern may depend also on the experimental approach. Blalock et al.[38] studied gene expression profiling in hippocampus in relation to cognitive impairment of rats. They reported that 29 genes were downregulated and 40 were upregulated; none of them was identical to those found altered in old rats hippocampi.

In cerebellum of old rats fed with fish oil for 1 month, 19 genes were overexpressed and 20 were repressed (Table 5.6). Out of the total 39 genes that exhibited significant alterations in their expression 15 coded for proteins with unknown or hypothetical functions. Unfortunately we do not have data on young rat cerebellum. However, the earlier data suggest that certain genes respond differently in distinct brain regions to the same fatty acid diet, which might be because of their specific functions or their different regulatory pathways.

SPECULATIONS ABOUT THE MODE OF ACTION OF n-3 POLYUNSATURATED FATTY ACIDS ON GENE EXPRESSION IN BRAIN

The exact mechanism of action of n-3 fatty acids is still far from being fully understood. Several lines of evidence, obtained on other tissues, suggest that

Table 5.6 Changes in Gene Expression in Old Rat Cerebellum

Δ-Fold	Gene product	Accession number
0.42	EST	AA409278
0.42	EST	AW544613
0.42	Human STS UT1919	AW544630
0.42	Splicing factor (45 kDa) (SPF45)	AW536113
0.42	Keratin type II (EndoA)	AW542449
0.42	EST	AA409520
0.42	EST	AA472919
0.42	EST	AA276873
0.42	CD2-associated protein (Cd2ap)	AW545446
0.42	EST	AW545785
0.42	Myeloblastosis oncogene	AA267899
0.42	Alpha-1,6-mannosyl-glycoprotein beta-1,2-*n*-acetylglucosaminyltransferase	AA260868
0.42	Carbon catabolite repression 4 homolog	AA404014
0.42	Galectin-6 (Lgals6)	AA404174
0.42	EST	AA270672
0.42	Gap junction membrane channel protein alpha 4	AA274472
0.42	Scavenger receptor class B type I (mSR-BI)	W82738
0.44	Adipocyte-specific mRNA	AA466094
0.46	Ferritin light chain 1 (Ftl1)	AW536161
0.47	Membrane protein TMS-2	AW542443
2.08	Cytochrome P450, 4a10	AA097980
2.08	EST	AA230682
2.09	Mad-related protein Smad7	AA068440
2.19	Similar to SRB7	AA067890
2.23	Integral membrane protein 2	AA066921
2.24	EST	AA080169
2.29	EST	AA065728
2.29	Keratin, type ii microfibrillar, component 7c	AA067967
2.30	Proteasome 28 kDa subunit 1	AA098608
2.40	Ajaba	AA106932
2.42	Similar to leucocyte antigen CD97 precursor	
2.45	EST	AA067324
2.53	Ganglioside-induced differentiation associated protein 2	AA068478
2.57	Stromal cell derived factor 1	AA068750
2.64	Hypothetical	AA068366
2.80	EST	AA065748
3.00	Hypothetical	AA087650
3.02	Heparan sulfate proteoglycan 1 (fibroglycan)	AA106952
3.23	EST	AA068436

both upregulation and repression can be realized involving peroxisome proliferator-activated receptor (PPAR) activation. PPARs are ligand-dependent transcription factors, belonging to the nuclear hormone receptor superfamily. They can modulate directly the gene activity: being first activated by specific ligands and then modulating DNA transcription by binding to defined nucleotide sequences in the promoter region of target genes PPAR responsive element (PPARE).[39] PPARs bind to DNA in the form of dimers, either homodimers or more often heterodimers with the receptor for 9-*cis* retinoic acid known as retinoid X receptor (RXR), which is the obligate partner of PPARs.[40] So far, three different isoforms of the PPARs have been isolated, showing differences either in the binding properties or in the distribution pattern.[41] Recent studies have shown that PPARs present a promiscuous binding pattern, they bind to saturated and unsaturated fatty acids as well as other lipid derivatives, such as eicosanoids and prostaglandins.[42] Interestingly α-linolenic acid had the highest binding constant among the rest of the fatty acids.[42] On the other hand, DHA has been shown to be a natural ligand for the RXR in brain.[40] The preference for these particular fatty acids might be one of the clues to understand the beneficial effects of the n-3 fatty acids on the nervous system.

Recently the role of fatty acids controlling gene expression has been reviewed[43] and it has been suggested that fatty acids can act also in PPAR independent pathways.[43] As mentioned earlier, n-3 fatty acids induce transthyretin transcription,[20] but other agents have also been shown to regulate transthyretin in mice and Li et al.[44] reported that nicotine enhanced the biosynthesis and secretion of this protein in the choroid plexus of rats. In addition to nicotine, a leaf extract from the *Gingko biloba* upregulated transthyretin by more than 15-fold in the hippocampus of mice.[45]

Therefore, it seems that depending on the cell-specific context and the target gene, fatty acids can take very different routes to alter transcription. Several transcription factors different from PPARs are likely candidates and it can be expected that new regulatory pathways of control will be soon discovered in this rapidly evolving field.[39]

These investigations highlight the complexity of how the n-3 polyunsaturated fatty acids affect cellular processes in brain and throw some light on molecular events supporting mental activity.

CONCLUSIONS

The data presented here and published elsewhere[18,20,21] along with those from the laboratory of Dr. A. Berger[32] are the first to show that n-3 polyunsaturated fatty acids affect the genetic machinery of the central nervous system and thus may open up new alleys to understand how these fatty acids control responses given by the brain to different challenges. However, there is much to learn concerning the speed of the response, the amount of the minimal dose, the best combinations of n-6 and n-3 fatty acids, and so on. Another exciting question is whether the different functional centers in the forebrain (like the visual, motoric, etc.), similar to the brain areas investigated here, respond differently to DHA in particular or to n-6 and n-3 fatty acids in general. In this respect, one can expect great plasticity and specialization within different brain areas and cellular structures. More recently, subregions' (dentate gyrus, CA1, and CA3) specificity or learning induced changes in gene expression of hippocampus were described,[46] indicating that the genetic machinery of the brain is highly and purposefully organized. A further hint for the plasticity of the genetic machinery of hippocampus came from the laboratory of Dr. Molteni who showed that acute and chronic exercises affect gene expression.[47]

ACKNOWLEDGMENTS

This work was supported by grants from the Hungarian Research Council (OTKA F 042850 and TS 044836) and EC QLRT-20010011172. K.K. and L.G.P. were supported by János Bólyai scholarship of the Hungarian Ministry of Education.

REFERENCES

1. Weisinger, H.S.; Vingrys, A.J.; Bui, B.V.; Sinclair, A.J. Effects of dietary n-3 fatty acid deficiency and repletion in the guinea pig retina. Invest. Ophthalmol. Vis. Sci. **1999**, *40*, 327–338.
2. Salem, N., Jr.; Niebylski, C.D. The nervous system has an absolute molecular species requirement for proper function. Mol. Membr. Biol. **1995**, *12*, 131–134.
3. Hussain, S.T.; Roots, B.I. Effects of essential fatty acid deficiency and immunopathological stresses on blood brain barrier (B-B-B) in Lewis rats: a biochemical study. Biochem. Soc. Trans. **1994**, *2*, 338s.
4. Salem, N., Jr.; Niebylski, C.D. *Essential Fatty Acids and Eicosanoids*; Sinclair, A., Gibson, R., Eds.; Amer. Oil Chem. Soc.: Champaign, IL, 1992; 844–47.
5. Bourre, J.-M.; Francois, M.; Youyou, A.; Dumont, O.; Piciotty, M.; Pascal, G.; Durand, G. The effects of dietary α-linolenic acid on the composition of nerve membranes, enzymatic activity, amplitude of electrophysiological parameters, resistance to poisons and learning tasks in rats. J. Nutr. **1989**, *119*, 1880–1892.
6. Vreugdenhil, M.; Bruehl, C.; Voskuyl, R.A.; Kang, J.X.; Leaf, A.; Waldman, W.J. Polyunsaturated fatty acids modulate sodium and calcium currents in CA1 neurons. Proc. Natl. Acad. Sci. USA **1996**, *93*, 12559–12563.
7. Ikemoto, A.; Nitta, A.; Furukava, S.; Ohisi, M.; Nakamura, A.; Fuji, I.; Okuyama, H. Dietary n-3 fatty acid deficiency decreases nerve growth factor content in rat hippocampus. Neursci. Lett. **2000**, *12*, 99–102.
8. Clarke, S.D.; Jump, D.B. Dietary polyunsaturated fatty acid regulation of gene transcription. Annu. Rev. Nutr. **1998**, *14*, 83–98.
9. Sessler, A.M.; Ntambi, J.M. Polyunsaturated fatty aid regulation of gene expression. J. Nutr. **1998**, *128*, 923–926.

10. Finstand, H.S.; Kolset, S.O.; Holme, J.A.; Wiger, R.; Ferrants, A.O.; Blomhoff, R.; Drevon, C.A. Effect of n-3 and n-6 fatty acids on proliferation and differentiation of promyelocytic leukemic HL-60 cells. Blood **1994**, *84*, 3799–3809.
11. Yehuda, S.; Rabinovitz, S.; Motsofsky, D.I. Essential fatty acids are the mediators of brain biochemistry and cognitive functions. J. Neurosci. **1999**, *56*, 565–570.
12. Gamoh, S.; Hashimoto, M.; Sugioka, K.; Shahdat Hossain, M.; Hata, N.; Misawa, Y.; Masumura, S. Chronic administration of docosahexaenoic acid improve reference memory-related learning ability in young rats. Neuroscience **1999**, *39*, 237–241.
13. Yehuda, S.; Rabinovitz, S. Essential fatty acid preparations (SR-3) improves Alzheimer's patients quality of life. Int. J. Neurosci. **1996**, *87*, 141–149.
14. Okaniwa, Y.; Yuasa, S.; Yamamoto, N.; Watanabe, N.; Kobayashi, T.; Okuyama, H.; Nomuras, M.; Nagata, Y. A high linoleate and a high α-linolenate diet induced changes in learning behavior of rats. Effects of a shift and reversal of training stimuli. Biol. Pharm. Bull. **1996**, *19*, 536–540.
15. Carriè, I.; Clèment, M.; Javel, de, A.; Frances, H.; Bourre, J-M. Phospholipid supplementation reverses behavioural and biochemical alterations induced by n-3 polyunsaturated fatty acid deficiency in mice. J. Lipid Res. **2000**, *41*, 473–480.
16. Kalmjin, S.; Feskens, E.J.M.; Launer, L.J.; Kromhout, D. Polyunsaturated fatty acids, anti-oxidants, and cognitive function in very old man. Am. J. Epidemiol. **1997**, *145*, 33–41.
17. Terano, T.; Fujishiro, S.; Ban, T.; Yamamoto, K.; Tanaka, T.; Noguchi, Y.; Tamura, Y.; Yazawa, K.; Hirayama, T. Docosahexaenoic acid supplementation improves the moderately severe dementia from thrombotic cerebrovascular diseases. Lipids **1999**, *34S* (suppl.), 345–346.
18. Takamura, H.; Kito, M. A highly sensitive method for quantitative analysis of phospholipid molecular species by high performance liquid chromatography. J. Chromatogr. **1991**, *109*, 436–439.
19. Kitajka, K.; Puskás, L.G.; Zvara, A.; Hackler, L., Jr.; Barceló-Coblijn, G.; Yeo, Y.K.; Farkas, T. The role of n-3 polyunsaturated fatty acids in brain: modulation of gene expression by dietary n-3 fatty acids. Proc. Natl. Acad. Sci. USA **2002**, *99*, 2619–2624.
20. Puskás, L.G.; Kitajka, K.; Nyakas, Cs.; Barceló Coblijn, G.; Farkas, T. Short term administration of omega 3 fatty acids from fish oil results in increased transthyretin transcription in old rat hippocampus. Proc. Natl. Acad. Sci. USA **2003**, *100*, 1580–1585.
21. Barceló-Coblijn, G.; Kitajka, K.; Puskás, L.G.; Hogyes, E.; Zvara, Á.; Hackler, L., Jr.; Farkas, T. Gene expression and molecular composition of phospholipids in rat brain in relation to dietary n-6 to n-3 fatty acid ratio. Biochim. Biophys. Acta **2003**, *1632*, 72–79.
22. Green, P.; Yavin, E. Fatty acid composition of late embryonic and early postnatal brain. Lipids **1999**, *31*, 859–865.
23. George, J.M.; Jim, H.; Woods, W.S.; Clayton, D.F. Characterization of a novel protein regulated during the critical period for song learning in the aebra finch. Neuron **1995**, *15*, 361–372.
24. Galvin, J.E.; Schuck, T.M.; Lee, V.M.; Trojanowski, J.Q. Differential expression and distribution of alpha-, beta-, and gamma-synuclein in the developing human substantia nigra. Exp. Neurol. **2001**, *168*, 347–355.
25. Hartman, V.N.; Miller, M.A.; Clayton, D.F.; Liu, W.C.; Kroodsma, D.E.; Brenowitz, E.A. Testosterone regulates alpha-synuclein mRNA in the avian song system. Neuroreport **2001**, *12*, 943–946.

26. Mauch, D.H.; Nägler, K.; Scumacher, S.; Göritz, Ch.; Müller, E.-Ch.; Otto, A.; Pfriger, F.W. CNS synaptogenesis promoted by glia-derived cholesterol. Science **2000**, *294*, 1354–1357.

27. Black, J.K.; Isaacs, K.R.; Anderson, B.J.; Alcantara, A.A.; Greenough. Learning causes synaptogenesis, whereas motor activity causes angiogenesis in cerebellar cortex of adult rats. Proc. Natl. Acad. Sci. USA **1990**, *87*, 5568–5572.

28. Carter, J.; Lippa, C.F. β-amyloid, neuronal death and Alzheimer's disease. Curr. Mol. Med. **2000**, *1*, 733–737.

29. Vara, H.; Martinez, B.; Santos, A.; Colino, A. Thyroid hormone regulates neurotransmitter release in neonatal rat hippocampus. Neuroscience **2002**, *11*, 19–23.

30. Favrelere, S.; Stadelman-Ingrand, S.; Huhuet, F.; Javel, D.; de Piriou, A.; Tallineau, C.; Durand, G. Age related changes in ethanolamine glycerophospholipid fatty acid level in rat frontal cortex and hippocampus. Neurobiol. Aging **2000**, *21*, 653–660.

31. Delion, S.; Chalon, S.; Guillotean, D.; Lejune, B.; Besnard, J.C.; Durand, G. Age-related changes in phospholipid fatty acid composition and monoaminergic neurotransmission in the hippocampus of rats fed a balanced or an n-3 polyunsaturated fatty acid-deficient diet. J. Lipid Res. **1997**, *38*, 680–689.

32. Berger, A.; Mutch, D.M.; Bruce-German, J.; Roberts, M.A. Dietary effects of arachidonate-rich fungal oil and fish oil on murine hepatic and hippocampal gene expression in old rat hippocampus. Lipids Health Dis. **2002**, 1.

33. Schwarzman, A.L.; Gregori, L.; Vitek, M.P.; Lyubski, S.; Strittmatter, W.J.; Enghilde, J.J.; Bhasin, R.; Silverman, J.; Weisgraber, K.H.; Coyle, P.K.; Zagorski, M.G.; Talafous, J.; Eisenberg, M.; Saunders, A.M.; Roses, A.D.; Goldgaber, D. Transthyretin sequesters amyloid beta protein and prevents amyloid formation. Proc. Natl. Acad. Sci. USA **1994**, *91*, 8368–8372.

34. Tsuzuki, K.; Fukatsu, R.; Yamaguchi, H.; Tateno, M.; Imai, K.; Fuji, N.; Yamauchi, T. Transthyretin binds amyloid beta peptides, Abeta 1-42 and Abeta 1-40 to form complex in the autopsied human kidney—possible role of transthyretin for Abeta sequestration. Neurosci. Lett. **2000**, *281*, 171–174.

35. Link, C.D. Expression of human beta-amyloid peptide in transgenic *Caenorhabditis elegans*. Proc. Natl. Acad. Sci. USA **2000**, *92*, 9368–9372.

36. Kuchler-Bopp, S.; Dietrich, J.-B.; Zaepfel, M.; Delaunoy, J.-P. Receptor-mediated endocytosis of transthyretin by ependymoma cells. Brain Res. **2000**, *7*, 185–194.

37. Peterson, S.A.; Klabunde, T.; Lashuel, H.A.; Purkey, H.; Sacchettini, J.C.; Kelly, J.W. Inhibiting transthyretin conformational changes that lead to amyloid formation. Proc. Natl. Acad. Sci. USA **1998**, *95*, 12956–12960.

38. Blalock, E.M.; Chen, K.-Ch.; Sharrow, K.; Herman, J.P.; Porter, N.M.; Foster, T.C.; Landfield, P.W. Gene microarrays in hippocampal aging: statistical profiling identifies novel processes correlated with cognitive impairment. J. Neurosci. **2003**, *23*, 3807–3819.

39. Wahli, W. Peroxisome proliferator-activated receptors (PPARs): from metabolic control to epidermal wound healing. Swiss Med. Wkly. **2002**, *132*, 83–91.

40. de Urquiza, A.M.; Liu, S.; Sjoberg, M.; Zetterstrom, R.H.; Griffiths, W.; Sjoval, J.; Perlmann, T. Docosahexaenoic acid, a ligand for the retinoid X receptor in mouse brain. Science **2000**, *290*, 2140–2144.

41. Cullingford, T.E.; Bhakoo, K.; Peuchen, S.; Dolphin, C.T.; Patel, R.; Clark, J.B. Distribution of mRNAs encoding the peroxisome proliferator-activated receptor alpha,

beta, and gamma and the retinoid X receptor alpha, beta, and gamma in rat central nervous system. J. Neurochem. **1998**, *70*, 1366–1375.

42. Xu, H.E.; Lambert, M.H.; Montana, V.G.; Parks, D.J.; Blanchard, S.G.; Brown, P.J-; Sternbach, D.D.; Lehmann, J.M.; Wisely, G.B.; Willson, T.M.; Kliewer, S.A.; Milburn, M.V. Molecular recognition of fatty acids by peroxisome proliferator-activated receptors. Mol. Cell. **1999**, *3*, 397–403.

43. Duplus, E.; Forest, C. Is there a single mechanism for fatty acid regulation of gene transcription? Biochem. Pharmacol. **2002**, *64*, 893–901.

44. Li, M.D.; Kana, J.K.; Matta, S.G.; Blanner, W.S.; Sharp, B.M. Nicotine enhances the biosynthesis and secretion of transthyretin from the choroids plexus in rats: implications for beta-amyloid formation. J. Neurosci. **2000**, *20*, 1318–1323.

45. Watanabe, C.M.; Wolffram, S.; Ader, P.; Rimbach, G.; Packer, L.; Maguire, J.J.; Schultz, P.G.; Gohil, K. The *in vivo* neuromodulatory effects of the herbal medicine gingko biloba. Proc. Natl. Acad. Sci. USA **2001**, *98*, 6577–6680.

46. Robles, V.; Vivas-Mejia, P.E.; Ortiz-Zuazaga, H.G.; Felix, J.; Ramos, X.; Pena de Ortiz, S. Hippocampal gene expression profiling in spatial discrimination learning. Neurobiol. Learn. Mem. **2003**, *8*, 80–95.

47. Molteni, R.; Ying, Z.; Gómez-Pinilla, F. Differential effects of acute and chronic exercise on plasticity-related genes in the rat hippocampus revealed by microarray. Eur. J. Neurosci. **2002**, *16*, 1107–1116.

6

Anterior and Posterior Cerebral Gene Expression in a Rat Model of Reduced Brain Docosahexaenoic Acid During Development

Judy B. Gaines, Nancy E. J. Berman, and Susan E. Carlson

University of Kansas Medical Center, Kansas City, KS, USA

INTRODUCTION

Docosahexaenoic acid (DHA) is a fatty acid found in high concentrations in brain[1] where it is concentrated in the phospholipids of synaptosomal membranes[2] and synaptic vesicles.[3] Animal studies conducted since the 1970s have investigated the effect of feeding an 18:3n-3-deficient diet during development on brain fatty acid composition and animal behavior. There is now an extensive literature on the biochemical and behavioral effects of 18:3n-3 deficiency. Brain phospholipid DHA decreases while at the same time there is an increase in a metabolite of linoleic acid (18:2n-6), docosapentaenoic acid (DPA, 22:5n-6).[4] Functionally, animals fed 18:3n-3-deficient diets have normal growth, reproduction, and lifespan. However, they have impaired visual development,[5,6] reduced pain threshold,[7] longer duration looking,[8] stereotyped behavior, and greater whole body activity.[9] The latter suggests alterations in brain dopamine function.[9] Indeed, dopamine neurotransmission is altered by several generations of an 18:3n-3-deficient diet[10] and dopamine-related behaviors and neurotransmission are influenced positively by intake of fish oil, which contains DHA.[11] Levant et al.[12] reported recently that even a modest reduction in rat brain

DHA ($\sim$25%) and the concomitant increase in brain DPA prior to weaning resulted in increased basal and amphetamine-stimulated activity, and reduced haloperidol-induced catalepsy, which evinced alterations in dopamine systems.

In the human fetus, brain DHA accumulates exponentially in the last trimester[13,14] and accumulation continues until at least 2 years of age.[14] Infants fed formulas that contain 18:3n-3 but no DHA, have significantly less circulating DHA than infants fed human milk. This was first reported over 20 years ago[15,16] and no exceptions have been reported despite similar comparisons done in many different centers worldwide. Lower circulating levels of DHA in formula-fed babies suggested less DHA accumulation in the infant brain and possible functional consequences analogous to those noted with 18:3n-3-deficient diets, for example, impaired visual acuity and altered attention. The first randomized trials compared human infants fed formulas with and without DHA to determine if DHA intake would enhance visual development and other neurologically related developmental outcomes in preterm and term infants.[17–22]

Results of these early trials showed that DHA-supplementated formula benefited visual acuity development[17–20] and attention,[21,22] evidence that brain DHA was increased by including DHA in infant formula that contained 18:3n-3. Later, two independent groups of researchers reported higher brain DHA levels in breast-fed infants compared with formula-fed infants.[23,24] Formula-fed term infants had about 25% less brain DHA than infants fed human milk,[23,24] whereas formula-fed preterm infants had about half as much brain DHA.[23] A number of clinical trials conducted in the intervening years have compared functional outcomes in infants fed diets with and without DHA (reviewed in Gibsen et al.[25] and Uauy et al.[26]). DHA was added to infant formulas in the US in 2002.

In the present study, we studied gene expression in the anterior and posterior cerebrum in Sprague–Dawley rats with physiologically lower brain DHA during development (exposed to 18:3n-3-deficient diets during gestation and weaning) to screen for early effects of low brain DHA accumulation. The degree of reduction in DHA and the developmental period during which the reduction occurred is analogous to the reduction in brain DHA of human infants born preterm and fed a postnatal diet without DHA.[23]

METHODS

Animals and Diets

Virgin Sprague–Dawley rats were obtained from Charles River Laboratories, Inc., Hartford, CT. The rats were housed individually and allowed to adapt to the new environment for at least 1 week before mating. The day they were sperm positive was defined as embryologic day 0 (E0). On E0, dams were changed from a chow diet to one of two nutritionally complete, semisynthetic diets (Table 6.1). Assignment to diet was random. The oil-free basal mix for AIN-93G was obtained from Harlan-Teklad, Madison, WI. Either soybean oil or sunflower oil was added to the dry mix as 7% of weight of the final diet. After the addition of soybean oil, the diet was identical in composition to AIN-93G.[27] Soybean oil contains a large amount of 18:3n-3, and animals fed this diet were the 18:3n-3-replete group. Sunflower oil contains much less 18:3n-3 and a much higher ratio of 18:2n-6 to 18:3n-3 than soybean oil (135 vs. 8). Sunflower oil replaced soybean oil as a dietary fat source in the 18:3n-3-deficient diet. The fatty acid composition of the two diets is shown in Table 6.2. Gene expression in the group fed the 18:3n-3-replete diet was compared to that of the group fed the 18:3n-3-deficient diet.

Six dams were assigned to each diet during pregnancy and lactation. Dams delivered on either embryological day 22 or 23 (E22 or E23). E23 was considered postnatal day 1 (P1). On E23 (P1), each litter was culled to eight pups. Two pups were killed at P7, P14, and P21. The two remaining pups were fed the maternal diet until >P56, when their brain visual evoked responses were studied. The rats were killed by decapitation and the brains were excised immediately. The brain

Table 6.1 Composition of 18:3n-3-Replete and 18:3n-3-Deficient Diets in kg/5 kg Diet

	18:3n-3-Replete	18:3n-3-Deficient
Basal mix formula[a]	4.65	4.65
Casein	1.00	1.00
L-Cystine	0.015	0.015
Corn starch	1.99	1.99
Maltodextrin	0.66	0.66
Sucrose	0.50	0.50
Cellulose	0.25	0.25
Mineral mix[b]	0.175	0.175
Vitamin mix[c]	0.050	0.050
Choline bitartrate	0.0125	0.0125
TBHQ	0.0007	0.0007
Soybean oil	0.35 kg	—
Sunflower oil	—	0.35 kg

[a]Harlan-Teklad TD00235. The individual basal mix components in kg/5 kg of diet are noted in the insert. The basal mix is modified from AIN-93G for use at the rate of 930 g/kg of diet prepared by adding 70 g/kg of selected fats and oils.
[b]Harlan-Teklad TD94046 (AIN 93G-MX).
[c]Harlan-Teklad TD 94047 (AIN-93-VX).

Table 6.2   Fatty Acid Composition of
α-Linolenic Acid-Replete and -Deficient
Diets (g/100 g Total Fatty Acids)

Fatty acid	Replete	Deficient
14:0[a]	0.3	0.3
14:1	0.0	0.0
16:0	4.8	6.3
16:1	0.1	0.1
18:0	3.4	4.4
18:1	22.3	20.0
18:2n-6	59.5	67.4
20:0	0.2	0.2
20:1n-9	0.2	0.1
18:3n-3	7.4	0.5
20:2n-6	0.3	0.2
22:0	0.2	0.2
24:0	0.1	0.1

[a]Number of carbons:number of double
bonds, carbons from CH_3 terminal of molecule to first double bond.

from one pup was placed in RNA*later*™ (Ambion, Austin, TX) and stored at
−20°C for later analysis. Prior to RNA extraction, the anterior and posterior cerebral sections were dissected by cuts through the brain immediately anterior to
the optic chiasm and through the cerebellum/medulla oblongata/pons posterior
to the mammillary bodies, respectively. The other brain from the same litter-age
was stored at −80°C until phospholipid fatty acid analysis. For analysis, the brain
was homogenized and lipids from a uniform aliquot were extracted. The phospholipids were isolated by TLC, and the phospholipid fatty acid methyl esters
prepared and quantified as previously described.[28]

cDNA Arrays

Gene expression was analyzed using BD Atlas™ Nylon Rat 1.2 cDNA
expression arrays representing 1185 genes including 9 housekeeping genes.
This is a broad coverage array profiling well-annotated cellular pathways and
functions. Genes are arrayed in six sections with genes of like function
grouped together. A list of genes included in this array, including array coordinates and GenBank accession numbers, is available at www.clontech.com/
atlas. Membranes were stripped according to the manufacturer's specifications
and probed up to three times. The intensities reported in Tables 6.3–6.6 for
each diet group were obtained by averaging the data from 12 individual

Table 6.3 Relative Expression of Anterior Cerebral Genes Expressed Higher in Rats with Normal Brain DHA (Soybean Oil) at all Ages (P1, P7, P14, and P21), $p < 0.05$

Clontech	GenBank accession no.	Gene	Soybean	Sunflower
B02h	**U53211**	**Degenerin channel MDEG; amiloride-sensitive brain Na channel BNAC1 (PM)**	**1.2 (0.6–1.9)**	**0.5 (0.0–1.5)**
B12d	**L19031**	**Organic anion transporter (PM)**	**1.3 (0.6–2.7)**	**0.7 (0.3–1.6)**
E03e	M13750	Prolactine like protein A (rPLP-A) (ECS)	1.1 (0.0–14.5)	0.7 (0.3–1.6)
E06a	X58023	Corticotropin-releasing factor binding protein (ECS)	1.0 (0.3–1.8)	0.4 (0.0–1.8)
A02g	D25290	Cadherin 6 precursor; kidney-cadherin (K-cadherin) (PM)	1.2 (0.6–1.9)	0.9 (0.4–2.2)
B12e	U67958	Urate transporter/ channel (C, ECS)	1.2 (0.8–2.4)	0.8 (0.3–2.0)
F11d	U27201	Metalloproteinase inhibitor 3 precursor; TIMP3 (ECM)	1.2 (0.7–2.4)	0.8 (0.4–1.6)
A01c	M10072	Leukocyte common antigen precursor (LCA); CD45 antigen; T200; PTPRC (PM)	1.1 (0.3–37.6)	1.0 (0.0–6.7)
E10b	L15618	Casein kinase II (alpha subunit) (C)	1.2 (0.5–2.7)	0.7 (0.2–1.4)
C04f	M32801+	3-Ketoacyl-CoA thiolase A + 3-ketoacyl-CoA thiolase B (P)	1.1 (0.8–1.4)	0.9 (0.5–1.9)
B06m	X16002	Potassium channel RCK4, subunit, putative (PM)	1.2 (0.4–3.1)	0.8 (0.4–1.5)
C01g	U20283	Syntaxin binding protein Munc18-2 (C)	1.1 (0.4–1.8)	0.8 (0.3–1.3)

Note: Analyzed with GeneSpring software. Each diet mean (range) represents analysis of pups from three different litters at each of the four ages. Proteins associated with C, cytoplasm; P, peroxisome; PM, plasma membrane; ECS, extracellular secreted; ECM, extracellular matrix. Genes shown in bold differ by diet, $p < 0.01$.

Table 6.4 Relative Expression of Anterior Cerebral Genes Expressed Higher in Rats with Lower Brain DHA (Sunflower Oil) at all Ages (P1, P7, P14, and P21)

Clontech	GenBank accession no.	Gene	Soybean	Sunflower
D03b	**L03202**	**Serotonin receptor; 5-hydroxytryptamine 6 receptor (5-HT-6) (PM)**	**0.2 (0.0–2.3)**	**1.1 (0.1–2.1)**
E02g	**X14232**	**Heparin-binding growth factor 1 (ECS)**	**0.5 (0.0–1.5)**	**1.3 (0.4–4.1)**
A07a	**D14014**	**G1/S-specific cyclin D1 (CCND1) (C)**	**0.6 (0.4–1.1)**	**1.2 (0.2–2.2)**
D13k	X97375	Nociceptin precursor; orphanin FQ; PPNOC; ORL1 agonist (ER, ECS)	0.8 (0.3–1.4)	1.1 (0.3–1.9)
A13c	S45392	*c*-Ros-1 proto-oncogene (PM)	0.7 (0.1–1.3)	1.1 (0.4–2.0)
A09a	X75305	Antigen peptide transporter 2; TAP2L; APT2; TAP2; MTP2 (PM)	0.6 (0.0–2.0)	1.2 (0.1–4.6)
C07n	Y08172	2-Arylpropionyl-CoA epimerase; alpha-methylacyl-CoA racemase (C)	0.5 (0.0–2.2)	1.1 (0.1–2.7)
A07h	D38560	GAK; cyclinG-associated kinase (C, N)	0.7 (0.2–3.2)	1.3 (0.5–3.5)
C06n	M33986	Cytochrome P-450 19; aromatase (PM)	0.6 (0.1–1.8)	1.1 (0.3–2.1)
A04c	M58587	Interleukin-6 receptor alpha precursor (IL-6R-alpha; IL6R) (PM, ECS)	0.5 (0.0–1.8)	1.0 (0.0–3.2)
A13d	J03823	Mas proto-oncogene; G-protein coupled receptor; Mas-1 (PM)	0.6 (0.1–1.9)	1.2 (0.7–2.2)
B11a	D17695	Water channel aquaporin 3 (AQP3) (PM)	0.8 (0.2–1.9)	1.2 (0.8–1.9)
A05a	S73583	Glutamyl aminopeptidase A (PM)	0.3 (0.0–1.7)	0.9 (0.0–3.9)

(continued)

Table 6.4   Continued

Clontech	GenBank accession no.	Gene	Soybean	Sunflower
D13h	X13722	Low-density lipoprotein receptor precursor (LDL receptor; LDLR) (PM)	0.5 (0.0–1.8)	0.9 (0.0–3.0)
D13i	J02762	Asialoglycoprotein receptor R2/3 (ASGPR); hepatic lectin 2/3; RHL-2 (PM)	0.2 (0.0–2.8)	1.0 (0.0–4.2)
D06g	U05239	Opioid receptor-like orphan receptor (PM, C)	0.2 (0.0–4.9)	0.9 (0.0–2.6)
D07k	U66274	Neuropeptide Y5 receptor (PM)	0.3 (0.0–3.7)	1.5 (0.5–4.5)

Note: Analyzed with GeneSpring software. Each diet mean (range) represents analysis of pups from three different litters at each of the four ages. Proteins associated with C, cytoplasm; P, peroxisome; PM, plasma membrane; ECS, extracellular secreted; ECM, extracellular matrix; N, nucleus. Genes shown in bold differ by diet, $p < 0.01$.

animals (three each from P1, P7, P14, and P21). The three pups chosen for analysis from each diet-age represented three of the six possible litters fed the diet.

RNA Extraction, Labeling, and Hybridization

Total RNA was extracted from anterior and posterior cerebral sections using Trizol Reagent (Gibco/BRL, Gaithersburg, MD) following the manufacturer's instructions. The quantity, quality, and integrity of the RNA were determined using UV absorption spectrophotometry and agarose gel electrophoresis. Radio-labeled probes were prepared using the BD Atlas™ Pure Total RNA Labeling System (BD Biosciences Clontech) according to the Clontech User's Manual. Briefly, total RNA was treated with DNAseI to minimize genomic contamination. To ensure removal of RNases, anti-RNase (Ambion) was used with this protocol at a concentration of 1 U/μL. Following DNase treatment, RNA integrity was again confirmed on an agarose gel. Poly A+ RNA was isolated using streptavidin-coated magnetic beads and biotinylated oligo(dT) and was subsequently reverse transcribed using MMLV reverse transcriptase in the presence of [α-^{32}P]dATP (Amersham) and primers (BD Bioscience Clontech) specific for the genes represented on the Rat 1.2 Array. The probes were purified and eluted by column chromatography.

Table 6.5 Relative Expression of Posterior Cerebral Genes Expressed Higher in Animals with Normal Brain DHA (Soybean Oil) Across all Ages Studied (P1, P7, P14, and P21), $p < 0.05$

Clontech	GenBank accession no.	Gene	Soybean	Sunflower
C10k	L12407	Dopamine beta-hydroxylase (PM)	1.5 (0.4–4.0)	0.8 (0.5–1.7)
C14g	L26110	Transforming growth factor beta receptor type 1 precursor (PM)	1.3 (0.6–2.4)	0.4 (0.0–1.9)
B10j	X70062	ATPase, sodium/potassium, gamma subunit (PM)	1.4 (0.3–2.2)	0.5 (0.1–1.8)
C14f	D38082	Bone morphogenetic protein type IA receptor (PM)	1.3 (1.0–2.4)	0.5 (0.0–1.9)
F12e	M94152	Adenosine A3 receptor (ADORA3); TGPCR1 (PM, ECM, ECS)	1.0 (0.4–2.3)	0.4 (0.0–1.2)
C13h	M77809	TGF-beta receptor type III; betaglycan; cand. tumor suppressor (PM)	1.2 (0.5–2.8)	0.4 (0.0–1.3)
C13k	D13566	Erythropoietin receptor precursor (EPOR) (PM)	1.5 (0.8–4.4)	0.7 (0.3–1.7)
C14k	U69278	Rek4 Eph-related receptor tyrosine kinase; type-A receptor 3 (PM)	1.4 (0.7–3.9)	0.5 (0.0–1.2)
C13l	M91599	Fibroblast growth factor receptor subtype 4 (PM)	1.4 (1.0–2.6)	0.8 (0.4–1.7)
B11f	Y09164	Voltage-gated sodium channel (atypical) (PM)	1.5 (0.8–2.6)	0.4 (0.0–1.2)
C14l	X78689	EHK1; ephrin type-A receptor 5 (EPHA5); tyrosine kinase (PM)	1.3 (0.8–2.5)	0.6 (0.3–1.1)
C13m	U97142	GFRα1, RET ligand 1 (PM)	1.3 (0.9–1.9)	0.8 (0.4–1.2)
C02l	M86240	Fructose-16-bisphosphatase, liver (C)	1.2 (0.6–4.2)	0.8 (0.2–2.0)

(continued)

Table 6.5 Continued

Clontech	GenBank accession no.	Gene	Soybean	Sunflower
C01m	M83678	Rab13, ras related GTPase (C)	1.2 (0.6–2.7)	0.9 (0.5–1.4)
C04j	L09216	Pancreatic lipase related protein 2 precursor; glycoprotein GP-3 (ECS)	1.2 (0.5–1.9)	0.8 (0.5–1.8)
C14h	M37394	Epidermal growth factor receptor (EGF receptor; EGFR) (PM)	1.4 (0.3–3.2)	0.4 (0.0–3.0)
F03l	D10393	Calcineurin B subunit, CALCINEURIN B SUBUNIT ISOFORM 1 (C,N)	1.3 (0.6–2.6)	0.7 (0.2–1.7)
C06e	X63410	Alcohol sulfotransferase A (EC 2.8.2.2); hydroxysteroid sulfotransferase (C)	0.7 (0.0–2.7)	0.3 (0.0–1.3)
C09m	J05030	Short chain acyl-CoA dehydrogenase precursor (SCAD; ACADS) (M)	1.3 (0.5–5.9)	0.8 (0.5–1.2)
E11b	L22294	Pyruvate dehydrogenase kinase kinase precursor (M)	1.3 (0.5–2.5)	0.7 (0.3–2.7)
F12b	U73458	Phosphotyrosine phosphatase 6 (PM)	1.1 (0.5–2.5)	0.5 (0.2–2.0)
C10f	U12268	Carbonic anhydrase 5 (M)	1.4 (0.6–3.2)	0.9 (0.4–2.4)
C10e	S68245	Carbonic anhydrase 4 (PM,C)	1.3 (0.8–2.7)	0.8 (0.4–2.3)
C12f	X83399	Initiation factor, eukaryotic, (eIF-4E) (C)	1.3 (0.8–2.9)	0.8 (0.2–2.9)
C06i	M20559	Annexin III (ANX3); lipocortin 3; placental anticoagulant protein III (C)	1.1 (0.4–1.6)	0.6 (0.2–1.7)
E05e	L26913	Interleukin 13 precursor (IL-13); T-cell activation protein P600 (ECS)	0.6 (0.0–2.4)	0.5 (0.0–1.1)

(*continued*)

Table 6.5   Continued

Clontech	GenBank accession no.	Gene	Soybean	Sunflower
C12m	D83697	Activator of apoptosis harakiri (HRK); neuronal death protein 5 (DP5) (C)	1.4 (0.8–2.8)	0.8 (0.3–1.5)
B11e	U77971	Urea transporter (PM)	1.3 (0.5–2.7)	0.7 (0.1–1.6)
B09k	D83598	Sulfonylurea receptor (PM)	1.0 (0.1–2.6)	0.5 (0.0–1.3)
C13i	U93306	KDR/flk1 vascular endothelial growth factor tyrosine kinase receptor (PM)	1.2 (0.6–2.5)	0.7 (0.1–1.5)
E11d	M94454	Cot proto-oncogene; Tpl-2 (C)	1.3 (0.8–2.2)	0.7 (0.3–1.7)
C12n	U01022	Huntington disease gene homolog (C,N)	1.2 (0.6–3.1)	0.7 (0.3–1.4)
C12j	L14680	bcl-2 (M, ER, N)	1.2 (0.3–3.1)	0.5 (0.1–1.3)

Note: Analyzed with GeneSpring software. Each diet mean (range) represents analysis of pups from three different litters at each of the four ages. Proteins associated with C, cytoplasm; P, peroxisome; PM, plasma membrane; ECS, extracellular secreted; ECM, extracellular matrix; N, nucleus; M, mitochondria. Genes shown in bold differ by diet, $p < 0.01$.

Probes were hybridized to the array membranes at 68°C overnight in a rotating hybridization oven. The membranes were washed the following day in low to high stringency buffers and exposed to a Phosphor screen for varying lengths of time.

Imaging and Data Analysis

Images were acquired using a Packard Cyclone Phosphorimager, and gene expression was quantified using Clontech's AtlasImage^{TM} 2.0 Software. Atlas-Image generates a color coded comparison view and numeric data in a tabular format. The numeric data were imported into GeneSpring (Silicon Genetics, Redwood City, CA) and measurements of individual cDNA intensities were normalized as follows: (1) values below 0 were set to 0 and (2) each measurement was divided by the 50th percentile of all measurements in the array. Finally, the 12 measurements obtained for each dietary treatment were combined and the results expressed as a median intensity and range for each group.

Within GeneSpring a nonparametric test (Wilcoxon–Mann–Whitney/ Kruksal–Wallis test) was used for statistical analysis to determine gene expression that was significantly affected by a diet across all ages studied. Results are shown only for genes that differed by diet ($p < 0.05$). We did not attempt to correct

Table 6.6 Relative Expression of Posterior Cerebral Genes Expressed Higher in Animals with Lower Brain DHA (Sunflower Oil) Across all Ages Studied (P1, P7, P14, and P21), $p < 0.05$

Clontech	GenBank accession no.	Gene	Soybean	Sunflower
A11m	M85214	trk, proto-oncogene, precursor (PM)	0.7 (0.2–2.2)	1.6 (0.7–2.8)
B01j	AB04559	Renal organic anion transporter (ROAT1) (PM)	0.6 (0.2–1.9)	1.1 (0.4–3.3)
E03l	M17523	Peptide YY precursor (PYY) (ECS)	0.0 (0.0–1.6)	0.4 (0.0–2.8)
A06b	X81193	LIM, muscle (N,CS)	0.1 (0.0–1.5)	0.6 (0.0–2.0)
C12d	K03502	Elongation factor 2 (EF2) (C)	0.7 (0.3–2.7)	1.3 (0.7–1.9)
A12d	L15619	Casein kinase II beta subunit (CKII; CSNK2B; CK2N); phosvitin (C)	0.7 (0.2–2.4)	1.4 (0.2–2.8)
C11h	X78327	Ribosomal protein L13 (C)	0.7 (0.2–1.1)	1.2 (0.2–2.4)
C02c	J04526	Type 1 hexokinase (HK1); brain hexokinase (M)	0.7 (0.3–1.8)	1.4 (0.7–2.5)
B12c	J05167	Band 3 (B3RP3), 3 Cl-HCO3-anion exchanger (PM)	0.7 (0.5–1.2)	1.1 (0.6–1.7)
D03i	U00475	Delta-type opioid receptor (DOR-1); opioid receptor A (PM)	0.1 (0.0–1.3)	0.6 (0.0–2.5)
D05m	L10072	5-Hydroxytryptamine 5A receptor (5HT5A; HTR5A) (PM)	0.1 (0.0–1.2)	0.6 (0.0–4.3)
D04h	U00935	Gonadotropin releasing hormone receptor (PM)	0.0 (0.0–0.5)	0.2 (0.0–1.7)
D09c	Y10369	GABA-B receptor1a (GABA-BR1A receptor) + GABA-B receptor1b (PM)	0.6 (0.3–1.3)	1.2 (0.3–2.1)
A10c	X13058	p53 Nuclear oncoprotein (N)	0.4 (0.1–1.4)	1.2 (0.5–2.2)

(*continued*)

Table 6.6 Continued

Clontech	GenBank accession no.	Gene	Soybean	Sunflower
A09c	**M34097**	**Natural killer (NK) cell protease 1 (RNKP-1) (ECS)**	**0.2 (0.0–2.1)**	**1.5 (0.7–2.4)**
A04h	**L33413**	**Advanced glycosylation end product-specific receptor (AGER) (PM)**	**0.4 (0.0–1.1)**	**1.4 (0.6–2.2)**
E05g	**J04486**	**Insulin-like growth factor binding protein 2 (IGF-BP 2) (ECS)**	**0.6 (0.3–1.5)**	**1.4 (0.6–2.7)**
A09a	**X75305**	**Antigen peptide transporter 2; TAP2L; APT2; TAP2; MTP2 (PM)**	**0.0 (0.0–1.2)**	**1.2 (0.5–3.4)**
A05e	**X54686**	**Jun-B; c-jun-related transcription factor (N)**	**0.3 (0.0–1.3)**	**1.4 (0.8–2.5)**
E06k	M91590	Beta-arrestin 2 (ARRB2) (C)	0.8 (0.3–1.5)	1.1 (0.5–1.9)
E14e	M83681	Rab-3a ras-related protein (PM, C)	0.9 (0.3–1.8)	1.1 (0.4–2.0)
G47	X59051	40S ribosomal protein S29 (RPS29) (C)	0.7 (0.2–3.0)	1.1 (0.7–1.9)
F07d	M76426	Dipeptidyl aminopeptidase related protein (DPP6) (PM)	0.7 (0.2–1.3)	1.0 (0.2–1.9)
D14f	M60525	VGF8A protein precursor (ECS)	0.7 (0.1–1.7)	1.2 (0.4–2.3)
B01a	U09540	P450 IB1; C3H cytochrome P450; CYP1B1 (PM, ER)	0.1 (0.0–1.4)	0.3 (0.0–2.1)
B02m	U12425	Cyclic nucleotide-gated channel, olfactory (PM)	0.1 (0.0–1.8)	0.4 (0.0–2.3)
B03h	X70521	Sodium channel, amiloride sensitive, alpha subunit; SCNEA (PM)	0.3 (0.0–1.1)	0.6 (0.0–2.4)
B01n	L15079	Multidrug resistance protein 2 (MDR2); P-glycoprotein (PGY2) (PM)	0.1 (0.0–1.3)	0.4 (0.0–2.1)

(*continued*)

Table 6.6 Continued

Clontech	GenBank accession no.	Gene	Soybean	Sunflower
D02b	J04811	Growth hormone receptor precursor (GH receptor; GHR) (PM)	0.3 (0.0–2.1)	0.2 (0.0–3.2)
D07l	U67863	Melanocortin receptor 4 (PM)	0.2 (0.0–1.4)	1.1 (0.5–2.3)
E04l	U77776	Interferon gamma inducing factor precursor (ECS)	0.1 (0.0–1.5)	0.5 (0.0–2.3)
E05b	M26744	Interleukin 6 (IL-6) (ECS)	0.2 (0.0–0.9)	0.7 (0.2–3.3)
D14i	U37101	Granulocyte colony stimulating factor (ECS)	0.1 (0.0–1.0)	0.4 (0.0–3.5)
D04l	U16523	Corticotropin-releasing factor receptor subtype 2 (CRF2R) (PM)	0.2 (0.0–3.1)	0.9 (0.0–3.8)
E06b	X06832	Chromogranin A (ECS)	0.9 (0.5–3.2)	1.3 (0.8–2.3)
E09i	M73808	Phosphorylase kinase, catalytic subunit (C)	0.9 (0.6–1.6)	1.3 (0.9–1.8)
E06j	M91589	Beta-arrestin 1 (C)	0.8 (0.3–1.7)	1.3 (0.7–2.9)
C11g	X62146	Ribosomal protein L11 (C)	0.8 (0.4–1.4)	1.3 (0.5–3.8)
E09b	M18332	Protein kinase C zeta type (PKC-zeta) (C, PM)	0.9 (0.6–1.8)	1.2 (0.9–2.2)
D03f	L20684	mu opioid receptor (MUOR1); mu-type opioid receptor (MOR-1) (PM)	0.0 (0.0–0.9)	0.3 (0.0–3.1)
D08d	M81766	RXR-beta *cis*-11-retinoic acid receptor; nuclear receptor co-regulator 1 (N)	0.5 (0.0–1.3)	1.1 (0.4–2.1)
A04i	X05111	CD2, membrane glycoprotein, T-cell marker (PM)	0.2 (0.0–1.8)	1.1 (0.4–2.3)
D01c	X64179	*c*-fms proto-oncogene; macrophage colony stimulating factor 1 recept. (PM)	0.4 (0.0–1.6)	0.8 (0.0–2.1)
A07c	L32591	G1/S-specific cyclin D3 (CCND3) (N, C)	0.7 (0.4–1.7)	1.1 (0.5–2.0)

(*continued*)

Table 6.6 Continued

Clontech	GenBank accession no.	Gene	Soybean	Sunflower
E08i	U37462	Dual-specificity mitogen-activated protein kinase kinase 5 (CS, C)	0.8 (0.4–1.7)	1.2 (0.7–1.7)
E03h	U25651	Muscle 6-phosphofructokinase (PFKM); phosphofructokinase 1 (C)	0.7 (0.3–2.4)	1.3 (0.5–3.0)
A14n	M12516	NADPH-cytochrome P450 reductase (CPR); POR (ER)	0.9 (0.4–1.5)	1.3 (0.8–2.4)
A13m	Z27118	Heat shock 70 kDa protein (HSP70) (C)	0.9 (0.5–1.2)	1.2 (0.7–2.4)
A09b	M58950	Plasma kallikrein (rPK) (ECS)	0.3 (0.0–2.1)	0.97 (0.0–2.4)
D06e	M89954	5-Hydroxytryptamine (serotonin) receptor 1B; 5-HT1B (PM)	0.2 (0.0–1.6)	0.98 (0.0–2.7)
D06f	M93273	Somatostatin receptor 2 (PM)	0.2 (0.0–3.5)	1.1 (0.0–3.4)
D02g	M85183	Angiotensin/vasopressin receptor (AII/AVP) (PM)	0.0 (0.0–0.7)	0.1 (0.0–2.0)
B03f	X95882	ATP ligand gated ion channel (PM)	0.1 (0.0–1.2)	0.4 (0.0–4.2)
D10f	M81142	Gamma-aminobutyric acid (GABA-A) receptor, gamma 3 subunit (PM)	0.0 (0.0–1.8)	0.3 (0.0–2.1)
A04e	AF007789	Urokinase receptor + GPI-anchored plasminogen activator (PM, ECS)	0.1 (0.0–2.8)	0.6 (0.0–3.1)
F08h	U61696	Aminipeptidase B (C)	0.8 (0.4–2.0)	1.2 (0.4–2.2)
F01c	D85760	Guanine nucleotide-binding protein alpha 12 subunit (GNA12) (PM,C)	0.9 (0.6–1.6)	1.2 (0.9–1.9)
D04f	M30705	Serotonin 5HT2 receptor (PM)	0.0 (0.0–1.1)	0.2 (0.0–1.4)

(*continued*)

Table 6.6 Continued

Clontech	GenBank accession no.	Gene	Soybean	Sunflower
E06c	L27867	NEUREXOPHILIN 1 (NEUROPHILIN) (ECS)	0.8 (0.4–1.3)	1.2 (0.5–1.6)
B02n	U32497	Purinergic receptor P2X4, ligand-gated ion channel (PM)	0.1 (0.0–1.9)	0.6 (0.0–2.7)
B01k	D79981	Kidney specific organic anion transporter OAT-K1 (PM)	0.0 (0.0–0.3)	0.2 (0.0–1.8)
A08f	D14051	Inducible nitric oxide synthase (iNOSl); type II NOS (NM, G, M)	0.1 (0.0–2.0)	1.0 (0.1–2.6)
A10l	L20681	*c*-ets-1 proto-oncogene protein; p54 (N)	0.9 (0.5–2.1)	1.2 (0.7–2.0)
D12f	M18642	Coagulation factor II (thrombin) receptor (CF2R) (PM)	0.2 (0.0–1.7)	0.9 (0.0–2.6)
D11n	AF018957	Neuropilin (PM)	0.8 (0.5–2.2)	1.2 (0.7–3.2)
A11n	Z14117	Platelet-derived growth factor B-chain (PDGFb); c-sis (ECS)	0.8 (0.5–1.8)	1.2 (0.8–1.6)
F03h	M96159	Adenylyl cyclase type V (PM)	0.9 (0.6–1.3)	1.2 (0.5–2.3)

Note: Analyzed with GeneSpring software. Each diet mean (range) represents analysis of pups from three different litters at each of the four ages. Proteins associated with C, cytoplasm; P, peroxisome; PM, plasma membrane; ECS, extracellular secreted; ECM, extracellular matrix; N, nucleus; CS, cytoskeleton; NM, nuclear membrane, M, mitochondria; G, golgi; ER, endoplasmic reticulum. Genes shown in bold differ by diet, $p < 0.01$.

statistically for number of genes compared, because the intention of these studies was to identify all possibly relevant genes affected by changing brain DHA content. The genes found to be significantly affected by diet using a more conservative measure of significance ($p < 0.01$) are noted in bold in each table.

RESULTS

Effect of Diet on Brain DHA

Figure 6.1 shows the effect of 18:3n-3-replete and -deficient diets on the relative percent of DHA in whole brain total phospholipids at various ages. With the

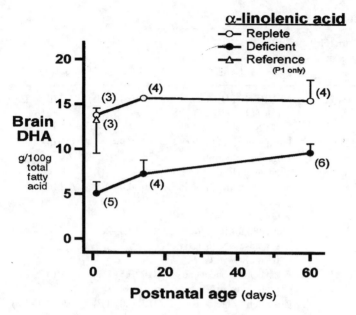

Figure 6.1 Brain DHA in the offspring of Sprague–Dawley rats fed AIN-73G (18:3n-3-replete diet) or modified AIN-73G containing sunflower oil in place of soybean oil (18:3n-3-deficient) diet from the start of pregnancy until the completion of weaning. At each age, individual brains from different litters were analyzed. The reference group shown for comparison at P1 was exposed to a typical, nutritionally undefined rat chow. After weaning, the pups remaining in each litter were raised to adulthood on the maternal diet.

reduction in brain DHA in the 18:3n-3-deficient group compared with the 18:3n-3 replete group, there was a reciprocal increase in total phospholipid DPA, 22:5n-6 (data not shown). No other brain fatty acids changed significantly.

Effect of Diet on Gene Expression Present in Anterior Cerebrum

Genes with greater expression in rats fed the 18:3n-3-replete compared with the 18:3n-3-deficient diet are shown in Table 6.3. The two genes that showed the most significant effect of diet ($p < 0.01$) were the amiloride-sensitive brain sodium channel (*BNaC1*; degenerin) and the organic anion transporter. Proteins from both of these genes are associated with the plasma membrane. Gene expression for a number of extracellularly secreted, extracellular matrix, cytosolic, peroxisomal, and plasma membrane proteins was significantly higher in rats exposed to the 18:3n-3-replete compared with the 18:3n-3-deficient diet (Table 6.3).

The genes shown in Table 6.4 had higher expression in the anterior cerebrum of rats fed the 18:3n-3-deficient compared with the 18:3n-3-replete diet. Genes that were most significantly elevated in association with lower DHA and increased DPA in brain ($p < 0.01$) were (1) the serotonin (5-hydroxytrypta-mine-6) receptor, (2) heparin-binding growth factor, and (3) G1/S-specific cyclin (Table 6.2). The products of these genes are associated with plasma membrane, extracellular secretion, and the cytoplasm, respectively. Gene expression for a number of other plasma membrane receptors (p value of $0.01-0.05$) was also significantly increased.

Effect of Diet on Gene Expression Present in Posterior Cerebrum

Twelve genes studied in the posterior cerebrum had higher expression in rats fed the 18:3n-3-replete compared with the 18:3n-3-deficient diet at a significance of $p < 0.01$ (Table 6.5). All 12 genes are associated with the plasma membrane. Among them were *GFRα1*, two ephrin-type A (*EphA3* and *EphA5*) receptors, two transforming growth factor beta receptors (*TGF-βI* and *-III*), and fibroblast growth factor 4.

Among animals fed an 18:3n-3-deficient compared with an 18:3n-3-replete diet, 19 genes were expressed in higher amounts during the first 21 days of age ($p < 0.01$) (Table 6.6). The genes coded for two nuclear proteins (Jun-B and p53 nuclear oncoprotein), three extracellular secreted proteins (IGF-BP2, peptide YY precursor, and natural killer cell protease 1), three cytosolic proteins (casein kinase II beta subunit, elongation factor 2, and ribosomal protein L12), and one mitochondrial protein (type 1 hexokinase). The remaining genes that had higher expression when brain DHA was lower were associated with the plasma membrane and included a number of hormone receptors and trk proto-oncogene precursor.

DISCUSSION

Our primary interest has been the effects and reversibility of brain changes caused by low accumulation of DHA during brain development. One approach we have used is to evaluate brain regional gene expression in response to 18:3n-3-deficient diets that reduce brain DHA. Because we were interested in developing testable hypotheses for mechanisms by which altered membrane DHA influences behavior, we considered genes of "high interest" to be those for which we found the same effect of diet across four preweaning ages. One advantage of this approach is that it has a statistically greater chance of identifying genes that are truly affected by changes in brain DHA. As can be seen in Tables 6.3–6.6, the expression of many genes appeared to be affected consistently by modifying brain DHA. A limitation of this approach is that it potentially ignores important genes that are affected by diet at a particular stage in development.

We have noted one particular pattern among these genes. Gene expression (gene, GenBank accession number) for three receptor tyrosine kinases [*EphA3* (Rek4), U69278; *EphA5* (EHK1), X78689; and *ErbB4*, U52531] was higher in rats fed the 18:3n-3-replete compared with the 18:3n-3-deficient diet. For ErbB4, expression was significantly increased in the posterior cerebrum at P1 and P7, and in the anterior cerebrum at 14 days (data not shown). These data suggest that *ErbB4* gene expression may be influenced by developmental stage. A fourth gene that showed higher expression in animals fed the soybean compared with the sunflower oil-containing diet was *GFRα1*. Although it is not itself a receptor tyrosine kinase, *GFRα1* must bind to glial cell line-derived neurotrophic factor (GDNF) before the receptor tyrosine kinase, c-RET, can bind to the complex and be activated.[29]

Tyrosine receptor kinases may (a) permanently associate with lipid rafts, (b) be recruited to lipid rafts after ligand binding, (c) be released from lipid rafts when activated, or (d) use lipid raft components for signal propagation or amplification without becoming associated with lipid rafts.[30] The three receptor tyrosine kinases that were upregulated in rats fed the 18:3n-3-replete compared with the 18:3n-3-deficient diet operate via the first mechanism and are attached to rafts by a GPI-anchor. The ephrin receptors, Rek4 (EphA3) and EHK1 (EphA5) bind to class A ephrins, and ErbB4 binds to neuroregulin in lipid rafts.[30] In the case of GFRα1, the protein is GPI-linked to the detergent-resistant portion of the plasma membrane. GDNF must translocate to the lipid raft and complex with GFRα1 before signaling can occur.[31]

EphA5 receptors are important in the development of topographic maps related to the developing visual system by directing axonal growth and/or reorganization between the retina and targets in the superior colliculus of the midbrain.[32,33] In addition, multiple Eph receptors and ligands are expressed in the hippocampus and its subcortical projection target, and expression of a truncated Eph receptor in mouse brain was shown to alter the hippocamposeptal topographic map.[34] Intrahippocampal infusion of an EphA antagonist immunoadhesin led to impaired performance on T-maze spontaneous alteration and context-dependent fear conditioning while activation by infusion of an agonist immunoadhesin resulted in enhanced performance on these tests.[35]

ErbB4 is a receptor for neuregulin.[36,37] ErbB4 is enriched at postsynaptic sites in the brain in PSD,[38] an electron-dense submembrane beneath post-synaptic membranes that is believed to be important for clustering of neurotrans-mitter receptors and regulation of receptor function.[39] Neuregulin-1 regulates survival of neurons, synaptogenesis, astrocyte differentiation, and microglial activation in the nervous and neuroendocrine systems.[40]

GDNF signaling requires c-RET and has been implicated in dopamine function at a number of different levels. GDNF is a well-known factor for survival of dopaminergic neurons,[41,42] increases expression of tyrosine hydroxylase, the rate-limiting enzyme for synthesis of dopamine,[43] and increases release of dopamine, possibly by increasing new functional synaptic terminals.[44] GFRα1 is highly expressed in dopamine neurons and in GABAergic neurons in substantia nigra and ventral tegmental areas.[45] Furthermore, adult midbrain dopamine neurons stimulated with GDNF were larger and had more dendritic extensions.[46] GDNF also has been shown to promote recovery from injury of the nigrostriatal dopamine system and to improve motor functions in animal models of Parkinson's disease.[46]

Chalon and coworkers[10] observed changes in brain dopamine systems in animals raised for several generations on an 18:3n-3-deficient diet. Conversely, behavioral changes in animals with lower brain DHA have been linked to dopaminergic function.[8,9] Moreover, such behavioral changes evincing dopaminergic function have been noted in rats exposed to 18:3n-3-deficient diets for relatively brief periods during development. In particular, Levant et al.[12] found abnormal dopamine-related behaviors such as less haloperidol-induced catalepsy and increased basal and amphetamine-stimulated activity in 18:3n-3-deficient compared with 18:3n-3-replete rats with our study design. Moreover, only one of these behavioral effects (less haloperidol-induced catalepsy) returned to normal when DHA was introduced into the diet at weaning and brain DHA and DPA were returned to normal[12] suggesting long-term programming of the dopaminergic system by lower brain DHA during development.

DHA has been shown to influence phase transition states of natural and artificial membranes, including detergent resistant regions of the plasma membranes. Salem et al.[47] showed that phospholipids with 22:5n-6 (DPA) in the sn-2 position have order parameters more like phospholipids with 18:1 in the sn-2 position than those with 22:6n-3 (DHA) in the sn-2 position. Shaikh et al.[48] investigated interactions of cholesterol and sphingomyelin in monolayers and bilayers containing either 18:1 or 22:6n-3 in the sn-2 position and concluded that lateral phase separation occurs in the presence of 22:6n-3 but not in the presence of 18:1-containing phosphatidylethanolamine. They suggested that DHA was needed to promote formation of stable lipid rafts in the plasma membrane and to phase separate from sphingomyelin–cholesterol microdomains.

The ability to promote phase separation from rafts may be critical to the mechanism by which DHA induced phospholipase D activation in lymphocytes. Diaz et al.[49] found that DHA activation of phospholipase D was inhibited by brefeldin A, suggesting ADP ribosylation (ARF)-dependence. They found a dose-dependent translocation of ARF to lipid soluble regions of the cell membranes in response to DHA. At the same time, the enzyme- and ARF-dependent phospholipase D activity were displaced from rafts toward detergent soluble membranes where ARF was present.

Not surprisingly, because DHA is localized in membrane phospholipids, many of the genes significantly affected by the amount of DHA in brain are associated with the plasma membrane. Quite a few of these are receptors, but some receptors had higher expression associated with normal brain DHA and others associated with lower brain DHA. The reader is referred to the tables for a complete list of genes in which expression was significantly affected by diets that changed brain DHA accumulation.

There are two prior reports of gene expression in rat brain after alterations in dietary n-3 fatty acid intake[49,50] Kitajka et al.[50] fed chow or chow with added perilla oil (high in 18:3n-3) or fish oil (high in DHA, 22:6n-3 and eicosa-pentaenoic acid or EPA, 20:5n-3) from conception to adulthood. All of the groups had a normal profile of total brain ethanolamine phosphoglyceride DHA and normal molecular species of DHA-containing ethanolamine phosphoglycer-ides.[50] Although changes in gene expression in the experimental groups when compared with the control group were found, these are better explained by the large increase in the proportion of energy intake from fat than by changes in either 18:3n-3 or n-3 long-chain polyunsaturated fatty acids (LC-PUFA) (DHA and EPA) intake.

In a different model that altered the ratios of brain n-6/n-3 LC-PUFA instead of the ratio of brain DHA (22:6n-3) to DPA (22:5n-6), Berger et al.[51] fed experimental diets containing arachidonic acid (AA, 20:4n-6) from fungal oil, DHA and EPA from fish oil or AA, DHA and EPA from both fungal oil and fish oil to mice. All diets were 18:2n-6- and 18:3n-3-replete and were con-trolled so that each contained the same amount of energy from fat. Although DPA (22:5n-6) was not reported, increases in DPA would not have been anti-cipated from the diets fed. Following 57 days on the various diets, some effects on gene expression in hippocampus were noted. Unlike our study, in which the majority of genes expressed differentially in relation to an 18:3n-3-deficient diet were associated with the plasma membrane, the genes noted to be influenced by changes in the n-6/n-3 LC-PUFA ratio were involved in nucleic acid binding.

In summary, we focused on genes upregulated or downregulated by dietary changes that influenced brain DHA across several postnatal ages prior to weaning. As far as we are aware, this is the first study that shows a cause and effect relationship between 18:3n-3-deficient diets that reduce brain DHA and increase brain n-3 DPA during development and gene expression in the anterior and posterior cerebral regions of the brain. Many of the proteins that showed differential expression have been associated with the plasma membrane and are known to be important for various types of signal transduction. One pattern observed was higher expression among genes for receptor tyrosine kinases that act as receptors for ligands located in detergent resistant membranes. These results suggest one possible mechanism by which alterations of DHA in plasma membranes may influence a large number of physiological responses, quite possibly including those related to organs and tissues besides the brain. Studies are currently underway to determine if the apparent changes in expres-sion of the brain tyrosine kinases and GFRα1 can be confirmed by RT-PCR or by quantification of the proteins encoded by the message.

CONCLUSIONS

Newborn infants fed diets without DHA (22:6n-3) have lower brain DHA than infants fed human milk, a dietary source of DHA. Results of randomized clinical studies have shown that infants fed DHA have more mature visual, attentional, and cognitive development. However, questions remain about the mechanism by which DHA influences these outcomes and the optimal timing of DHA accumulation. In an effort to target a likely mechanism, we screened for changes in gene expression in the anterior and posterior cerebrum of rats fed an alpha-linolenic acid (18:3n-3)-deficient diet compared with an 18:3n-3-replete diet. Clontech Rat 1.2 cDNA arrays were employed for the studies. Rat brain DHA was decreased by approximately 50% in the 18:3n-3-deficient compared with the 18:3n-3-replete group. The changes in gene expression reported here are limited to genes for which the change in expression was observed from birth to weaning. Many of the genes most significantly influenced by a diet that altered brain DHA ($p < 0.01$) were associated with the plasma membrane, and a large number of these were receptors for hormones and neurotransmitters. Four genes that had higher expression in rats with normal brain DHA during early development either were receptor tyrosine kinases (*EphA3, EphA5, ErbB4*) or were required for binding (*GFRα1*) and activation of a receptor tyrosine kinase (c-RET). The receptor tyrosine kinases identified in our studies operate through a common pattern (i.e., by attachment to a GPI-linked protein in detergent resistant regions of the plasma membrane also known as lipid rafts). GFRα1 is localized in the raft by virtue of a GPI anchor. Recruitment of the receptor tyrosine kinases to lipid rafts initiates the process that culminates in neurotrophic and other signaling. Studies are underway to determine if these changes can be confirmed by quantifying message and proteins.

ACKNOWLEDGMENTS

The authors which to acknowledge the Dr. Lisa Cui and Dr. Rajani Choudhuri for training in microarrays and Drs. Jeff Radel and Beth Levant for assistance with the animal model and helpful scientific discussions about the work.

REFERENCES

1. Svennerholm, L. Distribution and fatty acid composition of the major brain lipids in normal gray matter, white matter and myelin. J. Lipid Res. **1965**, *4*, 545–551.
2. Breckenridge, W.C.; Gombos, G.; Morgan, I.G. The lipid composition of adult rat bran synaptosomal membranes. Biochim. Biophys. Acta **1972**, *266*, 695–707.
3. Breckenridge, W.C.; Morgan, I.G.; Zanetta, J.P.; Vincendon, G. Adult rat brain synaptic vesicles II. Biochim. Biophys. Acta **1973**, *211*, 681–686.
4. Galli, C.; Trzeciak, H.I.; Paoletti, R. Effect of dietary fatty acids on the fatty acid composition of brain ethanolamine phosphoglyceride: reciprocal replacement of n-6 and n-3 polyunsaturated fatty acids. Biochim. Biophys. Acta **1971**, *248*, 449–254.
5. Neuringer, M.; Connor, W.E.; Lin, D.S.; Barstad, L.; Luck, S. Biochemical and functional effects of prenatal and postnatal omega-3 fatty acid deficiency on retina and brain in rhesus monkeys. Proc. Natl. Acad. Sci. USA **1986**, *83*, 4021–4025.
6. Neuringer, M.; Connor, W.E.; Van Petten, C.; Barstad, L. Dietary omega-3 fatty acid deficiency and visual loss in infant rhesus monkeys. J. Clin. Invest. **1984**, *73*, 272–276.

7. Yehuda, S. Behavioral effects of dietary fatty acids. In *Health Effects of Fish and Fish Oils*; Chandra, R.K., Ed.; ARTS Biomedical Publishers: St. John's, Newfoundland, 1989; 327–335.

8. Reisbick, S.W.; Neuringer, M.; Gohl, E.; Wald, R.; Anderson, G.J. Visual attention in infant monkeys: effects of dietary fatty acids and age. Dev. Psychol. **1997**, *33*, 387–395.

9. Reisbick, S.W.; Neuringer, M.; Hasnain, R.; Connor, W.E. Home cage behavior of rhesus monkeys with long-term deficiency of omega-3 fatty acids. Physiol. Behav. **1994**, *55*, 231–239.

10. Delion, S.; Chalon, S.; Herault, J.; Guilloteau, D.; Besnard, J.-C.; Durand, G. Chronic dietary α-linolenic acid deficiency alters dopaminergic and serotoninergic neurotransmission in rats. J. Nutr. **1994**, *124*, 2466–2476.

11. Chalon, S.; Delion-Vancassel, S.; Belzung, C.; Guilloteau, D.; Leguisquet, A.-M.; Besnard, J.-C.; Durand, G. Dietary fish oil induces changes in monoaminergic neurotransmission and behavior in rats. J. Nutr. **1998**, *128*, 2512–2519.

12. Levant, B.; Radel, J.D.; Carlson, S.E. Decreased brain docosahexaenoic acid during development alters dopamine-related behaviors in adult rats that are differentially affected by dietary remediation. Behav. Brain Res. **2004**, *152*, 49–59.

13. Clandinin, M.T.; Chappell, J.E.; Leong, S.; Heim, T.; Swyer, P.R.; Chance, G.W. Intrauterine fatty acid accretion rates in human brain: implications for fatty acid requirements. Early Hum. Dev. **1980**, *4*, 121–129.

14. Martinez, M. Developmental profiles of polyunsaturated fatty acids in the brain of normal infants and patients with peroxisomal diseases: severe deficiency of docosahexaenoic acid in Zellweger's and pseudo-Zellweger's syndromes. In *Health Effects of Three Polyunsaturated Fatty Acids in Seafoods*, World Review on Nutrition Dietetics; Simopoulos, A.P., Kifer, R.R., Martin, R.E., Barlow, S.M., Eds.; Karger: Basel, 1991; Vol. 66, 87–102.

15. Sanders, T.A.; Naismith, D.J. A comparison of the influence of breast-feeding and bottle-feeding on the fatty acid composition of the erythrocytes. Br. J. Nutr. **1979**, *41*, 619–623.

16. Putnam, J.C.; Carlson, S.E.; Devoe, P.W.; Barness, L.A. The effect of variations in dietary fatty acids on the fatty acid composition of erythrocyte phosphatidylcholine and phosphatidylethanolamine in human infants. Am. J. Clin. Nutr. **1982**, *36*, 106–114.

17. Carlson, S.E.; Werkman, S.H.; Rhodes, P.G.; Tolley, E.A. Visual-acuity development in healthy preterm infants: effect of marine oil supplementation. Am. J. Clin. Nutr. **1993**, *58*, 35–42.

18. Birch, E.E.; Birch, D.G.; Hoffman, D.R.; Uauy, R.D. Dietary essential fatty acid supply and visual acuity development. Invest. Ophthalmol. Vis. Sci. **1992**, *33*, 3242–3253.

19. Makrides, M.; Neumann, M.A.; Simmer, K.; Pater, J.; Gibson, R. Are long-chain polyunsaturated fatty acids essential nutrients in infancy? Lancet **1995**, *345*, 1463–146.

20. Carlson, S.E.; Ford, A.J.; Werkman, S.H.; Peeples, J.M.; Koo, W.W.K. Visual acuity and fatty acid status of term infants fed human milk and formulas with and without docosahexaenoate and arachidonate from egg yolk lecithin. Pediatr. Res. **1996**, *39*, 882–888.

21. Werkman, S.H.; Carlson, S.E. A randomized trial of visual attention of preterm infants fed docosahexaenoic acid until nine months. Lipids **1996**, *31*, 91–97.

22. Carlson, S.E.; Werkman, S.H. A randomized trial of visual attention of preterm infants fed docosahexaenoic acid until two months. Lipids **1996**, *31*, 85–90.

23. Farquharson, J.; Cockburn, F.; Patrick, W.A.; Jamieson, E.C.; Logan, R.W. Infant cerebral cortex phospholipid fatty-acid composition and diet. Lancet **1992**, *340*, 810–813.

24. Makrides, M.; Neumann, M.A.; Byard, R.W.; Simmer, K.; Gibson, R.A. Fatty acid composition of brain, retina, and erythrocytes in breast-fed and formula-fed infants. Am. J. Clin. Nutr. **1994**, *60*, 189–194.

25. Gibson, R.A.; Chen, W.; Makrides, M. Randomized trials with polyunsaturated fatty acid interventions in preterm and term infants: functional and clinical outcomes. Lipids **2001**, *36*, 873–883.

26. Uauy, R.; Hoffman, D.R.; Peirano, P.; Birch, D.G.; Birch, E.E. Essential fatty acids in visual and brain development. Lipids **2001**, *36*, 885–895.

27. Reeves, P.G.; Nielsen, F.H.; Fahey, G.C., Jr. AIN-93 purified diets for laboratory rodents: final report of the American Institute of Nutrition ad hoc writing committee on the reformulation of the AIN-76A rodent diet. J. Nutr. **1993**, *123*, 1939–1951.

28. Smuts, C.M.; Huang, M.; Mundy, D.; Plasse, T.; Major, S.; Carlson, S.E. A randomized trial of docosahexaenoic acid supplementation during the third trimester of pregnancy. Obstet. Gynecol. **2003**, *101*, 469–479.

29. Sanicola, M.; Hession, C.; Worley, D.; Carmillo, P.; Ehrenfels, C.; Walus, L.; Robinson, S.; Jaworski, G.; Wei, H.; Tizard, R.; Whitty, A.; Pepinsky, R.B.; Cate, R.L. Glial cell line-derived neurotrophic factor-dependent RET activation can be mediated by two different cell-surface accessory proteins. Proc. Natl. Acad. Sci. USA **1997**, *94*, 6238–6243.

30. Paratcha, G.; Ibanez, C.F. Lipid rafts and the control of neurotrophic factor signaling in the nervous system: variations on a theme. Curr. Opin. Neurobiol. **2002**, *12*, 542–549.

31. Tansey, M.G.; Baloh, R.H.; Milbrandt, J.; Johnson, E.M., Jr. GFR-α mediated localization of RET to lipid rafts is required for effective downstream signaling, differentiation and neuronal survival. Neuron **2000**, *2000*, 611–623.

32. Marcus, R.C.; Matthews, G.A.; Gale, N.W.; Yancoupolos, G.D.; Mason, C.A. Axon guidance in the mouse optic chiasm: retinal neurite inhibition by Ephrin "A"-expressing hypothalamic cells *in vitro*. Dev. Biol. **2000**, *221*, 132–147.

33. Brown, A.; Yates, P.A.; Burrola, P.; Ortuno, D.; Vaidya, A.; Jessell, T.M.; Pfaff, S.L.; Oleary, D.D.; Lemke, G. Topographic mapping from the retina to the midbrain is controlled by relative but not absolute levels of EphA receptor signaling. Cell **2000**, *102*, 77–88.

34. Yue, Y.; Chen, Z.-Y.; Gale, N.W.; Blair-Flynn, J.; Hu, T.-J.; Ye, X.; Cooper, M.; Crockett, D.P.; Yancopoulos, G.D.; Tessarollo, L.; Zhou, R. Mistargeting hippocampal axons by expression of a truncated Eph receptor. Proc. Natl. Acad. Soc. **2002**, *99*, 10777–10782.

35. Gerlai, R.; Shinsky, N.; Shih, A.; Williams, P.; Winer, J.; Armanini, M.; Cairns, B.; Winslow, J.; Gao, W.; Phillips, H.S. Regulation of learning by EphA receptors: a protein targeting study. J. Neurosci. **1999**, *19*, 9538–9549.

36. Ma, L.; Huang, Y.Z.; Pitcher, G.M.; Valtschanoff, J.G.; Ma, Y.H.; Feng, F.Y.; Lu, B.; Xiong, W.C.; Salter, M.W.; Weinberg, R.J.; Mei, L. Ligand-dependent recruitment of the ErbB4 signaling complex into neuronal rafts. J. Neurosci. **2003**, *23*, 3164–3175.

37. Yau, H.J.; Wang, H.F.; Lai, C.; Liu, F.C. Neural development of the neuregulin receptor ErbB4 in the cerebral cortex and the hippocampus: preferential expression of interneurons tangentially migrating from the ganglionic eminences. Cereb. Cortex **2003**, *13*, 252–264.

38. Huang, Y.Z.; Won, S.; Ali, D.W.; Wang, Q.; Tanowtiz, M.; Du, Q.S.; Pelkey, K.A.; Yang, D.J.; Xiong, W.C.; Salter, M.W.; Mei, L. Regulation of neuregulin signaling by PSD-95 interacting with ErbBr at CNS synapses. Neuron **2000**, *26*, 443–455.

39. Kennedy, M.B. The postsynaptic density at glutaminergic synapses. Trends Neurosci. **1997**, *20*, 264–268.

40. Crone, S.A.; Lee, K.F. Gene targeting reveals multiple essential functions of the neuregulin signaling system during development of the neuroendocrine and nervous systems. Ann. NY Acad. Sci. **2002**, *971*, 547–553.

41. Bowenkamp, K.E.; Hoffman, A.F.; Gerhardt, G.A.; Henry, M.A.; Biddle, P.T.; Hoffer, B.J.; Granholm, A.C. Glial cell line-derived neurotrophic factor supports survival of injured midbrain dopaminergic neurons. J. Comp. Neurol. **1995**, *355*, 479–489.

42. Granholm, A.C.; Reyland, M.; Albeck, D.; Sanders, L.; Gerhardt, G.A.; Hoernig, G.; Chen, L.; Westphal, H.; Hoffer, B. Glial cell line-derived neurotrophic factor is essential for postnatal survival of midbrain dopamine neurons. J. Neurosci. **2000**, *20*, 3182–3190.

43. Xiao, H.; Hirata, Y.; Isobe, K.; Kiuchi, K. Glial cell line-derived neurotrophic factor up-regulates the expression of tyrosine hydroxylase gene in human neuroblastoma cell lines. J. Neurochem. **2002**, *82*, 801–808.

44. Bourque, M.J.; Trudeau, L.E. GDNF enhances the synaptic efficacy of dopaminergic neurons in culture. Eur. J. Neurosci. **2000**, *12*, 3172–3180.

45. Sarabi, A.; Hoffer, B.J.; Olson, L.; Morales, M. GFRalpha-1 mRNA in dopaminergic and nondopaminergic neurons in the substantia nigra and ventral tegmental areas. J. Comp. Neurol. **2001**, *441*, 106–117.

46. Grondin, R.; Gash, D.M. Glial cell line-derived neurotrophic factor (GDNF): a drug candidate for the treatment of Parkinson's disease. J. Neurol. **1998**, *245*, 35–42.

47. Salem, N., Jr.; Litman, B.; Kim, H.-Y.; Gawrisch, K. Mechanisms of action of docosahexaenoic acid in the nervous system. Lipids **2001**, *36*, 945–959.

48. Shaikh, S.R.; Brzustowicz, M.R.; Gustafson, N.; Stillwell, W.; Wassall, S.R. Monounsaturated PE does not phase-separate from the lipid raft molecules sphingomyelin and cholesterol: role for polyunsaturation? Biochemistry **2002**, *41*, 10593–10602.

49. Diaz, O.; Berquand, A.; Dubois, M.; Di Agostino, S.; Sette, C.; Bourgoin, S.; Lagarde, M.; Nemoz, G.; Prigent, A.-F. The mechanism of docosahexaenoic acid-induced phospholipase D activation in human lymphocytes involves exclusion of the enzyme from lipid rafts. J. Biol. Chem. **2002**, *277*, 39368–39378.

50. Kitajka, K.; Puskas, L.G.; Zvara, A.; Hackler, L.; Barcelo-Coblijn, G.; Yeo, Y.K.; Farkas, T. The role of n-3 polyunsaturated fatty acids in brain: modulation of rat brain gene expression by dietary n-3 fatty acids. Proc. Natl. Acad. Sci. USA **2002**, *99*, 2619–2624.

51. Berger, A.; Mutch, D.M.; German, J.B.; Roberts, M.A. Dietary effects of arachidonate-rich fungal oil and fish oil on murine hepatic and hippocampal gene expression. Lipids in Health and Disease **2002**, *1*, 2 (http://www.Lipidworld.com/content/1/1/2).

7

A Genomic Approach to Studying Leptin Signaling in a Hypothalamic Cell Line

Wiweka Kaszubska*

*Metabolic Disease Research, Abbott Laboratories,
Abbott Park, IL, USA*

Paul Hessler and Paul E. Kroeger

*Genomics and Molecular Biology, Global Pharmaceutical Research Division,
Abbott Laboratories, Abbott Park, IL, USA*

INTRODUCTION

The prevalence of obesity in adults and children in the US and some European countries is growing at an alarming rate.[1] Several components of the mechanism that regulate energy homeostasis (the balance between food intake, storage and energy expenditure) have been identified.[2] One of the key elements is the satiety hormone, leptin.[3] Leptin, a 16 kDa hormone, is secreted from the adipocytes into circulation in proportion to body fat mass and reports to the brain on the status of energy reserves.[4] Even though the leptin receptor (lepR) is broadly expressed it was shown that its selective deletion in neurons, not in peripheral tissue, leads to obesity.[5] The activation of lepR, localized in the brain mainly in the hypothalamus, is believed to initiate a cascade of receptor-proximal and -distal neuropeptide-transmitted events that lead to the upregulation of anorexigenic (appetite suppressing) peptides and to the downregulation of orexigenic (appetite stimulating) ones.[6] In addition to its effects on food intake and body weight, leptin is known to regulate reproduction, carbohydrate metabolism, and bone formation.[7]

Current affiliation: Global Product Development, Serono International S.A., Geneva, Switzerland.

Lack of functional leptin produces an obese and diabetic phenotype in rodents that can be normalized by the addition of exogenous leptin.[8–10] The condition of human obesity, however, does not seem to be generally character-ized by leptin deficiency since obese individuals have elevated plasma leptin levels and are generally resistant to pharmacological treatment with leptin.[11] The molecular mechanism of leptin resistance in humans is not completely understood but may be due to a defect in leptin transport across the blood–brain barrier resulting in reduced circulating leptin level reaching its receptor in the hypothalamus or an unidentified defect in leptin signal transduction pathway.[12] Continuing efforts to identify the negative and positive regulators of leptin signaling should aid in defining the cause of leptin resistance and might lead to the development of more effective antiobesity therapeutics. In that regard, equally important, would be further understanding of which down-stream genes are regulated by leptin.

A number of microarray studies addressing the effects of leptin on gene expression in peripheral tissues have been reported. Differentially expressed genes were identified in white adipose tissue of wild type and ob/ob, genetically leptin-deficient mice treated with leptin.[13] Another study identified changes in gene expression levels in diet-induced obese rats that have elevated levels of circulating leptin because of increased fat mass.[14] Transcriptional profiling in ob/ob mouse liver tissue also revealed gene expression changes on leptin treat-ment.[15,16] Reports of genomic studies in the brain and especially in the hypo-thalamic tissue are more rare. It is commonly observed that genes of interest are localized to specific nuclei in the hypothalamus. If RNA samples are prepared in such a way as to include the surrounding tissue, then a dilution effect might occur with an unfavorable signal-to-noise ratio. Nonetheless, one study compared hypothalamic gene expression profiles of freely fed and fasted rats[17] and another compared hypoglycemic and normal mice.[18] We are not aware of any published reports of leptin-induced gene expression profiles specifically in the hypothalamus. However, leptin-induced transcripts were identified in neuronal GT1-7 cell line by subtractive hybridization.[19] In the study reported here we used GT1-7 cells to interrogate leptin-induced gene expression using microarray analysis. In addition, we determined the effects of a negative regulator, protein tyrosine phosphatase 1B (PTP1B), on the leptin signal transduction pathway using microarray analysis.

MATERIALS AND METHODS

DNA Constructs, Cell Culture, and Transfections

Expression vector encoding the full-length human lepR was previously described.[34] The full-length human PTP1B (a gift from Metabolex, Hayward, CA) was cloned into pcDNA3.1/Zeo(+). GT1-7 cells (licensed from the Salk Institute, San Diego, CA) were cultured in DMEM with 4.5 mg/mL glucose and 10% (v/v) fetal bovine serum (GIBCO BRL, Grand Island, NY). All trans-fections were performed with Lipofectamine 2000 (GIBCO BRL, Grand Island,

NY) according to the manufacturer's specifications. Generally, transfections were carried out in a six-well cell culture dish using 2 μg of lepR expression vector and varied amounts of PTP1B expression vector. Cells were incubated with the DNA/transfection reagent mix for 18–24 h. They were allowed to recover in regular growth media and 48 h posttransfection transferred to serum-free media containing indicated concentrations of leptin.

Western Blotting and Antibodies

For Western blot analysis, 48 h after transfection, cells were starved in a serum-free media for 18 h and stimulated with leptin for indicated length of time. Representative experiments of at least two independent transfections are shown. Transfected cells were harvested in 1× TBS, 10% triton X-100, 0.5% nonidet P-40, 0.25% sodium deoxycholate, 1 mM EGTA, 1 mM NaF, 0.2 mM NaV_3O_4, 1 μM microcystin, 1 mM AEBSF, and a complete EDTA-free protease inhibitor cocktail tablet (Roche, Indianapolis, IN). Protein concentration was determined using the BCA assay reagent (Pierce, Rockford, IL). Cell lysates containing equal amounts of protein were denatured and resolved by SDS–PAGE NuPage System (Invitrogen, Carlsbad, CA). Proteins were transferred to nitrocellulose membranes in 2× NuPage transfer buffer (Invitrogen) and 15% methanol. Membranes were blocked for 30 min at room temperature in 1× TBS containing 5% milk and incubated with a primary antibody at 4°C overnight. All subsequent washes were done in water and 1× TBS containing 0.1% Tween-20 (Sigma, St. Louis, MO) at room temperature. Membranes were incubated with HRP-conjugated secondary antibody in 1× TBS containing 5% milk for 1–2 h at room temperature and washed as mentioned earlier. The blots were developed with ECL-Plus reagents (Amersham-Pharmacia, Piscataway, NJ) according to the manufacturer's protocol. They were imaged and quantified by the STORM 860 software (Molecular Dynamics, Sunnyvale, CA). The following antibodies were purchased from Upstate Biotechnology (Lake Placid, NY): antiphospho-STAT3 (Y704), anti-STAT3, antiphospho-STAT5A/B (Y694/Y699), anti-STAT5. In addition, anti-JAK2 (pYpY 1007/1008) and anti-JAK2 were purchased from Biosource International (Camarillo, CA). Anti-PTP1B (AB-1) antibody was from Oncogene Research Products (San Diego, CA).

Microarray Analysis

Total RNA (5 μg) was converted into single-stranded cDNA and then to double-stranded cDNA using a Superscript Double Stranded cDNA Synthesis Kit (Invitrogen) with 100 pmol of an oligo(dT)24-primer containing a T7 RNA promoter site as described in the Affymetrix GeneChip Expression Analysis Technical Manual. The double-stranded cDNA was purified with phenol/chloroform extraction followed by ethanol precipitation and used to generate biotinylated cRNA using the Bioarray high-yield RNA transcript labeling kit with T7 polymerase (Enzo Diagnostics, Farmingdale, NY). Each cRNA sample (20 μg), purified with the RNeasy kit (Qiagen, Chatsworth, CA), was fragmented and hybridized to a single microarray chip as described in the Affymetrix GeneChip Expression Analysis Technical Manual. Each probe was tested in a GeneChip Test3 Array to ensure full-length transcripts and then hybridized to a Murine Genome U74Av2 array. Arrays were washed in a GeneChip Fluidics Station 400 and scanned in an Agilent GeneArray Scanner. Expression profiles were analyzed using Rosetta Resolver 3.2 (Rosetta Informatics, Kirkland, WA) analysis system. Gene specific annotation information regarding putative function was obtained from the NetAFFX web site maintained by Affymetrix (http://www. affymetrix.com) and is based on the gene annotation analysis of the GO consortium (http://www.geneontology.org).

Quantitative RT-PCR

The relative levels of SOCS3 and c-*fos* mRNAs were assessed by quantitative RT-PCR (TaqMan^{TM}) technology. Gene specific (GS) TaqMan^{TM} primers and probes were designed using the Primer Express program (Perkin–Elmer, Boston, MA) and synthesized with standard phosphoramidite chemistry. All Taq-Man^{TM} probes were labeled with the reporter fluorescein at the 5'-end and with the quencher tetramethylrhodamine at the 3'-end. The sequences of the forward and reverse primers and the probe for each gene were as follows: 5'-CTCCAAAAGCGAGTACCAGCT-3', 5'-CAGTAGAATCCGCTCTCCTGCA-3', 5'-TGGTGAACGCCGTGCGCA-3' for SOCS-3; 5'-CAACGAGCCCTCCT CCGACT-3', 5'-TGCCTTCTCTGACTGCTCACA-3', 5'-TGAGCTCACCCAC GCTGCTGGCC-3' for c-*fos*; 5'-TGTGTCCGTCGTGGATCTGA-3', 5'-CC TGCTTCACCACCTTCTTGA-3', 5'-GCCGCCTGGAGAAACCTGCCAA GTAT-3' for GAPDH. Each reaction was performed in triplicate in a final volume of 25 μL as follows: 1× Platinum Quantitative PCR Thermoscript buffer (Invitrogen), 5.5 mM MgCl$_2$, 400 nM GS primers, 100 nM GS probe, 100 nM 28S rRNA primers, 150 nM 28S rRNA probe, 0.5 U Thermoscript polymerase mix (Invitrogen), and 100 ng RNA previously treated with DnaseI. The reactions were incubated for 30 min at 50°C, followed by 5 min at 95°C, and then 40 cycles of 15 s at 95°C and 1 min at 60°C. Data was collected during each extension phase of the PCR reaction and analyzed with the ABI-7700 SDS software package. Each gene of interest was amplified separately and standardized to 28S rRNA copy number determined by the $\Delta\Delta$Ct method (Perkin–Elmer-ABI Prism 7700 Users Bulletin Number 2). The final fold differences in expression were calculated relative to the corresponding leptin stimulated or not stimulated vector control.

RESULTS AND DISCUSSION

Leptin Signaling Pathway

Structurally, lepR belongs to the class I cytokine-receptor superfamily.[20] Like other members of this family, it does not possess an intrinsic tyrosine kinase activity but recruits JAK2 to direct the phosphorylation of the proteins that comprise an active leptin-signaling complex.[21] Upon ligand binding and activation of JAK2, two tyrosine residues in the intracellular domain of lepR become phosphorylated and serve as docking sites for signaling molecules.[22,23] One of the phosphorylated residues (Tyr1138), binds STAT3, a member of the family of signal transducers and activators of transcription.[24] Although several STAT protein isoforms can be activated by leptin in different cell lines,[24,25] STAT3 is activated by leptin in the mouse hypothalamus.[26] Upon tyrosine phosphorylation by activated JAK2, STAT3 dimerizes and translocates to the nucleus where it regulates gene transcription.[21] One of the genes that is activated by STAT3 is its own negative feedback regulator, suppressor-of-cytokine-signaling-3 (SOCS3). Leptin induces the expression of SOCS3 mRNA in transfected cell lines[27,28] and *in vivo* in the mouse hypothalamus.[29] Furthermore, SOCS3 is a potential mediator of leptin resistance as its overexpression *in vitro* blocks lepR signaling.[29,30]

It is believed that the JAK–STAT pathway is responsible for the weight-reducing effects of leptin. A mutation in lepR resulting in expression of a truncated form of the receptor that is unable to activate the JAK–STAT pathway is responsible for the obese and diabetic phenotype of db/db mice.[25] Genetically engineered knock-in mice carrying a mutation in the STAT-binding site of lepR have also been shown to be hyperphagic and overweight.[31] In addition to the JAK–STAT pathway, recent evidence suggests that neuronal leptin signaling can utilize the insulin receptor substrate phosphatidylinositol-3 kinase (IRS-PI3K) pathway as well. There is a growing body of evidence that the leptin and insulin signaling pathways "cross-talk".[32]

Leptin Signaling in the GT1-7 Hypothalamic Cell Line

The leptin signaling pathway has been studied in a number of cell lines. However, the nonavailability of leptin-responsive hypothalamic cell lines has greatly impeded leptin receptor signaling studies in the context of the brain. A neuronal GnHR-secreting cell line, GT1-7, derived from an SV40 T-antigen expressing tumor of the hypothalamus has been described[33] and was chosen as an *in vitro* model for leptin signaling studies reported here. Although GT1-7 cells do not express an endogenous leptin receptor they do support the leptin-signaling pathway.[34,35] We have demonstrated that leptin stimulation of GT1-7 cells transfected with lepR resulted in selective phosphorylation of endogenous STAT3 (Fig. 7.1). Tyrosine phosphorylation of STAT3 reached a maximum between 10 and 30 min following leptin stimulation and declined thereafter. In addition, despite detectable endogenous levels, STAT5 was not significantly tyrosine phosphorylated in response to leptin. This finding is consistent with the observation that, in the mouse hypothalamus, leptin signals through STAT3.[26]

Upon tyrosine phosphorylation STAT3 dimerizes and translocates to the nucleus where it regulates gene transcription.[21] Another indication that

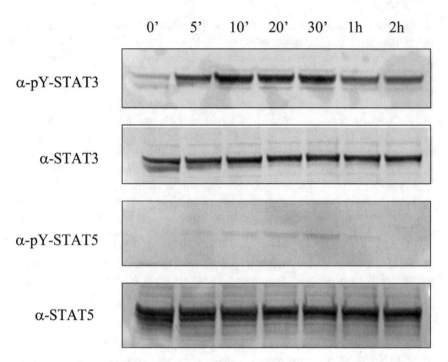

Figure 7.1   Time course of endogenous STAT3 and STAT5 tyrosine phosphorylation in leptin-stimulated GT1-7 cells. GT1-7 cells, transiently transfected with lepR, were stimulated with leptin (100 ng/mL) and harvested at indicated times. Cell lysates were normalized for protein content and resolved by SDS–PAGE. Western blot analysis was performed with listed antibodies. (Reprinted from Kaszubska et al.,[35] with permission from Elsevier.)

GT1-7 cells express appropriate leptin signaling molecules is the observation of leptin-dependent transcriptional regulation in these cells. GT1-7 cells were cotransfected with a vector encoding lepR and a STAT-responsive luciferase reporter gene. Stimulation with increasing concentrations of leptin resulted in a dose-dependent activation of the luciferase gene (data not shown). Thus GT1-7 cells transfected with lepR are capable of leptin-dependent STAT3 activation leading to transcriptional activation of a STAT-dependent reporter gene. These results indicate that GT1-7 cells constitute a suitable *in vitro* system for the study of leptin signaling.

Microarray Analysis of Leptin-Regulated Genes in GT1-7 Cells

GT1-7 cells transiently transfected with lepR were treated with leptin (10 ng/mL) for 1 h. The cells were harvested and total RNA was analyzed using Affymetrix microarrays (U74Av2) and the Rosetta Resolver system as described in the "Materials and Methods" section. Initial inspection of the leptin treated vs.

untreated control samples demonstrated that the expression of ∼300 sequences was affected, although many of the changes were relatively small (less than twofold). This represents immediate early transcriptional events and is consistent with previous studies demonstrating that gene expression changes in the hypothalamus are generally modest.[18] A plot of the +leptin/−leptin ratio vs. microarray probe intensity for each gene, provides an overview of the results and demonstrates that more genes were upregulated (229) than downregulated (75) by leptin treatment when a *p*-value of 0.05 is used as a cut-off (Fig. 7.2). An initial inspection of the data demonstrated that SOCS3, a known target of leptin signaling.[27,28] as well as SOCS1 and SOCS2, were some of the most highly regulated genes and provided confidence that we were indeed examining genes affected by leptin. Additionally, of the genes that were most affected, there was a preponderance of known transcriptional regulators including c-*fos*, Jun, C/EBP δ, ATF3, and bHLHb2, suggesting that the regulators of transcription are themselves targets of leptin signaling. Two of these factors, ATF3[36] and

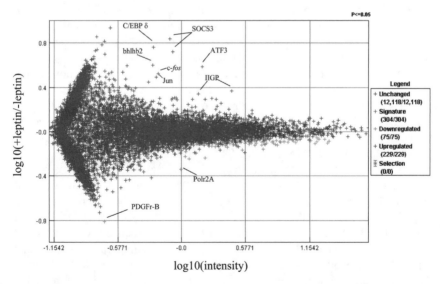

Figure 7.2 Microarray analysis of GT1-7 cells transfected with lepR and treated with leptin (10 ng/mL) for 1 h or not treated. For each sequence on the U74Av2 mouse microarray the +leptin/−leptin ratio is plotted against the fluorescent intensity of the probe set as an indication of abundance. The threshold for significance was set at $p \leq 0.05$. Specific genes identified include: CCAAT/enhancer binding protein (C/EBP δ), suppressor-of-cytokine-signaling-3 (SOCS3), activating transcription factor-3 (ATF3), basic helix loop helix binding protein 2 (bhlhb2), FBJ oncogene homologue (c-*fos*), interferon induced GTPase (IIGP), platelet derived growth factor receptor beta polypeptide (PDGFr-B), Jun, oncogene (Jun), and RNA polymerase polypeptide A (Polr2A).

bHLHb2,[37] have been identified primarily as repressors of transcription via their interaction with ATF/CRE and E-box sequences, respectively, suggesting transcriptional inhibition processes as well as activation are a result of leptin signaling. Further studies with extended treatments will elucidate the effects of these early regulatory events.

To understand the potential relationships between leptin-regulated genes we took advantage of the growing body of gene annotation information to determine if there were any regulatory themes that would permit the functional grouping of genes. A fold change cut-off of 1.5 with a *p*-value of ≤0.05 provided an acceptable number of sequences (Table 7.1) that were considered further. The Affymetrix probe set ID for each sequence was entered into the NetAFFX website and gene ontology (GO) annotation information was retrieved and organized. Though not every gene is annotated with respect to function there were several major groups of proteins found to be regulated by leptin treatment and these are arranged into functional groups in Table 7.2 (all available GO annotation information is in Table 7.1). Of the 91 sequences examined there were 19 related to transcriptional regulation and nucleic acid binding, 7 related to GTPase activity or GTP binding, and 13 related to the process of signal transduction. In many instances these proteins are considered members of multiple GO terms, but for the purpose of this discussion they have been grouped based on their primary functional classification. Given the small number of genes considered these groups represent significant functional associations.

The effects of leptin on transcriptional regulators suggested that one of the earliest events in leptin signaling is the activation of factors that can modulate other pathways through *de novo* synthesis or repression of mRNAs (Table 7.2). The JAK–STAT branch of the leptin signal transduction pathway is believed to be essential for mediating leptin effects on appetite and body weight.[25] Upon binding to pTyr1138 of lepR, STAT3 becomes tyrosine phosphorylated by JAK2 and translocates to the nucleus, where it regulates gene expression. In our microarray analysis STAT3 was modestly induced at the transcriptional level (+1.5-fold), and consistent with its activation via phosphorylation (Fig. 7.1) we observed that C/EBP δ, a known STAT3 target gene[38,39] was upregulated 5.7-fold. C/EBP δ activation has been associated with stress and inflammatory responses in a number of tissues including those of neuronal origin.[40] Glucocorticoids have been shown to activate C/EBP δ[41] and have been implicated in learning and memory processes.[42] ATF3 and bHLHb2 were induced +3.8- and +4.0-fold, respectively. Both of these factors are generally considered to be negative regulators of gene expression[36,37] and as such may repress gene expression in a leptin-dependent manner in GT1-7 cells.

In addition, there were a number of genes induced or repressed by leptin treatment that were identified in the GO annotation as signal transduction-related suggesting effects on multiple pathways. For example, we noted that the PDGF-beta receptor (−6.5-fold), oncostatin-M receptor (+1.8-fold), and insulin-receptor-substrate-3 (−4.1-fold) protein were regulated (Table 7.2).

Table 7.1 Leptin Regulated Genes

Affy sequence code	Sequence description	Fold change	p-Value	GO annotation
104689_at	E4F transcription factor 1	7.12	0.05	Regulation of transcription, DNA-dependent
92232_at	Suppressor of cytokine signaling 3	6.78	5.17×10^{-13}	Intracellular signaling cascade
104750_at	Interferon gamma inducible protein	6.47	0.022	GTP binding
160894_at	CCAAT/enhancer binding protein (C/EBP), delta	5.73	3.89×10^{-11}	Regulation of transcription, DNA-dependent
162206_f_at	Suppressor of cytokine signaling 3	5.21	1.38×10^{-19}	Intracellular signaling cascade
99955_at	Numb gene homolog (*Drosophila*)	4.52	0.016	Neurogenesis
104282_at	RIKEN clone:9930106P14	4.00	0.041	
97204_s_at	RIKEN cDNA 1110003P16 gene	3.97	0.047	
104701_at	Basic helix–loop–helix domain containing, class B2	3.95	0.000000317	Regulation of transcription, DNA-dependent
103696_r_at	RIKEN cDNA C330007P06 gene	3.94	0.016	
161371_r_at	Protein tyrosine phosphatase, receptor type, K	3.92	0.017	Protein amino acid dephosphorylation
104610_at	Similar to KIAA0595 protein	3.89	0.047	

(continued)

Table 7.1 Continued

Affy sequence code	Sequence description	Fold change	p-Value	GO annotation
104156_r_at	Activating transcription factor 3	3.88	0.00434	Regulation of transcription, DNA-dependent
96906_at	Ethanol induced 6	3.84	0.018	GTPase activator
98001_at	Rho guanine nucleotide exchange factor (GEF) 1	3.82	0.036	
104155_f_at	Activating transcription factor 3	3.81	1.54×10^{-24}	Regulation of transcription, DNA-dependent
103791_at	NMDA receptor-regulated gene 1	3.75	0.05	Transferase activity
98813_at	Reticuloendotheliosis oncogene	3.75	0.038	Regulation of transcription, DNA-dependent
161232_r_at	GPI-anchored membrane protein 1	3.65	0.021	
93430_at	Chemokine orphan receptor 1	3.59	0.036	G-protein coupled receptor protein signaling pathway
161317_r_at	Single WAP motif protein 2	3.58	0.049	
99475_at	Suppressor of cytokine signaling 2	3.49	0.016	Signal transduction
103393_at	RIKEN cDNA 5730470C09 gene	3.40	0.00752	
103495_at	ESTs	3.40	0.048	
160901_at	FBJ osteosarcoma oncogene	3.31	0.000000947	Regulation of transcription, DNA-dependent
94355_at	DNA segment, Chr 2, ERATO Doi 435, expressed	3.28	0.046	Regulation of transcription, DNA-dependent

Probe	Gene	Fold	p-value	Function
102362_i_at	Jun-B oncogene	3.08	0.035	Regulation of transcription, DNA-dependent
161567_r_at	Transducer of ErbB-2.1	3.04	0.029	Transferase activity
93557_at	Selenophosphate synthetase 2	2.95	0.0091	Helicase activity
95944_at	RIKEN cDNA 2810407E23 gene	2.95	0.02	
103015_at	B-cell leukemia/lymphoma 6	2.89	0.02	Regulation of transcription, DNA-dependent
103314_at	DNA segment, Chr 13, ERATO Doi 275, expressed	2.58	0.00771	Regulation of transcription, DNA-dependent
102384_at	RIKEN cDNA 2610209L14 gene	2.54	0.000695	
93104_at	B-cell translocation gene 1, antiproliferative	2.51	0.018	Cell growth and/or maintenance
161817_f_at	RIKEN cDNA 4930422J18 gene	2.45	0.012	Intracellular signaling cascade
98818_at	Nuclear receptor subfamily 3, group C, member 1	2.33	0.018	Regulation of transcription, DNA-dependent
96764_at	Interferon-inducible GTPase	2.31	2.7×10^{-13}	Cytokine and chemokine mediated signaling pathway
103963_f_at	Interferon-inducible GTPase	2.16	0.0000147	GTPase activity
92648_at	Syntaxin binding protein 3	2.09	0.043	Intracellular protein transport
97409_at	Interferon inducible protein 1	2.00	0.000773	

(*continued*)

Table 7.1 Continued

Affy sequence code	Sequence description	Fold change	p-Value	GO annotation
102363_r_at	Jun-B oncogene	1.99	0.019	Regulation of transcription, DNA-dependent
102381_at	Expressed sequence AU018108	1.98	0.000066	Fatty acid metabolism
102906_at	T-cell specific GTPase	1.98	1.64×10^{-8}	GTP binding
94820_r_at	Cyclin I	1.97	0.000546	Regulation of cell cycle
92534_at	GTP binding protein (gene overexpressed in skeletal muscle)	1.94	0.00000496	Small GTPase mediated signal transduction
100130_at	Jun oncogene	1.93	0.000748	Regulation of transcription, DNA-dependent
98569_at	RIKEN cDNA 1110030N17 gene	1.93	0.02	Mitochondrial inner membrane
92832_at	Suppressor of cytokine signaling 1	1.92	0.00644	Signal transduction
100074_at	RIKEN cDNA 2400003B06 gene	1.91	0.015	Intracellular protein transport
95577_at	Expressed sequence AI314180	1.91	0.032	
102315_at	Testis expressed gene 292	1.86	0.028	
103330_at	Spermatid perinuclear RNA binding protein	1.85	0.044	Single-stranded RNA binding
104316_at	*Mus musculus*, clone IMAGE:1379624, mRNA, partial cds	1.84	0.046	
104373_at	Expressed sequence AI646570	1.83	0.025	
101524_at	RIKEN cDNA 2610510D13 gene	1.82	0.017	

Probe ID	Gene	Ratio	p-value	Function
102255_at	Oncostatin receptor	1.81	0.05	Cell surface receptor linked signal transduction
160247_at	Ubiquitin-conjugating enzyme E2 variant 2	1.80	0.02	Ligase activity
160517_at	Lamin B1	1.78	0.0000017	Intermediate filament
160092_at	Interferon-related developmental regulator 1	1.75	0.027	
94382_at	RIKEN cDNA 1110057K04 gene	1.75	0.043	Lipid metabolism
92841_f_at	Chromogranin B	1.74	0.041	Extracellular space
93058_at	Eukaryotic translation initiation factor 1A	1.74	0.000463	Translation factor activity
160933_at	Interferon gamma induced GTPase	1.72	0.00281	GTPase activity
100088_at	Protein phosphatase 1, catalytic subunit, beta isoform	1.67	0.000000122	Protein serine/threonine phosphatase activity
160200_at	RIKEN cDNA 3230401D17 gene	1.66	0.0000913	
96777_at	Splicing factor 3b, subunit 1, 155 kDa	1.65	0.031	mRNA splicing
94433_at	RIKEN cDNA 5033402L14 gene	1.64	0.0002	
95705_s_at	Actin, beta, cytoplasmic	1.57	0.00000116	Cytoskeleton organization and biogenesis
98410_at	Interferon-g induced GTPase	1.57	0.00018	GTPase activity
98881_at	RIKEN cDNA 3110052N05 gene	1.57	0.016	
162172_f_at	Neural precursor cell expressed, developmentally downregulated gene 4	1.55	0.00032	Ubiquitin ligase complex
95914_at	RIKEN cDNA 6720461J16 gene	1.55	0.047	
100148_at	CCCTC-binding factor	1.54	0.019	DNA methylation
160397_at	IK cytokine	1.54	0.044	
94832_at	Heterogeneous nuclear ribonucleoprotein H2	1.53	0.00121	Nucleic acid binding

(continued)

Table 7.1 Continued

Affy sequence code	Sequence description	Fold change	p-Value	GO annotation
97689_at	Coagulation factor III	1.52	0.00433	Blood coagulation
99100_at	Signal transducer and activator of transcription 3	1.51	0.042	Regulation of transcription, DNA-dependent
99985_at	Thioredoxin reductase 1	1.50	0.04	Electron transport
102036_at	RIKEN cDNA 1810004B07 gene	−1.53	0.048	DNA binding
98552_at	RIKEN cDNA 2600009M07 gene	−1.69	0.021	
96293_at	RIKEN cDNA 2410015N17 gene	−1.75	0.00882	
96069_at	Aflatoxin B1 aldehyde reductase	−1.78	0.027	ATP binding
103054_at	Polymerase (RNA) II (DNA directed) polypeptide A	−2.20	0.027	DNA-directed RNA polymerase II
99433_at	Proline-serine-threonine phosphatase-interacting protein 2	−3.59	0.044	Cytoskeleton
93505_at	Methionine aminopeptidase 2	−3.81	0.028	Proteolysis and peptidolysis
102921_s_at	Tumor necrosis factor receptor superfamily, member 6	−3.84	0.024	Apoptosis
100445_f_at	Small proline-rich protein 1B	−3.85	0.022	Regulation of cell shape
94746_at	Histocompatibility 2, T region locus 24	−4.04	0.05	Immune response
94664_at	Insulin receptor substrate 3	−4.12	0.048	Insulin receptor binding
161285_r_at	Intergral membrane protein 1	−4.51	0.024	Protein amino acid glycosylation
161154_at	Platelet derived growth factor receptor, beta polypeptide	−6.51	0.014	Signal transduction

Table 7.2 Gene Ontology Biological Process Classification

Gene description	Sequence ID	Fold change	p-Value
Transcriptional regulation and nucleic acid binding			
E4F transcription factor 1	104689_at	7.1	0.05
CCAAT/enhancer binding protein (C/EBP), delta	160894_at	5.7	3.89×10^{-11}
Basic helix–loop–helix domain containing, class B2	104701_at	4.0	3.17×10^{-7}
Activating transcription factor 3	104156_r_at	3.9	0.00434
Similar to KIAA0595 protein	104610_at	3.9	0.047
Activating transcription factor 3	104155_f_at	3.8	1.54×10^{-24}
Reticuloendotheliosis oncogene	98813_at	3.8	0.038
RIKEN cDNA 5730470C09 gene	103393_at	3.4	0.00752
FBJ osteosarcoma oncogene	160901_at	3.3	9.47×10^{-7}
Jun-B oncogene	102362_i_at	3.1	0.035
B-cell leukemia/lymphoma 6	103015_at	2.9	0.02
Nuclear receptor subfamily 3, group C, member 1	98818_at	2.3	0.018
Jun-B oncogene	102363_r_at	2.0	0.019
Jun oncogene	100130_at	1.9	0.000748
Splicing factor 3b, subunit 1, 155 kDa	96777_at	1.6	0.031
CCCTC-binding factor	100148_at	1.5	0.019
Heterogeneous nuclear ribonucleoprotein H2	94832_at	1.5	0.00121
Signal transducer and activator of transcription 3	99100_at	1.5	0.042
Polymerase (RNA) II (DNA directed) polypeptide A	103054_at	−2.2	0.027
GTPase and GTP binding			
Interferon gamma inducible protein	104750_at	6.47	0.022
Interferon-inducible GTPase	96764_at	2.3	2.7×10^{-13}

(continued)

Table 7.2 Continued

Gene description	Sequence ID	Fold change	p-Value
Interferon-inducible GTPase	103963_f_at	2.2	1.47×10^{-5}
GTP binding protein (gene overexpressed in skeletal muscle)	92534_at	1.9	4.96×10^{-7}
Interferon gamma induced GTPase	160933_at	1.7	0.00281
Interferon-g induced GTPase	98410_at	1.6	0.00018
T-cell specific GTPase	102906_at	2.0	1.64×10^{-8}
Signal transduction			
Suppressor of cytokine signaling 3	92232_at	6.8	5.17×10^{-13}
Suppressor of cytokine signaling 3	162206_f_at	5.2	1.38×10^{-19}
Numb gene homolog (Drosophila)	99955_at	4.5	0.016
Protein tyrosine phosphatase, receptor type, K	161371_r_at	3.9	0.017
Rho guanine nucleotide exchange factor (GEF) 1	98001_at	3.8	0.036
Chemokine orphan receptor 1	93430_at	3.6	0.036
Suppressor of cytokine signaling 2	99475_at	3.5	0.016
Transducer of ErbB-2.1	161567_r_at	3.0	0.029
Suppressor of cytokine signaling 1	92832_at	1.9	0.00644
Oncostatin receptor	102255_at	1.8	0.05
Tumor necrosis factor receptor superfamily, member 6	102921_s_at	−3.8	0.024
Insulin receptor substrate 3	94664_at	−4.1	0.048
Platelet derived growth factor receptor, beta polypeptide	161154_at	−6.5	0.014

These proteins are members of distinct signaling pathways suggesting that leptin signaling can overlap with other pathways. In fact, a cross-talk between the leptin and insulin pathways has been suggested.[32] It is also possible that the regulation of these genes is a secondary event that occurs as a result of STAT3 activation requiring further analysis. We observed that the *Drosophila* Numb gene homologue was induced (+4.5-fold), suggesting that there is enhanced negative regulation of the Notch signaling pathway. The interplay of Notch and Numb is believed to control many aspects of neuronal cell fate[43] and the regulation of Numb by leptin suggests that pathways related to neuronal cell fate are possible targets of leptin signaling.

The GTP-related genes (Table 7.2) regulated in response to leptin treatment were only modestly changed (+1.7- to +2.3-fold) although the interferon gamma inducible protein was significantly induced (+6.5-fold). Based on DNA sequence analysis (data not shown) these GTPases are only partially related and thus represent a coordinate upregulation by leptin of a functionally related set of proteins that may modulate a variety of extracellular signals. One possibility is that if the leptin signaling pathway overlaps with that of insulin at the level of the IRS protein, one of the downstream targets of IRS is phosphatidylinositol 3-kinase (PI3K) which catalyzes the production of phosphatidylinositol trisphosphate (PIP3). PIP3, in turn, recruits and activates numerous other downstream molecules that contain the pleckstrin homology domain such as GTPases.[44]

PTP1B as a Negative Regulator of Leptin Signaling

Protein tyrosine phosphatase 1B (PTP1B) has been implicated in the regulation of leptin signaling. A surprising phenotype of PTP1B knockout (KO) mice was their resistance to diet-induced obesity.[45] The mechanism by which PTP1B affected body weight was not understood until a recent finding that PTP1B KO mice exhibit increased leptin sensitivity.[46] We and others have demonstrated that PTP1B regulates leptin signaling by dephosphorylating JAK2.[35,47] Again, we chose the GT1-7 cell line to study the effects of PTP1B overexpression. In the absence of exogenous PTP1B, leptin-dependent tyrosine phosphorylation of JAK2 and STAT3 was observed as expected [Fig. 7.3(a)]. Overexpression of PTP1B resulted in a dose-dependent reduction of JAK2 tyrosine phosphorylation. Quantification of the signal revealed a 55% reduction in pY-JAK2 when 5 μg of PTP1B expression plasmid was used in transfection relative to PTP1B-untransfected cells [Fig. 7.3(b)]. Western blotting confirmed that PTP1B protein expression level correlated with the amount of PTP1B plasmid transfected in the cells [Fig. 7.3(a)].

JAK2 contains a sequence motif [E/D]-pY-pY-[R/K] within its kinase activation loop that is a known substrate recognition site for PTP1B.[48] The two phosphotyrosines found at amino acid position 1007 and 1008 within the kinase activation loop are essential for the catalytic activity of JAK2. It would

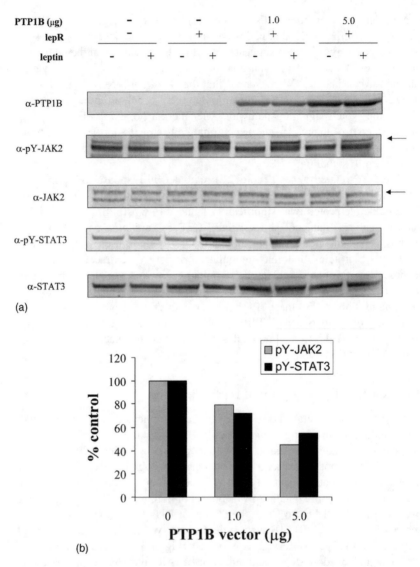

(a)

(b)

Figure 7.3 Overexpression of PTP1B results in decreased tyrosine phosphorylation of endogenous JAK2 and STAT3 in GT1-7 cells. Cells were transiently transfected with a lepR expression vector (+) or an equal amount of empty vector control (−) and an indicated amount of PTP1B expression vector. Cells were either stimulated (+) with leptin (10 ng/mL) for 30 min or not stimulated (−). (a) Cell lysates were normalized for protein content and resolved by SDS–PAGE. Western blot analysis was performed with indicated antibodies. (b) Quantification of the Western blot signal. The level of pY-JAK2 and pY-STAT3 was calculated as percent of leptin-induced tyrosine phosphorylation of these proteins in the absence of exogenous PTP1B. (Reprinted from Kaszubska et al.,[35] with permission from Elsevier.)

be predicted that dephosphorylation of these tyrosine residues would reduce JAK2 kinase activity and block the phosphorylation and activation of lepR and other downstream signaling molecules such as STAT3. Indeed we demonstrated that endogenous STAT3 tyrosine phosphorylation was reduced by 45% on PTP1B overexpression relative to untransfected GT1-7 cells [Fig. 7.3(b)]. The mechanism by which PTP1B, an endoplasmic reticulum-bound protein, comes in contact with its substrates within the cell is not well understood. However, a recent report demonstrates that receptor endocytosis might be required for PTP1B-catalyzed dephosphorylation.[49]

Effects of PTP1B on Leptin-Regulated Genes in GT1-7 Cells

In addition to the effects on JAK2 and STAT3 tyrosine phosphorylation, we tested the effects of PTP1B overexpression on other leptin-regulated genes. GT1-7 cells transiently cotransfected with lepR and two different amounts of

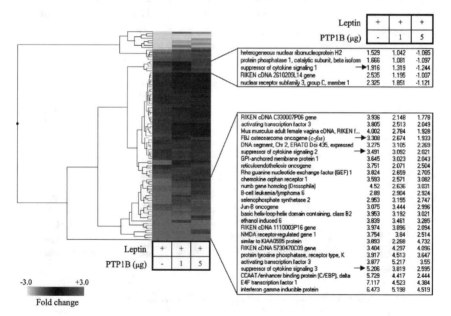

| Leptin | + | + | + |
| PTP1B (µg) | - | 1 | 5 |

heterogeneous nuclear ribonucleoprotein H2	1.529	1.042	-1.065
protein phosphatase 1, catalytic subunit, beta isoform	1.666	1.081	-1.097
suppressor of cytokine signaling 1 →	1.916	1.319	-1.244
RIKEN cDNA 2610209L14 gene	2.535	1.195	-1.007
nuclear receptor subfamily 3, group C, member 1	2.325	1.851	-1.121

RIKEN cDNA C330007P06 gene	3.936	2.148	1.778
activating transcription factor 3	3.805	2.513	2.049
Mus musculus adult female vagina cDNA, RIKEN f...	4.002	2.784	1.928
FBJ osteosarcoma oncogene (c-fos) →	3.308	2.674	1.933
DNA segment, Chr 2, ERATO Doi 435, expressed	3.275	3.105	2.269
suppressor of cytokine signaling 2 →	3.491	3.092	2.021
GPI-anchored membrane protein 1	3.645	3.023	2.043
reticuloendotheliosis oncogene	3.751	2.071	2.504
Rho guanine nucleotide exchange factor (GEF) 1	3.824	2.659	2.705
chemokine orphan receptor 1	3.593	2.571	3.082
numb gene homolog (Drosophila)	4.52	2.636	3.031
B-cell leukemia/lymphoma 6	2.89	2.904	2.924
selenophosphate synthetase 2	2.953	3.155	2.747
Jun-B oncogene	3.075	3.444	2.996
basic helix-loop-helix domain containing, class B2	3.953	3.192	3.021
ethanol induced 6	3.839	3.461	3.285
RIKEN cDNA 1110003P16 gene	3.974	3.896	2.094
NMDA receptor-regulated gene 1	3.754	3.84	2.514
similar to KIAA0595 protein	3.693	2.268	4.732
RIKEN cDNA 5730470C09 gene	3.404	4.297	4.096
protein tyrosine phosphatase, receptor type, K	3.917	4.513	3.647
activating transcription factor 3	3.877	5.217	3.55
suppressor of cytokine signaling 3 →	5.206	3.819	2.595
CCAAT/enhancer binding protein (C/EBP), delta	5.729	4.417	2.444
E4F transcription factor 1	7.117	4.523	4.384
interferon gamma inducible protein	6.473	5.198	4.519

| Leptin | + | + | + |
| PTP1B (µg) | - | 1 | 5 |

-3.0   +3.0

Fold change

Figure 7.4 Effects of PTP1B overexpression on leptin-regulated genes in GT1-7 cells. A 1D cluster was prepared in Spotfire 7.0 to visualize the 91 leptin-regulated sequences and the effect of PTP1B overexpression. For illustration purposes two regions of the cluster are highlighted with additional detail in the right as to the fold change of each gene as the expression level of PTP1B was increased. The SOCS genes and c-*fos* are noted with arrows as examples of genes for which the effects of leptin are reversed by PTP1B expression. The exact number of genes that met various filtering criteria is detailed in the "Results and Discussion" section.

PTP1B expression vector, were treated with leptin (10 ng/mL) for 1 h. The cells were harvested and the total RNA was analyzed using Affymetrix microarrays (U74Av2) and the Rosetta Resolver system as described in the "Materials and Methods" section. We expanded our analysis, described in the previous section, to assess how many of the leptin-regulated genes were affected by PTP1B. Of the 91 sequences regulated by leptin, 56 demonstrated a dose-dependent reversal in the level of expression as a result of increasing PTP1B expression (Fig. 7.4). If instead of a dose-response relationship we only consider whether the 5 μg PTP1B transfection had an effect, then 79 of the 91 sequences exhibit some reversal in expression.

To corroborate our findings by microarray analysis we performed quantitative RT-PCR to analyze the expression levels of two leptin-regulated genes. RNA samples that were analyzed by microarrays were used with primers specific to SOCS3 and c-*fos* (Fig. 7.5). In the absence of PTP1B, an approximately sixfold increase of SOCS3 mRNA and a twofold of c-*fos* over vector control were observed upon leptin stimulation. PTP1B overexpression caused a dose-dependent reduction in SOCS3 and c-*fos* mRNA. We conclude that microarray analysis is a valid approach by which the effects of PTP1B on leptin-regulated gene expression can be studied. Taken together these observations confirm that PTP1B negatively regulates leptin signal transduction.

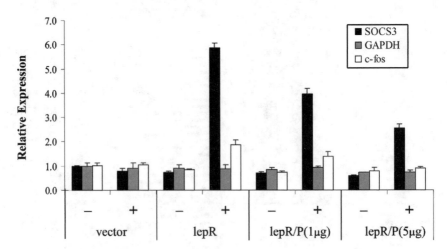

Figure 7.5 PTP1B decreases SOCS3 and c-*fos* mRNA accumulation in GT1-7 cells. Cells were transiently transfected with a lepR expression vector or an equal amount of empty vector control or cotransfected with lepR and an indicated amount of PTP1B expression vector (P). Cells were either stimulated (+) with leptin (10 ng/mL) for 1 h or not stimulated (−). RNA was isolated from cell lysates and corresponding mRNA analyzed by quantitative RT-PCR. Relative expression levels were determined in triplicate and standardized to 28S rRNA copy number determined by the ΔΔCt method. (Reprinted from Kaszubska et al.,[35] with permission from Elsevier.)

CONCLUSIONS

In summary, we have employed a hypothalamic GT1-7 cell line to study the leptin signaling pathway and leptin-regulated gene expression. We have shown that leptin signaling can be reconstituted in GT1-7 cell. As an *in vitro* model, this cell line represents a less complex system for genomic studies than a naturally heterogeneous hypothalamic tissue. Using microarray analysis we demonstrated that overexpression of PTP1B in GT1-7 cells leads to the reversal of mRNA levels of leptin-regulated genes, as would be predicted of a negative regulator of leptin signaling. In addition, we identified leptin-regulated genes that upon further validation might serve as targets for future antiobesity therapeutics.

ACKNOWLEDGMENTS

We thank the following individuals for their contributions to the work described here: Doug Falls, Verlyn Schaefer, Leigh Frost, Jim Trevillyan, Christine Collins, David White and Michael Jirousek.

REFERENCES

1. Kopelman, P.G. Obesity as a medical problem. Nature **2000**, *404*, 635–643.
2. Spiegelman, B.M.; Flier, J.S. Obesity and the regulation of energy balance. Cell **2001**, *104*, 531–543.
3. Friedman, J.M. Obesity in the new millennium. Nature **2000**, *404*, 632–634.
4. Houseknecht, K.L.; Baile, C.A.; Matteri, R.L.; Spurlock, M.E. The biology of leptin: a review. J. Anim. Sci. **1998**, *76*, 1405–1420.
5. Cohen, P.; Zhao, C.; Cai, X.; Montez, J.M.; Rohani, S.C.; Feinstein, P.; Mombaerts, P.; Friedman, J.M. Selective deletion of leptin receptor in neurons leads to obesity. J. Clin. Invest. **2001**, *108*, 1113–1121.
6. Schwartz, M.W.; Woods, S.C.; Porte, D., Jr.; Seeley, R.J.; Baskin, D.G. Central nervous system control of food intake. Nature **2000**, *404*, 661–671.
7. Elmquist, J.K. Anatomic basis of leptin action in the hypothalamus. Front. Horm. Res. **2000**, *26*, 21–41.
8. Halaas, J.L.; Gajiwala, K.S.; Maffei, M.; Cohen, S.L.; Chait, B.T.; Rabinowitz, D.; Lallone, R.L.; Burley, S.K.; Friedman, J.M. Weight-reducing effects of the plasma protein encoded by the obese gene. Science **1995**, *269*, 543–546.
9. Pelleymounter, M.A.; Cullen, M.J.; Baker, M.B.; Hecht, R.; Winters, D.; Boone, T.; Collins, F. Effects of the obese gene product on body weight regulation in ob/ob mice. Science **1995**, *269*, 540–543.
10. Rentsch, J.; Levens, N.; Chiesi, M. Recombinant ob-gene product reduces food intake in fasted mice. Biochem. Biophys. Res. Commun. **1995**, *214*, 131–136.
11. Heymsfield, S.B.; Greenberg, A.S.; Fujioka, K.; Dixon, R.M.; Kushner, R.; Hunt, T.; Lubina, J.A.; Patane, J.; Self, B.; Hunt, P.; McCamish, M. Recombinant leptin for weight loss in obese and lean adults: a randomized, controlled, dose-escalation trial. J. Am. Med. Assoc. **1999**, *282*, 1568–1575.

12. El-Haschimi, K.; Pierroz, D.D.; Hileman, S.M.; Bjorbaek, C.; Flier, J.S. Two defects contribute to hypothalamic leptin resistance in mice with diet-induced obesity. J. Clin. Invest. **2000**, *105*, 1827–1832.

13. Soukas, A.; Cohen, P.; Socci, N.D.; Friedman, J.M. Leptin-specific patterns of gene expression in white adipose tissue. Genes Dev. **2000**, *14*, 963–980.

14. Lopez, I.P.; Marti, A.; Milagro, F.I.; Zulet Md Mde, L.; Moreno-Aliaga, M.J.; Martinez, J.A.; De Miguel, C. DNA microarray analysis of genes differentially expressed in diet-induced (cafeteria) obese rats. Obes. Res. **2003**, *11*, 188–194.

15. Ferrante, A.W., Jr.; Thearle, M.; Liao, T.; Leibel, R.L. Effects of leptin deficiency and short-term repletion on hepatic gene expression in genetically obese mice. Diabetes **2001**, *50*, 2268–2278.

16. Liang, C.P.; Tall, A.R. Transcriptional profiling reveals global defects in energy metabolism, lipoprotein, and bile acid synthesis and transport with reversal by leptin treatment in ob/ob mouse liver. J. Biol. Chem. **2001**, *276*, 49066–49076.

17. Li, J.Y.; Lescure, P.A.; Misek, D.E.; Lai, Y.M.; Chai, B.X.; Kuick, R.; Thompson, R.C.; Demo, R.M.; Kurnit, D.M.; Michailidis, G.; Hanash, S.M.; Gantz, I. Food deprivation-induced expression of minoxidil sulfotransferase in the hypothalamus uncovered by microarray analysis. J. Biol. Chem. **2002**, *277*, 9069–9076.

18. Wurmbach, E.; Gonzalez-Maeso, J.; Yuen, T.; Ebersole, B.J.; Mastaitis, J.W.; Mobbs, C.V.; Sealfon, S.C. Validated genomic approach to study differentially expressed genes in complex tissues. Neurochem. Res. **2002**, *27*, 1027–1033.

19. White, D.W.; Zhou, J.; Stricker-Krongrad, A.; Ge, P.; Morgenstern, J.P.; Dembski, M.; Tartaglia, L.A. Identification of leptin-induced transcripts in the mouse hypothalamus. Diabetes **2000**, *49*, 1443–1450.

20. Tartaglia, L.A.; Dembski, M.; Weng, X.; Deng, N.; Culpepper, J.; Devos, R.; Richards, G.J.; Campfield, L.A.; Clark, F.T.; Deeds, J. Identification and expression cloning of a leptin receptor, OB-R. Cell **1995**, *83*, 1263–1271.

21. Carpenter, L.R.; Yancopoulos, G.D.; Stahl, N. General mechanisms of cytokine receptor signaling. Adv. Protein Chem. **1998**, *52*, 109–140.

22. Li, C.; Friedman, J.M. Leptin receptor activation of SH2 domain containing protein tyrosine phosphatase 2 modulates Ob receptor signal transduction. Proc. Natl. Acad. Sci. USA **1999**, *96*, 9677–9682.

23. Banks, A.S.; Davis, S.M.; Bates, S.H.; Myers, M.G., Jr. Activation of downstream signals by the long form of the leptin receptor. J. Biol. Chem. **2000**, *275*, 14563–14572.

24. Baumann, H.; Morella, K.K.; White, D.W.; Dembski, M.; Bailon, P.S.; Kim, H.; Lai, C.F.; Tartaglia, L.A. The full-length leptin receptor has signaling capabilities of interleukin 6-type cytokine receptors. Proc. Natl. Acad. Sci. USA **1996**, *93*, 8374–8378.

25. Ghilardi, N.; Ziegler, S.; Wiestner, A.; Stoffel, R.; Heim, M.H.; Skoda, R.C. Defective STAT signaling by the leptin receptor in diabetic mice. Proc. Natl. Acad. Sci. USA **1996**, *93*, 6231–6235.

26. Vaisse, C.; Halaas, J.L.; Horvath, C.M.; Darnell, J.E., Jr.; Stoffel, M.; Friedman, J.M. Leptin activation of Stat3 in the hypothalamus of wild-type and ob/ob mice but not db/db mice. Nat. Genet. **1996**, *14*, 95–97.

27. Emilsson, V.; Arch, J.R.; de Groot, R.P.; Lister, C.A.; Cawthorne, M.A. Leptin treatment increases suppressors of cytokine signaling in central and peripheral tissues. FEBS Lett. **1999**, *455*, 170–174.

28. Bjorbaek, C.; Lavery, H.J.; Bates, S.H.; Olson, R.K.; Davis, S.M.; Flier, J.S.; Myers, M.G., Jr. SOCS3 mediates feedback inhibition of the leptin receptor via Tyr985. J. Biol. Chem. **2000**, *275*, 40649–40657.

29. Bjorbaek, C.; Elmquist, J.K.; Frantz, J.D.; Shoelson, S.E.; Flier, J.S. Identification of SOCS-3 as a potential mediator of central leptin resistance. Mol. Cell **1998**, *1*, 619–625.

30. Bjorbaek, C.; El-Haschimi, K.; Frantz, J.D.; Flier, J.S. The role of SOCS-3 in leptin signaling and leptin resistance. J. Biol. Chem. **1999**, *274*, 30059–30065.

31. Bates, S.H.; Stearns, W.H.; Dundon, T.A.; Schubert, M.; Tso, A.W.; Wang, Y.; Banks, A.S.; Lavery, H.J.; Haq, A.K.; Maratos-Flier, E.; Neel, B.G.; Schwartz, M.W.; Myers, M.G., Jr. STAT3 signalling is required for leptin regulation of energy balance but not reproduction. Nature **2003**, *421*, 856–859.

32. Niswender, K.D.; Schwartz, M.W. Insulin and leptin revisited: adiposity signals with overlapping physiological and intracellular signaling capabilities. Front. Neuroendocrinol. **2003**, *24*, 1–10.

33. Mellon, P.L.; Windle, J.J.; Goldsmith, P.C.; Padula, C.A.; Roberts, J.L.; Weiner, R.I. Immortalization of hypothalamic GnRH neurons by genetically targeted tumorigenesis. Neuron **1990**, *5*, 1–10.

34. White, D.W.; Wang, D.W.; Chua, S.C., Jr.; Morgenstern, J.P.; Leibel, R.L.; Baumann, H.; Tartaglia, L.A. Constitutive and impaired signaling of leptin receptors containing the Gln → Pro extracellular domain fatty mutation. Proc. Natl. Acad. Sci. USA **1997**, *94*, 10657–10662.

35. Kaszubska, W.; Falls, H.D.; Schaefer, V.G.; Haasch, D.; Frost, L.; Hessler, P.; Kroeger, P.E.; White, D.W.; Jirousek, M.R.; Trevillyan, J.M. Protein tyrosine phosphatase 1B negatively regulates leptin signaling in a hypothalamic cell line. Mol. Cell. Endocrinol. **2002**, *195*, 109–118.

36. Hai, T.; Hartman, M.G. The molecular biology and nomenclature of the activating transcription factor/cAMP responsive element binding family of transcription factors: activating transcription factor proteins and homeostasis. Gene **2001**, *273*, 1–11.

37. St-Pierre, B.; Flock, G.; Zacksenhaus, E.; Egan, S.E. Stra13 homodimers repress transcription through class B E-box elements. J. Biol. Chem. **2002**, *277*, 46544–46551.

38. Yamada, T.; Tobita, K.; Osada, S.; Nishihara, T.; Imagawa, M. CCAAT/enhancer-binding protein delta gene expression is mediated by APRF/STAT3. J. Biochem. **1997**, *121*, 731–738.

39. Hutt, J.A.; O'Rourke, J.P.; DeWille, J. Signal transducer and activator of transcription 3 activates CCAAT enhancer-binding protein delta gene transcription in G0 growth-arrested mouse mammary epithelial cells and in involuting mouse mammary gland. J. Biol. Chem. **2000**, *275*, 29123–29131.

40. Cardinaux, J.R.; Allaman, I.; Magistretti, P.J. Pro-inflammatory cytokines induce the transcription factors C/EBPbeta and C/EBPdelta in astrocytes. Glia **2000**, *29*, 91–97.

41. Cao, Z.; Umek, R.M.; McKnight, S.L. Regulated expression of three C/EBP isoforms during adipose conversion of 3T3-L1 cells. Genes Dev. **1991**, *5*, 1538–1552.

42. Sterneck, E.; Paylor, R.; Jackson-Lewis, V.; Libbey, M.; Przedborski, S.; Tessarollo, L.; Crawley, J.N.; Johnson, P.F. Selectively enhanced contextual fear conditioning in mice lacking the transcriptional regulator CCAAT/enhancer binding protein delta. Proc. Natl. Acad. Sci. USA **1998**, *95*, 10908–10913.

43. Cayouette, M.; Raff, M. Asymmetric segregation of Numb: a mechanism for neural specification from Drosophila to mammals. Nat. Neurosci. **2002**, *5*, 1265–1269.
44. Shepherd, P.R.; Withers, D.J.; Siddle, K. Phosphoinositide 3-kinase: the key switch mechanism in insulin signalling (Erratum appears in Biochem. J. **1998**, *335* (Pt 3), 711). Biochem. J. **1998**, *333*, 471–490.
45. Elchebly, M.; Payette, P.; Michaliszyn, E.; Cromlish, W.; Collins, S.; Loy, A.L.; Normandin, D.; Cheng, A.; Himms-Hagen, J.; Chan, C.C.; Ramachandran, C.; Gresser, M.J.; Tremblay, M.L.; Kennedy, B.P. Increased insulin sensitivity and obesity resistance in mice lacking the protein tyrosine phosphatase-1B gene (Comment). Science **1999**, *283*, 1544–1548.
46. Zabolotny, J.M.; Bence-Hanulec, K.K.; Stricker-Krongrad, A.; Pierroz, D.D.; Mozell, R.; Haj, F.G.; Wang, Y.P.; Minokoshi, Y.; Kim, Y.B.; Flier, J.S.; Neel, B.G.; Kahn, B.B. Mice lacking protein tyrosine phosphatase 1B have increased leptin sensitivity. Diabetes **2001**, *50*, A58.
47. Zabolotny, J.M.; Bence-Hanulec, K.K.; Stricker-Krongrad, A.; Haj, F.; Wang, Y.; Minokoshi, Y.; Kim, Y.B.; Elmquist, J.K.; Tartaglia, L.A.; Kahn, B.B.; Neel, B.G. PTP1B regulates leptin signal transduction *in vivo* (Comment). Dev. Cell **2002**, *2*, 489–495.
48. Myers, M.P.; Andersen, J.N.; Cheng, A.; Tremblay, M.L.; Horvath, C.M.; Parisien, J.P.; Salmeen, A.; Barford, D.; Tonks, N.K. TYK2 and JAK2 are substrates of protein tyrosine phosphatase 1B. J. Biol. Chem. **2001**, *276*, 47771–47774.
49. Haj, F.G.; Verveer, P.J.; Squire, A.; Neel, B.G.; Bastiaens, P.I.H. Imaging sites of receptor dephosphorylation by PTP1B on the surface of the endoplasmic reticulum. Science **2002**, *295*, 1708–1711.

8

Characterizing Cholesterol Metabolism in Atherosclerosis Susceptible and Resistant Mouse Models Using DNA Microarrays

Laurent Vergnes, Jack Phan, and Karen Reue

*Department of Medicine and Department of Human Genetics,
University of California, Los Angeles, and Veterans Administration
Greater Los Angeles Healthcare System, Los Angeles, CA, USA*

GENETIC VARIATION IN CHOLESTEROL METABOLISM

Hypercholesterolemia has long been recognized as a major risk factor for atherosclerosis and cardiovascular diseases (reviewed in Steinberg[1]). Individuals in the human population exhibit a large range of serum cholesterol levels as a result of differences in several factors, including absorption of cholesterol from the diet, lipoprotein production and catabolism, and synthesis and excretion of cholesterol in the form of bile acids. An example of an extreme individual variation in regulation of cholesterol levels is the "Egg Man" described by Kern.[2] This 88 year old man habitually ate 25 eggs a day, thus consuming more than $20\times$ the recommended amount of cholesterol, yet maintained normal blood cholesterol levels and had no clinical evidence of atherosclerosis. This appeared to be explained by physiological compensatory mechanisms, including reduced absorption of dietary cholesterol in the intestine, and increased conversion of cholesterol to bile acids for elimination. This case illustrates that the study of individuals who exhibit aberrant responses to dietary cholesterol intake may provide insight into the metabolic processes involved in cholesterol homeostasis.

With the advent of tools such as DNA microarrays, the study of genetic variation becomes even more valuable, as the basis for individual differences can be assessed at the molecular level.

MOUSE MODELS THAT ARE SUSCEPTIBLE AND RESISTANT TO HYPERCHOLESTEROLEMIA AND ATHEROSCLEROSIS

Studies in human populations have revealed that response to dietary cholesterol is highly heritable, but known genetic variations explain only about one-third of the variation in serum cholesterol levels.[3] Owing to confounding factors such as genetic heterogeneity and lack of power to detect genes exerting small effects, the identification of genes controlling cholesterol homeostasis has been difficult in humans. However, it is feasible to identify genes contributing to complex traits, such as cholesterol metabolism, using the mouse as a genetic model. Inbred mouse strains are particularly well suited to genetic studies because each strain represents an unlimited source of genetically identical animals. The hundreds of available inbred strains represent a valuable resource of naturally occurring genetic mutations, including numerous mutations affecting lipid metabolism and atherosclerosis.[4−8] Furthermore, genetic manipulation in the mouse is easily accomplished through selective breeding or alteration of the germline via transgenic or gene knockout technology, and has been used extensively to study lipid metabolism genes (reviewed in Reue[7]). Together, these features have made the mouse the pre-eminent mammalian model for identification of genetic factors in complex diseases.

It should be noted, however, that the mouse and other rodents are not optimal models of human lipid metabolism as they differ from humans in the distribution of cholesterol among lipoprotein subclasses. Thus, whereas humans transport cholesterol primarily in the form of low density lipoproteins (LDL), mice consuming a typical low fat laboratory chow diet carry most cholesterol in the form of high density lipoproteins (HDL).[5,6] This limitation has been overcome to some extent by feeding mice diets containing cholesterol and the bile salt cholate, which cause a redistribution of cholesterol to the LDL fraction. Paigen and colleagues devised an atherogenic (Ath) diet containing 1.25% cholesterol, 15% fat, 30% protein, and 0.5% cholate, which produces a humanistic lipoprotein profile in some mouse strains, such as C57BL/6J.[5,6] When fed the Ath diet, C57BL/6J mice exhibit substantially elevated cholesterol levels and develop aortic lesions with characteristics of human atherosclerotic plaques including smooth muscle cell proliferation, fibrous caps, and calcification.[9] These qualities of the C57BL/6J strain have led to its adoption as a preferred genetic background on which to express transgenes and knockout gene alleles, and to identify novel genes that may have a role in lipid metabolism.[7]

Inbred mouse strains differ in their response to the Ath diet, and these genetic variations can be exploited to identify genes having roles in lipid metabolism.[4−9] For example, an intriguing genetic variation exists between the

C57BL/6J (B6J) strain, one of the strains most susceptible to Ath diet-induced hypercholesterolemia and atherosclerosis, and a closely related strain originally derived from the same breeding stock, known as C57BL/6ByJ (B6By). In contrast to B6J mice, B6By mice are resistant to both hypercholesterolemia and atherosclerotic lesion formation in response to the Ath diet.[10] Given their common origin, any differences between the strains are attributable to genetic variations that have accumulated since they were separated as individual breeding stocks (Fig. 8.1). Evidence supporting the near genetic identity of the two strains is provided by failure to detect any nucleotide differences in more than 120 microsatellite DNA markers and numerous gene sequences that we have analyzed in the two strains (Mouzeyan et al.,[10] Phan et al.,[11] and unpublished data). Despite the close genetic similarity, B6J and B6By mice exhibit a dramatic difference in their response to the Ath diet. Whereas B6J mice increase total cholesterol levels by 2.5-fold on the Ath diet, B6By mice exhibit no significant increase on the Ath diet [Fig. 8.2(a)]. The B6By mice are also resistant to atherosclerotic lesion development, showing virtually no lipid accumulation in the aorta [Fig. 8.2(b)]. We have determined that the resistance in B6By mice cannot be accounted for by reduced food intake, dietary cholesterol absorption, or endogenous cholesterol synthesis.[10]

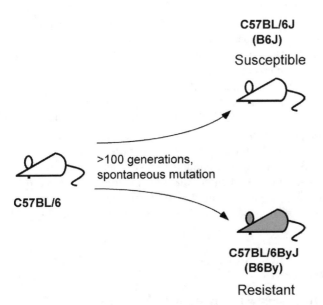

C57BL/6J
(B6J)
Susceptible

C57BL/6

>100 generations,
spontaneous mutation

C57BL/6ByJ
(B6By)

Resistant

Figure 8.1 Origin of C57BL/6 substrains. Mice from the original C57BL/6 stock were bred independently for many generations, providing the opportunity for spontaneous mutations. Although still very similar genetically, some divergence has occurred such that the resulting C57BL/6J strain is susceptible and the C57BL/6ByJ strain is resistant to hypercholesterolemia and atherosclerosis.

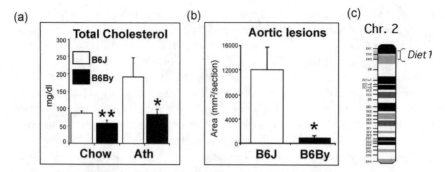

Figure 8.2   Resistance to hypercholesterolemia and atherosclerosis in B6By when compared with B6J mice. (a) Total plasma cholesterol levels in mice fed the chow diet or the Ath diet for 3 weeks. *$p < 0.01$, **$p < 0.001$. (b) Aortic lesions after 15 weeks on the Ath diet. (c) Position of the *Diet 1* gene locus on proximal chromosome 2 in mouse as determined by genetic mapping. [Panels (a) and (b) adapted from Mouzeyan et al.[10]]

Because of the common genetic origin of B6J and B6By mice, we hypothesized that the differences in response to the Ath diet must result from genetic variation in one or just a few genes. To map the responsible gene(s), we performed a genetic cross between B6By and an evolutionarily distant mouse strain, CAST/Ei. We detected a locus on proximal chromosome 2 that segregated with the resistance to elevated cholesterol levels, and called this locus *Diet1* [Fig. 8.2(c)[10]]. Within the *Diet1* interval, there are no candidate genes having known roles in cholesterol metabolism. We are therefore taking a combined genetic and molecular approach to identify the *Diet1* gene. DNA microarrays are a useful tool for several aspects of this work, and here we describe our use of expression profiling to address three different questions regarding the basis for the resistant phenotype in B6By mice: (A) What metabolic pathway(s) are altered in B6By when compared with B6J mice to account for the resistance to hypercholesterolemia? (B) Are there differential effects of atherogenic diet components on gene expression in B6J when compared with B6By mice? and (C) Which of the candidate genes located within the *Diet1* gene interval on chromosome 2 have altered expression in B6By when compared with B6J mice?

ANALYSIS OF CHOLESTEROL METABOLISM IN MOUSE STRAINS USING DNA MICROARRAYS

Metabolic Pathways Associated with Hypercholesterolemia Resistance in B6By Mice Revealed by Expression Profiling

To identify metabolic pathway differences that might distinguish B6J and B6By mice, we performed expression profiling using a commercial cDNA array containing 18,000 mouse cDNA sequences, one-third of which were known genes, and the remainder derived from expressed sequence tags (ESTs).[11] The array was hybridized to liver RNA samples from mice fed the Ath diet for 3 weeks, normalized using 30 double spotted controls, and ratios between the signal on the B6J and B6By arrays were calculated. On the basis of fold differences in expression levels, we selected the 100 mRNA species having highest B6J/B6By expression ratios, and the 100 having highest B6By/B6J expression ratios for in-depth analysis. Of these 200 genes, 29 were known mouse genes, and another 65 were ESTs that appeared homologous to known genes from human or other species. The remaining 106 were ESTs derived from genes of unknown function, and were therefore not useful in the analysis. Among the known genes, we recognized 17 as having roles in lipid metabolism (Fig. 8.3). These included 10 involved in cholesterol metabolism, 3 in phospholipid metabolism, and 2 in fatty acid synthesis. On the basis of the hypercholesterolemia resistance phenotype we had observed in B6By mice, we focused on the 10 genes implicated in cholesterol metabolism. These included seven genes with

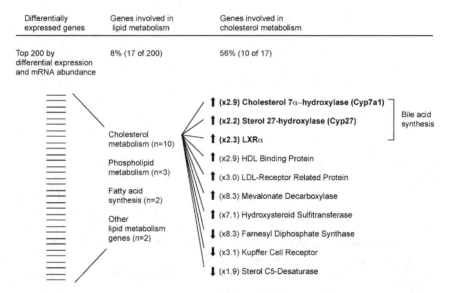

Figure 8.3 Identification of metabolic pathway altered in B6By mice by microarray expression profiling. Liver RNA samples from mice fed the Ath diet for 3 weeks were hybridized to a cDNA array. From the 200 genes showing the greatest expression difference between the two strains, 17 were involved in lipid metabolism. Of these 10 genes were specifically involved in cholesterol metabolism, 3 of which are key regulators of bile acid synthesis (shown in bold letters). The vertical black arrows indicate higher or lower expression in B6By when compared with B6J, and numbers in parentheses indicate the fold difference in expression levels. The elevated expression of enzymes that promote cholesterol conversion to bile acids in B6By mice revealed a mechanism for enhanced cholesterol elimination in that strain.

higher expression in B6By, and three genes with higher expression in B6J liver (Fig. 8.3, right column). Three of the genes with higher expression in B6By were known to have prominent roles in bile acid synthesis (bold letters in Fig. 8.3). Cholesterol 7α-hydroxylase (*Cyp7a1*) and sterol 27-hydroxylase (*Cyp27*) cata-lyze the rate-limiting steps in the conversion of cholesterol to bile acids via the classic and alternate pathways, respectively.[12] LXRα (liver X receptor α) is a nuclear receptor that is activated by oxysterols to stimulate *Cyp7a1* transcription, and thus, bile acid synthesis.[13] We confirmed the expression levels detected on the array using a competitive PCR assay or Northern blots with liver RNA from five mice of each strain.[11] Thus, B6By mice exhibited elevated expression levels for three key regulators of cholesterol conversion to bile acids, suggesting that increased bile acid synthesis may provide a mechanism for these mice to eliminate dietary cholesterol more rapidly than B6J mice do.

To further explore the hypothesis that B6By mice have enhanced bile acid metabolism, we performed Northern blot analysis of additional genes involved in bile acid transport and metabolism, and observed higher expression in B6By when compared with B6J, as summarized in Fig. 8.4 and detailed in Phan et al.[11] These genes included additional bile acid synthetic enzymes (*Cyp7b1*, oxysterol 7α-hydroxylase, increased 1.7-fold; *Cyp8b1*, sterol 12α-hydroxylase, increased 1.5-fold), bile acid transport proteins that reside on the sinusoidal mem-brane (NTCP, Na^+-taurocholate cotransporting polypeptide, increased twofold; OATP, organic anion transporting polypeptide, increased 2.5-fold) bile acid trans-port proteins that reside on the canalicular membrane (BSEP, bile salt export pump, increased 3.2-fold; Mdr2, multidrug-resistance-associated 2, increased 10.1-fold), and an intracellular bile acid binding protein (3α-HSD, 3α-hydroxysteroid dehydro-genase, increased 12.4-fold).[11] Given that the expression of rate-limiting enzymes involved in both bile acid synthesis (*Cyp7a1*) and secretion (BSEP) were elevated in B6By mice, we hypothesized that these mice should have enhanced rates of bile acid excretion. Indeed, fecal bile acid excretion was increased twofold and urinary bile acid excretion was increased 18-fold in B6By when compared with B6J mice [Fig. 8.4(b) and (c)].[11] We subsequently determined that bile acid levels in off-spring of the B6By × CAST/Ei genetic backcross segregate with the *Diet1* locus on chromosome 2.[11] These results suggest that genetic variation at the *Diet1* locus confers enhanced bile acid synthesis and secretion in B6By mice, leading to the observed lower blood cholesterol levels.

Ath Diet Components Elicit Different Gene Expression Profiles in B6J vs. B6By Mice

In contrast to humans, laboratory mice typically eat a diet that is low in fat and cholesterol, and do not spontaneously develop elevated serum cholesterol levels or atheromatous lesions in the aorta. However, hypercholesterolemia and aortic lesions can be induced in genetically susceptible mouse strains such as B6J by feeding the animals the Ath diet containing cholesterol, cholic acid, and fat.

(a)

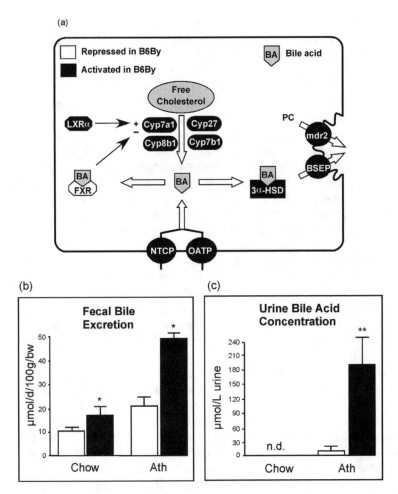

Figure 8.4 Bile acid metabolic pathway is enhanced in B6By mice. (a) Summary of cholesterol metabolism gene expression differences between B6By and B6J. Genes having higher expression levels in B6By when compared with B6J liver are shown with black shading (see text for full gene names and fold-differences in expression levels). B6By liver exhibited 50% reduced expression of FXR (farnesoid X-activated receptor), as indicated by the white box. Differences in gene expression were reflected in increased bile acid levels in (b) feces and (c) urine, where open bars correspond to B6J and black bars to B6By. $*p < 0.05$, $**p < 0.001$. [Adapted from Phan et al.[11]]

The Ath diet has been used for nearly 20 years to study lipid metabolism and atherogenesis in inbred mouse strains in numerous laboratories around the world, and alterations in serum lipids and lipoprotein levels that occur in several inbred strains in response to the Ath diet have been characterized in detail. In B6J mice, for example, the Ath diet produces a major change in plasma lipoproteins, with a shift from primarily high density lipoproteins to the large, cholesterol-rich low density and very low density lipoproteins.[5,14,15] In contrast, B6By mice do not exhibit this shift in lipoprotein profile in response to the Ath diet, and maintain low overall cholesterol levels.[10] It is well accepted that atherogenesis in humans results from a combination of increased circulating cholesterol levels and induction of an inflammatory response.[1] Consistent with this mechanism, susceptibility of B6J mice to atherosclerosis on the Ath diet is characterized by the induction of inflammatory and oxidative stress response genes in the liver, including serum amyloid A proteins, colony-stimulating factors, and heme oxygenase.[16] Lipid peroxidation products such as conjugated dienes also dramatically increase in liver in response to the Ath diet.[16] These observations indicate that diet induced atherogenesis in the mouse involves similar mechanisms as those thought to occur in humans, and suggest that a better understanding of the relationship between diet and disease may be derived from studying the mouse model.

There is limited information about how individual components of the Ath diet influence lipoproteins or gene expression. At the gene expression level, it is well established that cholesterol controls its own synthesis by acting through the sterol regulatory element-binding protein (SREBP) transcription factors,[17] and dietary taurocholic acid decreases the relative abundance of mRNA for 7-alpha-hydroxylase, the rate-limiting enzyme in the bile acid biosynthesis.[18,19] In B6J mice, taurocholate is required for the large increase in low density/very low density lipoprotein levels observed on the Ath diet.[20] The presence of 0.5% cholate in a high fat diet also represses acyl-CoA synthetase expression, and prevents high fat diet-induced hyperglycemia and obesity in B6J mice.[21] These studies suggest that the effects of the Ath diet are a composite of independent or synergistic effects of the individual diet components, but do not provide a global picture of what the nutrient-specific effects are. We therefore set out to answer the following questions: (1) Which genes are regulated specifically by the cholesterol, cholic acid, and fat components of the Ath diet? and (2) Do expression levels for these genes differ between the susceptible B6J and resistant B6By mouse strains?

We have applied microarray technology to address these questions. For this purpose, B6J and B6By mice were fed one of five diets: mouse chow, the Ath diet, or modified versions of the Ath diet in which either cholesterol, cholate, or fat were omitted (see Table 8.1). After 3 weeks on a specific diet, liver RNA was isolated and used for cRNA synthesis and hybridization to Affymetrix

Table 8.1 Composition of the Five Diets

	Chow	Ath	No cholate	No cholesterol	No Fat
Cholate	−	+	−	+	+
Cholesterol	−	+	+	−	+
Fat	−	+	+	+	−

MU11K oligonucleotide microarrays to assess the relative expression levels of more than 10,000 mouse genes and ESTs. We initially focused our analysis of diet components on the effects produced in B6J mice, which are susceptible to atherogenic changes on the Ath diet. A comparison of gene expression profiles between the chow and Ath diet showed that 6.4% of the genes present on the microarray were activated at least threefold and 3.4% were repressed to a similar degree.[22] A systematic comparison of expression levels among all five diets allowed the identification of genes that are regulated by each of the specific dietary components. Using the criterion of at least twofold induction or inhibition by a particular component across all five diets, we defined six gene expression patterns: cholesterol activated, cholesterol repressed, cholate activated, cholate repressed, fat activated, and fat repressed.

Representative expression profiles for each of the six groups are shown in Fig. 8.5. The numbers of genes falling into each category are shown in Fig. 8.6. For example, dietary cholesterol specifically regulated 38 genes, 25 of which were activated (Fig. 8.6, upper panel) and 13 of which were repressed by this component (Fig. 8.6, lower panel). The greatest number of genes was regulated by cholate (104 genes total), and the fewest showed regulation specifically by fat (15 total). The magnitude of expression of fat-regulated genes was modest (twofold to fourfold) when compared with cholesterol- or cholate-regulated genes, which varied up to 30-fold.[22] We confirmed and extended previous findings for several genes that have been shown to be responsive to the Ath diet. For example, expression of apolipoprotein A-IV, a protein constituent of various lipoprotein classes, was previously shown to be repressed by the Ath diet,[23] and our studies revealed that it is specifically cholesterol that is responsible for the effect (Fig. 8.5). The expression of *Cyp7b1* is known to be regulated through feedback control by bile acids.[12,24] In agreement with this, we detected repression of *Cyp7b1* specifically by the cholate component of the Ath diet (Fig. 8.5).

We also observed regulation by one of the Ath diet components of several other genes having roles in lipid metabolism, but which have not previously been shown to be regulated by diet. For example, cholesterol-repressed genes included aquaporin-8 (2.6-fold), which is involved in canalicular bile secretion from liver,[25] and *Cyp17* (9.6-fold), an enzyme involved in C-21 steroid synthesis.[26,27] Among the lipid metabolism genes induced by cholate were phospholipid transfer protein (4.1-fold), which is involved in lipoprotein remodeling in the circulation, the oxysterol binding nuclear hormone receptor LXRβ (fivefold), the phospholipid biosynthetic enzyme choline kinase (6.5-fold), and the intracellular lipid transporter protein, lipocalin 2 (13.5-fold).[22] There were also genes that have been previously shown to be regulated by diet components that did not show up in our analysis. In many cases this was due to extremely low expression levels or because they failed to meet the strict criteria of having at least twofold differential expression levels in a consistent pattern across all five diets. In addition, there were sets of genes that required two, or even all three, of the Ath diet components for activation or repression

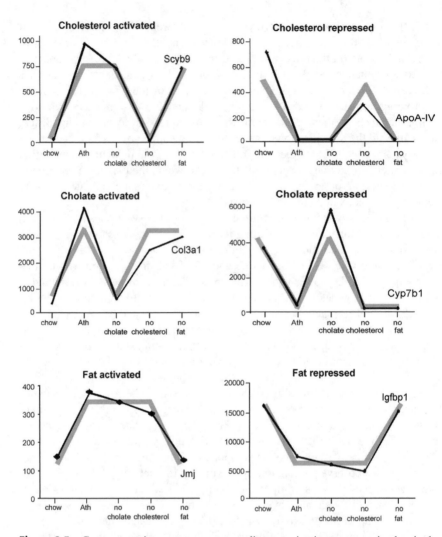

Figure 8.5 Gene expression patterns corresponding to activation or repression by single Ath diet components. The thick gray lines represent patterns of expression we expected for activation or repression of a gene specifically by cholesterol, cholate, or fat. The thin black lines show examples of genes following each of the patterns. Scyb9, Small inducible cytokine B subfamily (Cys-X-Cys); ApoA-IV, Apolipoprotein A-IV; Col3a1, procollagen type III alpha 1; Cyp7b1, Cytochrome P450, 7b1; Jmj, Jumonji; Igfbp1, Insulin-like growth factor binding protein 1.

(see intersections of Venn diagram circles in Fig. 8.6). For example, 100 genes were maximally activated only when both cholate and cholesterol were included. Among these were several genes having established roles in lipid metabolism, including lipoprotein lipase (3.8-fold), low density lipoprotein receptor related

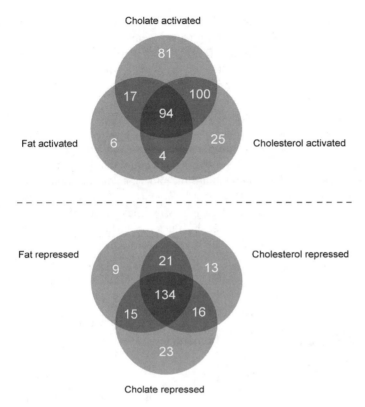

Cholate activated

81

17   100

94

Fat activated   6   25   Cholesterol activated

4

Fat repressed   Cholesterol repressed

9   21   13

134

15   16

23

Cholate repressed

Figure 8.6 Summary of gene expression response to components of the Ath diet. B6J mice were fed one of five diets containing various combinations of cholesterol, cholate, and fat (see Table 8.1), and liver RNA was analyzed by microarray hybridization. Venn diagram representation of genes activated or repressed by single diet components (light gray shading), by a combination of two components (intermediate shading), or requiring all three components for altered expression (dark shading in center).

protein associated protein 1 (3.6-fold), macrophage scavenger receptor AI (2.4-fold), and fatty acid binding protein 5 (3.8-fold) (L. Vergnes and K. Reue, unpublished data). A substantial number of genes also required all three components for activation (94 genes) or repression (134 genes). Although fewer in number, there were examples of genes that fell into each of the other categories as well, with a requirement for two of the three components for either activation or repression. These sets of genes most likely exhibit complex regulatory control. Further studies to examine the regulatory sequences associated with genes having a common expression pattern may lead to the identification of nutrient-responsive regulatory elements. In addition, since many of the genes in these groups are ESTs of unknown function, they represent interesting new candidates that may have a role in cholesterol homeostasis or atherogenesis.

Among the most valuable information provided by the gene array analysis of the Ath diet was the assignment of distinct proatherogenic effects to the cholesterol and cholate components of the diet (Fig. 8.7 and Vergnes et al.[22]). We found that cholesterol is the primary activator of genes involved in the acute inflammatory response, whereas cholate induced a program of gene expression that occurs in hepatic fibrogenesis, a response to inflammation in the liver. Thus, among the 25 cholesterol-activated genes, we detected 12 inflammatory/ immune response genes. These included members of the serum amyloid A (SAA) family (*Saa2*, *Saa3*, *Saa4*), histocompatibility antigens (H2-1A-α, H2-1A-β, H2-E-β, Ia-associated invariant chain), and additional inflammation/ immune associated genes such as interleukin-2 receptor γ, small inducible cytokine B9 (*Scyb9*), SAM domain and HD domain 1 (*Samhd1*), paired-Ig-like receptor A5 (*Pira5*), and galectin-3 (*Lgals3*) (see Fig. 8.7). These genes all required cholesterol for activation, as demonstrated by the drop to basal levels when

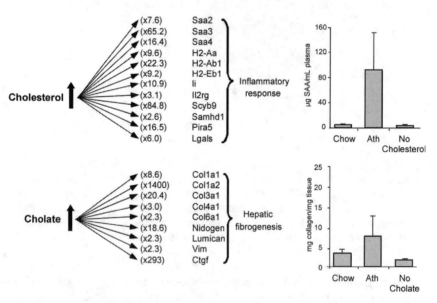

Figure 8.7 Activation of inflammatory response and fibrogenesis by distinct components of the Ath diet. B6J mice were fed one of five diets containing various combinations of cholesterol, cholate, and fat (see Table 8.1), and liver RNA was analyzed by microarray hybridization. As shown at left, cholesterol induced expression of inflammatory response genes, whereas cholate-activated genes involved in fibrogenesis. Fold induction of each gene by cholesterol or cholate is indicated in parentheses. As shown at right, the microarray analysis was followed by measurement of plasma serum amyloid A (SAA) protein levels and hepatic collagen content in mice fed the diets indicated. Results are consistent with the gene expression profiles, with SAA levels dramatically elevated by the Ath diet, but diminished by removal of cholesterol, and collagen levels exhibiting a requirement for dietary cholate.

cholesterol was omitted from the diet.[22] These results extend the previous observation that the Ath diet induces the acute inflammatory SAA genes[16] by demonstrating that cholesterol is the critical component responsible for this induction.

Among the genes induced by cholate, we detected several encoding collagen and related extracellular matrix proteins (Fig. 8.7 and Vergnes et al.[22]). The expression and excretion of extracellular matrix proteins is indicative of fibrogenesis, a process that occurs in response to inflammation induced by infectious or metabolic agents.[28,29] Five collagen genes were activated by cholate: *Col1a1* and *Col1a2* (which encode procollagen, type I, subunits alpha 1 and alpha 2), *Col3a1* (procollagen, type III, alpha 1), *Col4a1* (procollagen, type IV, alpha 1), and *Col6a1* (procollagen, type VI, alpha 1). Additional cholate-activated genes involved in extracellular matrix synthesis included nidogen and lumican, two proteins that have direct interactions with extracellular collagens, as well as vimentin, a cytoskeletal intermediate filament protein, and connective tissue growth factor (*Ctgf*), which modulates extracellular matrix secretion. These results demonstrate that cholate plays a key role in activating the expression of genes causing progression of inflammation to a chronic stage, leading to fibrosis in the liver and perhaps other sites, such as the vessel wall.

As the gene expression results showed clear cut differences in the effects of cholesterol and cholate on acute and chronic inflammatory responses, we sought to confirm these effects at the protein level. We determined that circulating SAA levels and hepatic collagen levels reflected the gene expression results, with elevated plasma SAA levels in response to cholesterol (Fig. 8.7, upper panel) and increased hepatic collagen deposition in response to cholate (Fig. 8.7, lower panel). These studies establish that cholesterol and cholate have distinct proatherogenic effects on gene expression, with cholesterol being required for activation of acute inflammatory genes, and cholate required for the fibrotic response to chronic liver inflammation. It is thought that both of these processes contribute to development of atherosclerotic lesions. The ability to "decouple" the two through dietary manipulation may allow further elucidation of the role that each process plays in atherogenesis.

Further support for the idea that the induction of inflammatory gene expression contributes to atherogenesis in B6J mice was derived from performing similar gene expression profiling in the atherosclerosis-resistant B6By strain. B6By mice were fed the five diets described earlier, and the same microarray experiment was performed. After similar computational analysis, we obtained a list of the six groups of genes induced or repressed by each diet component.[22] Overall, there were substantially fewer genes activated by cholesterol in B6By when compared with B6J mice (7 vs. 25). Some overlaps between the two strains were observed, such as cholesterol repression of *Cyp17* (3.5-fold) and cholate repression of *Cyp7b1* (5.8-fold). But the majority of genes found to be regulated by cholesterol and cholate in B6By were unique from those observed in B6J. The most striking difference was that the cholesterol induction of inflammatory genes and cholate induction of fibrogenic genes seen in B6J did not occur in B6By mice, except for a single collagen gene, *Col3a1*. The expression levels of the SAA gene family and histocompatibility antigens in B6By reached only 20% the levels of B6J, and many were not influenced by cholesterol at all.[22] Some of the fibrotic genes that were induced by cholate in B6J mice were activated slightly in B6By mice, but did not reach the criterion of twofold induction that we set for the analysis. Consistent with this result, when we measured plasma SAA levels and collagen in liver they were both significantly lower in B6By when compared with B6J (Fig. 8.8). The ability of B6By mice to more efficiently metabolize and excrete cholesterol[11] may prevent the accumulation of lipids in

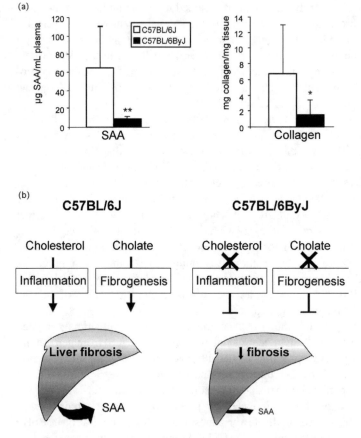

Figure 8.8 Attenuated inflammation and fibrogenesis in response to cholesterol and cholate in B6By mice. (a) Plasma serum amyloid A (SAA) and hepatic collagen levels were significantly lower in B6By vs. B6J after 3 weeks on the Ath diet. *t*-test, $^*p < 0.05$, $^{**}p < 0.01$. (b) Summary of the differential response to cholesterol and cholate in B6By and B6J mice. B6By showed less inflammation and less fibrotic response when compared with B6J.

the liver and attenuate the induction of inflammatory responses in liver which contribute to atherogenesis.

Analysis of *Diet1* Gene Candidates by Construction of a Locus-Specific DNA Microarray

As described earlier, we have learned about the metabolic basis for resistance to hypercholesterolemia in B6By mice and delineated the effects of specific Ath diet components on gene expression profiles using DNA microarrays. The primary goal at present is to isolate the *Diet1* gene and characterize its function. Known genes involved in cholesterol metabolism can be ruled out as candidates for *Diet1* based on the fact that they do not map within the *Diet1* gene region on chromosome 2. This suggests that *Diet1* is either a novel gene or a known gene that has a previously unrecognized role in lipid metabolism. The best strategy to isolate the gene is through a positional cloning approach, where identification of candidates is based on location rather than on the suspected function of the gene. With this strategy, the gene is mapped to a finite chromosomal region, and all genes residing there are evaluated as candidates. The ultimate screening of candidates involves sequencing the protein coding and gene regulatory regions to detect sequence variations that could alter protein structure or expression. As many mutant alleles underlying disease result in altered mRNA levels,[30-33] one strategy is to prioritize the candidates for sequencing by first screening for mRNA expression levels in the mutant when compared with wild-type tissue. This process is ideally suited to the use of DNA microarrays that contain the candidate genes under consideration. We are using such a strategy to screen positional candidates for *Diet1* (Fig. 8.9), and here we discuss some of the considerations that have gone into the design of the *Diet1* locus microarray.

We delimited the position of the *Diet1* gene to a region of approximately 5 Mb through genetic mapping. The identification of candidate genes in this interval has recently become considerably more straightforward than it was previously because of the availability of the draft mouse genome assembly, which contains sequence for all known genes as well as genes predicted by several algorithms.[34] These include Genscan, Geneid, and Ensembl, which predict genes on the basis of structural features such as open reading frames, intron/ exon boundaries, and CpG islands, as well as Twinscan and SGP, which rely upon mouse/human homology to assign genes.[35] We have taken a conservative approach and included nearly all gene predictions produced by the algorithms named earlier, including those obtained by a single algorithm, to ensure that genuine genes are not missed. Using these criteria, within the *Diet1* interval we identified 18 known genes, as well as another 50 putative genes predicted by algorithms (L. Vergnes and K. Reue, unpublished data).

Having compiled a catalog of known and predicted genes in the *Diet1* locus, it is feasible to begin screening candidates. The ultimate test will be to identify sequence variations between B6J and B6By mice that produce aberrant

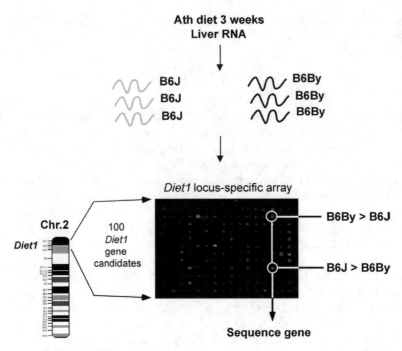

Figure 8.9 Strategy to analyze *Diet1* gene candidates using a locus-specific DNA micro-array. A DNA array corresponding to the *Diet1* locus was produced after amplification and printing cDNA fragments representing each candidate within the gene interval. Differentially labeled cDNA from liver of B6By and B6J mice fed the Ath diet were hybridized to the array. Examples of genes showing increased or decreased expression in B6By when compared with B6J represent good candidates for further analysis by sequencing.

protein structure or quantity. Because sequencing all 68 candidate genes in the two strains is not a trivial task, we decided to prioritize candidate genes for sequencing by first screening for variations between the two strains in mRNA expression levels. As we have described in the previous two sections, several differences in mRNA expression levels exist between B6J and B6By, most of which probably represent secondary effects of the *Diet1* mutation. However, we expect few expression differences among the 68 genes located within the *Diet1* gene interval, such that detection of expression differences will point to promising candidate genes within the interval. In order to include all known and predicted genes in the *Diet1* locus, some of which are not represented on commercial microarrays, we constructed a locus-specific cDNA microarray containing all of the candidate genes. We prepared cDNA fragments of 500–800 bp in length using reverse transcriptase-PCR for each of the candidate genes. In addition to the 68 gene candidates, we included a series of double-spotted control cDNAs, which included species that we previously determined to be induced, repressed, or expressed at similar levels in B6By and B6J liver. The cDNA fragments were purified and printed onto glass slides. These arrays can be hybridized to labeled cDNA samples from tissues of B6J and B6By mice to identify genes that are differentially expressed in the two strains. Such genes will represent excellent candidates for *Diet1*, and will receive top priority for sequence analysis to search for variations between the two strains that could account for the differences observed in cholesterol metabolism.

SUMMARY

The advent of DNA microarray technology has made it possible to approach the characterization of mutant phenotypes at the molecular level in an efficient manner. Here we have described the utility of this technology to characterize several aspects of a mouse mutation that affects susceptibility to diet-induced hypercholesterolemia and atherosclerosis. A major advance in our understanding of the resistant phenotype observed in B6By mice resulted from the detection of aberrant mRNA expression levels for several genes involved in bile acid synthesis and transport. Additional insight into differences between B6J and B6By responsiveness to the atherogenic diet was provided by analysis of the differential effects of individual diet components on proatherogenic gene expression in the two strains. Finally, generation of a custom locus-specific DNA microarray serves as a useful tool for efficient screening of positional candidate genes, and may accelerate identification of the *Diet1* gene. Similar strategies should be applicable to characterization of disease phenotypes in the mouse, human, and other organisms.

REFERENCES

1. Steinberg, D. Atherogenesis in perspective: hypercholesterolemia and inflammation as partners in crime. Nat. Med. **2002**, *8*, 1211–1217.
2. Kern, F., Jr. Normal plasma cholesterol in an 88-year-old man who eats 25 eggs a day. N. Engl. J. Med. **1991**, *324*, 896–899.
3. Lusis, A.J.; Weinreb, A.; Drake, T.A. Genetics of atherosclerosis. In *Textbook of Cardiovascular Disease*; Topol, E.J., Ed.; Lippincott-Raven Publishers: Philadelphia, 1998; 2389–2413.
4. Reue, K.; Doolittle, M.H. Naturally occurring mutations in mice affecting lipid transport and metabolism. J. Lipid Res. **1996**, *37*, 1387–1405.
5. Paigen, B.; Morrow, A.; Brandon, C.; Mitchell, D.; Holmes, P. Variation in susceptibility to atherosclerosis among inbred strains of mice. Atherosclerosis **1985**, *57*, 65–73.
6. Paigen, B.; Mitchell, D.; Reue, K.; Morrow, A.; Lusis, A.J.; LeBoeuf, R.C. *Ath-1*, a gene determining atherosclerosis susceptibility and high density lipoprotein levels in mice. Proc. Natl. Acad. Sci. USA **1987**, *84*, 3763–3767.

7. Reue, K. Mouse genetic models in atherosclerosis and lipid metabolism research. In *Genetic Models in Cardiorespiratory Biology*; Haddad, G.G., Xu, T., Eds.; Marcel Dekker, Inc.: New York, **2001**; 313–339.

8. Sheth, S.S.; Deluna, A.; Allayee, H.; Lusis, A.J. Understanding atherosclerosis through mouse genetics. Curr. Opin. Lipidol. **2002**, *13*, 181–189.

9. Qiao, J.H.; Xie, P.Z.; Fishbein, M.C.; Kreuzer, J.; Drake, T.; Demer, L.L.; Lusis, A.J. Pathology of atheromatous lesions in inbred and genetically engineered mice. Genetic determination of arterial calcification. Arterioscler. Throm. **1994**, *14*, 1480–1497.

10. Mouzeyan, A.; Choi, J.; Allayee, H.; Wang, X.; Sinsheimer, J.; Phan, J.; Castellani, L.; Reue, K.; Lusis, A.J.; Davis, R.C. A locus conferring resistance to diet-induced hypercholesterolemia and atherosclerosis on mouse chromosome 2. J. Lipid Res. **2000**, *41*, 573–582.

11. Phan, J.; Pesaran, T.; Davis, R.C.; Reue, K. The *Diet1* locus confers protection against hypercholesterolemia through enhanced bile acid metabolism. J. Biol. Chem. **2002**, *277*, 469–477.

12. Schwarz, M.; Lund, E.G.; Russell, D.W. Two 7alpha-hydroxylase enzymes in bile acid biosynthesis. Curr. Opin. Lipidol. **1998**, *9*, 113–118.

13. Repa, J.; Mangelsdorf, D. The role of orphan nuclear receptors in the regulation of cholesterol homeostasis. Annu. Rev. Cell. Dev. Biol. **2000**, *16*, 459–481.

14. LeBoeuf, R.C.; Doolittle, M.H.; Montcalm, A.; Martin, D.C.; Reue, K.; Lusis, A.J. Phenotypic characterization of the *Ath-1* gene controlling high density lipoprotein levels and susceptibility to atherosclerosis. J. Lipid Res. **1990**, *31*, 91–101.

15. Ishida, B.Y.; Blanche, P.J.; Nichols, A.V.; Yashar, M.; Paigen, B. Effects of atherogenic diet consumption on lipoproteins in mouse strains C57BL/6 and C3H. J. Lipid Res. **1991**, *32*, 559–568.

16. Liao, F.; Andalibi, A.; deBeer, F.C.; Fogelman, A.M.; Lusis, A.J. Genetic control of inflammatory gene induction and NF-kappa B-like transcription factor activation in response to an atherogenic diet in mice. J. Clin. Invest. **1993**, *91*, 2572–2579.

17. Horton, J.D.; Goldstein, J.L.; Brown, M.S. SREBPs: activators of the complete program of cholesterol and fatty acid synthesis in the liver. J. Clin. Invest. **2002**, *109*, 1125–1131.

18. Pandak, W.M.; Li, Y.C.; Chiang, J.Y.; Studer, E.J.; Gurley, E.C.; Heuman, D.M.; Vlahcevic, Z.R.; Hylemon, P.B. Regulation of cholesterol 7 alpha-hydroxylase mRNA and transcriptional activity by taurocholate and cholesterol in the chronic biliary diverted rat. J. Biol. Chem. **1991**, *266*, 3416–3421.

19. Shefer, S.; Nguyen, L.B.; Salen, G.; Ness, G.C.; Tint, G.S.; Batta, A.K.; Hauser, S.; Rani, I. Regulation of cholesterol 7 alpha-hydroxylase by hepatic 7 alpha-hydroxylated bile acid flux and newly synthesized cholesterol supply. J. Biol. Chem. **1991**, *266*, 2693–2696.

20. Dueland, S.; Drisko, J.; Graf, L.; Machleder, D.; Lusis, A.J.; Davis, R.A. Effect of dietary cholesterol and taurocholate on cholesterol 7 alpha-hydroxylase and hepatic LDL receptors in inbred mice. J. Lipid Res. **1993**, *34*, 923–931.

21. Ikemoto, S.; Takahashi, M.; Tsunoda, N.; Maruyama, K.; Itakura, H.; Kawanaka, K.; Tabata, I.; Higuchi, M.; Tange, T.; Yamamoto, T.T.; Ezaki, O. Cholate inhibits high-fat diet-induced hyperglycemia and obesity with acyl-CoA synthetase mRNA decrease. Am. J. Physiol. **1997**, *273*, E37–E45.

22. Vergnes, L.; Phan, J.; Strauss, M.; Tafuri, S.; Reue, K. Cholesterol and cholate components of an atherogenic diet induce distinct stages of hepatic inflammatory gene expression. J. Biol. Chem. **2003**, *278*, 42774–42784.

23. Baroukh, N.; Ostos, M.A.; Vergnes, L.; Recalde, D.; Staels, B.; Fruchart, J.; Ochoa, A.; Castro, G.; Zakin, M.M. Expression of human apolipoprotein A-I/C-III/A-IV gene cluster in mice reduces atherogenesis in response to a high fat-high cholesterol diet. FEBS Lett. **2001**, *502*, 16–20.

24. Pandak, W.M.; Hylemon, P.B.; Ren, S.; Marques, D.; Gil, G.; Redford, K.; Mallonee, D.; Vlahcevic, Z.R. Regulation of oxysterol 7alpha-hydroxylase (CYP7B1) in primary cultures of rat hepatocytes. Hepatology **2002**, *6*, 1400–1408.

25. Huebert, R.C.; Splinter, P.L.; Garcia, F.; Marinelli, R.A.; LaRusso, N.F. Expression and localization of aquaporin water channels in rat hepatocytes. Evidence for a role in canalicular bile secretion. J. Biol. Chem. **2002**, *277*, 22710–22717.

26. Lieberman, S.; Warne, P.A. 17-Hydroxylase: an evaluation of the present view of its catalytic role in steroidogenesis. J. Steroid Biochem. Mol. Biol. **2001**, *78*, 299–312.

27. Hartmann, R.W.; Ehmer, P.B.; Haidar, S.; Hector, M.; Jose, J.; Klein, C.D.; Seidel, S.B.; Sergejew, T.F.; Wachall, B.G.; Wachter, G.A.; Zhuang, Y. Inhibition of CYP 17, a new strategy for the treatment of prostate cancer. Arch. Pharm. (Weinheim) **2002**, *335*, 119–128.

28. Neubauer, K.; Saile, B.; Ramadori, G. Liver fibrosis and altered matrix synthesis. Can. J. Gastroenterol. **2001**, *15*, 187–193.

29. Li, D.; Friedman, S.L. Liver fibrogenesis and the role of hepatic stellate cells: new insights and prospects for therapy. J. Gastroenterol. Hepatol. **1999**, *14*, 618–633.

30. Péterfy, M.; Phan, J.; Xu, P.; Reue, K. Lipodystrophy in the *fld* mouse results from mutation of a new gene encoding a nuclear protein, lipin. Nat. Genet. **2001**, *27*, 121–124.

31. Zhang, Y.; Proenca, R.; Maffei, M.; Barone, M.; Leopold, L.; Friedman, J.M. Positional cloning of the mouse obese gene and its human homologue. Nature **1994**, *372*, 425–432.

32. Bultman, S.J.; Michaud, E.J.; Woychik, R.P. Molecular characterization of the mouse agouti locus. Cell **1992**, *71*, 1195–1204.

33. Bodnar, J.S.; Chatterjee, A.; Castellani, L.W.; Ross, D.A.; Ohmen, J.; Cavalcoli, J.; Wu, C.; Dains, K.M.; Chu, M.; Sheth, S.S.; Charugundla, K.; Demant, P.; West, D.B.; de Jong, P.; Lusis, A.J. Positional cloning of the combined hyperlipidemia gene *Hyplip1*. Nat. Genet. **2002**, *30*, 110–116.

34. The Mouse Genome Sequencing Consortium. Initial sequencing and comparative analysis of the mouse genome. Nature **2002**, *420*, 520–562.

35. Mathé, C.; Sagot, M.-F.; Schiex, T.; Rouzé, P. Current methods of gene prediction, their strengths and weaknesses. Nucl. Acids Res. **2002**, *30*, 4103–4117.

9

Microarray Analysis of Lipid Metabolism in Drug Abuse, Neurological and Psychiatric Disorders

E. Lehrmann, C.-T. Lee, J. Chen, and W. J. Freed

Cellular Neurobiology Research Branch, National Institute on Drug Abuse, NIH/DHHS, Baltimore, MD, USA

K. G. Becker

Gene Expression and Genomics Unit, Research Resources Branch, National Institute on Aging, NIH/DHHS, Baltimore, MD, USA

INTRODUCTION

Disturbances in lipid metabolism, lipid-dependent cellular processes, and lipid-mediated signal transduction pathways have been implicated in a number of human brain disorders, including multiple sclerosis (MS),[1] Alzheimer's disease (AD),[2,3] Down's syndrome,[4] Niemann-Pick type C disease,[5] and perhaps even schizophrenia[6–12] and drug abuse.[13–18] This chapter will discuss representative human brain disorders in which microarrays have been employed to analyze the underlying pathophysiology from a transcriptional perspective. Although this chapter is focused mainly on the postmortem analysis of human tissue, we will also consider studies in whole animals, primary cell cultures, and cell lines to show how these can be employed in a complementary manner to study human disorders.

To highlight the unique role of lipids in maintaining brain and neuronal functions, we briefly discuss four examples of lipid-dependent functions in the central nervous system (CNS). First, we illustrate the importance of oligodendrocyte-derived lipids for neuronal signal propagation by myelin formation and maintenance. Second, we illustrate the central role of membrane lipids as both a structural component and a reservoir of membrane-derived signaling molecules. Third, the integral role of cholesterol in synaptic function is highlighted by the requirement for astrocyte-derived cholesterol for neuronal synaptogenesis. Finally, endocannabinoids, endogenous lipid agonists at central cannabinoid receptors, exemplify a direct lipid-mediated signaling system. We also consider two conditions involving primary perturbation of lipid function in the brain: firstly, MS, which is functionally a disorder in which the integrity of CNS myelin is compromised and secondly, the central effects of cannabinoid substances. Although microarray studies of cannabinoid abuse in human postmortem brain have not been reported, the existing literature in animals and cell culture will be discussed. The remainder of the chapter will be devoted to two disorders, schizophrenia and drug abuse, for which some evidence of perturbed lipid metabolism and function has been found in postmortem tissue.

LIPID METABOLISM IN BRAIN FUNCTION

The brain is an extraordinarily lipid-rich structure, composed of $\sim$60% lipids.[19] Lipids provide a cellular boundary and structural integrity, supporting metabolic, signaling, and structural functions, and serve as an anchor and substrate for a wide range of biochemical processes in both neuronal and nonneuronal cells. Glycerophospholipids comprise the major fraction of brain lipids, accounting for 20–25% of the brain dry weight,[20] with phospholipids, cholesterol, and glycolipids accounting for $\sim$60% of the total neural membrane mass.[20] The phospholipid fraction is particularly rich in highly unsaturated essential fatty acids (HUEFAs), of which 75% consists of arachidonic acid (AA) and docosahexaenoic acid (DHA), with smaller amounts of linoleic acid (LA), alpha-LA (ALA), eicosapentaenoic acid (EPA), and the n-3 and n-6 forms of docosapentaenoic acid (DPA).[20] The essential fatty acids AA, DHA, EPA, and dihomogammalinolenic acid (DGLA) are either procured directly from the diet or supplied from metabolism of the main essential dietary fatty acids, LA and ALA. From the liver, they must be transported to the brain by gradient-dependent association/dissociation with plasma albumins, or transported in lipoprotein particles, from which they will be released primarily by the action of lipoprotein lipase from endothelial cells perfusing the tissue. Fatty acids can cross membranes by diffusion down a concentration gradient, created by the active removal of fatty acids from the cytosol by fatty acid binding proteins (FABPs), or may be actively transported by fatty acid transfer proteins (FATPs).[19,20]

Myelin Formation and Function

Rapid communication of information over long distances, for example, more than a few hundred microns, in both the CNS and the peripheral nervous system (PNS) is critically dependent on intact myelin. Myelin facilitates the propagation of action potentials along axons, by serving an insulating function much like the insulation of an electrical cable. The nodes of Ranvier are nonmyelinated intervals between myelinated axon segments, where sodium channels are concentrated and signal propagation occurs. Myelin is mainly composed of lipids and structural proteins, such as proteolipid protein (PLP1), myelin-associated glycoprotein (MAG), and myelin basic protein (MBP). As myelin matures, multiple membrane layers become tightly compacted into layers of juxtaposed membranes (Fig. 9.1). Schwann cells myelinate PNS axonal processes, whereas a cell type unique to the CNS, the oligodendrocyte, is responsible for CNS myelination. An essential difference is that each Schwann cell is devoted to myelinating one

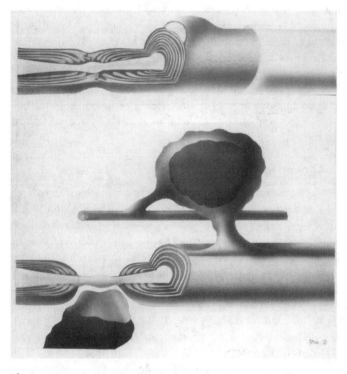

Figure 9.1 The oligodendrocyte-derived myelin sheath enables long-distance neural communication in the CNS. CNS myelin (lower picture) is organized differently than the Schwann-cell-derived PNS myelin (upper diagram). A specialized CNS cell, the oligodendrocyte, is responsible for myelinating axons in the brain. Myelin proteins, such as PLP1, MBP, MOBP, MPZ, MOG, and MAG, along with multiple layers of myelin membranes, which are extensions of the oligodendrocytic plasma membrane become compacted to form a myelin sheath that insulates axonal processes. The brief nonmyelinated intervals, the Nodes of Ranvier, allow for the saltatory signal conduction that permits the rapid transmission of signal over long distances in the CNS. Reproduced from Freed et al.[21]

axon, and in fact each axon will have many Schwann cells, whereas in the CNS each oligodendrocyte myelinates a number of axons (Fig. 9.1).

General Role of Lipids in Cell-to-Cell Signaling

As growth cones, synaptic and dendritic processes are composed of ~75% phospholipids,[22] it is not surprising that many proteins associated with lipid metabolism and organization are additionally involved in synaptogenesis and synaptic remodeling.[23] Bioactive lipids have been shown to modulate signaling in a number of different ways. In the developing brain, lysophosphatidic acid (LPA)-responsive cells emerge prior to gamma-aminobutyric acid (GABA) and glutamate (GLU) responsive cells,[24] and LPA continues to influence the developing brain as evidenced by the influence of LPA on the morphology and motility of immature postmitotic cortical neurons.[25] Membrane-derived signaling molecules, including AA, eicosanoids (such as prostaglandins, leukotrienes, thromboxanes, and hydroxyeicosatetraenoic acids), and platelet activating factor (PAF) are released by postsynaptic activation of phospholipases, and immediately affect ion channel activities, receptors, and enzymes. These signaling pathways subsequently alter gene expression via effects of lipid mediators on transcription. Also presynaptic neurotransmitter release or reuptake as well as neighboring neuronal or neuroglial cells may be modulated by the release of bioactive lipids into the extracellular space.[26]

Role of Astrocytic Cholesterol in Synaptogenesis

It has been reported that neurons in culture form only a few synapses if astrocytes are not present, and that the neuronal synapse formation coincides spatially and temporally with astrocyte development.[27,28] Later studies demonstrated that the synapse-inducing effect of astrocytes was due to the astrocytic secretion of cholesterol.[29] The role of astrocyte-derived cholesterol in neuronal function has been reviewed by Pfrieger.[30] In human brain disorders, cholesterol itself may be a risk factor for AD,[31] whereas Niemann-Pick disease type C involves a lysosomal accumulation of cholesterol.[5]

Cholesterol is synthesized from acetyl-Coenzyme A, and secreted in apolipoprotein (APO) E (APOE)-containing lipoprotein particles, which can be endocytosed via the low density lipoprotein (LDL) receptor. In that context, it is interesting to note that specific isoforms of the APOE allele appear to differentially affect neurite outgrowth and risk for several neurodegenerative disorders.[32] Cholesterol is further required for biogenesis of synaptic vesicles and binds to the integral synaptic vesicle protein synaptophysin.[33] Cholesterol and sphingolipid microdomains, called lipid rafts, on membranes are involved in a number of trafficking, signaling, and cytoskeletal functions, and are associated with postsynaptic proteins including surface alpha-amino-3-hydroxy-5-methyl-4-isoxazole-propionic acid (AMPA) receptors in dendrites. The importance of lipid rafts in maintaining normal synapse density and morphology was demonstrated by the gradual loss of synapses and dendritic spines after depletion of cholesterol or sphingolipid.[34] In addition to the specific role of cholesterol in synaptogenesis, fatty acids play many other important roles in neuronal function and plasticity. For example, sterol-regulatory element binding protein (SREBP) is a membrane-bound transcription factor which controls the expression of several genes involved in fatty acid metabolism, and in the brain it is expressed predominantly in neurons.[35] Glutamic acid decarboxylase-65, the rate-limiting enzyme in the biosynthesis of the neurotransmitter GABA, is targeted to cholesterol-rich membrane domains associated with synapses by palmitoylation of a region in the N-terminal domain of the protein.[36] Therefore, cholesterol and fatty acids are structural components of synaptic membranes that play essential regulatory roles in synapse formation and function.

Lipids in Cannabinoid Signaling

The endocannabinoid system is a prominent example of lipid-mediated information processing in the brain. Cannabinoid receptors are named for their affinity for delta9-tetrahydrocannabinol (delta9-THC), a ligand found in organic extracts from *Cannabis sativa*. The two types of cannabinoid receptors, CB1 and CB2, are coupled through $G_{i/o}$ proteins to signal transduction mechanisms, including inhibition of adenylyl cyclase (ADCY), activation of mitogen-activated protein kinases (MAPKs), and regulation of calcium and potassium channels.[37,38] The CB1 and CB2 cannabinoid receptors differ in their regional location and function. Whereas the CB1 receptor is predominantly expressed in CNS regions and in a variety of other organs, the CB2 receptor is predominantly associated with immune system functions.[39–41]

Endocannabinoids are the endogenous agonists for cannabinoid receptors and may mediate retrograde signaling at central synapses and other forms of short-range neuronal communication. They are widely distributed in the brain, are synthesized and released upon neuronal stimulation, undergo reuptake and are hydrolyzed intracellularly by fatty acid amide hydrolase (FAAH).[42] The endocannabinoids identified to date include arachidonoylethanolamide (anandamide), 2-arachidonylglyceryl ether (noladin ether) and 2-arachidonoylglycerol (2-AG).[38] The latter acts as a full agonist at both the CB1 and the CB2 cannabinoid receptors, and may be the most efficacious endogenous natural ligand for the cannabinoid receptors.[43] Anadamide and other *N*-acyl ethanolamines have been demonstrated to be elevated by dietary supplementation of the corresponding long-chain polyunsaturated fatty acids.[44]

MICROARRAYS AND GENE EXPRESSION PROFILING TECHNIQUES

The advent of microarray technologies has allowed for the measurement of large numbers of transcripts in minute amounts of brain tissue, thus permitting the characterization of transcriptional patterns and individual transcripts associated with the pathophysiology of human brain disorders.[7,15–17,45–51] Currently, two major types of microarrays predominate, oligonucleotide-based arrays (such as Affymetrix™ or Code-Link™ arrays) and a large number of commercial or in-house prepared cDNA-based arrays.[52,53] One significant difference between these platforms is probe length. The shorter oligonucleotide probes theoretically enable a high degree of stringency, which may require a highly intact RNA input.

Microarrays also differ in whether they are focused on tissue-specific or more universal transcripts, and hence the genes contained. The new NIA Mammalian Gene Collection (MGC) array (Fig. 9.2) is constructed predominantly from clones with validated high accuracy sequences and identified complete open reading frames,[54] exemplifies the focused array format (K.G. Becker, unpublished data). A neuronal/stem cell-focused array, the NIA NSC1, comprising 2666 neuronal, stem cell, and drug abuse-related cDNAs is also being constructed (C.-T. Lee et al., unpublished). Both arrays contain genes related to lipid catabolism, transport, biosynthesis, metabolism, and myelin structure (Table 9.1), and may be used to study lipid metabolism and function in parallel with changes in other functional groups.

Microarrays provide analog data related to the intensity of hybridization signals, but quantitative transcriptional profiles can be performed by simple sequencing of large numbers of individual transcripts. As it is impractical to routinely sequence large numbers of complete RNA molecules, an alternate approach to this problem involves sequencing smaller tags of 10–20 bases.

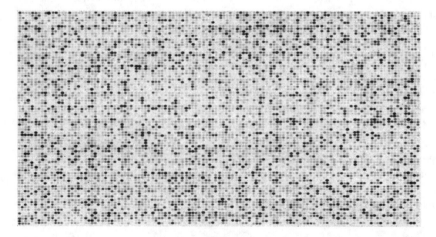

Figure 9.2 Example of a CNS-focused microarray, the NIA MGC array. The Mammalian Gene Collection (MGC) array is a cDNA array containing 9216 elements, and is constructed from validated high accuracy sequences and identified open reading frames.[54] The array is printed on a nylon membrane and utilizes ^{33}P-radiolabeled cDNA, which has a dynamic range spanning several logarithmic units of signal intensity.

Table 9.1 Examples of Lipid Metabolism and Myelination Genes on Human Neuronal/Stem Cell Array (NSC1), Broadly Annotated Using Gene Ontology™ (GO)-Annotation

Gene ontology (GO)-classification/ gene name	Unigene	LLID	Symbol
Acetyltransferase activity			
Bile acid CoA/glycine N-choloyltransferase	Hs.159440	570	*BAAT*
Carnitine acetyltransferase	Hs.12068	1384	*CRAT*
Glycine B complementing	Hs.18508	10249	*GLYAT*
Lecithin–cholesterol acyltransferase	Hs.348401	3931	*LCAT*
Cholesterol biosynthesis			
3-Hydroxy-3-methylglutaryl-CoA synthase 1 (sol.)	Hs.77910	3157	*HMGCS1*
3-Hydroxy-3-methylglutaryl-CoA synthase 2 (mit.)	Hs.59889	3158	*HMGCS2*
7-Dehydrocholesterol reductase	Hs.11806	1717	*DHCR7*
Farnesyl-diphosphate farnesyltransferase 1	Hs.48876	2222	*FDFT1*
Isopentenyl-diphosphate delta isomerase	Hs.76038	3422	*IDI1*
Cholesterol metabolism			
Sterol regulatory element binding TF 1	Hs.166	6720	*SREBF1*
Sterol regulatory element binding TF 2	Hs.108689	6721	*SREBF2*
ATP-binding cassette, sub-family A (ABC1), member 1	Hs.211562	19	*ABCA1*
Apolipoprotein A-I	Hs.93194	335	*APOA1*
Apolipoprotein A-II	Hs.237658	336	*APOA2*
Apolipoprotein C-I	Hs.268571	341	*APOC1*
Apolipoprotein D	Hs.75736	347	*APOD*
Apolipoprotein E	Hs.169401	348	*APOE*
Apolipoprotein J/clusterin	Hs.75106	1191	*CLU*
(Similar to) Apolipoprotein L	Hs.114309	8542	*APOL1*
HDL binding protein (vigilin)	Hs.177516	3069	*HDLBP*
LDL receptor-related protein 8, APOE receptor	Hs.54481	7804	*LRP8*
(Similar to) LDL receptor	Hs.213289	3949	*LDLR*
Fatty acid biosynthesis			
Acetyl-CoA carboxylase beta	Hs.183857	32	*ACACB*
Prostaglandin–endoperoxide synthase 2	Hs.196384	5743	*PTGS2*

(*continued*)

Table 9.1 Continued

Gene ontology (GO)-classification/ gene name	Unigene	LLID	Symbol
Protein kinase, AMP-activated, β 2 noncatalytic	Hs.50732	5565	*PRKAB2*
Protein kinase, AMP-activated, γ 1 noncatalytic	Hs.3136	5571	*PRKAG1*
Triosephosphate isomerase 1	Hs.83848	7167	*TPI1*
Fatty acid metabolism			
Acyl-CoA dehydrogenase, C-4-C-12 straight chain	Hs.79158	34	*ACADM*
Acyl-CoA dehydrogenase, long chain	Hs.1209	33	*ACADL*
Acyl-CoA dehydrogenase family, member 8	Hs.14791	27034	*ACAD8*
(Similar to) acyl-CoA oxidase 1, palmitoyl	Hs.379991	51	*ACOX1*
Degenerative spermatocyte hom./lipid desaturase	Hs.185973	8560	*DEGS*
Fatty acid binding protein 7, brain	Hs.26770	2173	*FABP7*
Fatty-acid-CoA ligase, long-chain 6	Hs.14945	23305	*FACL6*
Galactosylceramidase (Krabbe disease)	Hs.273	2581	*GALC*
Short-chain dehydrogenase/ reductase 1	Hs.17144	9249	*SDR1*
Phospholipid biosynthesis			
Phosphatidylinositol 4-kinase, catalytic, beta	Hs.154846	5298	*PIK4CB*
Phospholipid metabolism			
Glutathione peroxidase 4 (phospholipid hydroperoxidase)	Hs.2706	2879	*GPX4*
Nuclear receptor subfamily 1, group I, member 2	Hs.118138	8856	*NR1I2*
PYK2 N-terminal domain-interacting receptor 1	Hs.301272	83394	*PITPNM3*
Phospholipase A2, group V	Hs.290	5322	*PLA2G5*
Phospholipase A2, group VI (cytoplasmic, Ca^{2+}-independent)	Hs.120360	8398	*PLA2G6*
Phospholipase C, beta 2	Hs.355888	5330	*PLCB2*
Phospholipase C, delta 1	Hs.80776	5333	*PLCD1*
Phospholipase C, epsilon	Hs.153322	5334	*PLCE*

(continued)

Table 9.1 Continued

Gene ontology (GO)-classification/ gene name	Unigene	LLID	Symbol
Phospholipase C, γ 2 (phosphatidylinositol-specific)	Hs.75648	5336	*PLCG2*
Phospholipase D1	Hs.82587	5337	*PLD1*
Phospholipase D2	Hs.104519	5338	*PLD2*
Sphingolipid metabolism			
Sialyltransferase 8 A	Hs.82527	6489	*SIAT8A*
Sphingomyelin phosphodiesterase 1, acid lysosomal	Hs.77813	6609	*SMPD1*
Myelin sheath			
Myelin basic protein	Hs.69547	4155	*MBP*
Myelin protein zero (Charcot– Marie–Tooth neuropathy 1B)	Hs.93883	4359	*MPZ*
Myelin transcription factor 1	Hs.279562	4661	*MYT1*
Proteolipid protein 1 (Pelizaeus- Merzbacher disease)	Hs.1787	5354	*PLP1*

Note: The GO^{TM} terms were extracted from LocusLink (http://www.ncbi.nlm.nih.gov/LocusLink) and the linked GO^{TM} Consortium AmiGO browser (http://www.geneontology.org/). Phospholipases form a well-represented group of transcripts on the NSC1 array. Phospholipases have been demonstrated to be differentially altered postmortem in the brains of humans who suffered from schizophrenia[10] or abused cocaine.[13,15,18,55] PLA2, induced by synaptic activity resulting from the actions of a number of neurotransmitters, preferentially targets the Sn2-position in which most highly HUFAs are located, with the resultant genesis of lysophospholipids, the parent HUFA and oxidation products, all of which may have bioactivity. PLC and PLD preferentially target the Sn3 position of glycerophosphatidates, resulting in the formation of a phosphorylated nitrogenous base such as choline and in the formation of bioactive compounds 1,2-diacylglycerol for PLC and a phosphatidic acid for PLD. Phospholipid catabolism by phospholipase activity therefore specifically targets different regions of phospholipid, resulting in different catabolites, which subsequently initiate distinct signaling cascades.[6,10,26]

Serial analysis of gene expression (SAGE) involves sequential sequencing from 10 base tags[56] to 17–20 base tags (Long SAGE).[57] MicroSAGE utilizing small amounts of starting material has recently been evaluated to be representative, reproducible, and accurate.[58] Massively parallel signature sequencing (MPSS)[59] allows sequencing on the order of 1 million 20 base tags per sample, sufficient to provide a complete and quantitative picture of gene expression, including rare transcripts, within a sample. Data are obtained independent of technique and provide information on the total number of specific RNA transcripts per million, and can be accumulated over time for the development of large databases, relating differences in gene expression to genetic variation, or to patient characteristics.

CONSIDERATIONS IN USING POSTMORTEM BRAIN TISSUE

Logistics of Human Brain Studies

The use of human postmortem brain tissue is generally the only means to directly investigate the pathogenesis of human brain disorders. Obtaining, cataloging, and preservation of human brain tissue in an organized manner, however, require a major investment of resources. It is essential to obtain patient histories, perform neuropathological examinations, to preserve and catalog material in a systematic manner, and to identify any comorbid disease and drug and alcohol abuse. Some approaches to brain bank operation have been described in the literature.[60–63] Using postmortem tissue for gene expression studies with microarrays also requires that the tissue be relatively well preserved and appropriately stored.

Lipid metabolism is likely to be disturbed in a number of relatively common human disorders which are described elsewhere in this chapter, and which are the subject of collection by brain banks in the USA and elsewhere. There are many disorders, ranging from various forms of drug abuse to less common conditions, for which postmortem brain tissue may not be available. Appropriate controls, which are comparable to the condition being studied, must also be available. For example, for disorders that affect children, control brain samples will often not be possible to obtain.

Problems and Limitations

Ante- and postmortem factors including patient history, comorbid drug abuse, medication history, individual genetic and environmental variations, and life history may influence brain gene expression. Also, ante- and postmortem factors such as agonal state, brain pH, postmortem interval, and tissue storage may affect RNA quality and quantity.[64–66] Some of these factors can be dealt with by matching control and experimental samples, but it will not usually be possible to match samples for all possible confounding variables. For example, subjects and controls might be matched for age, postmortem interval, and gender. The potential influence of other possible factors, such as medications present, alcohol present at time of death, brain pH, manner of death, and duration of storage might be examined *post hoc*. When compared with cells or animal tissue samples, human postmortem brain tissue will be less readily available and not as experimentally "clean", but provides the opportunity to study human disorders directly and in considerable detail.

Tissue Samples vs. Captured Identified Cells

A major obstacle in deciphering gene expression profiles from microarray studies of brain tissue is rooted in the heterogeneity and multitude of cellular components that populate the CNS. Analyzing gene expression profiles at the level of individual cell types or even single cells has now become feasible with RNA amplification protocols that allow RNA from single cells to be amplified for gene expression profiling.[67] There are clear similarities between single cells of a specific phenotype,[68,69] but while methodology is reproducible, it also underscores that a single cell type is not homogeneous at the gene expression level.[67] A logical strategy for studies of bran function is to intially analyze gene expression in tissue samples. From studies in tissues, identification of transcripts that display altered regulation may allow a determination of which cell types are involved in the particular disorder in question. This information may then allow for studies in specific cell types by capturing RNA from identified single cells.

Postarray Analysis

Currently, there is no gold standard by which data are extracted from microarray studies nor on what constitutes significant gene expression changes. Gene regulation data are likely to be obtained using different experimental procedures and data processing methods, and thus a standard for the minimal reporting needed to allow for replication and evaluation of microarray data, the "Minimum information about a microarray experiment (MIAME, http://www.mged.org/index.html) was developed,[70] and has been adopted by a number of scientific journals. Matching samples for known confounds such as age, postmortem interval, brain pH, known neurological disease, or substance abuse is paramount. When designing studies using human subjects, it may also be desirable to build in the ability to *post hoc* subdivide or exclude samples, such that valuable information is not lost if a control or experimental sample later is found to be unusable.[15,66]

MICROARRAY STUDIES IN HUMAN BRAIN AND EXPERIMENTAL MODELS: DRUGS OF ABUSE, SCHIZOPHRENIA, AND MS

Studies of Drugs of Abuse in Human Postmortem Brain Samples

Effects of drugs of abuse such as alcohol, cannabis, cocaine, or amphetamines are manifest in a number of biochemical, signaling, structural, and morphological parameters.[66,71] Altered presence and function of lipids appear to be involved in damage to both gray and white matter in human alcoholics,[14] whereas alterations in phospholipid turnover,[55] myelination pattern,[13] and regional activity of phospholipase A2 (PLA2) were demonstrated in cocaine abusers.[18]

Abused substances, including both alcohol and cocaine, have been demonstrated to produce altered expression of messenger RNA encoding a number of lipid-associated molecules in human brain.[15–17] For example, metabolic enzymes, such as the rate-limiting enzyme for cholesterol synthesis, 3-hydroxy-3-methylglutaryl-Coenzyme A reductase (HMGCR) was decreased in alcohol abuse,[17] whereas CYP51, the rate-limiting first step in the conversion of lanosterol to cholesterol[72] was altered in the prefrontal cortex of cocaine abusers.[15] Phosphatidic acid phosphatase type 2C (PPAP2C), altered in the prefrontal cortex in cocaine abusers,[15] converts phosphatidic acid to diacylglycerol (DAG), and functions in *de novo* synthesis of glycerolipids as well as in receptor-activated signal transduction mediated by phospholipase D (PLD). Apolipoproteins are important for various lipid transport and exchange functions, and altered expressions of APO A1 (APOA1),[17] APOD,[16,17] and APOE[16] were found in microarray studies in postmortem brain tissue from alcohol abusers.

A large number of myelin-related gene expression changes have been demonstrated in postmortem brains from humans who abused cocaine or alcohol, including PLP1,[15–17] glycoproteins M6A and M6B (GPM6A and GPM6B),[15,17] MAG and MBP, myelin oligodendrocyte glycoprotein (MOG), myelin protein zero (MPZ), OMGP, peripheral myelin protein 22 (PMP22), and myelin and lymphocyte protein/T-cell differentiation protein (MAL).[16] Further, the proteolipid promoter-binding transcription factor, myelin transcription factor 1 (MYT1), was found to be differentially regulated in the two different cohorts studied.[16]

It is notable that though the direction of regulation of myelin-associated oligodendrocyte basic protein (MOBP) differed between the latter two studies, almost all myelin- and APO-related genes were downregulated in alcohol abuse brains,[16,17] which may reflect the transcriptional equivalent of the neuropathological findings of white matter loss and demyelination in alcoholics.[14] In contrast, a number of transcripts encoding proteins that may affect lipid metabolism and signaling were upregulated, including calmodulin (CALM), calcineurin Abeta (PPP3CB), PLA2, PAF acetylhydrolase, isoform Ib, alpha (PAFAH1B1), the small GTP binding protein ras-related nuclear protein (RAN), regulator of G-protein signaling 4 (RGS4), and the PLD-inhibitor synuclein alpha (SNCA).

Changes in gene expression profiles in animal models of abuse have also demonstrated altered lipid metabolism and structural genes. Cocaine was demonstrated to alter the expression of lipid-related signal transduction cascades, including upregulation of the adenosine A1 receptor (ADORA1), which interacts with G-proteins to inhibit adenylate cyclase,[73] whereas phosphatidylinositol 4-kinase (Pik4ca) was downregulated after 3 days, but not after 1 day of "binge" cocaine administration.[74] Freeman et al.[75] reported that self-administration of cocaine resulted in upregulation of ADCY8 at both the mRNA and the protein levels whereas protein kinase C alpha (PRKCA) mRNA and protein were induced in the rat hippocampus after "binge" administration of cocaine.[76] Morphine was reported to induce changes in FABP, glyceraldehyde 3 phosphate dehydrogenase (Gapdh), acid ceramidase (Asah), saposin (SAP), and ApoE.[77]

Cannabinoids

Although there has not been a microarray study of human cannabinoid abuse, several studies have examined the effects of cannabinoids on gene expression in animals and cell cultures. Kittler et al.[78] administered delta9-THC to rats, and observed changes in the expression of a number of genes involved in cannabinoid synthesis or second messenger systems [prostaglandin D synthase (Ptgds) and

Calm], as well as prominent changes in the expression of a number of genes related to myelination or structural plasticity. Lipid-related changes included phospholipid glutathione peroxidase (Gpx4), Calm, Mobp, Plp, and Fabp7. Neural cell adhesion molecule (NCAM) expression was increased after 24 h, but this change had disappeared by 1 week of delta9-THC administration. A decrease in myelin basic protein (Mbp) expression was present after both 1 and 3 weeks of delta9-THC administration, but not after 24 h. Increased expressions of myelin proteolipid protein (Plp) at all time points and Mobp at 7 days were also found.

Grigorenko et al.[79] performed microarray experiments to compare the effects of delta9-THC *in vivo* with those seen using an *in vitro* model, in which a pretreatment with the delta9-THC agonist WIN-55212-2 attenuated the neurodegenerative effect of *N*-methyl-D-aspartate (NMDA) on primary rat hippocampal neuronal cultures. Several genes showed changes in expression *in vitro* comparable to those induced by delta9-THC *in vivo*, including NCAM and Mbp. Neuroprotective WIN-55212-2 pretreatment reversed the expressional changes in NCAM and Plp seen with NMDA alone.

Finally, Parmentier-Batteur et al.[80] examined gene expression in the mouse forebrain 12 h after administration of delta9-THC or WIN-55212-2. Both the drugs changed the expression of 20 genes in a similar way, including *Ptgds* and several genes associated with neuronal signaling and differentiation. Interestingly, Plp expression was decreased by delta9-THC, but not by WIN-55212-2. Thus, cannabinoid receptor stimulation produces complex and time-dependent changes in the expression of a wide range of genes, many with no known direct connection to cannabinoid signaling pathways.

Schizophrenia

Several studies have indicated that there may be significant changes in lipid metabolism and function in schizophrenia.[6,10,19] Microarray studies of schizophrenia have, in most cases, suggested that this disorder principally involves changes in the expression of genes related to synaptic function, glutamate and GABA-mediated neurotransmission, signaling, and certain metabolic pathways.[9,46,48,49] There are, nonetheless, some indications that aberrant myelin-related functions are present in schizophrenia. Hakak et al.[7] reported decreased expression of several myelin-related genes in schizophrenia. A subsequent study[81] identified a decrease in number of oligodendrocytes in schizophrenia as a possible basis for the decreased expression of myelin-related genes, citing a number of additional studies that were generally consistent with this hypothesis. This concept also has some support from a study by Pongrac et al.,[82] showing a decrease in the expression of the major myelin protein PLP1 gene in some schizophrenic subjects. Notably, decreased expression of PLP1 was seen in five of ten schizophrenic subjects, with increases in two subjects and essentially no change in the remaining three subjects. Therefore, although loss of myelination is not likely to be the principal cause of schizophrenia, impaired myelin function may contribute to the impaired thought processes, which are the essence of this disorder.

One further study indicating abnormal lipid function in schizophrenia employed specialized microarrays to examine a relatively small number of transcripts.[45] This study identified mRNA encoding apolipoproteins APOL1, APOL2, and APOL4 as being substantially increased in the prefrontal cortex in schizophrenia. The authors note that APOL genes are clustered in the 22q-11–22q-13 chromosomal region, which has been linked to susceptibility for schizophrenia, and contains the catechol *o*-methyltransferase (COMT) gene that strongly affects prefrontal function.[83] The physiological significance of this finding is not clear; however, the important role of apolipoproteins in brain lipid function and synapse formation has been mentioned previously.

Multiple Sclerosis

An autoimmune process in MS may target proteins of the myelin sheath, oligodendrocytes, and neurons, producing oligodendrocytic and neuronal loss and demyelinated axons. Oligodendrocytic progenitor cells can migrate in and may develop into myelinating oligodendrocytes, thus, in part resolving the white matter axonal demyelination by remyelinating denuded axons.[84] However, MS patients possessing the APOE epsilon2 allele are characterized by the absence of remyelination.[1] An additional feature of MS is the presence of axonal loss,[85] and in that context, it is interesting to note that humans who lack PLP1 develop length-dependent axonal degeneration in the absence of demyelination and inflammation.[86]

MS disease activity differs with disease stage. Echoing the different pathology and activity in the center and edges of lesions in chronic MS,[87] it was demonstrated that more genes display differences in expression at the lesion margin than in the lesion centers for both active and inactive lesions.[88] The difference between lesion center and lesion edge appears more pronounced for silent lesions, whereas the average magnitude of gene expression difference was two- to threefold higher in both the center and edge of active lesions.[88] Microarray studies of acute lesions with inflammation or silent lesions without inflammation have demonstrated a number of changes in lipid metabolism associated with MS. The initial study by Whitney et al.[50] studied acute lesions from one MS patient, and found expression changes in the retinoic acid receptor alpha-1 (RARA) and a gene highly similar to MYT1.[50] Studying acute and silent MS lesions, Lock et al.[89] found reduced expression of key myelin-associated proteins and enzymes, including MAG, peripheral myelin protein 2 (PMP2), PLP1, OMGP and MOBP, and UDP-galactose ceramide galactosyl transferase (CGT),[90] as well as reduced expression of neuronal markers such as pentraxin 1 (PTX1), 43 kDa growth-associated protein (GAP43), neurofilaments light and medium (NEFL and NEF3) and synaptic proteins such as synaptotagmin (SYT), synaptobrevin 1 and 2 (VAMP1 and VAMP2), synapsin (SYN), and the neuronal glycoproteins GPM6A and GPM6B.

Comparing gene expression changes in human MS lesions with those obtained using the predominant animal model system, murine experimental allergic encephalitis (EAE), Whitney et al.[51] demonstrated upregulation of 5-lipoxygenase (5-LO) in both, underscoring the involvement of lipid mediators in the inflammatory reactions associated with both MS and EAE.[51] A general downregulation of myelination genes such as Mog and Plp and trend to downregulation of Mbp, Mobp, and Mag were further demonstrated in both the active phase of EAE and in the recovery phase.[91] As a result, and consistent with the pathophysiology of different phases of MS,[84,87] microarray studies of MS and the murine model EAE suggest that there are different and reduced levels of myelin-associated transcripts as well as a reduced capacity for myelin repair, both in the acute phase and in the recovery phase of MS. In both conditions, inflammatory lipid mediators such as 5-LO appear strongly upregulated.

CONCLUSIONS

Microarray analysis of disorders of the brain have demonstrated an important role played by lipid metabolism, structure, and signaling in several very different afflictions including MS, drug abuse, and schizophrenia. Lipids play a number of critical and unique roles in the functioning of the nervous system. These include unique structural roles (myelin), regulatory (synapse formation) and signaling processes (cannabinoids, second messengers, and membrane function). These processes are related to functions that lipids perform in other organs, but also have special features that are unique to the brain. A few disorders are directly related to lipid functions in the brain, and include disorders of lipid metabolism, especially lysosomal storage disorders, and conditions related to loss of myelin integrity or function, such as MS.

The recent advent of microarray technology, which can now be performed readily in many laboratories, allows for broad surveys of gene expression changes to be performed in brain disorders of unknown pathophysiology. Studies of gene expression in the brain in human disorders, and in animal models, have resulted in the identification of a range of possible lipid-related dysfunctions in brain disorders not directly associated with aberrations in lipid metabolism. For example, the expression of CYP51, the rate-limiting first step in the conversion of lanosterol to cholesterol[72] was altered in the prefrontal cortex of cocaine abusers.[66]

Several changes in the expression of genes related to lipid function have been unexpectedly found in brain disorders. Apolipoproteins are important for lipid transport and exchange functions. Altered expressions of APOA1, APOD, and APOE were found in microarray studies in postmortem brain tissue from alcohol abusers.[16,17] In schizophrenia, altered expressions of APOL1, APOL2, and APOL4 were substantially increased in the prefrontal cortex.[45] The authors note that each of these APOL genes are clustered in the 22q-11–22q-13 chromosomal region. Expression of the major myelin proteolipid protein PLP1 gene appears to be changed in a number of human brain disorders, including MS lesions,[89] schizophrenia,[7,82] alcoholism,[16,17] and cocaine abuse.[15]

The role of these changes in lipid-related genes in the pathogenesis of brain disorders is entirely unknown at present. These conditions, such as alcohol abuse, cocaine abuse, and schizophrenia, have not classically been ascribed to altered lipid function. It is likely that altered lipid-mediated functions are not primary features of these disorders, but may be altered in response to other events. Future studies are likely to explore metabolic pathways that are responsible for these changes and further exploration of the role of these transcripts in brain function. The contribution of changes in lipid-mediated functions to pathophysiology may be explored using other methods, such as gene targeting techniques in mouse models. At present, these studies are only in the early stages and changes in gene expression in human brain disorders have been explored only in a very few studies. Nonetheless, the common observations of, for example, alterations in PLP1 and apolipoprotein expression in a number of brain disorders strongly suggest that these changes are of significant functional importance. Thus, it appears likely that aberrations in lipid-mediated functions are present in a number of conditions, such as schizophrenia and drug abuse, which are characterized by alterations in higher order brain functions.

REFERENCES

1. Carlin, C.; Murray, L.; Graham, D.; Doyle, D.; Nicoll, J. Involvement of apolipoprotein E in multiple sclerosis: absence of remyelination associated with possession of the APOE epsilon2 allele. J. Neuropathol. Exp. Neurol. **2000**, *59*, 361–367.
2. Papassotiropoulos, A.; Streffer, J.R.; Tsolaki, M.; Schmid, S.; Thal, D.; Nicosia, F.; Iakovidou, V.; Maddalena, A.; Lutjohann, D.; Ghebremedhin, E.; Hegi, T.; Pasch, T.; Traxler, M.; Bruhl, A.; Benussi, L.; Binetti, G.; Braak, H.; Nitsch, R.M.; Hock, C. Increased brain beta-amyloid load, phosphorylated tau, and risk of Alzheimer disease associated with an intronic CYP46 polymorphism. Arch. Neurol. **2003**, *60*, 29–35.
3. Roher, A.E.; Weiss, N.; Kokjohn, T.A.; Kuo, Y.M.; Kalback, W.; Anthony, J.; Watson, D.; Luehrs, D.C.; Sue, L.; Walker, D.; Emmerling, M.; Goux, W.; Beach, T. Increased A beta peptides and reduced cholesterol and myelin proteins characterize white matter degeneration in Alzheimer's disease. Biochemistry **2002**, *41*, 11080–11090.
4. Murphy, E.J.; Schapiro, M.B.; Rapoport, S.I.; Shetty, H.U. Phospholipid composition and levels are altered in Down syndrome brain. Brain Res. **2000**, *867*, 9–18.
5. Carstea, E.D.; Morris, J.A.; Coleman, K.G.; Loftus, S.K.; Zhang, D.; Cummings, C.; Gu, J.; Rosenfeld, M.A.; Pavan, W.J.; Krizman, D.B.; Nagle, J.; Polymeropoulos, M.H.; Sturley, S.L.; Ioannou, Y.A.; Higgins, M.E.; Comly, M.; Cooney, A.; Brown, A.; Kaneski, C.R.; Blanchette-Mackie, E.J.; Dwyer, N.K.; Neufeld, E.B.; Chang, T.Y.; Liscum, L.; Strauss, J.F., III; Ohno, K.; Zeigler, M.; Carmi, R.; Sokol, J.; Markie, D.; O'Neill, R.R.; van Diggelen, O.P.; Elleder, M.; Patterson, M.C.; Brady, R.O.; Vanier, M.T.; Pentchev, P.G.; Tagle, D.A. Niemann-Pick C1 disease gene: homology to mediators of cholesterol homeostasis. Science **1997**, *277*, 228–31.

6. Bennett, C.N.; Horrobin, D.F. Gene targets related to phospholipid and fatty acid metabolism in schizophrenia and other psychiatric disorders: an update. Prostaglandins Leukot. Essent. Fatty Acids **2000**, *63*, 47–59.
7. Hakak, Y.; Walker, J.R.; Li, C.; Wong, W.H.; Davis, K.L.; Buxbaum, J.D.; Haroutunian, V.; Fienberg, A.A. Genome-wide expression analysis reveals dysregulation of myelination-related genes in chronic schizophrenia. Proc. Natl. Acad. Sci. USA **2001**, *98*, 4746–4751.
8. Jensen, J.E.; Al-Semaan, Y.M.; Williamson, P.C.; Neufeld, R.W.; Menon, R.S.; Schaeffer, B.; Densmore, M.; Drost, D.J. Region-specific changes in phospholipid metabolism in chronic, medicated schizophrenia: (31)P-MRS study at 4.0 Tesla. Br. J. Psychiat. **2002**, *180*, 39–44.
9. Middleton, F.A.; Mirnics, K.; Pierri, J.N.; Lewis, D.A.; Levitt, P. Gene expression profiling reveals alterations of specific metabolic pathways in schizophrenia. J. Neurosci. **2002**, *22*, 2718–2729.
10. Ross, B.M.; Turenne, S.; Moszczynska, A.; Warsh, J.J.; Kish, S.J. Differential alteration of phospholipase A2 activities in brain of patients with schizophrenia. Brain Res. **1999**, *821*, 407–413.
11. Yao, J.K.; Leonard, S.; Reddy, R.D. Membrane phospholipid abnormalities in postmortem brains from schizophrenic patients. Schizophr. Res. **2000**, *42*, 7–17.
12. Puri, B.K. Impaired phospholipid-related signal transduction in advanced Huntington's disease. Exp. Physiol. **2001**, *86*, 683–685.
13. Bartzokis, G.; Beckson, M.; Lu, P.H.; Edwards, N.; Bridge, P.; Mintz, J. Brain maturation may be arrested in chronic cocaine addicts. Biol. Psychiat. **2002**, *51*, 605–611.
14. Kril, J.J.; Halliday, G.M. Brain shrinkage in alcoholics: a decade on and what have we learned? Prog. Neurobiol. **1999**, *58*, 381–387.
15. Lehrmann, E.; Oyler, J.; Vawter, M.P.; Hyde, T.M.; Kolachana, B.; Kleinman, J.E.; Huestis, M.A.; Becker, K.G.; Freed, W.J. Transcriptional profiling in the human prefrontal cortex: evidence for two activational states associated with cocaine abuse. Pharmacogenomics J. **2003**, *3*, 27–40.
16. Lewohl, J.M.; Wang, L.; Miles, M.F.; Zhang, L.; Dodd, P.R.; Harris, R.A. Gene expression in human alcoholism: microarray analysis of frontal cortex. Alcohol Clin. Exp. Res. **2000**, *24*, 1873–1882.
17. Mayfield, R.D.; Lewohl, J.M.; Dodd, P.R.; Herlihy, A.; Liu, J.; Harris, R.A. Patterns of gene expression are altered in the frontal and motor cortices of human alcoholics. J. Neurochem. **2002**, *81*, 802–813.
18. Ross, B.M.; Moszczynska, A.; Peretti, F.J.; Adams, V.; Schmunk, G.A.; Kalasinsky, K.S.; Ang, L.; Mamalias, N.; Turenne, S.D.; Kish, S.J.SJ. Decreased activity of brain phospholipid metabolic enzymes in human users of cocaine and methamphetamine. Drug Alcohol Depend. **2002**, *67*, 73–79.
19. Horrobin, D.F. Lipid metabolism, human evolution and schizophrenia. Prostaglandins Leukot. Essent. Fatty Acids **1999**, *60*, 431–437.
20. Farooqui, A.A.; Horrocks, L.A.; Farooqui, T. Glycerophospholipids in brain: their metabolism, incorporation into membranes, functions, and involvement in neurological disorders. Chem. Phys. Lipids **2000**, *106*, 1–29.
21. Freed, W.J.; de Medinaceli, L.; Wyatt, R.J. Promoting functional plasticity in the damaged nervous system. Science **1985**, *227*, 1544–1552.

22. Pfenninger, K.H.; Ellis, L.; Johnson, M.P.; Friedman, L.B.; Somlo, S. Nerve growth cones isolated from fetal rat brain: subcellular fractionation and characterization. Cell **1983**, *35*, 573–584.

23. Goritz, C.; Mauch, D.H.; Nagler, K.; Pfrieger, F.W. Role of glia-derived cholesterol in synaptogenesis: new revelations in the synapse–glia affair. J. Physiol. Paris **2002**, *96*, 257–263.

24. Dubin, A.E.; Bahnson, T.; Weiner, J.A.; Fukushima, N.; Chun, J. Lysophosphatidic acid stimulates neurotransmitter-like conductance changes that precede GABA and L-glutamate in early, presumptive cortical neuroblasts. J. Neurosci. **1999**, *19*, 1371–1381.

25. Fukushima, N.; Weiner, J.A.; Kaushal, D.; Contos, J.J.; Rehen, S.K.; Kingsbury, M.A.; Kim, K.Y.; Chun, J. Lysophosphatidic acid influences the morphology and motility of young, postmitotic cortical neurons. Mol. Cell. Neurosci. **2002**, *20*, 271–282.

26. Bazan, N.G.; Tu, B.; Rodriguez de Turco, E.B. What synaptic lipid signaling tells us about seizure-induced damage and epileptogenesis. Prog. Brain Res. **2002**, *135*, 175–185.

27. Pfrieger, F.W.; Barres, B.A. Synaptic efficacy enhanced by glial cells *in vitro*. Science **1997**, *277*, 1684–1687.

28. Ullian, E.M.; Sapperstein, S.K.; Christopherson, K.S.; Barres, B.A. Control of synapse number by glia. Science **2001**, *29*, 657–661.

29. Mauch, D.H.; Nagler, K.; Schumacher, S.; Goritz, C.; Muller, E.C.; Otto, A.; Pfrieger, F.W. CNS synaptogenesis promoted by glia-derived cholesterol. Science **2001**, *294*, 1354–1357.

30. Pfrieger, F.W. Outsourcing in the brain: do neurons depend on cholesterol delivery by astrocytes? BioEssays *25*, 72–78.

31. Burns, M.; Duff, K. Cholesterol in Alzheimer's disease and tauopathy. Ann. NY Acad. Sci. **2002**, *977*, 367–375.

32. Teter, B.; Xu, P.T.; Gilbert, J.R.; Roses, A.D.; Galasko, D.; Cole, G.M. Defective neuronal sprouting by human apolipoprotein E4 is a gain-of-negative function. J. Neurosci. Res. **2002**, *68*, 331–336.

33. Thiele, C.; Hannah, M.J.; Fahrenholz, F.; Huttner, W.B. Cholesterol binds to synaptophysin and is required for biogenesis of synaptic vesicles. Nat. Cell. Biol. **2002**, *2*, 42–49.

34. Hering, H.; Lin, C.C.; Sheng, M. Lipid rafts in the maintenance of synapses, dendritic spines, and surface AMPA receptor stability. J. Neurosci. **2003**, *23*, 3262–3271.

35. Ong, W.Y.; Hu, C.Y.; Soh, Y.P.; Lim, T.M.; Pentchev, P.G.; Patel, S.C. Neuronal localization of sterol regulatory element binding protein-1 in the rodent and primate brain: a light and electron microscopic immunocytochemical study. Neuroscience **2000**, *97*, 143–153.

36. Kanaani, J.; el-Hussuini, A.-D.; Aguilera-Moreno, A.; Diacovo, J.M.; Bredt, D.S.; Baekkeskov, S. A combination of three distinct trafficking signals mediates axonal targeting and presynaptic clustering of GAD65. J. Cell Biol. **2002**, *158*, 1229–1238.

37. Matsuda, L.A. Molecular aspects of cannabinoid receptors. Crit. Rev. Neurobiol. **1997**, *11*, 143–66.

38. Howlett, A.C. The cannabinoid receptors. Prostaglandins Other Lipid Mediat. **2002**, *68–69*, 619–631.

39. Breivogel, C.S.; Childers, S.R. The functional neuroanatomy of brain cannabinoid receptors. Neurobiol. Dis. **1998**, *5*, 417–431.

40. Munro, S.; Thomas, K.L.; Abu-Shaar, M. Molecular characterization of a peripheral receptor for cannabinoids. Nature **1993**, *365*, 61–65.
41. Pettit, D.A.; Harrison, M.P.; Olson, J.M.; Spencer, R.F.; Cabral, G.A. Immunohisto-chemical localization of the neural cannabinoid receptor in rat brain. J. Neurosci. Res. **1998**, *51*, 391–402.
42. Fride, E. Endocannabinoids in the central nervous system—an overview. Prostaglandins Leukot. Essent. Fatty Acids **2002**, *66*, 2212–2233.
43. Sugiura, T.; Kobayashi, Y.; Oka, S.; Waku, K. Biosynthesis and degradation of anandamide and 2-arachidonoylglycerol and their possible physiological significance. Prostaglandins Leukot. Essent. Fatty Acids **2002**, *66*, 173–192.
44. Berger, A.; Crozier, G.; Bisogno, T.; Cavaliere, P.; Innis, S.; Di Marzo, V. Anandamide and diet: inclusion of dietary arachidonate and docosahexaenoate leads to increased brain levels of the corresponding N-acylethanolamines in piglets. Proc. Natl. Acad. Sci. USA **2001**, *98*, 6402–6406.
45. Mimmack, M.L.; Ryan, M.; Baba, H.; Navarro-Ruiz, J.; Iritani, S.; Faull, R.L.; McKenna, P.J.; Jones, P.B.; Arai, H.; Starkey, M.; Emson, P.C.; Bahn, S. Gene expression analysis in schizophrenia: reproducible up-regulation of several members of the apolipoprotein L family located in a high-susceptibility locus for schizophrenia on chromosome 22. Proc. Natl. Acad. Sci. USA **2002**, *99*, 4680–4685.
46. Mirnics, K.; Middleton, F.A.; Marquez, A.; Lewis, D.A.; Levitt, P. Molecular characterization of schizophrenia viewed by microarray analysis of gene expression in prefrontal cortex. Neuron **2000**, *28*, 53–67.
47. Tang, W.X.; Fasulo, W.H.; Mash, D.C.; Hemby, S.E. Molecular profiling of midbrain dopamine regions in cocaine overdose victims. J. Neurochem. **2003**, *85*, 911–924.
48. Vawter, M.P.; Barrett, T.; Cheadle, C.; Sokolov, B.P.; Wood, W.H., III; Donovan, D.M.; Webster, M.; Freed, W.J.; Becker, K.G. Application of cDNA microarrays to examine gene expression differences in schizophrenia. Brain Res. Bull. **2001**, *55*, 641–650.
49. Vawter, M.P.; Crook, J.M.; Hyde, T.M.; Kleinman, J.E.; Weinberger, D.R.; Becker, K.G.; Freed, W.J. Microarray analysis of gene expression in the prefrontal cortex in schizophrenia: a preliminary study. Schizophr. Res. **2002**, *58*, 11–20.
50. Whitney, L.W.; Becker, K.G.; Tresser, N.J.; Caballero-Ramos, C.I.; Munson, P.J.; Prabhu, V.V.; Trent, J.M.; McFarland, H.F.; Biddison, W.E. Analysis of gene expression in mutiple sclerosis lesions using cDNA microarrays. Ann. Neurol. **1999**, *46*, 425–428.
51. Whitney, L.W.; Ludwin, S.K.; McFarland, H.F.; Biddison, W.E. Microarray analysis of gene expression in multiple sclerosis and EAE identifies 5-lipoxygenase as a component of inflammatory lesions. J. Neuroimmunol. **2001**, *121*, 40–48.
52. Shilling, P.D.; Kelsoe, J.R. Functional genomics approaches to understanding brain disorders. Pharmacogenomics **2002**, *3*, 21–45.
53. Watson, S.J.; Meng, F.; Thompson, R.C.; Akil, H. The "chip" as a specific genetic tool. Biol. Psychiat. **2000**, *48*, 1147–1156.
54. Strausberg, R.L.; Feingold, E.A.; Grouse, L.H.; Derge, J.G.; Klausner, R.D.; Collins, F.S.; Wagner, L.; Shenmen, C.M.; Schuler, G.D.; Altschul, S.F.; Zeeberg, B.; Buetow, K.H.; Schaefer, C.F.; Bhat, N.K.; Hopkins, R.F.; Jordan, H.; Moore, T.; Max, S.I.; Wang, J.; Hsieh, F.; Diatchenko, L.; Marusina, K.; Farmer, A.A.; Rubin, G.M.; Hong, L.; Stapleton, M.; Soares, M.B.; Bonaldo, M.F.; Casavant, T.L.; Scheetz, T.E.;

Brownstein, M.J.; Usdin, T.B.; Toshiyuki, S.; Carninci, P.; Prange, C.; Raha, S.S.; Loquellano, N.A.; Peters, G.J.; Abramson, R.D.; Mullahy, S.J.; Bosak, S.A.; McEwan, P.J.; McKernan, K.J.; Malek, J.A.; Gunaratne, P.H.; Richards, S.; Worley, K.C.; Hale, S.; Garcia, A.M.; Gay, L.J.; Hulyk, S.W.; Villalon, D.K.; Muzny, D.M.; Sodergren, E.J.; Lu, X.; Gibbs, R.A.; Fahey, J.; Helton, E.; Ketteman, M.; Madan, A.; Rodrigues, S.; Sanchez, A.; Whiting, M.; Madan, A.; Young, A.C.; Shevchenko, Y.; Bouffard, G.G.; Blakesley, R.W.; Touchman, J.W.; Green, E.D.; Dickson, M.C.; Rodriguez, A.C.; Grimwood, J.; Schmutz, J.; Myers, R.M.; Butter-field, Y.S.; Krzywinski, M.I.; Skalska, U.; Smailus, D.E.; Schnerch, A.; Schein, J.E.; Jones, S.J.; Marra, M.A. Mammalian Gene Collection Program Team. Gener-ation and initial analysis of more than 15,000 full-length human and mouse cDNA sequences. Proc. Natl. Acad. Sci. **2002**, *99*, 16899–16903.

55. MacKay, S.; Meyerhoff, D.J.; Dillon, W.P.; Weiner, M.W.; Fein, G. Alteration of brain phospholipid metabolites in cocaine-dependent polysubstance abusers. Biol. Psychiat. **1993**, *34*, 261–264.

56. Velculescu, V.E.; Zhang, L.; Vogelstein, B.; Kinzler, K.W. Serial analysis of gene expression. Science **1995**, *270*, 484–487.

57. Saha, S.; Sparks, A.B.; Rago, C.; Akmaev, V.; Wang, C.J.; Vogelstein, B.; Kinzler, K.W.; Velculescu, V.E. Using the transcriptome to annotate the genome. Nat. Biotechnol. **2002**, *20*, 508–512.

58. Blackshaw, S.; Kuo, W.P.; Park, P.J.; Tsujikawa, M.; Gunnersen, J.M.; Scott, H.S.; Boon, W.M.; Tan, S.S.; Cepko, C.L. MicroSAGE is highly representative and repro-ducible but reveals major differences in gene expression among samples obtained from similar tissues. Genome Biol. **2003**, *4*, R17.

59. Brenner, S.; Johnson, M.; Bridgham, J.; Golda, G.; Lloyd, D.H.; Johnson, D.; Luo, S.; McCurdy, S.; Foy, M.; Ewan, M.; Roth, R.; George, D.; Eletr, S.; Albrecht, G.; Vermaas, E.; Williams, S.R.; Moon, K.; Burcham, T.; Pallas, M.; DuBridge, R.B.; Kirchner, J.; Fearon, K.; Mao, J.; Corcoran, K. Gene expression analysis by massively parallel signa-ture sequencing (MPSS) on microbead arrays. Nat. Biotechnol. **2000**, *18*, 630–634.

60. Bell, J.E.; Ironside, J.W. Principles and practice of 'high risk' brain banking. Neuro-pathol. Appl. Neurobiol. **1997**, *23*, 281–288.

61. Jendroska, K.; Patt, S.; Janisch, W.; Cervos-Navarro, J.; Poewe, W. How to run a "brain bank"? Clinical and institutional requirements for "brain banking". J. Neural. Transm. Suppl. **1993**, *39*, 71–75.

62. McKee, A.C. Brain banking: basic science methods. Alzheimer Dis. Assoc. Disord. **1999**, *13* (suppl 1), S39–S44.

63. Sarris, M.; Garrick, T.M.; Sheedy, D.; Harper, C.G. Banking for the future: an Australian experience in brain banking. Pathology **2002**, *34*, 225–229.

64. Barton, A.J.; Pearson, R.C.; Najlerahim, A.; Harrison, P.J. Pre- and postmortem influ-ences on brain RNA. J. Neurochem. **1993**, *61*, 1–11.

65. Leonard, S.; Logel, J.; Luthman, D.; Casanova, M.; Kirch, D.; Freedman, R. Biological stability of mRNA isolated from human postmortem brain collections. Biol. Psychiat. **1993**, *33*, 456–466.

66. Lehrmann, E.; Hyde, T.M.; Vawter, M.P.; Becker, K.G.; Kleinman, J.E.; Freed, W.J. The use of microarrays to characterize neuropsychiatric disorders: postmortem studies of substance abuse and schizophrenia. Curr. Mol. Med. **2003**, *3*, 437–445.

67. Luo, L.; Salunga, R.C.; Guo, H.; Bittner, A.; Joy, K.C.; Galindo, J.E.; Xiao, H.; Rogers, K.E.; Wan, J.S.; Jackson, M.R.; Erlander, M.G. Gene expression profiles of laser-captured adjacent neuronal subtypes. Nat. Med. **1999**, *5*, 117–122.
68. Kamme, F.; Salunga, R.; Yu, J.; Tran, D.T.; Zhu, J.; Luo, L.; Bittner, A.; Guo, H.Q.; Miller, N.; Wan, J.; Erlander, M. Single-cell microarray analysis in hippocampus CA1: demonstration and validation of cellular heterogeneity. J. Neurosci. **2003**, *23*, 3607–3615.
69. Tietjen, I.; Rihel, J.M.; Cao, Y.; Koentges, G.; Zakhary, L.; Dulac, C. Single-cell transcriptional analysis of neuronal progenitors. Neuron **2003**, *38*, 161–175.
70. Brazma, A.; Hingamp, P.; Quackenbush, J.; Sherlock, G.; Spellman, P.; Stoeckert, C.; Aach, J.; Ansorge, W.; Ball, C.A.; Causton, H.C.; Gaasterland, T.; Glenisson, P.; Holstege, F.C.; Kim, I.F.; Markowitz, V.; Matese, J.C.; Parkinson, H.; Robinson, A.; Sarkans, U.; Schulze-Kremer, S.; Stewart, J.; Taylor, R.; Vilo, J.; Vingron, M. Minimum information about a microarray experiment (MIAME)-toward standards for microarray data. Nat. Genet. **2001**, *29*, 365–371.
71. Koob, G.F.; Nestler, E.J. The neurobiology of drug addiction. J. Neuropsychiat. Clin. Neurosci. **1997**, *9*, 482–497.
72. Gibbons, G.F. The role of cytochrome P450 in the regulation of cholesterol biosynthesis. Lipids **2002**, *37*, 1163–1170.
73. Toda, S.; McGinty, J.F.; Kalivas, P.W. Repeated cocaine administration alters the expression of genes in corticolimbic circuitry after a 3-week withdrawal: a DNA macroarray study. J. Neurochem. **2002**, *82*, 1290–1299.
74. Yuferov, V.; Kroslak, T.; Laforge, K.S.; Zhou, Y.; Ho, A.; Kreek, M.J. Differential gene expression in the rat caudate putamen after "binge" cocaine administration: advantage of triplicate microarray analysis. Synapse **2003**, *48*, 157–169.
75. Freeman, W.M.; Brebner, K.; Patel, K.M.; Lynch, W.J.; Roberts, D.C.; Vrana, K.E. Repeated cocaine self administration causes multiple changes in rat frontal cortex gene expression. Neurochem. Res. **2002**, *27*, 1181–1192.
76. Freeman, W.M.; Brebner, K.; Lynch, W.J.; Robertson, D.J.; Roberts, D.C.; Vrana, K.E. Cocaine-responsive gene expression changes in rat hippocampus. Neuroscience **2001**, *108*, 371–380.
77. Loguinov, A.V.; Anderson, L.M.; Crosby, G.J.; Yukhananov, R.Y. Gene expression following acute morphine administration. Physiol. Genomics **2001**, *6*, 169–181.
78. Kittler, J.T.; Grigorenko, E.V.; Clayton, C.; Zhuang, S.Y.; Bundey, S.C.; Trower, M.M.; Wallace, D.; Hampson, R.; Deadwyler, S. Large-scale analysis of gene expression changes during acute and chronic exposure to [delta]9-THC in rats. Physiol. Genomics **2000**, *3*, 175–185.
79. Grigorenko, E.; Kittler, J.; Clayton, C.; Wallace, D.; Zhuang, S.; Bridges, D.; Bundey, S.; Boon, A.; Pagget, C.; Hayashizaki, S.; Lowe, G.; Hampson, R.; Deadwyler, S. Assessment of cannabinoid induced gene changes: tolerance and neuroprotection. Chem. Phys. Lipids **2002**, *121*, 257–266.
80. Parmentier-Batteur, S.; Jin, K.; Xie, L.; Mao, X.O.; Greenberg, D.A. DNA microarray analysis of cannabinoid signaling in mouse brain *in vivo*. Mol. Pharmacol. **2002**, *62*, 828–35.
81. Hof, P.R.; Haroutunian, V.; Copland, C.; Davis, K.L.; Buxbaum, J.D. Molecular and cellular evidence for an oligodendrocyte abnormality in schizophrenia. Neurochem. Res. **2002**, *27*, 1193–1200.

82. Pongrac, J.; Middleton, F.A.; Lewis, D.A.; Levitt, P.; Mirnics, K. Gene expression profiling with DNA microarrays: advancing our understanding of psychiatric disorders. Neurochem. Res. **2002**, *27*, 1049–1063.
83. Egan, M.F.; Goldberg, T.E.; Kolachana, B.S.; Callicott, J.H.; Mazzanti, C.M.; Straub, R.E.; Goldman, D.; Weinberger, D.R. Effect of COMT Val108/158 Met genotype on frontal lobe function and risk for schizophrenia. Proc. Natl. Acad. Sci. USA **2001**, *98*, 6917–6922.
84. Hemmer, B.; Archelos, J.J.; Hartung, H.P. New concepts in the immunopathogenesis of multiple sclerosis. Nat. Rev. Neurosci. **2002**, *3*, 291–301.
85. Bjartmar, C.; Wujek, J.R.; Trapp, B.D. Axonal loss in the pathology of MS: consequences for understanding the progressive phase of the disease. J. Neurol. Sci. **2003**, *206*, 165–171.
86. Garbern, J.Y.; Yool, D.A.; Moore, G.J.; Wilds, I.B.; Faulk, M.W.; Klugman, M.; Nave, K.A.; Sistermans, E.A.; van der Knaap, M.S.; Bird, T.D.; Shy, M.E.; Kamholz, J.A.; Griffiths, I.R. Patients lacking the major CNS myelin protein, proteolipid protein 1, develop length-dependent axonal degeneration in the absence of demyelination and inflammation. Brain **2002**, *125*, 551–561.
87. Raine, C.S. The neuropathology of multiple sclerosis. In *Multiple Sclerosis: Clinical and Pathogenetic Basis*; Raine, C.S., McFarland, H.F., Tourtelotte, W.W., Eds.; Chapman and Hall: London, 1997; 151–171.
88. Mycko, M.P.; Papoian, R.R.; Boschert, U.; Raine, C.S.; Selmaj, K.W. cDNA microarray analysis in multiple sclerosis lesions: detection of genes associated with disease activity. Brain **2003**, *126*, 1048–1057.
89. Lock, C.; Hermans, G.; Pedotti, R.R.; Brendolan, A.; Schadt, E.; Garren, H.; Langer-Gould, A.; Strober, S.; Cannella, B.; Allard, J.; Klonowski, P.; Austin, A.; Lad, N.; Kaminski, N.; Galli, S.J.; Oksenberg, J.R.; Raine, C.S.; Heller, R.; Steinman, L. Gene-microarray analysis of multiple sclerosis lesions yields new targets validated in autoimmune encephalomyelitis. Nat. Med. **2002**, *8*, 500–508.
90. Dupree, J.L.; Coetzee, T.; Blight, A.; Suzuki, K.; Popko, B. Myelin galactolipids are essential for proper node of Ranvier formation in the CNS. J. Neurosci. **1998**, *18*, 1642–1649.
91. Carmody, R.J.; Hilliard, B.; Maguschak, K.K.; Chodosh, L.A.; Chen, Y.H. Genomic scale profiling of autoimmune inflammation in the central nervous system: the nervous response to inflammation. J. Neuroimmunol. **2002**, *133*, 95–107.

10

Microarray Analyses of SREBP-1 Target Genes

Naoya Yahagi

*Department of Internal Medicine, Graduate School of Medicine,
University of Tokyo, Tokyo, Japan*

Hitoshi Shimano

*Department of Internal Medicine, Institute of Clinical Medicine,
University of Tsukuba, Ibaraki, Japan*

INTRODUCTION

Microarray analyses are becoming increasingly prevalent and are being used in various research fields in life science. Application of this technique for the quantification of mRNA expression levels has enabled us to obtain vast amount of information on the expression levels of numerous genes at once. This is of great help especially for the elucidation of transcriptionally regulated pathways. In this chapter, we present our recent studies on a transcription factor, sterol regulatory element-binding factor (SREBP)-1, as an example where we applied the microarray technique for research on the nutritional regulation of gene expression.

ADVANTAGE OF MICROARRAY ANALYSES FOR THE RESEARCH OF LIPOGENIC GENE REGULATION

Mammals store their body energy in the form of glycogen and triglycerides. The former is stored chiefly in liver and skeletal muscle and the latter in liver and adipose tissue. Whereas glycogen synthesis is regulated by protein phosphory-lation cascades, the synthesis of fatty acids and triglycerides (lipogenesis) is mainly controlled at the transcriptional level.[1] This is presumably relevant to the differences in turnover rate between glycogen and triglycerides; glycogen is synthesized or degraded in minutes to hours while for triglycerides it is in hours to days. Transcriptional regulation is best suited for these relatively slow responses.

Another feature of lipogenic responses is that each enzyme involved in the metabolic pathway is coordinately regulated.[2] The fatty acid biosynthetic pathway is composed of about 25 enzymes. Among these enzymes, the following are of particular importance: fatty acid synthase (FAS), the main synthetic enzyme that catalyzes the condensation of malonyl-CoA to produce the 16-carbon saturated fatty acid palmitate; acetyl-CoA carboxylase (ACC), which synthesizes malonyl-CoA from acetyl-CoA; and ATP citrate lyase (ACL), which synthesizes acetyl-CoA from citrate provided through glycolysis. These enzymes are known to be coordinately upregulated 10–50-fold at the tran-scriptional level when lipogenesis is accelerated in hepatocytes and adipocytes, for instance, when fasted animals are refed.[1]

As lipogenic enzyme genes are coordinately upregulated upon refeeding, it has been postulated that these genes have regulatory sequences in their promoter that interact with common trans-acting factors. Currently, the most important reg-ulator is considered to be SREBP-1.[3,4]

SREBPs ARE MEMBRANE-BOUND TRANSCRIPTION FACTOR OF bHLH SUPERFAMILY

SREBPs are transcription factors that belong to the basic helix–loop–helix (bHLH) leucine zipper family. Unlike other transcription factors, SREBP pro-teins are initially bound to the rough endoplasmic reticulum (rER) membrane and form a complex with SREBP-cleavage activating protein (SCAP), a sterol-sensing molecule. Upon sterol deprivation from the cells, SREBP/SCAP travels from rER to Golgi where SREBP is cleaved to liberate the amino-terminal portion containing a bHLH leucine zipper domain (nuclear SREBP), and enters the nucleus where it can bind to specific sterol response elements (SRE) in the promoters of target genes (reviewed in Brown and coworkers[5–7]). Meanwhile, many cellular sterols retain the SREBP/SCAP complex on the rER where insulin inducible genes (INSIGs) play an important role in the regulation of SCAP con-figuration and the cleavage of SREBP, and thus, cholesterol synthesis does not occur. To date, three isoforms of SREBP-1a, -1c, and -2 have been found and characterized. In the differentiated tissues, SREBP-2 plays a crucial role in the regulation of cholesterol synthesis[6,7], whereas SREBP-1c controls the transcrip-tion of lipogenic enzymes involved in the synthesis of fatty acids and triglycer-ides.[8–11] SREBP-1c is drastically induced when fasted animals are refed.[12] SREBP-1a is expressed in actively growing cells, and activates both lipogenic and cholesterogenic genes. Thus, SREBP-1a has the most potent and global range of target genes among the isoforms.

SREBP-1a TARGET GENES ON A DNA MICROARRAY ANALYSIS

In an attempt to comprehensively understand SREBP targets and potentially identify new target genes, we performed microarray analysis (randomly selected some 9000 genes on an array manufactured by IncyteGenomics) for mRNA from livers of a transgenic mouse model that overexpresses a nuclear form of SREBP-1a protein in liver. This mouse exhibited markedly elevated expression of enzymes involved in fatty acid and cholesterol biosynthetic pathways, and thereby developed massive enlargement of the liver, owing to the engorgement of hepatocytes with cholesterol and triglycerides (hepatic steatosis).[13] From the microarray data we calculated the fold changes in mRNA expression of each gene in the transgenic mice livers compared with wild-type, and listed them in descending order in Table 10.1.

Consistency and Reliability of the Data

There has been much discussion about the problems in the quantification and reproducibility of the results from DNA chips. It could be technical issues related to cDNA or oligoprobes, or sample problems. However, fortunately, the data obtained from DNA microarrays of SREBP1a-transgenic/wild-type liver RNAs, exhibited perfect reliability and consistency; we had a chance to compare the extent of SREBP-1a-induction estimated by DNA microarrays and by Northern blot analysis in some known SREBP target genes such as FAS and ACL, and the fold-induction was highly consistent between the two different methods convincing us about the validity of the rest of DNA microarray data. Although rare, some EST clones used in microarrays were identical to provide us two results for the same gene, the data were very similar again. For example, secreted modular calcium-binding protein 2 listed in Table 10.1 had two independent EST clones (GenBank accession nos.: AA272826 and AA059909) on the array, and the fold-increase obtained from them were 9.3 and 8.2, respectively (only the former value is presented in Table 10.1).

Most of the Inducible Genes were Lipogenic and Cholesterogenic Genes, as Predicted

In the top 25 genes whose expression was the most strongly increased in transgenic mice, 9 were lipogenic genes and 5 were cholesterogenic. The most highly ranked gene was FAS and the second was ACL, both were well characterized as SREBP targets in the promoter analyses. 6-Phosphogluconate dehydrogenase (6PGD) was a key enzyme, so was glucose-6-phosphate dehydrogenase, an enzyme in the pentose pathway, which produces NADPH for reducing potential for lipid synthesis, and was considered as one of lipogenic enzyme members, although the promoter analysis was not yet done in the context of the SREBP target.

Table 10.1   Microarray Analysis of Mice Overexpressing SREBP-1a in Liver

Fold increase	Gene name	Function	Accession no.
19.5	Fatty acid synthase (*FAS*)	Lipogenesis	AA116513
19.5	Long chain fatty-acyl CoA elongase (*FACE*)	Lipogenesis	AA239254
16.0	Membrane-associated protein 17	Unknown	AA274146
15.1	ATP citrate lyase (*ACL*)	Lipogenesis	W33415
14.8	Lymphocyte antigen 6 complex, locus D	Unknown	AA163336
14.6	Diphosphomevalonate decarboxylase (*MVD*)	Cholesterol synthesis	AI510113
14.4	Secretory leukocyte protease inhibitor	Unknown	AA200339
13.7	6-Phosphogluconate dehydrogenase (*6PGD*)	Lipogenesis	AI893710
12.6	Squalene epoxidase (*SQLE*)	Cholesterol synthesis	AA268608
12.4	Acetyl-CoA synthetase (*ACAS*)	Lipogenesis	AA537637
12.1	Isopentenyl-diphosphate delta-isomerase 1 (*IDI1*)	Cholesterol synthesis	AA237469
11.1	Proline-serine-threonine phosphatase-interacting protein 2 mitochondrial glycerol-3-phosphate acyltransferase (*GPAT*)	Unknown Lipogenesis	AA390032 AA209041
10.4	HMGCoA synthase (*HMGCS*) 1	Cholesterol synthesis	AI892192
9.4	Acetyl-CoA carboxylase (*ACC*) 1	Lipogenesis	AA014384
9.3	Secreted modular calcium-binding protein 2	Unknown	AA272826
7.9	Delta-5 fatty acid desaturase (*D5D*)	Lipogenesis	AA068575
7.6	7-Dehydrocholesterol reductase (*DHCR7*)	Cholesterol synthesis	AA003001
7.3	UDP-glucose 4-epimerase	Conversion from galactose to glucose	AA386807
7.1	p21Waf1	Cell cycle arrest	W88005

Note: The fold changes in mRNA expression of each gene in the transgenic mice livers compared with wild-type are listed in descending order. Methods: PolyA RNAs were extracted from livers of transgenic or wild-type mice ($n = 4$) and pooled. The microarray analysis was performed by IncyteGenomics with their standard protocol. Briefly, each set of polyA RNA was labeled with fluorescence (Cy3 for wild-type and Cy5 for transgenic mice). Then they were mixed and hybridized onto the microarray plate containing approximately 9000 clones. After washing, the plate was scanned by fluorescence imager and signals were quantified. The same method was used for all microarray data shown in this chapter.

Mitochondrial glycerol-3-phosphate acyltransferase was also identified as a highly inducible gene. Many of cholesterolgenic genes were also detected, which, along with established genes, led us to conclude that SREBP can activate an entire pathway of cholesterol synthesis.[14]

SREBP-1 PLAYS THE CENTRAL ROLE IN REFEEDING RESPONSES

To clarify the positioning of SREBP-1 activation in the refeeding responses in liver, we also performed microarray analysis on gene expression profiles in refed animals in a similar way. First we quantified each gene expression level (for some 9000 genes on the IncyteGenomics microarray) in liver and adipose tissue from fasted or refed mice in order to obtain the entire list of genes that are nutritionally regulated at mRNA level. The calculated fold increases in refed state vs. fasted state were displayed in the scatter plot in logarithmic scale (data from liver are plotted along the horizontal axis and those from adipose tissue are along vertical axis) (Fig. 10.1). As shown here, both in liver and in adipose tissue, lipogenic enzymes such as FAS, ACL, and ACC are markedly induced upon refeeding and are revealed to be the representative genes of refeeding responses.

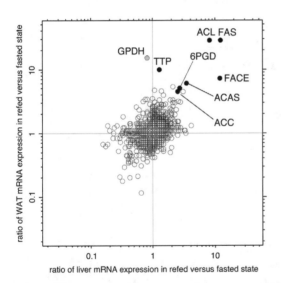

Figure 10.1 Lipogenic genes are the most strongly upregulated genes when fasted animals are refed. Increases in mRNA expression in liver (on horizontal axis) and adipose (on vertical axis) after refeeding fasted mice quantified by microarray analysis are plotted in two dimension in logarithmic scale. Lipogenic genes under SREBP-1 control (shown as solid circles) are coordinately elevated both in liver and in adipose tissue. GPDH does not seem to be an SREBP-1 target gene because it was not increased in SREBP-1 transgenic mice. Methods: Mice ($n = 8$ for each group) were 24 h fasted or 12 h refed on a high-sucrose fat free diet after 24 h starvation. PolyA RNA samples extracted from livers and white adipose were pooled within each group and analyzed with IncyteGenomics DNA microarray. *ACAS*, acetyl-CoA synthase; *ACC*, acetyl-CoA carboxylase; *ACL*, ATP citrate lyase; *FACE*, fatty-acyl-CoA elongase; *FAS*, fatty acid synthase; *GPDH*, glycerophosphate dehydrogenase; *6PGD*, 6-phosphogluconate dehydrogenase; *TTP*, tricarboxylate transport protein; WAT, white adipose tissue.

Then we compared these refeeding-responsive genes with SREBP-1 target genes. In Fig. 10.2, mRNA quantification results by microarray analysis including those of lipogenic and cholesterogenic genes are plotted in a 2D field of refed-vs.-fasted changes on the horizontal axis and transgenic changes on the vertical axis. This scatter diagram demonstrates that target genes of SREBP-1 are activated the most vigorously in refed mice livers and indicates that SREBP-1 plays the central role in refeeding responses.

SREBP-1 IS INVOLVED IN SYNTHESIS OF PHOSPHOLIPIDS

It has been reported that phosphocholine cytidylyltransferase, a rate-controlling enzyme in the phosphatidylcholine biosynthesis pathway is an SREBP target.[15,16] Thus, SREBPs could also be involved in phospholipid metabolism,

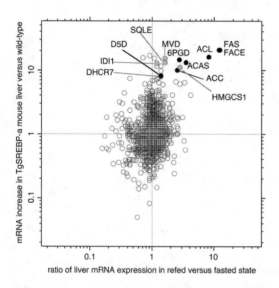

Figure 10.2   SREBP-1 plays the pivotal role in refeeding responses. Increases in mRNA expression in liver after refeeding fasted mice (on horizontal axis) and those in SREBP-1a transgenic mice compared with wild-type mice (on vertical axis) quantified by microarray analysis are plotted in 2D in logarithmic scale. Lipogenic and cholesterogenic genes are shown as black and gray circles, respectively. Methods: For the fasting–refeeding experiment, mice (*n* = 8 for each group) were 24 h fasted or 12 h refed on a high-sucrose fat free diet after 24 h starvation. PolyA RNA samples extracted from livers were pooled within each group and analyzed with IncyteGenomics DNA microarray. *ACAS*, acetyl-CoA synthase; *ACC*, acetyl-CoA carboxylase; *ACL*, ATP citrate lyase; *D5D*, delta-5 fatty acid desaturase; *FACE*, fatty-acyl-CoA elongase; *FAS*, fatty acid synthase; *6PGD*, 6-phosphogluconate dehydrogenase; *DHCR7*, 7-dehydrocholesterol reductase; *HMGCS1*, hydroxymethylglutaryl-CoA synthase 1; *IDI1*, isopentenyl-diphosphate delta-isomerase 1; *MVD*, diphosphomevalonate decarboxylase; *SQLE*, squalene epoxidase.

which potentially places SREBPs in the regulation of entire lipid synthesis. Intriguingly, in *Drosophila*, the SREBP/SCAP system is regulated by phosphatidylethanolamine,[17,18] suggesting that the SREBP/SCAP system conserved in a wide range of species plays a more diverse role in lipid synthesis by regulating physicochemical functions of biomembranes. Although it has been published that there are some SRE sites in the promoters of some phospholipid genes, fold changes of activation by SREBP-1a is not marked in those genes. SREBPs could influence, but not dominate, the expression of these genes in mammalians.

NEWLY IDENTIFIED SREBP-1 TARGET GENES WHOSE FUNCTIONS HAVE BEEN ELUCIDATED

The most intriguing and exciting part of DNA microarray analysis is identification of new target genes of a transcription factor. The listing of SREBP-1a-induced EST clones offered us an opportunity to characterize in detail these genes whose functions were not yet well documented. These are acetyl-CoA synthetase (ACAS),[19] fatty-acyl-CoA elongase (FACE),[20] and delta-5 fatty acid desaturase (D5D)[21] as discussed.

Acetyl-CoA Synthetase

One of the highly inducible genes (12-fold increase by SREBP-1 overexpression) was designated as a gene highly similar to bacterial ACAS. We and another group[22] cloned the whole cDNA and confirmed that it has activity for mammalian ACAS and that its regulation is in a manner similar to that of lipogenic genes, and controlled by SREBPs. This enzyme catalyzes production of acetyl-CoA from acetate and CoA. Acetyl-CoA is a hub molecule for energy metabolism and nutrition. SREBP regulation of this enzyme could implicate that mammalian lipogenic organs have a pathway for energy conversion from free acetate as well as glucose. ACAS activity has also been well known among researchers in ruminology to play a crucial role in energy production on ruminants as volatile fatty acids (also known as short chain fatty acids) produced through fermentation of cellulose and other fibers in the rumen are their main source of energy.[23]

Long Chain Fatty-Acyl-CoA Elongase; LCE

Pursuing another SREBP inducible EST helped us clone a novel elongase which was revealed to catalyze the last elongation step in mammalian fatty acid synthesis,[20,24] and was named FACE or LCE. The process was as follows: after sequencing a long 3' UTR cloned by rapid amplification of cDNA ends (RACE), we found that this gene encodes an integral membrane protein similar to Cig30 and SSC1 in mammalians and ELO2 and ELO3 in yeast. It was well known that FAS covers synthesis from malonyl-CoA to palmitate (C16:0), and stearoyl-CoA desaturase (SCD)-1 desaturates stearate to oleate

(C18:0 to C18:1), but the enzyme responsible for the conversion of C16:0 to C18:0 was missing. So, we intensively investigated the enzymatic activity of this new protein on various substrates, and identified it as the missing elongase. This enzyme also has an activity for elongation of monounsaturated fatty acids and could be involved in regulation of ratio of saturated/monounsaturated fatty acids with a link to SCD-1, and might potentially play a role in obesity and insulin sensitivity.

Delta-5 Fatty Acid Desaturase

D5D was also identified as an SREBP target gene on the basis of our microarray data,[21] and so was D6D (delta-6 fatty acid desaturase). They are involved in the production of polyunsaturated fatty acids (PUFA). As PUFA inhibit SREBP-1c and lipogenesis, it is hard to explain the physiological meaning of SREBP induction of these PUFA relating enzymes. D5D and D6D are also regulated by peroxisome proliferator-activated receptor (PPAR) alpha, and hence the expression of D5D and D6D could be regulated by two distinct mechanisms; one through SREBP-1c, as is shown by its induction during refeeding, and the other through PPAR alpha, which is inducible by fasting. Dual regulation of D5D and D6D by the two factors SREBP-1 and PPAR alpha that possess opposing activation properties depending upon nutritional status caused no remarkable change in the overall expression of these enzymes between fasting and refeeding. This dual regulation of both desaturases contribute to a stable production of PUFA that are essential for cellular functions regardless of the energy state.

Other Potential SREBP Target Genes

SREBP and Other Carbohydrate Metabolism

UDP-glucose (galactose) 4-epimerase which involves conversion from galactose to glucose is a potential SREBP target gene. SREBP-1a transgenic liver express this enzyme 5.7-fold higher than wild-type, suggesting that SREBP-1 could be involved in galactose metabolism and possibly in lactose production. SREBP-1c could play a role in suckling mammary glands, in which FAS activity is high. UDP-glucose dehydrogenase also showed 5.0-fold induction in SREBP-1a overproduction. This enzyme catalyzes a step which is important for the uronic acid pathway. Physiological consequences of this pathway are the production of pentose and UDP-glucronate. UDP-glucuronate is used for production of proteoglycans and conjugates with steroid hormones, drugs, and bilirubin. These data could implicate roles of SREBP-1 in galactose, lactose, and glucuronate metabolism. Recently, we and another group found that hepatic SREBP-1c could be induced by ingestion of fructose as well as glucose. Considering that the pentose pathway is important in lipogenesis for production of NADPH, SREBP-1 is likely to be highly involved in sugar metabolism, in which transcriptional regulation of related genes are not fully understood.

However, further investigation on promoters of these and other enzymes in the pathways are required.

SREBP-1 AND LIPOGENIC GENES ARE DOWNREGULATED IN ADIPOCYTES OF OBESE ANIMALS

To gain insight into the pathogenesis of obesity, we examined, with the same microarray technique discussed earlier, the gene expression profile of *ob/ob* mouse, a genetically obese mouse model formed from a nonsense mutation in appetite-suppressing hormone leptin. Quite unexpectedly, as shown in Fig. 10.3, we found that the mRNA expression levels of lipogenic enzymes were lowered. Later we demonstrated that refeeding responses of SREBP-1 and its downstream lipogenic enzymes were markedly impaired in adipocytes

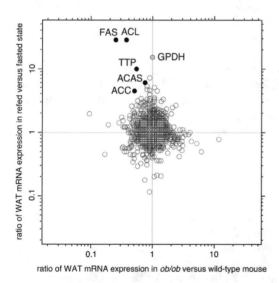

Figure 10.3 Lowered expression of SREBP-1-regulated lipogenic genes in *ob/ob* adipose. Increases in mRNA expression in adipose tissue after refeeding fasted mice (on vertical axis) and those in *ob/ob* adipose compared with wild-type (on horizontal axis) are plotted in 2D in logarithmic scale. The mRNA expression levels of lipogenic enzymes under SREBP-1 control (FAS, ACL, TTP, ACAS, and ACC; shown as solid circles) were lowered in *ob/ob* mice adipose tissue. GPDH that did not seem to be an SREBP-1 target gene because of no alteration in SREBP-1 transgenic mice was not suppressed in *ob/ob* adipose. Methods: For the *ob/ob* mice experiment, wild-type and *ob/ob* mice were sacrificed in a fed state and RNA was extracted from adipose tissue. RNA samples were pooled within each group and analyzed with IncyteGenomics DNA microarray. *ACAS*, acetyl-CoA synthase; *ACC*, acetyl-CoA carboxylase; *ACL*, ATP citrate lyase; *FAS*, fatty acid synthase; *GPDH*, glycerophosphate dehydrogenase; *TTP*, tricarboxylate transport protein; WAT, white adipose tissue.

of *ob*/*ob* mouse.[25] The similar results from microarray analyses performed elsewhere were reported by others.[26,27] Notably, this downregulation of SREBP-1 and lipogenic genes is also documented in humans[28–30] and seems to be an important characteristic of adipocytes in obesity.

In an attempt to search for the upstream regulator for this lipogenic gene suppression, we focused on the data by Soukas et al.[26] that p21[Waf1/CIP1] and Bax alpha were elevated two to threefold in *ob*/*ob* adipose tissue. Both of these genes are well-known p53 targets, so we hypothesized that p53 is activated in *ob*/*ob* adipocytes, leading to the activation of target genes such as p21 and at the same time to the suppression of SREBP-1 and its downstream lipogenic genes. The tumor suppressor p53 not only activates transcription of target genes through its response element, but also represses genes lacking the element by binding to and sequestering essential transcription factors such as TATA-binding protein.[31,32] To test this hypothesis, we intercrossed *ob*/*ob* and p53-knockout mice and demonstrated that the disruption of p53 in *ob*/*ob* mice completely suppressed the p53-regulated genes to wild-type levels and partially restored expression of lipogenic enzymes;[33] thus the hypothesis was proven.

This is an example where microarray analysis provides clues from which we can speculate upstream regulators. Microarray data can help us not only to search the downstream genes of a transcription factor but also to explore for upstream regulators. As we show here, gene expression profiling with microarray technique is of great assistance in the elucidation of transcriptionally regulated pathways.

REFERENCES

1. Goodridge, A.G. Dietary regulation of gene expression: enzymes involved in carbohydrate and lipid metabolism. Annu. Rev. Nutr. **1987**, *7*, 157–185.

2. Goodridge, A.G. *Fatty Acid Synthesis in Eucaryotes*; Elsevier Science: Amsterdam, 1991.

3. Shimomura, I.; Shimano, H.; Korn, B.S.; Bashmakov, Y.; Horton, J.D. Nuclear sterol regulatory element-binding proteins activate genes responsible for the entire program of unsaturated fatty acid biosynthesis in transgenic mouse liver. J. Biol. Chem. **1998**, *273*, 35299–35306.

4. Shimano, H.; Yahagi, N.; Amemiya-Kudo, M.; Hasty, A.H.; Osuga, J.; Tamura, Y.; Shionoiri, F.; Iizuka, Y.; Ohashi, K.; Harada, K.; Gotoda, T.; Ishibashi, S.; Yamada, N. Sterol regulatory element-binding protein-1 as a key transcription factor for nutritional induction of lipogenic enzyme genes. J. Biol. Chem. **1999**, *274*, 35832–35839.

5. Brown, M.S.; Goldstein, J.L. The SREBP pathway: regulation of cholesterol metabolism by proteolysis of a membrane-bound transcription factor. Cell **1997**, *89*, 331–340.

6. Brown, M.S.; Goldstein, J.L. A proteolytic pathway that controls the cholesterol content of membranes, cells, and blood. Proc. Natl. Acad. Sci. USA **1999**, *96*, 11041–11048.

7. Brown, M.S.; Ye, J.; Rawson, R.B.; Goldstein, J.L. Regulated intramembrane proteolysis: a control mechanism conserved from bacteria to humans. Cell **2000**, *100*, 391–398.
8. Shimano, H. Sterol regulatory element-binding protein-1 as a dominant transcription factor for gene regulation of lipogenic enzymes in the liver. Trends Cardiovasc. Med. **2000**, *10*, 275–278.
9. Shimano, H. Sterol regulatory element-binding proteins (SREBPs): transcriptional regulators of lipid synthetic genes. Prog. Lipid Res. **2001**, *40*, 439–452.
10. Shimano, H. Sterol regulatory element-binding protein family as global regulators of lipid synthetic genes in energy metabolism. Vitam. Horm. **2002**, *65*, 167–194.
11. Horton, J.D.; Goldstein, J.L.; Brown, M.S. SREBPs: activators of the complete program of cholesterol and fatty acid synthesis in the liver. J. Clin. Invest. **2002**, *109*, 1125–1131.
12. Horton, J.D.; Bashmakov, Y.; Shimomura, I.; Shimano, H. Regulation of sterol regulatory element binding proteins in livers of fasted and refed mice. Proc. Natl. Acad. Sci. USA **1998**, *95*, 5987–5992.
13. Shimano, H.; Horton, J.D.; Hammer, R.E.; Shimomura, I.; Brown, M.S.; Goldstein, J.L. Overproduction of cholesterol and fatty acids causes massive liver enlargement in transgenic mice expressing truncated SREBP-1a. J. Clin. Invest. **1996**, *98*, 1575–1584.
14. Sakakura, Y.; Shimano, H.; Sone, H.; Takahashi, A.; Inoue, N.; Toyoshima, H.; Suzuki, S.; Yamada, N.; Inoue, K. Sterol regulatory element-binding proteins induce an entire pathway of cholesterol synthesis. Biochem. Biophys. Res. Commun. **2001**, *286*, 176–183.
15. Lagace, T.A.; Storey, M.K.; Ridgway, N.D. Regulation of phosphatidylcholine metabolism in Chinese hamster ovary cells by the sterol regulatory element-binding protein (SREBP)/SREBP cleavage-activating protein pathway. J. Biol. Chem. **2000**, *275*, 14367–14374.
16. Kast, H.R.; Nguyen, C.M.; Anisfeld, A.M.; Ericsson, J.; Edwards, P.A. CTP: phosphocholine cytidylyltransferase, a new sterol- and SREBP-responsive gene. J. Lipid Res. **2001**, *42*, 1266–1272.
17. Seegmiller, A.C.; Dobrosotskaya, I.; Goldstein, J.L.; Ho, Y.K.; Brown, M.S.; Rawson, R.B. The SREBP pathway in *Drosophila*: regulation by palmitate, not sterols. Dev. Cell **2002**, *2*, 229–238.
18. Dobrosotskaya, I.Y.; Seegmiller, A.C.; Brown, M.S.; Goldstein, J.L.; Rawson, R.B. Regulation of SREBP processing and membrane lipid production by phospholipids in *Drosophila*. Science **2002**, *296*, 879–883.
19. Sone, H.; Shimano, H.; Sakakura, Y.; Inoue, N.; Amemiya-Kudo, M.; Yahagi, N.; Osawa, M.; Suzuki, H.; Yokoo, T.; Takahashi, A.; Iida, K.; Toyoshima, H.; Iwama, A.; Yamada, N. Acetyl-coenzyme A synthetase is a lipogenic enzyme controlled by SREBP-1 and energy status. Am. J. Physiol. Endocrinol. Metab. **2002**, *282*, E222–E230.
20. Matsuzaka, T.; Shimano, H.; Yahagi, N.; Yoshikawa, T.; Amemiya-Kudo, M.; Hasty, A.H.; Okazaki, H.; Tamura, Y.; Iizuka, Y.; Ohashi, K.; Osuga, J.; Takahashi, A.; Yato, S.; Sone, H.; Ishibashi, S.; Yamada, N. Cloning and characterization of a mammalian fatty acyl-CoA elongase as a lipogenic enzyme regulated by SREBPs. J. Lipid Res. **2002**, *43*, 911–920.

21. Matsuzaka, T.; Shimano, H.; Yahagi, N.; Amemiya-Kudo, M.; Yoshikawa, T.; Hasty, A.H.; Tamura, Y.; Osuga, J.; Okazaki, H.; Iizuka, Y.; Takahashi, A.; Sone, H.; Gotoda, T.; Ishibashi, S.; Yamada, N. Dual regulation of mouse Delta(5)- and Delta(6)-desaturase gene expression by SREBP-1 and PPARalpha. J. Lipid Res. **2002**, *43*, 107–114.

22. Luong, A.; Hannah, V.C.; Brown, M.S.; Goldstein, J.L. Molecular characterization of human acetyl-CoA synthetase, an enzyme regulated by sterol regulatory element-binding proteins. J. Biol. Chem. **2000**, *275*, 26458–26466.

23. Murray, R.K.; Granner, D.K.; Mayes, P.A.; Rodwell, V.W. Biosynthesis of fatty acids. In *Harper's Biochemistry*, 25th Ed.; McGraw-Hill, 1999; Chap. 23.

24. Moon, Y.A.; Shah, N.A.; Mohapatra, S.; Warrington, J.A.; Horton, J.D. Identification of a mammalian long chain fatty Acyl elongase regulated by sterol regulatory element-binding proteins. J. Biol. Chem. **2001**, *276*, 45358–45366.

25. Yahagi, N.; Shimano, H.; Hasty, A.H.; Matsuzaka, T.; Ide, T.; Yoshikawa, T.; Amemiya-Kudo, M.; Tomita, S.; Okazaki, H.; Tamura, Y.; Iizuka, Y.; Ohashi, K.; Osuga, J.; Harada, K.; Gotoda, T.; Nagai, R.; Ishibashi, S.; Yamada, N. Absence of sterol regulatory element-binding protein-1 (SREBP-1) ameliorates fatty livers but not obesity or insulin resistance in Lep(ob)/Lep(ob) mice. J. Biol. Chem. **2002**, *277*, 19353–19357.

26. Soukas, A.; Cohen, P.; Socci, N.D.; Friedman, J.M. Leptin-specific patterns of gene expression in white adipose tissue. Genes Dev. **2000**, *14*, 963–980.

27. Nadler, S.T.; Stoehr, J.P.; Schueler, K.L.; Tanimoto, G.; Yandell, B.S.; Attie, A.D. The expression of adipogenic genes is decreased in obesity and diabetes mellitus. Proc. Natl. Acad. Sci. USA **2000**, *97*, 11371–11376.

28. Ducluzeau, P.H.; Perretti, N.; Laville, M.; Andreelli, F.; Vega, N.; Riou, J.P.; Vidal, H. Regulation by insulin of gene expression in human skeletal muscle and adipose tissue. Evidence for specific defects in type 2 diabetes. Diabetes **2001**, *50*, 1134–1142.

29. Oberkofler, H.; Fukushima, N.; Esterbauer, H.; Krempler, F.; Patsch, W. Sterol regulatory element binding proteins: relationship of adipose tissue gene expression with obesity in humans. Biochim. Biophys. Acta **2002**, *1575*, 75–81.

30. Diraison, F.; Dusserre, E.; Vidal, H.; Sothier, M.; Beylot, M. Increased hepatic lipogenesis but decreased expression of lipogenic gene in adipose tissue in human obesity. Am. J. Physiol. Endocrinol. Metab. **2002**, *282*, E46–E51.

31. Seto, E.; Usheva, A.; Zambetti, G.P.; Momand, J.; Horikoshi, N.; Weinmann, R.; Levine, A.J.; Shenk, T. Wild-type p53 binds to the TATA-binding protein and represses transcription. Proc. Natl. Acad. Sci. USA **1992**, *89*, 12028–12032.

32. Mack, D.H.; Vartikar, J.; Pipas, J.M.; Laimins, L.A. Specific repression of TATA-mediated but not initiator-mediated transcription by wild-type p53. Nature **1993**, *363*, 281–283.

33. Yahagi, N.; Shimano, H.; Matsuzaka, T.; Najima, Y.; Sekiya, M.; Nakagawa, Y.; Ide, T.; Tomita, S.; Okazaki, H.; Tamura, Y.; Iizuka, Y.; Ohashi, K.; Gotoda, T.; Nagai, R.; Kimura, S.; Ishibashi, S.; Osuga, J.; Yamada, N. p53 Activation in adipocytes of obese mice. J. Biol. Chem. **2003**, *278*, 25395–25400.

11

Understanding the Coordinated Effects of PPARs on Lipid Metabolism Using Microarrays

Sander Kersten

Nutrition, Metabolism and Genomics group, Division of Human Nutrition, Wageningen University, Wageningen, The Netherlands

Pascal Escher and Sherrie Tafuri

Pfizer Global Research & Development, Ann Arbor Laboratories, Ann Arbor, MI, USA

Walter Wahli

Center for Integrative Genomics, University of Lausanne, Lausanne, Switzerland

INTRODUCTION

The application of high throughput gene expression profiling via microarrays is slowly causing a revolution in biomedical science. Until recently determining the expression of numerous genes at once was a wild fantasy; nowadays, however, the expression monitoring of thousands of genes simultaneously has become a routine procedure. Despite its relatively high cost, an increasing number of researchers, in both industry and academia, are able to benefit from the power of this technology. One important area that was predicted to profit immensely from microarray technology is drug development and validation. Although within the pharmaceutical industry huge progress has been made in the implementation of DNA microarrays into drug development, many obstacles still need to be overcome, some of which are alluded to in this paper.

One important set of molecular targets for drug discovery is a group of receptors called the peroxisome proliferator-activated receptors (PPARs). PPARs are ligand-activated transcription factors that belong to the superfamily of nuclear hormone receptors.[1] They mediate part of the effects of fatty acids and fatty acid-derived molecules (eicosanoids) on DNA transcription. In addition, they serve as receptors for several compounds used in the treatment of dyslipidemia and type 2 diabetes and are therefore of great interest to the pharmaceutical industry.

This chapter will discuss the specific advancements made in PPAR research, thanks to microarray, and includes a detailed discussion of specific data obtained. Furthermore, it will present an evaluation of the future benefits of microarrays to the field of PPARs.

PEROXISOME PROLIFERATOR-ACTIVATED RECEPTORS

PPARs are transcription factors of the nuclear hormone receptor superfamily, which bind to a specific sequence within the promoter region of a target gene (reviewed in Desvergne and Wahli[2]). This sequence, called a peroxisome proliferator-activated receptor response element or PPRE, usually consists of a direct repeat of the sequence AGGTCA separated by A. As with many other nuclear hormone receptors, binding to the PPRE is achieved via heterodimerization with the common binding partner, the retinoid X receptor. Binding of ligand induces a conformational change, leading to recruitment of coactivator proteins and subsequent activation of DNA transcription.

Three different PPAR genes can be distinguished, PPARα (*NR1C1*), PPARβ (also called delta; *NR1C2*), and PPARγ (*NR1C3*), which are expressed at different levels in various tissues.[2] The PPARα isotype is mainly expressed in brown adipose tissue, liver, heart, and kidney, and governs the expression of numerous genes involved in fatty acid metabolism, hepatocyte proliferation, and inflammation. In contrast, PPARγ mRNA is most abundant in adipose tissue, and to a lesser extent in colon and immune cells. It is considered the master regulator of adipocyte differentiation and stimulates the expression of genes involved in adipo- and lipogenesis, such as lipoprotein lipase and adipocyte fatty acid binding protein. The third PPAR isotype, PPARβ, is expressed ubiquitously and is currently the subject of intense investigation. The most compelling studies indicate a role for PPARβ in keratinocyte differentiation and apoptosis.[3]

PPARs are responsible for (part of) the effects of a variety of fatty acids and eicosanoids on DNA transcription. In addition, they serve as intracellular receptors for two important classes of drugs. Whereas PPARγ binds the insulin sensitizing thiazolidinediones that are used in the treatment of type 2 diabetes, PPARα binds hypolipidemic fibrates, which are used in the treatment of dyslipidemia. Currently, pharmaceutical companies are investing heavily in the development of novel PPAR ligands, with emphasis on dual PPARα and PPARγ agonists that combine the benefits of both PPARα and PPARγ activation.

OBJECTIVE OF THE APPLICATION OF MICROARRAY INTO PPAR RESEARCH

The application of microarray into research on PPARs can be driven by numerous interests. Overall, expression profiling of thousands of genes via microarray can promote increased understanding of which metabolic pathways and which specific set of genes are regulated by PPARs and PPAR ligands. This information in turn can be used for several different purposes.

First, microarrays can be an extremely useful tool to elucidate the function of PPARs in different organs. Inasmuch as the function of a PPAR is directly coupled to the identity and function of its target genes, the identification of genes regulated by PPARs by expression profiling provides important information about the biological role of the receptor. Several groups have successfully used this approach to establish a link between PPAR and pathways that previously had not been associated with PPARs, such as amino acid and urea metabolism.[4] By identifying novel PPAR-regulated pathways, novel clinical applications of PPAR agonists may emerge.

Second, by comparing the profiles of genes regulated by the different nuclear hormone receptors via microarray, important information can be extracted about potential cross-talk between PPARs and other receptors such as the bile acid receptor, FXR, and liver X receptor. In order to maximize the power of this analysis, collaboration among research groups with access to different research tools, such as knock-out animals, should be promoted. Although this type of analysis has great potential, there are no published studies available as yet.

Ideally, the clinical and metabolic effects of PPAR agonists can be linked to the induction or repression of specific genes. By comparing the expression profiles induced by a diverse range of PPAR agonists that have similar clinical effects, genes that are regulated in common and whose regulation is correlated with a clinical or metabolic outcome can be defined. The identification of these types of biomarkers would be expected to be very useful in drug development to evaluate the clinical potential of new compounds at an early stage.

In the pharmaceutical industry every year a large number of promising compounds feed into the pipeline of clinical testing, yet most eventually have to be abandoned because of deleterious side effects. Huge sums of money could be saved and health hazards minimized if the (likely) toxicity of a compound can be determined at an earlier stage. For this reason pharmaceutical companies are eager to link the toxicological effects (side effects) of PPAR agonists to changes in expression of specific genes. Future compounds can then be investigated for their effect on the expression of these particular genes in a cell-based assay or in animal models.

It is well known that the clinical and toxicological responses to a drug may differ a lot among individuals. One of the hopes is that in the future microarray, as well as proteomics and metabolomics, will assist in treating patients on a more individual basis. By examining a patient's response to a drug at the gene, protein, or metabolite level, the dose or type of drug can be adjusted. Currently, this application of microarrays remains largely theoretical, but one should not be surprised if in 10–15 years time transcriptomics and particularly metabolomics will have become important tools in clinical diagnostics and treatment.

OVERVIEW MICROARRAY STUDIES PPARα

Although the earlier discussion may make it sound like the actual implementation of microarrays is still in its infancy, in actuality much has already been achieved. This is partly because of the availability of some powerful tools, which have served as the basis of a number of microarray studies that have yielded some interesting new insights into PPARα biology. These tools include mice that lack PPARα (PPARα null mice), as well as several specific high affinity agonists.

In one of the first microarray studies on PPARα, one pathway that emerged to be regulated by PPARα was amino acid metabolism. Using Affymetrix oligonucleotide microarrays in combination with subtractive hybridization, it was observed that several genes involved in amino acid and urea metabolism were markedly upregulated in the liver of PPARα null mice, including the genes encoding aspartate amino transferase and agininosuccinate lyase.[4] Although the mechanism by which PPARα downregulates expression of these genes remains elusive, the importance of this regulation in mice was illustrated by the increased plasma urea levels in fasted PPARα null mice. On the basis of these and additional data, it is clear that, rather than merely regulating fatty acid catabolism, PPARα serves an integrated role in hepatic nutrient metabolism (Fig. 11.1).

This kind of comparative analysis of global gene expression has the ability to generate large sets of potential new targets of PPARα *in vivo*, both positive and negative. For example, in our studies it was found that a number of genes involved in heme synthesis were diminished in fasted PPARα null mice compared with fasted wild-type mice, suggesting that PPARα may govern heme synthesis (Fig. 11.1). Furthermore, several genes involved in biotransformation were decreased in the PPARα null mice, suggesting that PPARα might also positively regulate this pathway. Interestingly, our experiments have also revealed that many of the downstream effects of PPARα activation may actually be due to downregulation of gene expression. Table 11.1 shows the complete list of genes that are increased in livers of fasted PPARα null mice compared with wild-type mice, divided into functional groups, as determined using the Mu6500 Affymetrix GeneChip. This analysis confirmed that PPARα suppresses the hepatic acute phase and inflammatory response, and yielded numerous novel negative targets of PPARα, involved in such diverse pathways as growth factor signaling, extra- and intracellular matrix and intracellular protein transport (Table 11.1). On the basis of these studies it can be concluded that PPARα has a much broader role than previously envisioned, and may exert many of its effects by downregulating gene expression.

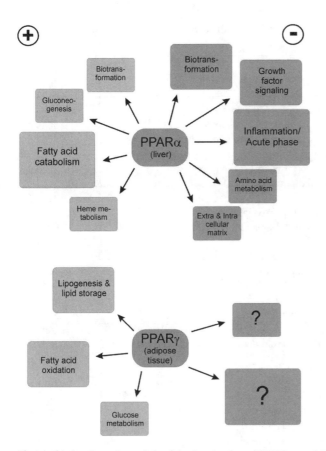

Figure 11.1 Overview of the functional roles of PPARα and PPARγ based on micro-arrays. Pathways that are upregulated are indicated in light grey, pathways that are down-regulated, in dark grey. The size of the boxes and of the text is commensurate with the number of genes belonging to the box.

Another way to study PPARα responsiveness is by exposing cells or animals to PPARα activators. According to oligonucleotide microarray analysis, exposure of rat hepatoma FAO cells to 50 μM of the synthetic PPARα agonist WY14643 for 6 h altered the expression of numerous genes, including many well-known PPARα target genes.[5] Interestingly, in these cells a relatively large number of protein kinases, phosphatases, and signaling molecules were regulated by WY14643, which was confirmed for some at the protein level. In transient reporter assay the promoter of MAPK phosphatase 1 was found to be responsive to PPARα and WY14643, although effects were marginal. At least two of the kinases that were downregulated by WY14643 in FAO cells, casein kinase I and phosphatidylinositol 3-kinase regulatory subunit (p85 beta), were upregulated in livers of PPARα null mice (Table 11.1), confirming their negative

Table 11.1 Genes Increased in Fasted PPARα Null Mice Compared with Fasted Wild-Type Mice

Accession no.	Gene	Fold-increase	Average difference
Biotransformation/cytochrome p450 system			
J03549	*Cyp2a4* (testosterone 15-alpha hydroxylase)	8.5	8089.5
L06463	*Cyp2a1* (testosterone 7-alpha hydroxylase)	1.5	5251.9
D17674	*Cyp2c29* (microsomal aldehyde oxygenase)	2.2	8391.7
X60452	*Cyp3a11* (testosterone 6-beta hydroxylase)	2.7	8876.4
X63023	*Cyp3a13* (nifedipine oxidase ?)	2.9	2764.6
D26137	*Cyp3a16*	4.9	3125.6
M64863	*Cyp17* (17-alpha hydroxylase/C17–20 lyase)	1.6	655.5
U36993	*Cyp7b1* (oxysterol 7-alpha hydroxylase)	4	6952.8
AA109782	Amine *N*-sulfo transferase	11.9	3014.7
J04696	Gluthatione *S*-transferase class mu	2	343.5
L06047	Gluthatione *S*-transferase class alpha	1.7	236.2
D17571	NADP cytochrome p450 oxidoreductase	1.9	4948
U31966	Carbonyl reductase	1.9	1605.7
X06358	UDP-glucuronosyltransferase	2	5704.8
AA117223	3-Alpha hydroxysteroid dehydrogenase (*AKR1C13*)	2.5	453.4
X95685	17-Beta hydroxysteroid dehydrogenase type II	2.2	1779.7
Plasma proteins (acute phase)			
X03208	Alpha 2 microglobulin	1.5	4601.4
U49430	Ceruloplasmin	1.8	5758.2
M13522	Serum amyloid A isoform 2	3.3–9	680.4
M13521	Serum amyloid A isoform 1	3.3–3.6	2446.2
U02554	Serum amyloid A isoform 4	3.3	667.6
Y00426	Serum amyloid P component	3	1830.7
X81579	IGF binding protein	2	2808.9
X70533	Corticosteroid-binding globulin	2.1	1661.2
W07946	Alpha-1-antitrypsin	4	153.2
Growth factor signaling			
X62600	C/EBPbeta	1.8	954.7
U09507	p21 (*Waf1*)	1.9	527.8
Y07836	Stra13 helix loop helix protein	4.3	254.4
U08378	Stat3	2	913.3
J04115	*c-Jun*	18.4	152.5
AA048098	*c-Myc*	4.6	51.5
J03236	*Jun-B*	4.4	52
J05205	*Jun-D*	4.1	1196.6
U74359	*Smad1*	2.7	145.8
Z19579	Seven in absentia homolog (*siah-1*)	2.2	341.9
L20315	Mitogen activated protein kinase MPS1	3.4	96.5

(*continued*)

Table 11.1 Continued

Accession no.	Gene	Fold-increase	Average difference
U39066	Mitogen activated protein kinase kinase	8	595.3
X61940	Growth factor ind. immediate early gene	4.3	528.6
M59821	Pip92 growth factor immediate early protein	2	181.9
W61805	Insulin ind. growth response prot., immediate early protein CL-6	1.8	2418.7
AA030483	Insulin ind. growth response prot. 2	2.2	2669.8
AA016564	Insulin-induced growth response protein CL-4	2.8	430.6
AA138777	Growth arrest and DNA-damage inducible protein GADD45	2.1	177.5
X69062	PCTAIRE-3 protein kinase (cdc2 related)	6.4	354.6
U50413	Phosphoinisitide 3-kinase regulatory subunit p85 alpha	6.2	142.5
W65513	Casein-kinase I, delta isoform	2.7–5.2	151.1
M27844	Calmodulin	1.8	641.3
Extracellular and intracellular matrix			
Z30970	Tissue inhibitor of metalloproteinase 3 (*timp-3*)	3.3	129.9
U74079	ERM (ezrin, radixin, moesin) binding phosphoprotein	2.2	769.2
X60672	Radixin	1.6	903.2
L00919	Band 4.1	2.3	117.3
Y07711	Zyxin	3.2	180
L33779	Desmocollin	2.5	203.2
L22355	Vascular cell adhesion molecule 1	3.5	63.5
Z22532	Syndecan	4.4	195.5
M21495	Cytoskeletal gamma actin	5.5	2522.9
X84325	Rhod GTPase	2.2	290.6
D31842	Cytoskeleton associated PTP non receptor type 14	2.2	112.8
X58251	*COL1A2* pro-alpha (2) collagen	9.4	204.4
Infection/inflammation			
M61909	NF-kappaB p65	4.7	327.6
W29669	Interferon related developmental regulator 1 (*TIS7*)	3.1	46.3
U73037	Interferon regulatory factor 7	4.4	261.9
D49949	Interleukin-18 (interferon gamma inducing factor)	3.8	64.6
AA120109	ERG2 interferon induced protein	4.7	1008.5
X53802	Interleukin-6 receptor	2.3	153.8
M20658	Interleukin-1 receptor	9.8	475
U15209	Macrophage inflammatory protein-1 gamma	2.2	813.6
M58691	*TIS11* (inhibits TNFalpha production)	1.6	597.7
X05475	Complement component C9	2.2	3446.9
M29008	Complement factor H-related factor	2.1	484.4

(*continued*)

Table 11.1   Continued

Accession no.	Gene	Fold-increase	Average difference
M35525	Complement component 5D	1.7	1660.5
AA059550	Phosphodiesterase I	2	461.3
Intracellular protein transport			
AA028877	KDEL receptor, ER protein retention receptor 2	2.1	520.7
AA023107	Vesicle trafficking protein SEC22b	2.2	104.3
AA111168	SEC61gamma, protein transport across ER	2.9	1456.9
Chromosomal proteins			
X12944	HMG-17 chromosomal protein	1.7	806.9
X13171	Histone H1	2.4	732.3
J03482	Histone H1	4.1	432.4
X13605	Histone H3.3	8.7	4134.6
Other metabolic enzymes			
X75129	Xanthine dehydrogenase	1.8	336.5
X56824	Heme oxygenase	2.8	468.4
D42048	Squalene epoxidase	3.2	204.7
L41631	Glucokinase	2.1	148.4
M27796	Carbonic anhydrase III	2.7	1369.9
Miscellaneous			
AA139907	*spot14*	12.8	3122.8
M64292	*BTG2/TIS21*	7.3	219.6
Z72000	*BTG3/tob5*	6.8	225.6
M73696	*Glvr-1* retrovirus receptor, Na+ depend. phosphate transporter	6.6	1274.2
J00621	Renin	5.4	313.1
M60523	Inhibitor of DNA binding HLH protein	4.9	802.1
AA060167	Pre beta cell colony enhancing factor	4.7	823.1
X14432	Thrombomodulin	4.1	86.5
X65627	p68 RNA helicase	4	745.7
U17132	Zinc transporter ZnT-1	3.7	327
L29006	Cationic amino acid transporter CAT2	3.7	246.1
W13646	Polyubiquitin	3.3	208.8
M13685	Prion protein PrP	3.3	267.4
D78135	Cold-inducible RNA binding protein CIRP	3.2	3.2
AA162233	*DEAH9* putative RNA helicase	2.8	218.5
AA002458	*Hsp22* stress protein	2.8	922.7
M32032	Selenium binding liver protein (acetaminophen-binding protein)	2.4	4077.4
U14172	Eukaryotic initiation factor 3 subunit 10 p162	2.2	465
M55154	Transglutaminase	2.1	1604.1

(continued)

Table 11.1 Continued

Accession no.	Gene	Fold-increase	Average difference
M17327	Leukemia virus envelope protein	2	202.1
L76567	*shp* (small heterodimer partner, nuclear hormone receptor)	2	795.7
X96639	Exostosin-1 (putative tumor suppressor)	1.9	366.4
W41883	Thymosin B4 (thymic peptide)	1.9	588.8
X06086	Cathepsin L (lysosomal protease)	1.8	8580.3
Z36270	GC binding protein Sp1-like (TGFbeta inducible early protein)	1.8	227.6
Y08640	Nuclear hormone receptor RORalpha	1.8	297.1
U36220	FK506 binding protein 51 (immunophilin, neg. target of leptin)	1.7	259.5
M13990	Leukemia virus gene fragment	1.5	313.1

Note: RNA was prepared from livers of wild-type and PPARα null mice fasted for 24 h and used for oligonucleotide microarray (Affymetrix). Differentially expressed genes are grouped into functional clusters.

regulation by PPARα *in vivo*. Using the same analysis, a much smaller number of genes was altered in human hepatoma HepG2 cells, with hardly any overlap with the set of genes regulated in FAO cells. The reason behind this is unclear but may reflect cell line- or species-specific regulation.

Instead of exposing cultured cells to a PPARα agonist *in vitro*, Cherkaoui-Malki et al.[6] treated mice with WY14643 for 2 weeks, and examined changes in hepatic gene expression using microarray. It is important to realize that treatment of mice for 2 weeks with a potent PPARα agonist causes massive liver enlargement and peroxisome proliferation and therefore observed changes in gene expression, although dependent on PPARα, may not directly follow ligand-activation of PPARα. Regardless of what cut-off criteria was used, many more genes were found to be down- than upregulated by WY14643 treatment. This corroborates other microarray experiments and suggests that although little mechanistic information is available about transcriptional suppression by PPARα, many of the downstream effects of PPARα activation may actually be accounted for by downregulation of gene expression. Besides the well-known positive PPARα targets such as Cyp4A1 and 4A3, HMG-CoA synthase, and pyruvate dehydrogenase kinase 4, several putative novel direct targets of PPARα were identified, the most interesting of which were a set of liver cell surface marker genes that included CD24, CD39, retinoic acid early transcript γ, and Ly-6D. As the authors point out, additional studies are necessary to ascertain whether these changes occur directly via PPARα, or are secondary effects of sustained hepatocyte and peroxisome proliferation.

Part of the answers may be found in a study in which mice were treated with WY14643 for a much shorter period of time (2 or 3 days) and changes in hepatic gene expression analyzed via oligonucleotide microarray.[7] To determine PPARα-agonist specific changes in hepatic gene expression, the effects of fenofibrate were studied in parallel. In general, the results were very similar to the previous study,[6] including upregulation of many genes involved in fatty acid catabolism. Scatter-plot analysis indicated that the pattern of changes in gene expression induced by WY14643 and fenofibrate were very similar, with the notable exception of metallothionein 1 and 2, which were upregulated by WY14643 but not fenofibrate. Perhaps the most interesting observation in this study is the pronounced downregulation of hepatic serum amyloid 2 expression by both PPARα agonists, which is in perfect agreement with the observed upregulation of serum amyloid expression in livers of PPARα null mice compared with wild-type mice under fasting conditions. These data demonstrate that serum amyloid is a negative target of PPARα *in vivo* and provide supporting evidence for a role of PPARα in suppressing the hepatic acute phase response.

As mentioned previously, by comparing the expression profiles induced by a diverse range of PPAR agonists that have similar clinical effects, genes that are regulated in common and whose regulation is correlated with a clinical, metabolic, or toxicological outcome can be defined. In one study[8] microarray was used to compare the gene expression response elicited by three PPARα agonists, clofibrate, gemfibrozil, and WY14643, and an unrelated compound, phenobarbital, in rat livers. It was observed that the PPARα agonists induced gene expression profiles that bore more similarity to each other than to the pattern induced by phenobarbital. Still, significant differences among the PPARα agonists were observed. For example, whereas WY14643 strongly upregulated expression of tripeptidylpeptidase II, this response was not observed for gemfibrozil and clofibrate. Part of the differential response among PPARα agonists may be attributed to the concept of SPARM, where different PPAR agonists may regulate expression of only partially overlapping sets of genes via differential recruitment of coactivators.

Although this chapter focuses on the implementation of microarray into research on PPARs, some interesting proteomics data have recently been published that are worthwhile to discuss in the context of the global effects of PPAR and/or PPAR agonists. Obese ob/ob mice were treated with either WY14643 or rosiglitazone and the effects examined via 2D gel electrophoresis.[9] In line with published data at the mRNA level, WY14643 influenced expression of several proteins involved in fatty acid binding and oxidation, ketogenesis, and amino acid metabolism. Interestingly, WY14643 diminished expression of proteins involved in homocysteine metabolism, suggesting that PPARα might serve a regulatory role in this pathway. In concert with the low expression of PPARγ in liver, rosiglitazone altered the abundance of a much smaller number of proteins. Surprisingly, several proteins involved in fatty acid oxidation in both mitochondria and peroxisomes, which are considered as classical PPARα regulated pathways, were induced by rosiglitazone. This suggests that PPARγ agonists may stimulate hepatic fatty acid oxidation. The mechanism behind this effect is not exactly clear. Several explanations are possible, including cross-activation of PPARα or lack of specificity of the PPAR response elements present in the gene promoters.

Similarly, male Sprague–Dawley rats were treated with agonists for PPARγ (troglitazone), PPARα (WY14643), or both (SB-219994), and the effects on liver protein abundance analyzed using proteomics.[10] The protein abundance of classical PPAR target genes, including HMG-CoA synthase and long chain acyl-CoA thioesterase, was observed to change after WY14643 treatment. Remarkably, the protein that responded most to WY14643 was stathmin, a protein involved in microtubule dynamics. Because of the small number of proteins that were regulated it is impossible to discern a pattern pointing toward regulation of a specific pathway. Overall, these proteomics data are well in agreement with comparable microarray data and suggest that changes in expression at the mRNA level are most often translated into changes at the protein level.

OVERVIEW MICROARRAY STUDIES PPARγ

Unlike for PPARα, presently there are no published studies that have used gene targeting in combination with microarray to study the effect of PPARγ on global gene expression. It is expected that these will be out soon. However, a limited number of studies did investigate the effects of PPARγ ligand-activation on global gene expression, both *in vitro* and *in vivo*. Way et al.[11] treated Zucker diabetic rats with the specific PPARγ agonist GW1929 and examined the effects on gene expression in white and brown adipose tissue, skeletal muscle, and liver. As expected, the largest effect of GW1929 was observed in white and brown adipose tissue, where PPARγ is most highly expressed. In these tissues it was estimated that about 10% of the expressed genes were altered by GW1929 treatment. In contrast, in liver and skeletal muscle the numbers were only 2% and 1%, respectively, indicating the preferential effect of PPARγ agonists on adipose tissue. Interestingly, in white adipose tissue GW1929 not only increased expression of genes involved in adipo- and lipogenesis, but also that of several genes involved in fatty acid catabolism, an effect which is considered to be reserved to PPARα. Conversely, in skeletal muscle GW1929 decreased mRNA levels of genes involved in fatty acid oxidation. Together with the GW1929-induced downregulation of pyruvate dehydrogenase kinase 4, which phosphorylates and thereby inactivates pyruvate kinase, these data suggest that PPARγ agonists may cause a shift in fuel utilization in muscle from fatty acids to glucose, whereas fatty acid utilization in white fat is enhanced. One of the problems of *in vivo* studies is that it is hard to determine whether the observed changes in gene expression are a direct effect or are secondary to metabolic perturbations elicited by the compound. Thus, despite the (weak) presence of PPARγ in skeletal muscle, it is entirely possible that the observed changes in gene expression in skeletal muscle are a consequence of decreased plasma free fatty acid levels. This may also be true in liver, where three genes involved in gluconeogenesis were suppressed by the PPARγ agonist. It should be noted that PPARγ is hardly expressed in normal liver, but becomes upregulated in fatty livers, such as observed in Zucker diabetic rats.

The *in vitro* system that has been most extensively employed to study the function of PPARγ is the 3T3-L1 adipogenesis model. In this model, 3T3-L1 fibroblasts are induced to differentiate into adipocytes by a cocktail of hormones. Taking advantage of oligonucleotide microarray, the effects of PPARγ activation on gene expression were studied in fibroblasts, preadipocytes (defined as cells treated with the differentiation cocktail for 6 h), and adipocytes.[12] For reasons that are not specified, the authors used an unknown PPARγ agonist. Genes regulated were divided into specific clusters depending on their response pattern. Interestingly, whereas numerous genes were upregulated during adipocyte differentiation, few genes were induced by the PPARγ agonist in differentiated adipocytes, where PPARγ is most highly expressed. Curiously, adipogenic marker genes such as lipoprotein lipase, adipsin and ACRP30/adiponectin adipoQ were not upregulated by PPARγ agonists in differentiated adipocytes. Though it is perfectly possible that the induction of these genes by PPARγ observed in other systems is not reproducible in 3T3-L1 adipocytes, it is also conceivable that the time window of 6 h taken in these studies was insufficient to detect certain effects. The β-catenin gene was signaled out for detailed analysis in this study, as it was downregulated during adipogenesis as well as by the PPARγ agonist in adipocytes. Treatment of db/db mice with rosiglitazone also reduced β-catenin mRNA levels in white adipose tissue, leading the authors to conclude that β-catenin may play an important role in adipocyte differentiation.

Though it is apparent that PPARγ plays an important role in adipocyte differentiation, until recently it was not exactly clear whether PPARγ upregulation *per se* is sufficient to drive adipogenesis. Li and Lazar[13] showed that infection of 3T3-L1 fibroblasts with a constitutively highly active form of PPARγ, VP16-PPARγ, is able to cause adipogenesis and used microarray to demonstrate that this effect was associated with upregulation of many adipogenic genes (adipsin, aP2, and ACRP30/adiponectin/adipoQ). Overall, the effect of VP16-PPARγ on 3T3-L1 adipogenesis was comparable to that of rosiglitazone although particular genes showed a marked difference in fold-induction. Most notably, whereas resistin was 147-fold induced during VP16-PPARγ-induced adipogenesis, it was upregulated only 40-fold during rosiglitazone-induced adipogenesis. Resistin is a small protein produced by adipocytes, whose regulation at the mRNA level has been the source of considerable debate. Interestingly, rosiglitazone reduced expression of resistin in VP16-PPARγ-induced adipocytes, which may indicate that rosiglitazone may influence resistin mRNA independently of PPARγ. Unfortunately, the authors limit the presentation of the microarray data to a small subset of mostly known PPARγ target and/or adipogenic marker genes. Clearly, the power of microarray has not been fully exploited as yet, and hopefully the remainder of the data will greet us in a different format in the near future.

Other groups have studied differential gene expression during 3T3-L1 adipogenesis, and though many are candidate PPARγ target genes, this requires confirmation in a different model in which the level of PPARγ or its ligand-activation can be modulated.[14,15] These studies also highlight one of the dangers of microarray in that authors should avoid presenting huge lists of merely descriptive data. This can be achieved by performing a detailed and clever follow-up analysis, or by carrying out advanced statistical analysis to present the data in a comprehensible and meaningful manner.

Ligand activation of PPARγ inhibits cell proliferation and stimulates differentiation of several epithelial derived cell lines. Accordingly, there has been an interest to find out what genes may mediate the antiproliferative effects of PPARγ ligands. To this end, Gupta et al.[16] treated the MOSER S colon carcinoma cell line with rosiglitazone and examined the effects on global gene expression by microarray. For reasons that are not clear, the expression of only nine genes, involved in several different processes, was presented. On the basis of the observed upregulation of the carcinoembryonic antigen (CEA) by rosiglitazone, other members of the same family were signaled out for further investigation. Both nonspecific cross-reacting antigen and biliary glycoprotein were found to be induced by rosiglitazone and by the PPARγ agonist GW7845. As CEA proteins have been implicated in aggregation of colon cancer cells, the effect of rosiglitazone on CEA-dependent aggregation in MOSER S cells was studied. A positive effect of rosiglitazone was observed, suggesting that PPARγ may regulate epithelial cell adhesion via CEA proteins. Although more elaborate follow-up experiments are required to verify this notion, the study nicely illustrates the potential of microarray to yield novel exciting data that can provide key insights into major biological processes.

CONCLUSIONS

As reviewed earlier, the implementation of microarrays into research on PPARs and their ligands has already led to important new insights into the biological function of PPARs, and has given rise to large lists of genes that are (either directly or indirectly) regulated by PPARs or their ligands. One of the major challenges ahead is how to interpret this huge amount of descriptive data and decide what type of follow-up analysis, both analytical and experimental, is required to exploit the benefits of this approach fully. The type of follow-up obviously depends on the objectives that have been defined and may range from real-time PCR analysis to obtain a more quantitative picture of the regulation, to elaborate cross-comparisons among numerous PPAR agonists to establish common targets and thus common mechanisms of action and/or toxicity. Because of the high price tag attached, this latter type of approach will mostly remain the exclusive realm of pharmaceutical companies, which are willing to invest heavily in order to develop the best PPAR agonists or antagonists.

ACKNOWLEDGMENTS

The original microarray data presented in Table 11.1 were generated in a collaborative effort with Pascal Escher and Sherrie Tafuri from Pfizer. The assistance and support of Steven Madore, Sandra Rojas-Caro, Tim Jatkoe, Steve Hunt, and Gary McMaster is gratefully acknowledged. Work done in the author's laboratories was supported by the Swiss National Science Foundation, Human Frontier Science Program, the Royal Netherlands Academy of Art and Sciences (KNAW), the Netherlands Organization for Scientific Research (NWO), and the Dutch Diabetes Foundation.

REFERENCES

1. Kersten, S.; Desvergne, B.; Wahli, W. Roles of PPARs in health and disease. Nature **2000**, *405*, 421–424.
2. Desvergne, B.; Wahli, W. Peroxisome proliferator-activated receptors: nuclear control of metabolism. Endocr. Rev. **1999**, *20*, 649–688.
3. Di-Poi, N.; Tan, N.S.; Michalik, L.; Wahli, W.; Desvergne, B. Antiapoptotic role of PPARbeta in keratinocytes via transcriptional control of the Akt1 signaling pathway. Mol. Cell **2002**, *10*, 721–733.
4. Kersten, S.; Mandard, S.; Escher, P.; Gonzalez, F.J.; Tafuri, S.; Desvergne, B.; Wahli, W. The peroxisome proliferator-activated receptor alpha regulates amino acid metabolism. FASEB J. **2001**, *15*, 1971–1978.
5. Van den Heuvel, J.P.; Kreder, D.; Belda, B.; Hannon, D.B.; Nugent, C.A.; Burns, K.A.; Taylor, M.J. Comprehensive analysis of gene expression in rat and human hepatoma cells exposed to the peroxisome proliferator WY14,643. Toxicol. Appl. Pharmacol. **2003**, *188*, 185–198.
6. Cherkaoui-Malki, M.; Meyer, K.; Cao, W.Q.; Latruffe, N.; Yeldandi, A.V.; Rao, M.S.; Bradfield, C.A.; Reddy, J.K. Identification of novel peroxisome proliferator-activated receptor alpha (PPARalpha) target genes in mouse liver using cDNA microarray analysis. Gene Expr. **2001**, *9*, 291–304.
7. Yamazaki, K.; Kuromitsu, J.; Tanaka, I. Microarray analysis of gene expression changes in mouse liver induced by peroxisome proliferator-activated receptor alpha agonists. Biochem. Biophys. Res. Commun. **2002**, *290*, 1114–1122.
8. Hamadeh, H.K.; Bushel, P.R.; Jayadev, S.; Martin, K.; DiSorbo, O.; Sieber, S.; Bennett, L.; Tennant, R.; Stoll, R.; Barrett, J.C.; Blanchard, K.; Paules, R.S.; Afshari, C.A. Gene expression analysis reveals chemical-specific profiles. Toxicol. Sci. **2002**, *67*, 219–231.
9. Edvardsson, U.; Brockenhuus Von Lowenhielm, H.; Panfilov, O.; Nystrom, A.C.; Nilsson, F.; Dahllof, B. Hepatic protein expression of lean mice and obese diabetic mice treated with peroxisome proliferator-activated receptor activators. Proteomics **2003**, *3*, 468–478.

10. White, I.R.; Man, W.J.; Bryant, D.; Bugelski, P.; Camilleri, P.; Cutler, P.; Hayes, W.; Holbrook, J.D.; Kramer, K.; Lord, P.G.; Wood, J. Protein expression changes in the Sprague–Dawley rat liver proteome following administration of peroxisome proliferator activated receptor alpha and gamma ligands. Proteomics **2003**, *3*, 505–512.

11. Way, J.M.; Harrington, W.W.; Brown, K.K.; Gottschalk, W.K.; Sundseth, S.S.; Mansfield, T.A.; Ramachandran, R.K.; Willson, T.M.; Kliewer, S.A. Comprehensive messenger ribonucleic acid profiling reveals that peroxisome proliferator-activated receptor gamma activation has coordinate effects on gene expression in multiple insulin-sensitive tissues. Endocrinology **2001**, *142*, 1269–1277.

12. Gerhold, D.L.; Liu, F.; Jiang, G.; Li, Z.; Xu, J.; Lu, M.; Sachs, J.R.; Bagchi, A.; Fridman, A.; Holder, D.J.; Doebber, T.W.; Berger, J.; Elbrecht, A.; Moller, D.E.; Zhang, B.B. Gene expression profile of adipocyte differentiation and its regulation by peroxisome proliferator-activated receptor-gamma agonists. Endocrinology **2002**, *143*, 2106–2118.

13. Li, Y.; Lazar, M.A. Differential gene regulation by PPARgamma agonist and constitutively active PPARgamma2. Mol. Endocrinol. **2002**, *16*, 1040–1048.

14. Okuno, M.; Arimoto, E.; Nishizuka, M.; Nishihara, T.; Imagawa, M. Isolation of up- or down-regulated genes in PPARgamma-expressing NIH-3T3 cells during differentiation into adipocytes. FEBS Lett. **2002**, *519*, 108–112.

15. Burton, G.R.; Guan, Y.; Nagarajan, R.; McGehee, R.E. Microarray analysis of gene expression during early adipocyte differentiation. Gene **2002**, *293*, 21–31.

16. Gupta, R.A.; Brockman, J.A.; Sarraf, P.; Willson, T.M.; DuBois, R.N. Target genes of peroxisome proliferator-activated receptor gamma in colorectal cancer cells. J. Biol. Chem. **2001**, *276*, 29681–29768.

12

Microarray Analysis of Changes Induced by Peroxisome Proliferator-activated Receptor α Agonists in the Expression of Genes Involved in Lipid Metabolism

**Kazuto Yamazaki, Taro Hihara, Hideki Yoshitomi,
Junro Kuromitsu, and Isao Tanaka**

*Tsukuba Research Laboratories, Eisai Co., Ltd., Tokodai,
Tsukuba, Ibaraki, Japan*

PEROXISOME PROLIFERATOR-ACTIVATED RECEPTORS (PPARs)

A family of transcriptional factors known as peroxisome proliferator-activated receptors (PPARs) plays a crucial role in lipid and lipoprotein metabolism, and glucose homeostasis. There are three PPAR subtypes, which are encoded by different genes and are commonly designated PPARα (NR1C1), PPARγ (NR1C3), and PPARδ (NR1C2, NUC1, PPARβ, or FAAR).[1] PPARs possess a domain structure that is common to other members of the nuclear receptor gene family: N-terminus, DNA-binding domain, and ligand-binding domain. PPARs form heterodimers with another nuclear receptor, 9-*cis*-retinoic receptor (RXR or NR2B).[1] PPAR/RXR heterodimers bind to DNA sequences containing direct repeats of the hexanucleotide sequence AGGTCA separated by one nucleotide, known as the peroxisome proliferator response element (PPRE).

PPARα regulates fatty acid and lipoprotein metabolism, and is highly expressed in the liver, heart, kidney, and muscle.[2] It is also known that

PPARα downregulates inflammatory response.[3] Saturated and unsaturated fatty acids, such as palmitic acid, oleic acid, linoleic acid, and arachidonic acid, activate PPARα as natural ligands.[4] The lipoxygenase metabolite, 8(S)-hydroxyeicosatetraenoic acid (HETE) was identified as a high-affinity PPARα ligand.

PPARγ exists in two distinct isoforms, termed PPARγ1 and PPARγ2, which have different tissue distributions and functions. PPARγ1 is found in the liver and, to a lesser degree, in other organs, including adipose tissue. On the other hand, PPARγ2 is expressed exclusively in adipose tissue. PPARγ2 activation leads to adipocyte differentiation, lipid storage, and inhibition of inflammation. Essential fatty acids, such as linoleic acid, linolenic acid, and arachidonic acid, and eicosapentaenoic acid (EPA), bind to PPARγ at micromolar concentrations.[5] The 15-lipoxygenase metabolites of linoleic acid, 9-hydroxyoctadecadienoic acid (HODE), and 13-HODE, are also active at micromolar concentrations.[6] In addition, 15-deoxy-$\Delta^{12,14}$-prostaglandin J_2 (15d-PGJ$_2$) activates PPARγ at low micromolar concentrations.[7,8]

Different from PPARα and PPARγ, PPARδ is ubiquitously expressed. Although its physiological functions are yet to be fully determined, it was reported that PPARδ is involved in reverse cholesterol transport via regulation of the ATP-binding cassette A1 gene expression.[9] Both saturated and unsaturated fatty acids are bound to PPARδ.[10] Among them, dihomo-γ-linolenic acid, arachidonic acid, and EPA bind at low micromolar concentrations.[10,11] Several eicosanoids, such as PGA$_1$ and PGD$_2$, activate PPARδ.[6]

Table 12.1 shows the EC$_{50}$ values of transactivation of several synthetic PPAR agonists. The hypolipidemic fibrate drugs are an important class of PPARα ligands. Wy-14,643 at micromolar concentrations is an activator of murine PPARα. These drugs, such as fenofibrate, bezafibrate, clofibrate, and gemfibrozil, have been used worldwide.[14,15] They activate human and murine PPARα at high micromolar concentrations. More potent compounds, such as GW9578, have been developed with EC$_{50}$ values at the nanomolar levels, and these are expected to be utilized as novel antihyperlipidemic drugs.[16] GW501506 is a PPARδ-selective agonist.[11] Thiazolidinedione compounds, including pioglitazone and rosiglitazone, are high-affinity PPARγ agonists, which are a class of antidiabetic agents.[15,17] A more potent PPARγ agonist, GI262570, was synthesized, and proved to be ~30-fold more active than rosiglitazone. Moreover, PPAR α/γ dual agonists including KRP-297 have been developed, and exert both antidiabetic and antihyperlipidemic effects.[18–21] Figure 12.1 shows the chemical structures of the compounds listed in Table 12.1. It can be seen that PPARα has a characteristic species-specificity, compared with PPARδ and PPARγ (Table 12.1). Thus, the EC$_{50}$ value of Wy-14,643 for PPARα is ~40× larger in humans than in mice. Conversely, the EC$_{50}$ value of KRP-297 for human PPARα is ~20× smaller in murine PPARα. GI262570 activates human PPARα with an EC$_{50}$ of 0.85 μM, but does not activate murine PPARα until about 30 μM. The other PPARs do not show such dramatic species differences.

Table 12.1   Activity of PPAR Agonists in PPAR-GAL4 Transactivation Assay

Compound[a]	Human receptor, EC$_{50}$ (μM)			Murine receptor, EC$_{50}$ (μM)		
	PPARα	PPARδ	PPARγ	PPARα	PPARδ	PPARγ
Wy-14,643[b]	4.2	NA	NA	0.11	NA	NA
Fenofibric acid[c]	13	NA	73	19	NA	95
Bezafibrate	39	120	44	85	110	61
Gemfibrozil[c]	61	NA	57	45	NA	75
GW2331	0.064	1.2	0.16	0.0015	0.72	0.20
GW9578	0.067	0.89	0.72	0.0051	0.99	0.88
GW501516	0.99	0.0013	1.4	1.5	0.034	1.5
Pioglitazone[a]	NA	NA	0.18	12	NA	0.32
Rosiglitazone[a]	NA	NA	0.016	11	NA	0.025
GI262570[a]	0.85	NA	0.00050	NA	NA	0.00094
KRP-297	0.67	3.8	0.088	12	2.4	0.16

[a]Compound structures are shown in Fig. 12.1. These data were generated using the PPAR-GAL4 transactivation assay[12] with a PLAP reporter[13] in transiently transfected CV-1 cells. EC$_{50}$ = the concentration of test compound that gave 50% of the maximal reporter activity, $n = 3$.
[b]NA = not activated at 30 μM.
[c]NA = not activated at 300 μM.

MICROARRAY ANALYSIS OF HEPATIC GENE EXPRESSION OF MICE TREATED WITH PPARα AGONISTS

We have recently analyzed changes of gene expression in the mouse liver induced by PPARα agonists, using microarray technology.[22] In this chapter, we outline and update our results, focusing on expression of genes involved in lipid metabolism. In addition to our goal of finding novel genes affected by PPARα, we selected two PPARα agonists, Wy-14,643 as a strong murine PPARα agonist, and fenofibrate as a marketed and high micromolar PPARα agonist with activity in the high micromolar concentration range, in order to examine whether or not there are differences in gene-regulatory activity between them. Wy-14,643 is ∼170-fold more potent than fenofibric acid as a murine PPARα transactivator (Table 12.1). We orally treated 8-week-old male C57BL/6N mice with fenofibrate at 100 mg/kg, or Wy-14,643 at 30 or 100 mg/kg daily, and collected the liver on day 2 or 3 to detect early and primary changes of gene expression caused by these PPARα agonists. In a preliminary examination, we identified the doses of fenofibrate and Wy-14,643 which produced similar degrees of induction of the palmitoyl acyl-CoA oxidase 1 (ACOX1) gene, a well-known marker of peroxisome proliferation, on day 2. On the basis of the results, the doses of fenofibrate and Wy-14,643 were set at 100 and 30 mg/kg, respectively. An Affymetrix Murine Genome U74Av2 GeneChip was used as a high-density oligonucleotide microarray; it is able to screen ∼6000 full-length cDNAs and ∼6000 expressed sequence tags (ESTs).

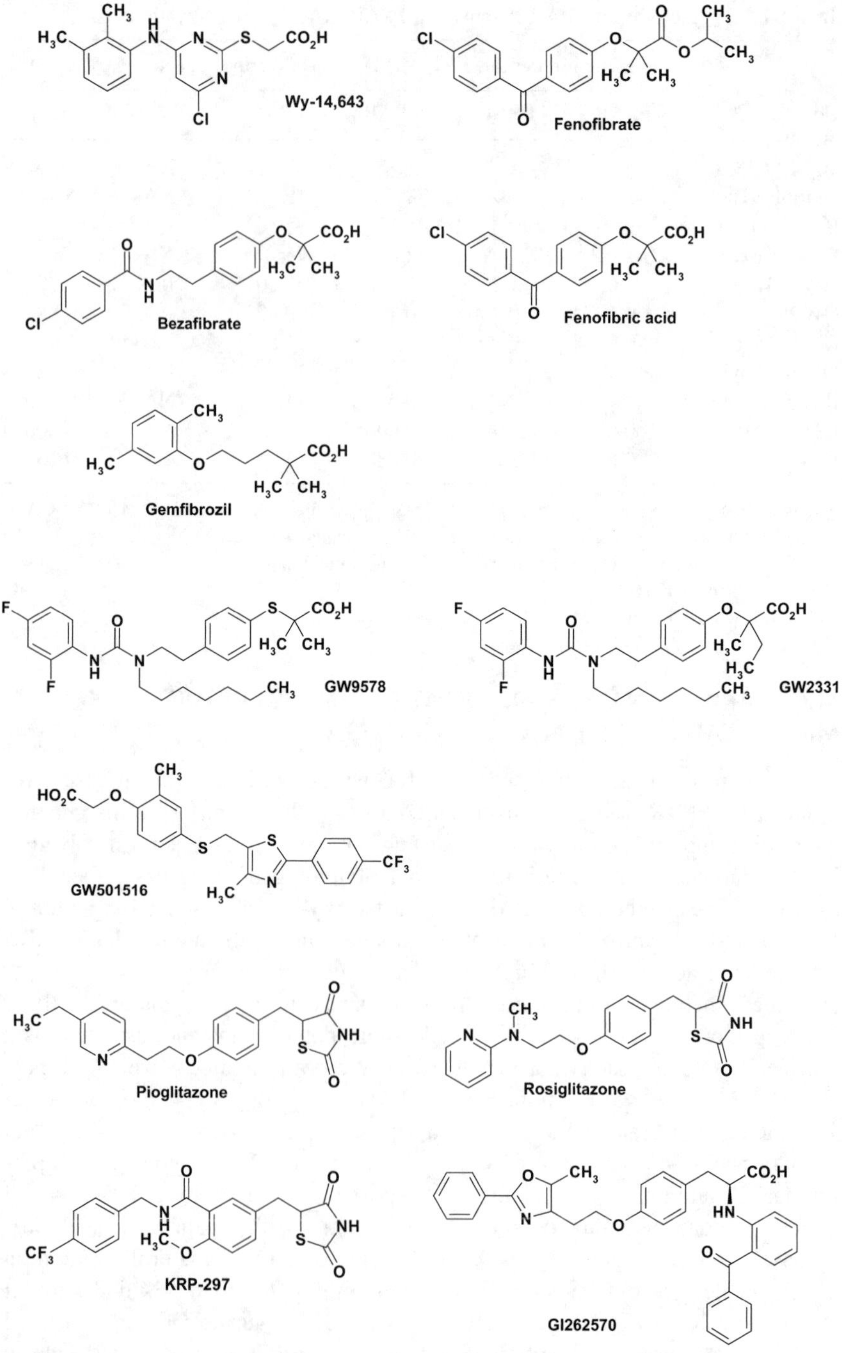

Figure 12.1 Chemical structures of the PPAR agonists listed in Table 12.1.

Data were analyzed with the Affymetrix GeneChip Expression Analysis Software (version 3.1).[23]

Representative scatter plots with regression curves are shown in Fig. 12.2. The scatter plots between control and PPARα agonists show that they induce both upregulation and downregulation of expression of several genes. The regression curves and r^2 values between the two PPARα agonists indicate that the gene regulation patterns by fenofibrate and Wy-14,643 are basically similar, in spite of the difference of *in vitro* mouse PPARα transactivation potency. We picked up changes of gene expression involved in lipid metabolism, and the upregulated and downregulated gene groups are summarized in Figs. 12.3

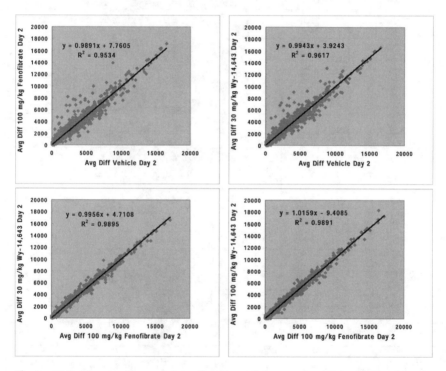

Figure 12.2 Scatter plots and regression lines of gene expression in the liver of male C57BL/6N mice treated with vehicle or a PPARα agonist, between vehicle and fenofibrate, between vehicle and Wy-14,643, and between fenofibrate and Wy-14,643 treatment. Mice were given fenofibrate (100 mg/kg), Wy-14,643 (30 or 100 mg/kg), or vehicle (0.5% methyl cellulose) by gavage once a day, and the liver was collected at 24 h after the last treatment on day 2 or 3 under nonfasting conditions. After extraction of poly(A)+ mRNA from the liver, equal quantities of poly(A)+ mRNA from mice in each group were pooled ($n = 3$ except for the 100 mg/kg Wy-14,643 group, $n = 2$), and 5 μg of the pooled poly(A)+ mRNA was subjected to microarray analysis. Avg Diff indicates average difference of intensity of gene expression determined by Affymetrix microarray analysis. The scatter plots were made with Microsoft Excel 2000.

Gene name	Gene Symbol	Accession number	Vehicle Day 3	Fenofibrate 100 mg/kg Day 2	Day 3	Wy-14,643 30 mg/kg Day 2	Day 3	Wy-14,643 100 mg/kg Day 2	Gene ontology classifications (biological process)
cytosolic acyl-CoA thioesterase 1	Cte1	Y14004	-1.3	6.3	8.8	4.3	2.6	6.7	acyl-CoA metabolism/long-chain fatty acid metabolism
peroxisomal acyl-CoA thioesterase 1	Pte1	AW046123	-1.1	1.4	2.1	1.3	1.3	1.3	acyl-CoA metabolism
hydroxyacyl-CoA dehydrogenase/3-ketoacyl-CoA thiolase/enoyl-CoA hydratase (trifunctional protein), β subunit	Hadhb	AW122615	-1.1	1.7	2	1.6	1.6	2	fatty acid β-oxidation/fatty acid metabolism
acetyl-CoA acyltransferase 1	Acaa1	AI530403	-1.1	3.6	4.3	3.2	2.9	3.9	fatty acid metabolism
enoyl-CoA hydratase/3-hydroxyacyl-CoA dehydrogenase	Ehhadh	AJ011864	-1.1	5.9	6.8	5.9	5.7	6.7	acetyl-CoA metabolism/fatty acid β-oxidation/fatty acid metabolism/metabolism
dodecenoyl-CoA Δ-isomerase (3,2 trans-enoyl-CoA isomerase)	Dci	Z14050	1	3.4	4	2.8	3.1	3.5	fatty acid metabolism/ metabolism
enoyl-CoA hydratase 1, peroxisomal	Echl	AF030343	-1.1	1.7	2.1	1.8	1.7	1.9	fatty acid metabolism/metabolism
carnitine acetyltransferase	Crat	X85983	1.2	6.8	10.7	4	3.4	7.8	fatty acid metabolism/transport
2,4-dienoyl-CoA reductase 1, mitochondrial	Decr1	AI844846	1.1	2.2	2.5	2	1.9	2.5	metabolism
acetyl-CoA dehydrogenase, medium chain	Acadm	U07159	1	2	2.2	2	2	2.3	electron transport/fatty acid metabolism
acyl-CoA oxidase 1, palmitoyl	Acox1	AF006688	1.1	1.7	1.9	1.5	1.5	1.7	electron transport/fatty acid β-oxidation/fatty acid metabolism/spermatogenesis
fatty acid CoA ligase, long chain 5	Facl5	AI838021	1.1	1.4	1.7	1.3	1.3	1.7	fatty acid metabolism/metabolism
cytochrome P450, family 4, subfamily a, polypeptide 10	Cyp4a10	AB01821	-1.4	2.2	2.2	2	2	2.3	—
cytochrome P450, family 4, subfamily a, polypeptide 12	Cyp4a12	Y10221	-1.3	3.3	4.3	3.5	3.5	4.6	electron transport
cytochrome P450, family 4, subfamily a, polypeptide 14	Cyp4a14	Y11638	-1.2	2	2.1	2.2	1.8	2.1	electron transport
epoxide hydrolase 2, cytoplasmic	Ephx2	Z37107	-1.1	1.6	1.5	1.5	1.4	1.7	aromatic compound metabolism/metabolism/response to toxin/xenobiotic metabolism
aldehyde dehydrogenase family 3, subfamily A2	Aldh3a2	U14390	-1.1	2.5	3.1	1.8	1.8	2.8	metabolism
lipoprotein lipase	Lpl	AA726364	-1.2	3.9	4.6	3.3	3.4	3.3	lipid catabolism/lipid metabolism
monoglyceride lipase	Mgll	AI846600	-1.1	1.9	2.4	1.4	1.3	2.2	aromatic compound metabolism/lipid metabolism/proteolysis and peptidolysis
phospholipid transfer protein	Pltp	U28960	1.1	1.3	3.3	1.6	2.7	1.8	lipid transport
solute carrier family 27 (fatty acid transporter), member 1	Slc27a1	U15976	1.1	5.7	5.7	4.2	4	5.4	lipid transport/metabolism/transport
fatty acid binding protein 2, intestinal	Fabp2	M65034	1.2	1.7	1.7	2.1	1.6	1.7	transport
fatty acid binding protein 4, adipocyte	Fabp4	M20497	-1.3	1.4	2	1.8	2.4	2.9	transport
malic enzyme, supernatant	Mod1	J02652	1.4	2.5	3.5	2.1	1.8	3.3	malate metabolism
glycerol-3-phosphate acyltransferase, mitochondrial	Gpam	U11680	1.4	1.4	1.7	1.2	1.5	2	fatty acid metabolism/metabolism/phospholipid biosynthesis/positive regulation of body size
glycerol-3-phosphate dehydrogenase 1 (soluble)	Gpd1	M25558	-1	1.6	1.8	1.4	1.2	1.8	carbohydrate metabolism/glycerol-3-phosphate metabolism
hydroxysteroid (17β) dehydrogenase 4	Hsd17b4	X89998	1	1.8	2.1	1.7	1.6	1.9	metabolism/protein targeting/steroid biosynthesis
hydroxysteroid (17β) dehydrogenase 11 (retinal short-chain dehydrogenase/reductase 2)	Retsdr2	AA822174	-1	2.5	2.6	2.2	2.3	2.3	metabolism
cytochrome P450, family 2, subfamily b, polypeptide 9	Cyp2b9	M21855	1.5	3.5	6.8	2.7	5.3	2.6	electron transport
cytochrome P450, family 2, subfamily b, polypeptide 10	Cyp2b10	M21856	1.5	2.5	3.4	2.9	4.1	2.5	electron transport
NAD(P)-dependent steroid dehydrogenase-like	Nsdhl	AW106745	1.3	1.3	1.8	-1.1	1.8	1.1	cholesterol biosynthesis/cholesterol metabolism/steroid biosynthesis
3-hydroxy-3-methylglutaryl-CoA reductase	Hmgcr	M62766	1.1	-1.1	1.4	1.2	1.7	1.4	biosynthesis/cholesterol biosynthesis
3-hydroxy-3-methylglutaryl-CoA lyase	Hmgcl	U49878	1.1	1.4	1.7	1.4	1.4	1.4	—
7-dehydrocholesterol reductase	Dhcr7	AF057368	1.4	1.2	1.6	1.3	2	1.9	cholesterol biosynthesis/sterol biosynthesis
cytochrome P450, family 8, subfamily b, polypeptide 1	Cyp8b1	AF090317	1.4	1.4	1.7	1.1	1.5	1.3	electron transport
squalene epoxidase	Sqle	D42048	1.3	1.5	1.7	1.1	2.3	1.8	aromatic compound metabolism/electron transport/metabolism
sterol-C_4-methyl oxidase-like	Sc4mol	AI848668	1.6	1.4	1.5	1.8	2.9	2.6	fatty acid metabolism/metabolism/steroid metabolism/sterol biosynthesis
retinol dehydrogenase 6	Rdh6	AF030513	1.1	2.6	3.8	2.8	3.4	3.7	metabolism
glutathione S-transferase, θ2	Gstt2	X98056	1.1	2	1.9	1.6	1.4	1.9	—
microsomal glutathione S-transferase 3	Mgst3	AI843448	1.3	1.8	2.2	1.9	2	1.6	biological process unknown

Legend:
- -1.5< <1.5
- >1.5
- >2
- >4
- >6
- >8
- >10

and 12.4, respectively, with greater than 1.5-fold or less than -1.5-fold being the cut-off relative to gene expression of the vehicle-treated group on day 2.

In a separate study, we determined blood triglyceride, total cholesterol, high-density lipoprotein cholesterol, and non-esterified fatty acid in 8-week-old male C57BL/6N mice treated with fenofibrate (100 mg/kg) or Wy-14,643 (30 or 100 mg/kg) on days 0, 2, and 3. On day 2, blood triglyceride levels were already decreased significantly in all PPARα agonist-treated groups to similar extents [Fig. 12.5(a)]. The PPARα agonist treatment seemed to reduce blood non-esterified fatty acid levels, but this reduction was not significant [Fig. 12.5(b)]. No significant change was observed in total cholesterol or high-density lipoprotein cholesterol. Similar results were obtained on day 3.

OXIDATION OF FATTY ACID

There are several lines of evidence that various genes involved in β- and ω-oxidation pathways are under the transcriptional control of PPARα.[24–26] We observed induction of several genes involved in fatty acid oxidation. Acyl-CoA thioesterases cleave thioester bonds and function in fatty acid β-oxidation, regulating intracellular levels of CoA esters and CoA. This function may be very important during high β-oxidation and fatty acid overload, to generate the free CoA necessary for fatty acid β-oxidation to proceed.[27] Cytosolic acyl-CoA thioesterase 1 (CTE1) expression was strongly induced, particularly by 100 mg/kg fenofibrate treatment (6.3–8.8-fold). On the other hand, induction of peroxisomal acyl-CoA thioesterase 1 (PTE1) was observed only in the 100 mg/kg fenofibrate-treated group on day 3 (2.1-fold).

Carnitine acyltransferase (CRAT) catalyzes the reversible transfer of acyl groups from an acyl-CoA thioester to carnitine, and is thought to play an important role in the rapid transfer of activated long-chain fatty acids into mitochondria for β-oxidation.[28] CRAT gene expression was strongly induced, especially by feno-fibrate treatment, by ~7-fold on day 2 and by ~11-fold on day 3. Wy-14,643 at 30 mg/kg increased CRAT mRNA at most by fourfold, though a higher dose caused an approximately eightfold elevation.

Figure 12.3 Upregulated genes and expressed sequence tags (ESTs) in the liver of male C57BL/6N mice treated with fenofibrate, Wy-14,643, or vehicle (0.5% methyl cellulose) on day 2 or 3. The liver was collected at 24 h after the last treatment under nonfasting conditions. After extraction of poly(A)+ mRNA from the liver, equal quantities of poly(A)+ mRNA from mice in each group were pooled ($n = 3$ except for the 100 mg/kg Wy-14,643 group, $n = 2$), and 5 μg of the pooled poly(A)+ mRNA was subjected to microarray analysis. The value in each box is the fold change relative to that of the group of vehicle-treated mice on day 2. ESTs were surveyed on the following Web site: http://www.ncbi.nlm.nih.gov/Unigene/. Mouse gene symbols and gene ontology classifications were retrieved from the Jackson Laboratory homepage: http://www.informatics.jax.org/. —, not described. (Adapted from Yamazaki et al.[20].)

Gene name	Gene symbol	Accession number	Vehicle Day 3	Fenofibrate 100 mg/kg		Wy-14,643 30 mg/kg		Wy-14,643 100 mg/kg	Gene ontology classifications (biological process)
				Day 2	Day 3	Day 2	Day 3	Day 2	
hydroxysteroid 11β dehydrogenase 1	Hsd11b1	X83202	1.1	-1.6	-1.9	-1.8	-1.7	-2.1	lung development/metabolism/steroid metabolism
hydroxysteroid dehydrogenase 5, Δ⁵-3β	Hsd3b5	L41519	-1.1	-2.5	-2.6	-1.5	-2.2	-2.1	C₂₁-steroid hormone biosynthesis/steroid biosynthesis
arachidonate lipoxygenase 3	Alox3	Y14695	-1.3	-1.1	-2.3	-1.1	-1.2	-1	electron transport/leukotriene biosynthesis/metabolism
cytochrome P450, family 2, subfamily c, polypeptide 37	Cyp2c37	AF047542	1	-2	-2	-2	-2	-2	electron transport
cytochrome P450, family 2, subfamily c, polypeptide 40	Cyp2c40	AF47727	1	-1.5	-1.4	-1.4	-1.2	-1.5	electron transport
apolipoprotein A-IV	Apoa4	M64248	1.4	-2.8	-4.9	-3.6	-2.1	-3.8	lipid transport/regulation of cholesterol absorption
apolipoprotein A-V	Apoa5	AA674450	1	-1.9	-1.2	-1.1	-1.1	-2.1	biological process unknown/lipid transport
glycosylphosphatidylinositol specific phospholipase D1	Gpld1	AF050666	1	-1.6	-1.7	1.6	-1.3	-1.8	cell adhesion/cell-matrix adhesion/GPI anchor release
cytochrome P450, family 7, subfamily b, polypeptide 1	Cyp7b1	U36993	-1	-1.5	-1.3	-1	-1.3	-1.2	cholesterol metabolism/electron transport
serum amyloid A1	Saa1	M13521	1	-1.2	-2.9	-1.5	-1.7	-2.2	acute-phase response
serum amyloid A2	Saa2	U60438	1.1	-3.4	-4.1	-3.7	-6.6	-6.6	acute-phase response
serum amyloid A3	Saa3	X03505	-1.4	-1.6	-1.8	-1.5	-1.8	-2.2	acute-phase response

-1.5< <1.5
<-1.5
<-2
<-4
<-6

Figure 12.4 Downregulated genes and expressed sequence tags (ESTs) in the liver of male C57BL/6N mice treated with fenofibrate, Wy-14,643, or vehicle (0.5% methyl cellulose) on day 2 or 3. The liver was collected at 24 h after the last treatment under non-fasting conditions. After extraction of poly(A)+ mRNA from the liver, equal quantities of poly(A)+ mRNA from mice in each group were pooled ($n = 3$ except for the 100 mg/kg Wy-14,643 group, $n = 2$), and 5 μg of the pooled poly(A)+ mRNA was subjected to microarray analysis. The value in each box is the fold change relative to that of the group of vehicle-treated mice on day 2. ESTs were surveyed on the following Web site: http://www.ncbi.nlm.nih.gov/Unigene/. Mouse gene symbols and gene ontology classifications were retrieved from the Jackson Laboratory homepage: http://www. informatics.jax.org/. (Adapted from Yamazaki et al.[20].)

Hydroxyacyl-CoA dehydrogenase/3-ketoacyl-CoA thiolase/enoyl-CoA hydratase (trifunctional protein), β subunit (HADHB) and α subunit (HADHA) are the mitochondrial trifunctional proteins. The heterocomplex containing four α and four β subunits catalyzes three steps in β-oxidation of fatty acid, including the long-chain 3-hydroxy-CoA dehydrogenase step. In this study, approximately twofold increase was seen in all compound-treated groups on days 2 and 3.

Acyl-CoA dehydrogenase, medium-chain (ACADM) is specific to C_4-C_{12} straight-chain fatty acids. This enzyme catalyzes the initial step of the mitochondrial fatty acid β-oxidation pathway. ACADM gene expression was doubled on days 2 and 3 after all treatments. Mitochondrial 2,4-dienoyl-CoA reductase 1 (DECR1) functions as an auxiliary β-oxidation enzyme, metabolizing unsaturated fatty enoyl-CoA esters that have double bonds in both odd- and even-numbered positions. PPARα agonist administration led to a three- to fourfold increase of DECR1 mRNA. Dodecenoyl-CoA Δ-isomerase (DCI) transforms 3-*cis* and 3-*trans* intermediates to 2-*trans*-enoyl-CoA compounds, functioning

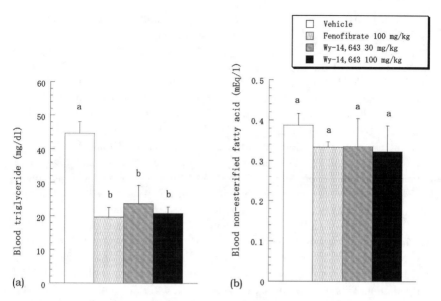

Figure 12.5 Blood triglyceride (a) and non-esterified fatty acid levels (b) of 8-week-old male C57BL/6N mice orally treated with fenofibrate (100 mg/kg), Wy-14,643 (30 or 100 mg/kg), or vehicle (0.5% methyl cellulose) on day 2. Blood was drawn from the caudal vein under nonfasting conditions on days 0, 2, and 3. Values are expressed as means ± SEM. For multiple comparison, a one-way analysis of variance (ANOVA) was carried out, followed by Scheffé's test as a *post hoc* test. Bars with the same letter are not statistically significantly different by Scheffé's test ($p > 0.05$), $n = 4$–5.

in unsaturated fatty acid β-oxidation. DCI gene expression was elevated by three- to fourfold in all treated groups.

ACOX1 is the first enzyme of fatty acid β-oxidation in peroxisomes, catalyzing the desaturation of acyl-CoAs to 2-*trans*-enoyl-CoAs. ACOX1 mRNA induction was about twofold by both PPARα agonists. Enoyl-CoA hydratase/ 3-hydroxyacyl-CoA dehydrogenase (EHHADH) is the second enzyme of the peroxisomal β-oxidation system. The EHHADH gene was strikingly induced by both fenofibrate and Wy-14,643, being increased six- to sevenfold. Acetyl-CoA acyltransferase 1 (ACAA1) is the final enzyme of β-oxidation, transforming 3-ketoacyl-CoAs in the presence of CoAs into acyl-CoAs and acetyl-CoAs. Both PPARα agonists caused three- to fourfold elevation of ACAA1 mRNA levels. Peroxisomal enoyl-CoA hydratase 1 (ECH1) also functions in the auxiliary step of fatty acid β-oxidation. The mRNA level of ECH1 was approximately doubled by both PPARα agonists.

Fatty acid-CoA ligase, long-chain 5 (FACL5) converts free long-chain fatty acids into fatty acyl-CoA esters, playing a role not in only fatty acid β-oxidation, but also in lipid biosynthesis. A small increase of hepatic FACL5 gene expression

(~1.7-fold) was seen on day 3 in the 100 mg/kg fenofibrate-treated group, and on day 3 on the 100 mg/kg Wy-14,643-treated group.

Our study showed that both Wy-14,643 and fenofibrate increase hepatic mRNA of CYP4A10, CYP4A12, and CYP4A14 by two- to fourfold from day 2. The CYP4A enzyme catalyzes the ω-oxidation of medium- and long-chain fatty acids and prostaglandins. It is well-known that peroxisome proliferators induce the expression of several CYP4A enzymes in various species. For example, clofibrate regulates hepatic CYP4A1, CYP4A2, and CYP4A3 mRNA coordinately in rats;[29] clofibrate induces hepatic CYP4A10 and CYP4A14 in mice.[30] Bell et al.[31] observed that methylclofenapate treatment for 4 days increased hepatic expression of CYP4A10, but not CYP4A12 in mice. CYP4A12 mRNA was increased by more than threefold already on day 2 in our study. This may be due to the difference of the peroxisome proliferators used.

The mRNA of aldehyde dehydrogenase 3 family, member A2 (ALDH3A2) was increased by two- to threefold, particularly in fenofibrate-treated mice from day 2. This aldehyde dehydrogenase isozyme is thought to play a major role in the detoxification of aldehydes generated by alcohol metabolism and lipid peroxidation. This gene product catalyzes the oxidation of long-chain aliphatic aldehydes to fatty acids. ALDH3A2 is also involved in the degradation of leukotriene B_4, which is a proinflammatory mediator synthesized from arachidonic acid.[32]

KETOGENESIS

We found a slight increase of 3-hydroxy-3-methylglutaryl-CoA (HMG-CoA) lyase (HMGCL) (1.4–1.7-fold), which is involved in the final step of ketogenesis: the conversion of HMG-CoA into acetyl-CoA and acetoacetate. This increase may be due to accelerated fatty acid oxidation in the liver caused by PPARα agonists. On the other hand, we found no change in the mRNA level of HMG-CoA synthase (data not shown), which produces HMG-CoA from acetyl-CoA and acetoacetyl-CoA, although its induction by Wy-14,643 was reported in HepG2 cells expressing human PPARα.[33]

FATTY ACID TRANSPORT

We observed two- to threefold increases of mRNAs of two fatty acid binding proteins (FABP): adipocyte-type (FABP2) and intestinal-type (FABP4). The increase was higher in the case of FABP4. FABPs are a family of cytoplasmic proteins that bind long-chain fatty acids. It is considered that roles of FABPs include fatty acid uptake, transport, and metabolism.[34] Motojima[35] showed that Wy-14,643 induces ectopic expression of heart- and adipose-type FABP mRNAs in the mouse liver.

Solute carrier family 27 (fatty acid transporter), member 1 [SLC27A1, fatty acid transport protein 1 (FATP1)] mRNA was prominently increased, particularly by fenofibrate (>5.5-fold from day 2). SLC27A1 is involved in cellular uptake and metabolism of very long-chain fatty acids, and was demonstrated to be a very long-chain acyl-CoA synthetase, suggesting that a potential mechanism for facilitating fatty acid uptake is via esterification-coupled influx.[36] It was reported that hepatic SLC27A1 expression is induced by PPARα agonists, Wy-14,643 and clofibrate, in mice.[37]

Phospholipid transfer protein (PLTP) is a member of the lipid transfer/lipopolysaccharide binding protein gene family. PLTP transfers phospholipids from triglyceride-rich lipoproteins to high-density lipoproteins during lipolysis, modulating the levels, the size, and the composition of high-density lipoproteins.[38] We observed an approximately threefold increase in hepatic PLTP mRNA by day 3. Bouly et al.[39] reported the induction of PLTP gene expression by fenofibrate.

Lecithin:cholesterol acyltransferase (LCAT) converts free cholesterol of lipoprotein into esterified cholesterol, playing an important role in the reverse cholesterol transport system. Rat LCAT was reported to be downregulated to <50% of the control at the mRNA level by fenofibrate treatment for 14 days.[40] We did not recognize any clear change of LCAT mRNA levels at least until day 3 (range of change was −1- to 1.1-fold). Bouly et al.[39] showed that 2 week fenofibrate treatment caused a marginal decrease (20–30%) in LCAT mRNA in C57BL/6 mice and did not affect LCAT mRNA level in human apolipoprotein A-I (apo A-I) transgenic mice.

LIPOLYSIS

Lipoprotein lipase (LPL) mRNA in the liver was increased by greater than threefold from day 2. LPL catalyzes the hydrolysis of triglycerides in lipoprotein particles into fatty acids and monoacylglycerol, being involved in the clearance of triglyceride-rich lipoprotein particles in the liver. Together with suppression of apo C-III, enhanced LPL production may contribute to the hypolipidemic action of PPARα agonists.[41] LPL activation seemed to result in a significant reduction in blood triglyceride from day 2 as described earlier.

A more than twofold increase of monoglyceride lipase (MGLL) was seen on day 3 in 100 mg/kg fenofibrate-treated mice, and on day 2 in 100 mg/kg Wy-14,643-treated mice. The major role of MGLL is to complete hydrolysis of monoglycerides formed during lipolysis of stored triglycerides of the adipocytes. Another role is considered to be to catalyze the hydrolysis of 2-monoglycerides produced as a result of LPL-catalyzed hydrolysis of triglycerides from chylomicrons and very low-density lipoproteins.[42] MGLL upregulation by Wy-14,643 was reported by Cherkaoui-Malki et al.[43]

GLYCERONEOGENESIS

Biosynthesis of triacylglycerol proceeds by way of the conversion of dihydroxyacetone phosphate to *sn*-glycerol-3-phosphate by glycerol-3-phosphate dehydrogenase, and the conversion of *sn*-glycerol-3-phosphate to 1-acylglycerol-3-phosphate by glycerol-3-phosphate acyltransferase. Malic enzyme generates NADPH for fatty acid biosynthesis. We observed an increase of cytosolic malic enzyme (MOD1) (two- to threefold) in both groups from day 2, and an increase of cytoplasmic glycerol-3-phosphate dehydrogenase 1 (GPD1) (twofold) in the 100 mg/kg fenofibrate-treated group from day 2 and in the 100 mg/kg Wy-14,643-treated group on day 2. Glycerol-3-phosphate acyltransferase (GPAM) mRNA was increased by 1.5–1.7-fold on day 3 in both groups, and by twofold on day 2 in the 100 mg/kg Wy-14,643-treated group, but it should be considered that its mRNA was also increased by 1.4-fold in the control group on day 3. Das et al.[44] reported induction of hepatic GPAM activity by clofibrate, bezafibrate, gemfibrozil, and Wy-14,643 in mice, and Hertz et al.[45] noted PPARα-mediated activation of malic enzyme transcription.

METABOLISM OF ARACHIDONIC ACID AND EICOSANOIDS

Arachidonic acid (5,8,11,14-eicosatetraenoic acid) is a precursor of many active endogenous compounds, which are formed by prostaglandin endoperoxide synthases, lipoxygenases, and CYP monooxygenases. The CYP-derived arachidonic acid metabolites have been shown to have important biological functions, including modulation of fluid and electrolyte transport, modulation of vascular tone, and release of neuropeptides. CYPs involved in the biotransformation of arachidonic acid include the CYP1A, CYP2B, CYP2C, CYP2E, CYP2J, and CYP4A subfamilies.[46,47] Reduction of CYP2C37 gene expression by twofold was identified in all PPARα-treated groups from day 2. In addition, a slight decrease (~1.5-fold) of CYP2C40 mRNA was seen on day 2 in the 100 mg/kg fenofibrate-treated group and on day 2 in the 100 mg/kg Wy-14,643-treated group. CYP2C37 produces 12-HETE, and CYP2C40 produces an unidentified metabolite that coeluted with 16-, 17-, and 18-HETE.[48] On day 3 in the 100 mg/kg fenofibrate-administered group, arachidonate lipoxygenase 3 (ALOXE3, epidermis-type lipoxygenase 3) expression was reduced to less than half. ALOXE3 has a homology with 12(*R*)-lipoxygenase, which metabolizes arachidonic acid to 12(*R*)-HETE.[49] The *S*- and *R*-enantiomers of 12-HETE are generated in the liver by P450 monooxygenase,[50] and 12(*R*)-HETE is involved in inflammation in the cornea and skin.[51,52] Reduction of CYP2C37 and ALOXE3 expression may be beneficial in damping down the inflammation process.

Grant et al.[53] and Johansson et al.[54] observed induction of hepatic soluble epoxide hydrolase mRNA in mice by clofibrate. We detected an ~1.5-fold increase of soluble epoxide hydrolase 2 (EPHX2) expression from day 2. Soluble epoxide hydrolase degrades not only potentially toxic epoxides, but also endogenous chemical mediators, including some of the active *cis*-epoxyeicosatrienoic acids produced by CYPs from arachidonic acid.[55] Thus, EPHX2 may play a role in xenobiotic metabolism, and in the regulation of physiological mediators derived from arachidonic acid.

Microsomal glutathione S-transferase 3 (MGST3) expression was increased by about twofold in both groups from day 2. MGST3 is a member of the MAPEG (membrane-associated proteins in eicosanoid and glutathione metabolism) family that is involved in the production of leukotrienes and prostaglandin E, important mediators of inflammation. The MGST3 gene encodes an enzyme which catalyzes the conjugation of leukotriene A_4 and reduced glutathione to produce leukotriene C_4. This enzyme also exhibits glutathione-dependent peroxidase activity toward lipid hydroperoxides, reducing 5(S)-hydroperoxyeicosatetraenoic acid (5-HPETE) to 5-HETE.[56] It was proposed that MGST3 is involved in the detoxication of xenobiotics and/or protection against harmful metabolites generated during oxidative stress.

In addition, we observed an increase in glutathione S-transferase $θ2$ (GSTT2) by about twofold from day 2. Human GSTT2 is active toward unsaturated aldehyde compounds, *trans,trans*-alka-2,4-dienals and *trans*-alk-2-enals, and may play a significant role in protection against the products of lipid peroxidation.[57]

METABOLISM OF CHOLESTEROL, STEROIDS, AND RETINOIDS

Bile acid is synthesized in the liver from cholesterol, and this is an important pathway for elimination of cholesterol from the body. The rate-limiting enzyme of this step is cholesterol $7α$-hydroxylase (CYP7A1), catalyzing the $7α$-hydroxylation of cholesterol. Cholesterol $12α$-hydroxylase (CYP8B1) is a branch-point enzyme in bile acid biosynthesis, catalyzing the conversion of $7α$-hydroxy-4-cholesten-3-one to $7α,12α$-dihydroxy-4-cholesten-3-one. It was reported that 3 day ciprofibrate treatment suppresses bile acid synthesis via PPARα-mediated downregulation of CYP7A1 expression in the mouse liver.[58] On the other hand, other studies indicated that clofibrate treatment does not affect CYP7A1 in rats.[59,60] In our analysis, no significant change of CYP7A1 mRNA was detected until day 3. Hunt et al.[61] showed that Wy-14,643 induces the expression of CYP8B1 in mice. We found a slight increase of CYP8B1 by ~1.5-fold on day 3 in both the fenofibrate- and Wy-14,643-treated groups.

We detected an increase in 7-hydrocholesterol reductase (DHCR7) by approximately twofold on day 3 in the 100 mg/kg fenofibrate-treated group and the 30 mg/kg Wy-14,643-treated group, and on day 2 in the 100 mg/kg Wy-14,643-treated group. DHCR7 is the final enzyme in cholesterol biosynthesis. A similar induction was observed for squalene epoxidase (SQLE), which converts squalene into 2,3-epoxysqualene in the biosynthetic route to cholesterol. We detected an increase in mRNA of the C_4-sterol-methyl oxidase-like (SC4MOL) gene after PPARα agonist treatment, especially on day 3 in the case of Wy-14,643 treatment (about threefold). This gene is the murine homologue of *ERG25*, which was originally cloned from *Saccharomyces cerevisiae*. The C_4-methyl oxidase removes two C_4-methyl groups in the biosynthetic pathway to cholesterol.[62]

The expression of NAD(P)-dependent steroid dehydrogenase-like (NSDHL) gene was increased by approximately twofold on day 3 in 100 mg/kg fenofibrate-treated and 30 mg/kg Wy-14,643-treated mice. NSDHL contains a 3β-hydroxysteroid dehydrogenase (3β-HSD) domain. Therefore, NSDHL has a key role in cholesterol synthesis. In fact, disturbances of cholesterol biosynthesis were reported in NSDHL mutant mice [bare patches (*Bpa*) and striated (*Str*)].[63] Several genes involved in glyceroneogenesis and cholesterol biosynthesis were activated by PPARα agonists. These observations seem paradoxical, because PPARα agonists decrease circulating triglyceride and cholesterol levels. PPARα agonists increase the liver weight and proliferation of peroxisomes and smooth endoplasmic reticulum, which may require an increase of synthesis of glycerolipids and cholesterol.

Dramatic increases of phenobarbital-inducible CYP2B9 and CYP2B10 mRNAs were detected, by three- to sevenfold and three- to fourfold, respectively, from day 2. These proteins exhibit testosterone 16α-hydroxylase activity, and are believed to oxidize testosterone to an inactive form.[64] Prominent increases in both CYP2B9 and CYP2B10 expression in the liver of Wy-14,643-treated mice were also reported by Cherkaoui-Malki et al.[43] 3β-Hydroxysteroid dehydrogenase 5 (3β-HSD5, HSD3B5) functions as a 3-ketosteroid reductase.[65] This enzyme converts an active androgen, dihydrotestosterone (5α-androstan-17β-ol-3-one) into an inactive androgen (5α-androstane-3β,17β-diol), and therefore may serve an important function in regulating the availability of dihydrotestosterone in the male liver. The expression of 3β-HSD5 was decreased by 1.5–2.5-fold from day 2.

We found a three- to fourfold increase of retinol dehydrogenase type 6 (RDH6) (or *cis*-retinol/3α-hydroxysterol short-chain dehydrogenase) mRNA. RDH6 could initiate the pathway by catalyzing the conversion of 9-*cis*-retinol into 9-*cis*-retinal, which is then converted into 9-*cis*-retinoic acid by retinal dehydrogenase type 1 abundant in the liver. Moreover, RDH6 may catalyze the conversion of inactive 5α-androstane-3α,17β-diol into active dihydrotestosterone, and of androsterone (5α-androstan-3α-ol-17-one) into androstanedione (5α-androstane-3,17-dione), which can then be converted into dihydrotestosterone.[66]

A two- to threefold increase of 17β-hydroxysteroid dehydrogenase type 11 (17β-HSD11, HSD17B11) expression was observed in both groups from day 2. The mouse 17β-HSD11 is the homolog of human retinal short-chain dehydrogenase/reductase (RETSDR2).[67] Accordingly, 17β-HDS11 may be involved in the biosynthesis of 9-*cis*-retinoic acid, as in the case of RDH6. 9-*cis*-Retinoic acid is known to be an RXR ligand. PPARα, PPARδ, and PPARγ form heterodimers with RXR to activate transcription of target genes. Therefore, the upregulation of RDH6 and 17β-HSD11 genes by PPARα agonists may affect the potency of not only PPARα itself, but also the other PPARs.

11β-Hydroxysteroid dehydrogenase type 1 (11β-HSD1, HSD11B1) catalyzes the conversion of inert 11-dehydrocorticosterone into active corticosterone, which amplifies local glucocorticoid action, particularly in the liver. Excess glucocorticoid action may result in dyslipidemia, insulin resistance, and impaired glucose tolerance. This idea was supported by the finding that 11β-HSD1-null mice have improved glucose tolerance.[68] Hermanowski-Vosatka et al.[69] reported that Wy-14,643 represses 11β-HSD1 gene expression, and further, Guerre-Millo, et al.[70] showed that fenofibrate, ciprofibrate and GW9578 improve insulin resistance in animal models. We confirmed the reduction of 11β-HSD1 by twofold from day 2 of both fenofibrate and Wy-14,643 treatment, by microarray analysis.

About twofold elevation of 17β-hydroxysteroid dehydrogenase type 4 (17β-HSD4, HSD17B4) was seen from day 2. This enzyme is unique in having a multidomain structure containing dehydrogenase, hydratase, and lipid transfer domains. 17β-HSD4 not only inactivates estradiol by conversion to estrone, but also its three domains participate in the peroxisomal β-oxidation of long- and branched-chain fatty acids.[71] Breitling et al.[71] suggested that the main function of human 17β-HSD4 is in fatty acid metabolism, while its steroid conversion activity is a secondary and maybe minor role *in vivo*. It is of interest that 17β-HSD4 gene expression is under PPARα, as is the case for other genes involved in peroxisomal β-oxidation.

METABOLISM OF APOLIPOPROTEIN

Plasma lipoprotein metabolism is regulated and controlled by the specific apolipoprotein constituents of the various lipoprotein classes. The major apolipoproteins include apo A-I, apo A-II, apo A-IV, apo C-I, apo C-III, apo B, and apo E. Specific apolipoproteins function in the regulation of lipoprotein metabolism through their involvement in the transport and redistribution of lipids among various cells and tissues, or through their roles as cofactors for enzymes of lipid metabolism, or through their maintenance of the structure of the lipoprotein particles. Fibrates downregulate apo C-III[72,73] and apo C-II, but do not affect apo C-I.[74] On the other hand, fibrates upregulate apo A-I,[75] apo A-II[76] and apo A-IV,[77] leading to increased serum high-density lipoprotein cholesterol levels in humans. In rodents, however, fibrate treatment reduces apo A-I[78] and apo A-IV mRNA levels.[79] In our study, hepatic apo A-IV expression was drastically reduced to about a third or less from day 2. In contrast, the changes of apo A-I and apo C-II were between -1.1- and 1.2-fold and between -1.3- and -1.2-fold, respectively, in any treatment up to day 3.

We observed a slight reduction by 2–1.5-fold of glycosylphosphatidylinositol-specific phospholipase D1 (GPLD1) mRNA. It was reported that the majority of GPLD1 appears to be associated with a fraction of lipoprotein containing apo A-I and apo A-IV.[80] We showed that apo A-IV mRNA was drastically reduced. Therefore, the decrease in GPLD1 may be a result of the reduction of apo A-IV production.

A transient reduction in apo A-V expression was detected on day 2 in the 100 mg/kg fenofibrate-treated group, and on day 2 in the 100 mg/kg Wy-14, 643-treated group. This novel apolipoprotein is upregulated after hepatectomy, and was identified in a plasma fraction containing high-density lipoprotein particles, though it was a minor constituent.[81] Its precise function is still unknown, but it may possibly act as an antagonist of lipid uptake by the liver to protect the reduced-sized liver against a possible lipid overload in the early phase of liver regeneration. van der Vliet et al.[82] reported that adenoviral over-expression of apo A-V leads to reductions of serum triglyceride and cholesterol, suggesting that apo A-V regulates circulating triglyceride and cholesterol levels.

The most drastic and prominent change of gene expression was the depression of serum amyloid A (SAA) 2 mRNA from day 2. Hepatic SAA2 mRNA was reduced to one-fourth by 100 mg/kg fenofibrate, and to one-seventh by 30 mg/kg Wy-14,643 on day 3. SAA3 mRNA was also reduced on day 2 by 100 mg/kg Wy-14,643. For SAA1, decreases of 2.9- and 1.7-fold by 100 mg/kg fenofibrate and 30 mg/kg Wy-14,643, respectively, on day 3, and 2.2-fold by 100 mg/kg Wy-14,643 on day 2 were found. SAA is an acute-phase protein and comprises a family of apolipoproteins encoded by at least three genes.[83,84] Among this family, SAA1 and SAA2 are major acute-phase proteins expressed in the liver. We were able to demonstrate a marked decrease in plasma SAA concentrations by fenofibrate treatment in mice challenged with acetaminophen.[22] In addition, it was shown that fenofibrate inhibits experi-mental amyloidosis by reducing SAA levels in mice.[85]

The acute-phase proteins, including SAA, interact with plasma lipoproteins and cause increases in serum triglyceride and triglyceride-rich very low-density lipoprotein, and a decrease in high-density lipoprotein.[86] High-density lipo-protein plays an important role in reverse cholesterol transport, and therefore is protective against atherosclerosis. Apo A-I, the predominant high-density lipoprotein–apolipoprotein mediating cholesterol efflux from the peripheral tissues, is displaced by acute-phase SAA. This displacement and subsequent remodeling of high-density lipoprotein impair the cholesterol efflux.[87]

PERSPECTIVES IN PPAR RESEARCH USING MICROARRAYS

We have outlined the PPARα agonist-induced hepatic gene expression changes in mice, focusing on genes involved in lipid metabolism, as evaluated with a microarray technique. We not only confirmed previously reported induction and reduction of expression of various genes by PPARα agonist(s), but also identified several new genes that are affected. After microarray analysis, it is important to distinguish which are primary and which are secondary changes of gene expression, or which are PPARα-dependent or -independent. For this purpose, detailed time course and dose-dependency studies are necessary as a matter of course. Additionally, the use of PPARα gene-disrupted mice,[88] as well as mice targeted for certain genes affected by PPARα, such as ACOX1,[89] is a powerful tool. As of 2002, the mouse genome project has nearly been finished and the mouse genome sequence is available in public data-bases.[90] Thus, we can directly look at promoter [41,91] or intron [92] sequences of a target gene to examine whether virtual PPREs exist there or not. Furthermore, we can perform a reporter assay to confirm whether or not the PPREs are functional.

It would be intriguing to know whether and, if so, to what degree the changes of gene expression are of physiological significances *in vivo*. In particular, the following points are noteworthy: (1) Several genes in the cholesterol biosynthesis pathway are activated by PPARα agonists (DHCR7, SQLE, SC4MOL, and NSDHL), although fibrates are known clinically to lower serum cholesterol. We did not observe a significant change of blood cholesterol levels at least until day 3. What are the roles of these activated genes in the liver? (2) Opposite pathways of androgen metabolism are both promoted by PPARα agonists: increase in CYP2B9, CYP2B10, and RDH6 (accelerated inactivation of active androgen) vs. decrease in 3β-HSD5 (downregulation of the conversion into active androgen). Which is (are) the primary or counteracting change(s)? (3) Induction of ALDH3A2, EPHX2, and MGST3, and suppression of CYP2C37 involved in eicosanoid metabolism may participate in the inhibition of inflammation by PPARα agonists. (4) Induction of ALDH3A2, MGST3, and GSTT2 genes may play a role in protection against products of lipid peroxidation, because peroxisome proliferation is associated with induction of an H_2O_2-generating enzyme and the resulting increase of oxidative stress. (5) Enhanced expression of genes that probably promote production of 9-*cis*-retinoic acid may affect PPARα-mediated changes of gene expression.

In this study, we picked up the changes, based on the criteria of >1.5-fold and <−1.5-fold, as well as utilizing the judgment of "increased" or "decreased" by the Affymetrix Analysis Software.[20] In another study, we treated human apo A-I genomic transgenic mice[93] with KRP-297 (30 mg/kg) or fenofibrate (100 mg/kg) for 7 days to examine the effects on blood high-density lipoprotein cholesterol and human apo A-I levels. In addition, we investigated hepatic human and murine apo A-I mRNA expression in the transgenic mice after 7 day fenofibrate treatment (100 mg/kg) by Northern blot analysis. Fenofibrate induced elevation of both blood high-density lipoprotein cholesterol and human apo A-I concentrations, whereas KRP-297 did not elevate blood human apo A-I levels, inspite of an increase in blood high-density lipoprotein cholesterol [Fig. 12.6(A) and (B)]. Like fenofibrate, GW2331 and GW9578 increased both blood high-density lipoprotein cholesterol and human apo A-I concentrations (data not shown).

By Northern blot analysis, we found a 20% increase in human apo A-I mRNA and a 20–30% decrease in murine apo A-I mRNA in the liver of fenofibrate-treated mice [Fig. 12.6(C) and (D)]. The increase and decrease were not statistically significantly different ($p > 0.05$). This finding suggests that a 20% increase of gene expression may be enough to induce a significant increase of protein levels and significant *in vivo* changes. Therefore, we could overlook subtle, but significant changes of gene expression by using microarray analysis. More intriguingly, fenofibrate and ciprofibrate posttranscriptionally

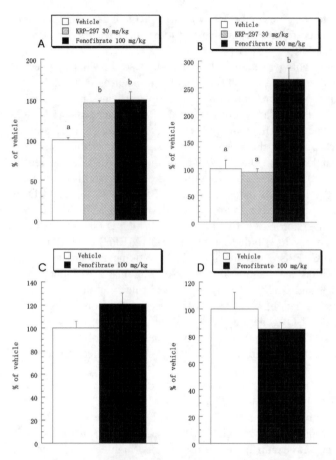

Figure 12.6 Blood high-density lipoprotein cholesterol levels (A) and human apolipo-protein A-I (apo A-I) levels (B) of transgenic male C57BL/6 mice containing an 11 kb human apo A-I genomic fragment [C57BL/6-TgN(APOA1)1Rub][91] treated with KRP-297 or fenofibrate. The mice were orally given KRP-297 (30 mg/kg), fenofibrate (100 mg/kg), or vehicle (0.5% methyl cellulose) for 7 days, and blood was drawn from the caudal vein under nonfasting conditions. Blood human apo A-I concentrations were determined by ELISA using monoclonal antibody against human apo A-I. Hepatic mRNA levels of human apo A-I (C) and murine apo A-I (D) of the transgenic mice treated with fenofibrate. The mice were orally given fenofibrate (100 mg/kg) or vehicle (0.5% methyl cellulose) for 7 days, and the liver was collected under nonfasting conditions. Northern membrane was made after extraction of total RNA from the liver. The membrane was hybridized with ^{32}P-labled human apo A-I, or mouse apo A-I, or β-actin cDNA probe. The human and murine apo A-I intensities were normalized to the β-actin intensity. Values are expressed as means $\pm$ SEM. For multiple comparison of blood high-density lipoprotein cholesterol and human apo A-I levels, a one-way analysis of variance (ANOVA) was carried out, followed by Fisher's protected least significant difference (PLSD) test as a *post hoc* test. Bars with the same letter are not statistically significantly different by Fisher's PLSD test ($p > 0.05$) (A and B), $n = 4$ in (A) and (B), $n = 3$ in (C) and (D).

downregulate hepatic scavenger receptor class B type I protein expression in mice, without affecting the mRNA level.[94] To resolve this problem, proteome analysis including posttranscriptional modification,[95] combined with transcriptome analysis, will be a crucial step.

PPAR α/γ dual agonists, including KRP-297 and AZ242 have been developed for treatment of diabetes, in expectation of an augmented improvement of both glucose and lipid metabolism.[18–21] It is therefore very interesting to examine additive and/or synergic effects of PPAR α/γ dual agonists on gene expression in target tissues including the liver, adipose tissue, and muscle, compared with the effects of PPARα- and PPARγ-specific agonists, by using the microarray technique. Among PPAR α/γ dual agonists, KRP-297 is a unique compound. Murakami et al. [21] suggested that improvement of lipid abnormalities by KRP-297 in Zucker fatty rats is a consequence of its PPARα agonism. In their paper, induction of ACOX1 mRNA was demonstrated in the liver of Zucker fatty rats and in primary rat hepatocytes treated with KRP-297. However, hepatomegaly and peroxisome proliferation were not observed. KRP-297 treatment of human apo A-I genomic transgenic mice resulted in an increase in blood high-density lipoprotein cholesterol levels, without change in blood human apo A-I concentrations, although other compounds with PPARα activity increased both high-density lipoprotein cholesterol and human apo A-I levels [Fig. 12.6(A) and (B)]. In the case of pioglitazone treatment (30 mg/kg for 7 days), increased blood high-density lipoprotein cholesterol and unchanged human apo A-I levels were also observed, as with KRP-297 (data not shown). Whether or not human apo A-I is exceptional is unknown. Microarray analysis of changes of gene expression induced by KRP-297, particularly in the liver, will be an interesting subject.

Rodents (such as mice and rats) and nonrodents (humans and guinea pigs) respond differently to PPARα agonists. PPARα treatment of rodents causes hepatic peroxisome proliferation and liver enlargement, and eventually hepatocarcinoma, whereas humans and guinea pigs are refractory to the adverse effects.[96] As PPARα agonists manifest hypotriglyceridemic effects clinically by suppression of apo C-III expression in humans, as well as in rodents,[73] the mechanism of the species-specific differences cannot be simply explained. The following main possibilities have been proposed: (1) Low hepatic levels of PPARα. Hepatic expression of PPARα is 10-fold lower in humans than in mice and rats.[97] (2) The presence of a human alternatively spliced variant of PPARα, exhibiting a dominant negative activity toward wild-type PPARα.[98] (3) Disruption of response elements in promoter regions in nonrodent species. For example, human ACOX1 promoter contains a disrupted PPRE sequence.[99] Lawrence et al.[100] proposed that other factors besides PPARα levels determine the species-specific response to PPARα agonists, and thus genomic differences between species may be more important. Coupling of data from the human[101,102] and mouse genome projects[90] with microarray results should be informative.

Roglans et al.[103] investigated hepatic mRNAs of humans treated with fibrates. The liver specimens were collected by biopsy from patients who had received fenofibrate, bezafibrate, or gemfibrozil at pharmacological doses, and showed a significant reduction in plasma triglyceride levels.[103] They found no increase in ACOX1 expression by any fibrate at clinical doses. Comparison of hepatic gene expression between rodents and humans will be necessary to understand the species-specific differences and clinical actions of fibrates. It will be particularly important to investigate human samples, as in the study by Roglans et al. Microarray analysis of human specimens might be suitable, because this technique provides a great deal of information on gene expression patterns from a few samples.

There has been a microarray study on gene expression changes in mouse liver induced by adenoviral PPARγ1 overexpression, in which PPARα-disrupted mice were used to eliminate PPARα effects.[104] PPARγ1 overexpression resulted in adipogenic hepatic steatosis through activation of adipocyte-specific and lipogenesis-related genes. It was suggested that PPARα expression in normal liver serves as a key regulator of fatty acid catabolism, keeping PPARγ1 and adipogenesis in check.[105] The authors of this study pointed out potential clinical implications, for example, individuals with hepatic low PPARα expression and hyperactive PPARγ1 could develop adipogenic hepatic steatosis. It would be of interest to conduct a microarray analysis on hepatic gene expression of wild-type mice treated with extremely potent PPARγ agonists, taking the effects of PPARα into consideration. Such studies would provide information on the interaction between PPARα and PPARγ, and on PPARs-mediated gene changes in the liver.

Microarray analysis offers a rapid and efficient method for large-scale profiling of gene expression. Similar to what we have done Cherkaoui-Malki et al.[43] have analyzed gene expression changes by PPARα agonists in the mouse liver using a cDNA microarray. They employed a UniGene mouse cDNA microarray from InCyte Genomics, consisting of a total of 7483 unigene/EST clusters. By combining their results with ours, we can accumulate information on genes affected by PPARα. Furthermore, changes of gene expression can be scrutinized by using Northern blot analysis, quantitative polymerase chain reaction method, promoter–reporter assay, and gene-engineered mice. The knowledge obtained by the combination of all these technologies should reveal new aspects of the physiological roles of PPARα, and be particularly valuable with respect to development of novel therapeutic applications of fibrates and more potent PPARα agonists, as well as miminizing clinical side effects caused by PPARα agonists.

REFERENCES

1. Nuclear Receptors Nomenclature Committee. A unified nomenclature system for the nuclear receptor superfamily. Cell **1999**, *97*, 161–163.
2. Braissant, O.; Foufelle, F.; Scotto, C.; Dauca, M.; Wahli, W. Differential expression of peroxisome proliferator-activated receptors (PPARs): tissue distribution of PPAR-α, -β, and -γ in the adult rat. Endocrinology **1996**, *137*, 354–366.

3. Devchand, P.R.; Keller, H.; Peters, J.M.; Vazquez, M.; Gonzalez, F.J.; Wahli, W. The PPARα-leukotriene B_4 pathway to inflammation control. Nature **1996**, *384*, 39–43.

4. Göttlicher, M.; Widmark, E.; Li, Q.; Gustafsson, J.A. Fatty acids activate a chimera of the clofibric acid-activated receptor and the glucocorticoid receptor. Proc. Natl. Acad. Sci. USA **1992**, *89*, 4653–4657.

5. Kliewer, S.A.; Sundseth, S.S.; Jones, S.A.; Brown, P.J.; Wisely, G.B.; Koble, C.S.; Devchand, P.; Wahli, W.; Willson, T.M.; Lenhard, J.M.; Lehmann, J.M. Fatty acids and eicosanoids regulate gene expression through direct interactions with peroxisome proliferator-activated receptors α and γ. Proc. Natl. Acad. Sci. USA **1997**, *94*, 4318–4323.

6. Yu, K.; Bayona, W.; Kallen, C.B.; Harding, H.P.; Ravera, C.P.; McMahon, G.; Brown, M.; Lazar, M.A. Differential activation of peroxisome proliferator-activated receptors by eicosanoids. J. Biol. Chem. **1995**, *270*, 23975–23983.

7. Forman, B.M.; Tontonoz, P.; Chen, J.; Brun, R.P.; Spiegelman, B.M.; Evans, R.M. 15-Deoxy-$\Delta^{12,14}$-prostaglandin J_2 is a ligand for the adipocyte determination factor PPARγ. Cell **1995**, *83*, 803–812.

8. Kliewer, S.A.; Lenhard, L.M.; Willson, T.M.; Patel, I.; Morris, D.C.; Lehmann, J.M. A prostaglandin J_2 metabolite binds peroxisome proliferator-activated receptor γ and promotes adipocyte differentiation. Cell **1995**, *83*, 813–819.

9. Oliver, W.R., Jr.; Shenk, J.L.; Snaith, M.R.; Russell, C.S.; Plunket, K.D.; Bodkin, N.L.; Lewis, M.C.; Winegar, D.A.; Sznaidman, M.L.; Lambert, M.H.; Xu, H.E.; Sternbach, D.D.; Kliewer, S.A.; Hansen, B.C.; Willson, T.M. A selective peroxisome proliferator-activated receptor δ agonist promotes reverse cholesterol transport. Proc. Natl. Acad. Sci. USA **2001**, *98*, 5306–5311.

10. Xu, H.E.; Lambert, M.H.; Montana, V.G.; Parks, D.J.; Blanchard, S.G.; Brown, P.J.; Sternbach, D.D.; Lehmann, J.M.; Wisely, G.B.; Willson, T.M.; Kliewer, S.A.; Milburn, M.V. Molecular recognition of fatty acids by peroxisome proliferator-activated receptors. Mol. Cell. **1999**, *3*, 397–403.

11. Forman, B.M.; Chen, J.; Evans, R.M. Hypolipidemic drugs, polyunsaturated fatty acids, and eicosanoids are ligands for peroxisome proliferator-activated receptors α and δ. Proc. Natl. Acad. Sci. USA **1997**, *94*, 4312–4317.

12. Lehmann, J.M.; Moore, L.B.; Smith-Oliver, T.A.; Wilkison, W.O.; Willson, T.M.; Kliewer, S.A. An antidiabetic thiazolidinedione is a high affinity ligand for peroxisome proliferator-activated receptor γ (PPARγ). J. Biol. Chem. **1995**, *270*, 12953–12956.

13. Yoshitomi, H.; Yamazaki, K.; Tanaka, I. Mechanism of ubiquitous expression of mouse uncoupling protein 2 mRNA: control by *cis*-acting DNA element in 5′-flanking region. Biochem. J. **1999**, *340*, 397–404.

14. Farmer, J.A.; Gotto, A.M., Jr. Currently available hypolipidaemic drugs and future therapeutic developments. Baillière's Clin. Endocrinol. Metab. **1995**, *9*, 825–847.

15. Vosper, H.; Khoudoli, G.A.; Graham, T.L.; Palmer, C.N.A. Peroxisome proliferator-activated receptor agonists, hyperlipidaemia, and atherosclerosis. Pharmacol. Ther. **2002**, *95*, 47–62.

16. Brown, P.J.; Winegar, D.A.; Plunket, K.D.; Moore, L.B.; Lewis, M.C.; Wilson, J.G.; Sundseth, S.S.; Koble, C.S.; Wu, Z.; Chapman, J.M.; Lehmann, J.M.; Kliewer, S.A.; Willson, T.M. A ureido-thioisobutyric acid (GW9578) is a subtype-selective

PPARα agonist with potent lipid-lowering activity. J. Med. Chem. **1999**, *42*, 3785–3788.

17. Stumvoll, M.; Häring, H.U. Glitazones: clinical effects and molecular mechanisms. Ann. Med. **2002**, *34*, 217–224.

18. Lohray, B.B.; Lohray, V.B.; Bajji, A.C.; Kalchar, S.; Poondra, R.R.; Padakanti, S.; Chakrabarti, R.; Vikramadithyan, R.K.; Misra, P.; Juluri, S.; Mamidi, N.V.S.R.; Rajagopalan, R. (−)3-[4-[2-(Phenoxazin-10-yl)ethoxy]phenyl]-2-ethoxypropanoic acid [(−)DRF 2725]: a dual PPAR agonist with potent antihyperglycemic and lipid modulating activity. J. Med. Chem. **2001**, *44*, 2675–2678.

19. Etgen, G.J.; Oldham, B.A.; Johnson, W.T.; Broderick, C.L.; Montrose, C.R.; Brozinick, J.T.; Misener, E.A.; Bean, J.S.; Bensch, W.R.; Brooks, D.A.; Shuker, A.J.; Rito, C.J.; McCarthy, J.R.; Ardecky, R.J.; Tyhonas, J.S.; Dana, S.L.; Bilakovics, J.M.; Paterniti, J.R., Jr.; Ogilvie, K.M.; Liu, S.; Kauffman, R.F. A tailored therapy for the metabolic syndrome. The dual peroxisome proliferator-activated receptor-α/γ agonist LY465608 ameliorates insulin resistance and diabetic hyperglycemia while improving cardiovascular risk factors in preclinical models. Diabetes **2002**, *51*, 1083–1087.

20. Ljung, B.; Bamberg, K.; Dahllöf, B.; Kjellstedt, A.; Oakes, N.D.; Östling, J.; Svensson, L.; Camejo, G. AZ 242, a novel PPARα/γ agonist with beneficial effects on insulin resistance and carbohydrate and lipid metabolism in ob/ob mice and obese Zucker rats. J. Lipid Res. **2002**, *43*, 1855–1863.

21. Murakami, K.; Tobe, K.; Ide, T.; Mochizuki, T.; Ohashi, M.; Akanuma, Y.; Yazaki, Y.; Kadowaki, T. A novel insulin sensitizer acts as a coligand for peroxisome proliferator-activated receptor-α (PPAR-α) and PPAR-γ. Effect of PPAR-α activation on abnormal lipid metabolism in liver of Zucker fatty rats. Diabetes **1998**, *47*, 1841–1847.

22. Yamazaki, K.; Kuromitsu, J.; Tanaka, I. Microarray analysis of gene expression changes in mouse liver induced by peroxisome proliferator-activated receptor α agonists. Biochem. Biophys. Res. Commun. **2002**, *290*, 1114–1122.

23. Lipshutz, R.J.; Fodor, S.P.; Gingeras, T.R.; Lockhart, D.J. High density synthetic oligonucleotide arrays. Nat. Genet. **1999**, *21* (suppl 1), 20–24.

24. Schoonjans, K.; Staels, B.; Auwerx, J. The peroxisome proliferator activated receptors (PPARs) and their effects on lipid metabolism and adipocyte differentiation. Biochim. Biophys. Acta **1996**, *1302*, 93–109.

25. Johnson, E.F.; Palmer, C.N.A.; Griffin, K.J.; Hsu, M.H. Role of the peroxisome proliferator-activated receptor in cytochrome P450 4A gene regulation. FASEB J. **1996**, *10*, 1241–1248.

26. Latruffe, N.; Cherkaoui Malki, M.; Nicolas-Frances, V.; Jannin, B.; Clemencet, M.C.; Hansmannel, F.; Passilly-Degrace, P.; Berlot, J.P. Peroxisome-proliferator-activated receptors as physiological sensors of fatty acid metabolism: molecular regulation in peroxisomes. Biochem. Soc. Trans. **2001**, *29*, 305–309.

27. Hunt, M.C.; Solaas, K.; Kase, B.F.; Alexson, S.E. Characterization of an acyl-CoA thioesterase that functions as a major regulator of peroxisomal lipid metabolism. J. Biol. Chem. **2002**, *277*, 1128–1138.

28. Ramsay, R.R. The carnitine acyltransferases: modulators of acyl-CoA-dependent reactions. Biochem. Soc. Trans. **2000**, *28*, 182–186.

29. Kimura, S.; Hardwick, J.P.; Kozak, C.A.; Gonzalez, F.J. The rat clofibrate-inducible CYP4A subfamily. II. cDNA sequence of IVA3, mapping of the Cyp4a locus to mouse chromosome 4, and coordinate and tissue-specific regulation of the CYP4A genes. DNA **1989**, *8*, 517–525.

30. Barclay, T.B.; Peters, J.M.; Sewer, M.B.; Ferrari, L.; Gonzalez, F.J.; Morgan, E.T. Modulation of cytochrome P-450 gene expression in endotoxemic mice is tissue specific and peroxisome proliferator-activated receptor-α dependent. J. Pharmacol. Exp. Ther. **1999**, *290*, 1250–1257.

31. Bell, D.R.; Plant, N.J.; Rider, C.G.; Na, L.; Brown, S.; Ateitalla, I.; Acharya, S.K.; Davies, M.H.; Elias, E.; Jenkins, N.A.; Gilbert, D.J.; Copeland, N.G.; Elcombe, C.R. Species-specific induction of cytochrome P-450 4A RNAs: PCR cloning of partial guinea-pig, human and mouse CYP4A cDNAs. Biochem. J. **1993**, *294*, 173–180.

32. Willemsen, M.A.A.P.; Rotteveel, J.J.; de Jong, J.G.N.; Wanders, R.J.A.; IJlst, L.; Hoffmann, G.F.; Mayatepek, E. Defective metabolism of leukotriene B$_4$ in the Sjögren-Larsson syndrome. J. Neurol. Sci. **2001**, *183*, 61–67.

33. Hsu, M.H.; Savas, Ü.; Griffin, K.J.; Johnson, E.F. Identification of peroxisome proliferator-responsive human genes by elevated expression of the peroxisome proliferator-activated receptor α in HepG2 cells. J. Biol. Chem. **2001**, *276*, 27950–27958.

34. Huang, H.; Starodub, O.; McIntosh, A.; Kier, A.B.; Schroeder, F. Liver fatty acid-binding protein targets fatty acids to the nucleus. Real time confocal and multiphoton fluorescence imaging in living cells. J. Biol. Chem. **2002**, *277*, 29139–29151.

35. Motojima, K. Differential effects of PPARα activators on induction of ectopic expression of tissue-specific fatty acid binding protein genes in the mouse liver. Int. J. Biochem. Cell. Biol. **2000**, *32*, 1085–1092.

36. Coe, N.R.; Smith, A.J.; Frohnert, B.I.; Watkins, P.A.; Bernlohr, D.A. The fatty acid transport protein (FATP1) is a very long chain acyl-CoA synthetase. J. Biol. Chem. **1999**, *274*, 36300–36304.

37. Motojima, K.; Passilly, P.; Peters, J.M.; Gonzalez, F.J.; Latruffe, N. Expression of putative fatty acid transporter genes are regulated by peroxisome proliferator-activated receptor α and γ activators in a tissue- and inducer-specific manner. J. Biol. Chem. **1998**, *273*, 16710–16714.

38. Jauhiainen, M.; Metso, J.; Pahlman, R.; Blomqvist, S.; van Tol, A.; Ehnholm, C. Human plasma phospholipid transfer protein causes high density lipoprotein conversion. J. Biol. Chem. **1993**, *268*, 4032–4036.

39. Bouly, M.; Masson, D.; Gross, B.; Jiang, X.C.; Fievet, C.; Castro, G.; Tall, A.R.; Fruchart, J.C.; Staels, B.; Lagrost, L.; Luc, G. Induction of the phospholipid transfer protein gene accounts for the high density lipoprotein enlargement in mice treated with fenofibrate. J. Biol. Chem. **2001**, *276*, 25841–25847.

40. Staels, B.; van Tol, A.; Skretting, G.; Auwerx, J. Lecithin: cholesterol acyltransferase gene expression is regulated in a tissue-selective manner by fibrates. J. Lipid Res. **1992**, *33*, 727–735.

41. Schoonjans, K.; Peinado-Onsurbe, J.; Lefebvre, A.M.; Heyman, R.A.; Briggs, M.; Deeb, S.; Staels, B.; Auwerx, J. PPARα and PPARγ activators direct a distinct tissue-specific transcriptional response via a PPRE in the lipoprotein lipase gene. EMBO J. **1996**, *15*, 5336–5348.

42. Karlsson, M.; Contreras, J.A.; Hellman, U.; Tornqvist, H.; Holm, C. cDNA cloning, tissue distribution, and identification of the catalytic triad of monoglyceride lipase. Evolutionary relationship to esterases, lysophospholipases, and haloperoxidases. J. Biol. Chem. **1997**, *272*, 27218–27223.

43. Cherkaoui-Malki, M.; Meyer, K.; Cao, W.Q.; Latruffe, N.; Yeldandi, A.V.; Rao, M.S.; Bradfield, C.A.; Reddy, J.K. Identification of novel peroxisome proliferator-activated receptor α (PPARα) target genes in mouse liver using cDNA microarray analysis. Gene Expr. **2001**, *9*, 291–304.

44. Das, A.K.; Aquilina, J.W.; Hajra, A.K. The rapid induction of liver glycerophosphate acyltransferase in mice by clofibrate, a hypolipidemic agent. J. Biol. Chem. **1983**, *258*, 3090–3093.

45. Hertz, R.; Nikodem, V.; Ben-Ishai, A.; Berman, I.; Bar-Tana, J. Thyromimetic mode of action of peroxisome proliferators: activation of 'malic' enzyme gene transcription. Biochem. J. **1996**, *319*, 241–248.

46. Rifkind, A.B.; Lee, C.; Chang, T.K.; Waxman, D.J. Arachidonic acid metabolism by human cytochrome P450s 2C8, 2C9, 2E1, and 1A2: regioselective oxygenation and evidence for a role for CYP2C enzymes in arachidonic acid epoxygenation in human liver microsomes. Arch. Biochem. Biophys. **1995**, *320*, 380–389.

47. Zeldin, D.C.; Moomaw, C.R.; Jesse, N.; Tomer, K.B.; Beetham, J.; Hammock, B.D.; Wu, S. Biochemical characterization of the human liver cytochrome P450 arachidonic acid epoxygenase pathway. Arch. Biochem. Biophys. **1996**, *330*, 87–96.

48. Luo, G.; Zeldin, D.C.; Blaisdell, J.A.; Hodgson, E.; Goldstein, J.A. Cloning and expression of murine CYP2Cs and their ability to metabolize arachidonic acid. Arch. Biochem. Biophys. **1998**, *357*, 45–57.

49. Kinzig, A.; Heidt, M.; Fürstenberger, G.; Marks, F.; Krieg, P. cDNA cloning, genomic structure, and chromosomal localization of a novel murine epidermis-type lipoxygenase. Genomics **1999**, *58*, 158–164.

50. Capdevila, J.; Yadagiri, P.; Manna, S.; Falck, J.R. Absolute configuration of the hydroxyeicosatetraenoic acids (HETEs) formed during catalytic oxygenation of arachidonic acid by microsomal cytochrome P-450. Biochem. Biophys. Res. Commun. **1986**, *141*, 1007–1011.

51. Conners, M.S.; Stoltz, R.A.; Webb, S.C.; Rosenberg, J.; Dunn, M.W.; Abraham, N.G.; Laniado-Schwartzman, M. A closed eye contact lens model of corneal inflammation. Part 1: increased synthesis of cytochrome P450 arachidonic acid metabolites. Invest. Ophthalmol. Vis. Sci. **1995**, *36*, 828–840.

52. Woollard, P.M. Stereochemical difference between 12-hydroxy-5,8,10,14-eicosatetraenoic acid in platelets and psoriatic lesions. Biochem. Biophys. Res. Commun. **1986**, *136*, 169–176.

53. Grant, D.F.; Storms, D.H.; Hammock, B.D. Molecular cloning and expression of murine liver soluble epoxide hydrolase. J. Biol. Chem. **1993**, *268*, 17628–17633.

54. Johansson, C.; Stark, A.; Sandberg, M.; Ek, B.; Rask, L.; Meijer, J. Tissue specific basal expression of soluble murine epoxide hydrolase and effects of clofibrate on the mRNA levels in extrahepatic tissues and liver. Arch. Toxicol. **1995**, *70*, 61–63.

55. Chacos, N.; Capdevila, J.; Falck, J.R.; Manna, S.; Martin-Wixtrom, C.; Gill, S.S.; Hammock, B.D.; Estabrook, R.W. The reaction of arachidonic acid epoxides (epoxyeicosatrienoic acids) with a cytosolic epoxide hydrolase. Arch. Biochem. Biophys. **1983**, *223*, 639–648.

56. Jakobsson, P.J.; Mancini, J.A.; Riendeau, D.; Ford-Hutchinson, A.W. Identification and characterization of a novel microsomal enzyme with glutathione-dependent transferase and peroxidase activities. J. Biol. Chem. **1997**, *272*, 22934–22939.

57. Tan, K.L.; Board, P.G. Purification and characterization of a recombinant human Theta-class glutathione transferase (GSTT2-2). Biochem. J. **1996**, *315*, 727–732.

58. Post, S.M.; Duez, H.; Gervois, P.P.; Staels, B.; Kuipers, F.; Princen, H.M. Fibrates suppress bile acid synthesis via peroxisome proliferator-activated receptor-α-mediated downregulation of cholesterol 7α-hydroxylase and sterol 27-hydroxylase expression. Arterioscler. Thromb. Vasc. Biol. **2001**, *21*, 1840–1845.

59. Angelin, B.O.; Björkhem, I.; Einarsson, K. Effects of clofibrate on some microsomal hydroxylations involved in the formation and metabolism of bile acids in rat liver. Biochem. J. **1976**, *156*, 445–448.

60. Ståhlberg, D.; Angelin, B.; Einarsson, K. Effects of treatment with clofibrate, bezafibrate, and ciprofibrate on the metabolism of cholesterol in rat liver microsomes. J. Lipid Res. **1989**, *30*, 953–958.

61. Hunt, M.C.; Yang, Y.Z.; Eggertsen, G.; Carneheim, C.M.; Gåfvels, M.; Einarsson, C.; Alexson, S.E.H. The peroxisome proliferator-activated receptor α (PPARα) regulates bile acid biosynthesis. J. Biol. Chem. **2000**, *275*, 28947–28953.

62. Bard, M.; Bruner, D.A.; Pierson, C.A.; Lees, N.D.; Biermann, B.; Frye, L.; Koegel, C.; Barbuch, R. Cloning and characterization of *ERG25*, the *Saccharomyces cerevisiae* gene encoding C-4 sterol methyl oxidase. Proc. Natl. Acad. Sci. USA **1996**, *93*, 186–190.

63. Liu, X.Y.; Dangel, A.W.; Kelley, R.I.; Zhao, W.; Denny, P.; Botcherby, M.; Cattanach, B.; Peters, J.; Hunsicker, P.R.; Mallon, A.M.; Strivens, M.A.; Bate, R.; Miller, W.; Rhodes, M.; Brown, S.D.; Herman, G.E. The gene mutated in bare patches and striated mice encodes a novel 3β-hydroxysteroid dehydrogenase. Nat. Genet. **1999**, *22*, 182–187.

64. Noshiro, M.; Lakso, M.; Kawajiri, K.; Negishi, M. Rip locus: regulation of female-specific isozyme (I-P-450$_{16α}$) of testosterone 16α-hydroxylase in mouse liver, chromosome localization, and cloning of P-450 cDNA. Biochemistry **1988**, *27*, 6434–6443.

65. Abbaszde, I.G.; Clarke, T.R.; Park, C.H.J.; Payne, A.H. The mouse 3β-hydroxysteroid dehydrogenase multigene family includes two functionally distinct groups of proteins. Mol. Endocrinol. **1995**, *9*, 1214–1222.

66. Chai, X.; Zhai, Y.; Napoli, J.L. cDNA cloning and characterization of a *cis*-retinol/3α-hydroxysterol short-chain dehydrogenase. J. Biol. Chem. **1997**, *272*, 33125–33131.

67. Haeseleer, F.; Palczewski, K. Short-chain dehydrogenases/reductases in retina. Methods Enzymol. **2000**, *316*, 372–383.

68. Morton, N.M.; Holmes, M.C.; Fievet, C.; Staels, B.; Tailleux, A.; Mullins, J.J.; Seckl, J.R. Improved lipid and lipoprotein profile, hepatic insulin sensitivity, and glucose tolerance in 11β-hydroxysteroid dehydrogenase type 1 null mice. J. Biol. Chem. **2001**, *276*, 41293–41300.

69. Hermanowski-Vosatka, A.; Gerhold, D.; Mundt, S.S.; Loving, V.A.; Lu, M.; Chen, Y.; Elbrecht, A.; Wu, M.; Doebber, T.; Kelly, L.; Milot, D.; Guo, Q.; Wang, P.R.; Ippolito, M.; Chao, Y.S.; Wright, S.D.; Thieringer, R. PPARα agonists reduce 11β-hydroxysteroid dehydrogenase type 1 in the liver. Biochem. Biophys. Res. Commun. **2000**, *279*, 330–336.

70. Guerre-Millo, M.; Gervois, P.; Raspé, E.; Madsen, L.; Poulain, P.; Derudas, B.; Herbert, J.M.; Winegar, D.A.; Willson, T.M.; Fruchart, J.C.; Berge, R.K.; Staels, B. Peroxisome proliferator-activated receptor α activators improve insulin sensitivity and reduce adiposity. J. Biol. Chem. **2000**, *275*, 16638–16642.

71. Breitling, R.; Marijanovic, Z.; Perovic, D.; Adamski, J. Evolution of 17β-HSD type 4, a multifunctional protein of β-oxidation. Mol. Cell. Endocrinol. **2001**, *171*, 205–210.

72. Haubenwallner, S.; Essenburg, A.D.; Barnett, B.C.; Pape, M.E.; DeMattos, R.B.; Krause, B.R.; Minton, L.L.; Auerbach, B.J.; Newton, R.S.; Leff, T.; Bisgaier, C.L. Hypolipidemic activity of select fibrates correlates to changes in hepatic apolipoprotein C-III expression: a potential physiologic basis for their mode of action. J. Lipid Res. **1995**, *36*, 2541–2551.

73. Staels, B.; Vu-Dac, N.; Kosykh, V.A.; Saladin, R.; Fruchart, J.C.; Dallongeville, J.; Auwerx, J. Fibrates downregulate apolipoprotein C-III expression independent of induction of peroxisomal acyl coenzyme A oxidase. A potential mechanism for the hypolipidemic action of fibrates. J. Clin. Invest. **1995**, *95*, 705–712.

74. Andersson, Y.; Majd, Z.; Lefebvre, A.M.; Martin, G.; Sechkin, A.V.; Kosykh, V.; Fruchart, J.C.; Najib, J.; Staels, B. Developmental and pharmacological regulation of apolipoprotein C-II gene expression. Comparison with apo C-I and apo C-III gene regulation. Arterioscler. Thromb. Vasc. Biol. **1999**, *19*, 115–121.

75. Berthou, L.; Duverger, N.; Emmanuel, F.; Langouet, S.; Auwerx, J.; Guillouzo, A.; Fruchart, J.C.; Rubin, E.; Denèfle, P.; Staels, B.; Branellec, D. Opposite regulation of human versus mouse apolipoprotein A-I by fibrates in human apolipoprotein A-I transgenic mice. J. Clin. Invest. **1996**, *97*, 2408–2416.

76. Vu-Dac, N.; Schoonjans, K.; Kosykh, V.; Dallongeville, J.; Fruchart, J.C.; Staels, B.; Auwerx, J. Fibrates increase human apolipoprotein A-II expression through activation of the peroxisome proliferator-activated receptor. J. Clin. Invest. **1995**, *96*, 741–750.

77. Bovard-Houppermans, S.; Ochoa, A.; Fruchart, J.C.; Zakin, M.M. Fenofibric acid modulates the human apolipoprotein A-IV gene expression in HepG2 cells. Biochem. Biophys. Res. Commun. **1994**, *198*, 764–769.

78. Staels, B.; van Tol, A.; Andreu, T.; Auwerx, J. Fibrates influence the expression of genes involved in lipoprotein metabolism in a tissue-selective manner in the rat. Arterioscler. Thromb. **1992**, *12*, 286–294.

79. Staels, B.; van Tol, A.; Verhoeven, G.; Auwerx, J. Apolipoprotein A-IV messenger ribonucleic acid abundance is regulated in a tissue-specific manner. Endocrinology **1990**, *126*, 2153–2163.

80. Deeg, M.A.; Bierman, E.L.; Cheung, M.C. GPI-specific phospholipase D associates with an apoA-I- and apoA-IV-containing complex. J. Lipid Res. **2001**, *42*, 442–451.

81. van der Vliet, H.N.; Sammels, M.G.; Leegwater, A.C.J.; Levels, J.H.M.; Reitsma, P.H.; Boers, W.; Chamuleau, R.A.F.M. Apolipoprotein A-V: a novel apolipoprotein associated with an early phase of liver regeneration. J. Biol. Chem. **2001**, *276*, 44512–44520.

82. van der Vliet, H.N.; Schaap, F.G.; Levels, J.H.; Ottenhoff, R.; Looije, N.; Wesseling, J.G.; Groen, A.K.; Chamuleau, R.A. Adenoviral overexpression of apolipoprotein A-V reduces serum levels of triglycerides and cholesterol in mice. Biochem. Biophys. Res. Commun. **2002**, *295*, 1156–1159.

83. Malle, E.; Steinmetz, A.; Raynes, J.G. Serum amyloid A (SAA): an acute phase protein and apolipoprotein. Atherosclerosis **1993**, *102*, 131–146.

84. Rokita, H.; Shirahama, T.; Cohen, A.S.; Meek, R.L.; Benditt, E.P.; Sipe, J.D. Differential expression of the amyloid SAA 3 gene in liver and peritoneal macrophages of mice undergoing dissimilar inflammatory episodes. J. Immunol. **1987**, *139*, 3849–3853.

85. Murai, T.; Yamada, T.; Miida, T.; Arai, K.; Endo, N.; Hanyu, T. Fenofibrate inhibits reactive amyloidosis in mice. Arthritis Rheum. **2002**, *46*, 1683–1688.

86. Sammalkorpi, K.; Valtonen, V.; Kerttula, Y.; Nikkilä, E.; Taskinen, M.R. Changes in serum lipoprotein pattern induced by acute infections. Metabolism **1988**, *37*, 859–865.

87. Pussinen, P.J.; Malle, E.; Metso, J.; Sattler, W.; Raynes, J.G.; Jauhiainen, M. Acute-phase HDL in phospholipid transfer protein (PLTP)-mediated HDL conversion. Atherosclerosis **2001**, *155*, 297–305.

88. Lee, S.S.; Pineau, T.; Drago, J.; Lee, E.J.; Owens, J.W.; Kroetz, D.L.; Fernandez-Salguero, P.M.; Westphal, H.; Gonzalez, F.J. Targeted disruption of the α isoform of the peroxisome proliferator-activated receptor gene in mice results in abolishment of the pleiotropic effects of peroxisome proliferators. Mol. Cell. Biol. **1995**, *15*, 3012–3022.

89. Fan, C.Y.; Pan, J.; Chu, R.; Lee, D.; Kluckman, K.D.; Usuda, N.; Singh, I.; Yeldandi, A.V.; Rao, M.S.; Maeda, N.; Reddy, J.K. Hepatocellular and hepatic peroxisomal alterations in mice with a disrupted peroxisomal fatty acyl-coenzyme A oxidase gene. J. Biol. Chem. **1996**, *271*, 24698–24710.

90. Waterston, R.H.; Lindblad-Toh, K.; Birney, E.; Rogers, J.; Abril, J.F.; Agarwal, P.; Agarwala, R.; Ainscough, R.; Alexandersson, M.; An, P.; Antonarakis, S.E.; Attwood, J.; Baertsch, R.; Bailey, J.; Barlow, K.; Beck, S.; Berry, E.; Birren, B.; Bloom, T.; Bork, P.; Botcherby, M.; Bray, N.; Brent, M.R.; Brown, D.G.; Brown, S.D.; Bult, C.; Burton, J.; Butler, J.; Campbell, R.D.; Carninci, P.; Cawley, S.; Chiaromonte, F.; Chinwalla, A.T.; Church, D.M.; Clamp, M.; Clee, C.; Collins, F.S.; Cook, L.L.; Copley, R.R.; Coulson, A.; Couronne, O.; Cuff, J.; Curwen, V.; Cutts, T.; Daly, M.; David, R.; Davies, J.; Delehaunty, K.D.; Deri, J.; Dermitzakis, E.T.; Dewey, C.; Dickens, N.J.; Diekhans, M.; Dodge, S.; Dubchak, I.; Dunn, D.M.; Eddy, S.R.; Elnitski, L.; Emes, R.D.; Eswara, P.; Eyras, E.; Felsenfeld, A.; Fewell, G.A.; Flicek, P.; Foley, K.; Frankel, W.N.; Fulton, L.A.; Fulton, R.S.; Furey, T.S.; Gage, D.; Gibbs, R.A.; Glusman, G.; Gnerre, S.; Goldman, N.; Goodstadt, L.; Grafham, D.; Graves, T.A.; Green, E.D.; Gregory, S.; Guigó, R.; Guyer, M.; Hardison, R.C.; Haussler, D.; Hayashizaki, Y.; Hillier, L.W.; Hinrichs, A.; Hlavina, W.; Holzer, T.; Hsu, F.; Hua, A.; Hubbard, T.; Hunt, A.; Jackson, I.; Jaffe, D.B.; Johnson, L.S.; Jones, M.; Jones, T.A.; Joy, A.; Kamal, M.; Karlsson, E.K.; Karolchik, D.; Kasprzyk, A.; Kawai, J.; Keibler, E.; Kells, C.; Kent, W.J.; Kirby, A.; Kolbe, D.L.; Korf, I.; Kucherlapati, R.S.; Kulbokas, E.J., III; Kulp, D.; Landers, T.; Leger, J.P.; Leonard, S.; Letunic, I.; Levine, R.; Li, J.; Li, M.; Lloyd, C.; Lucas, S.; Ma, B.; Maglott, D.R.; Mardis, E.R.; Matthews, L.; Mauceli, E.; Mayer, J.H.; McCarthy, M.; McCombie, W.R.; McLaren, S.; McLay, K.; McPherson, J.D.; Meldrim, J.; Meredith, B.; Mesirov, J.P.; Miller, W.; Miner, T.L.; Mongin, E.; Montgomery, K.T.; Morgan, M.; Mott, R.; Mullikin, J.C.; Muzny, D.M.; Nash, W.E.; Nelson, J.O.; Nhan, M.N.; Nicol, R.; Ning, Z.; Nusbaum, C.; O'Connor, M.J.; Okazaki, Y.; Oliver, K.; Overton-Larty, E.; Pachter, L.; Parra, G.; Pepin, K.H.; Peterson, J.; Pevzner, P.; Plumb, R.; Pohl, C.S.; Poliakov, A.; Ponce, T.C.; Ponting, C.P.;

Potter, S.; Quail, M.; Reymond, A.; Roe, B.A.; Roskin, K.M.; Rubin, E.M.; Rust, A.G.; Santos, R.; Sapojnikov, V.; Schultz, B.; Schultz, J.; Schwartz, M.S.; Schwartz, S.; Scott, C.; Seaman, S.; Searle, S.; Sharpe, T.; Sheridan, A.; Shownkeen, R.; Sims, S.; Singer, J.B.; Slater, G.; Smit, A.; Smith, D.R.; Spencer, B.; Stabenau, A.; Stange-Thomann, N.; Sugnet, C.; Suyama, M.; Tesler, G.; Thompson, J.; Torrents, D.; Trevaskis, E.; Tromp, J.; Ucla, C.; Ureta-Vidal, A.; Vinson, J.P.; von Niederhausern, A.C.; Wade, C.M.; Wall, M.; Weber, R.J.; Weiss, R.B.; Wendl, M.C.; West, A.P.; Wetterstrand, K.; Wheeler, R.; Whelan, S.; Wierzbowski, J.; Willey, D.; Williams, S.; Wilson, R.K.; Winter, E.; Worley, K.C.; Wyman, D.; Yang, S.; Yang, S.P.; Zdobnov, E.M.; Zody, M.C.; Lander, E.S. Initial sequencing and comparative analysis of the mouse genome. Nature **2002**, *420*, 520–562.

91. Frohnert, B.I.; Hui, T.Y.; Bernlohr, D.A. Identification of a functional peroxisome proliferator-responsive element in the murine fatty acid transport protein gene. J. Biol. Chem. *274*, 3970–3977.

92. Helledie, T.; Grøntved, L.; Jensen, S.S.; Kiilerich, P.; Rietveld, L.; Albrektsen, T.; Boysen, M.S.; Nøhr, J.; Larsen, L.K.; Fleckner, J.; Stunnenberg, H.G.; Kristiansen, K.; Mandrup, S. The gene encoding the acyl-CoA-binding protein is activated by peroxisome proliferator-activated receptor γ through an intronic response element functionally conserved between humans and rodents. J. Biol. Chem. **2002**, *277*, 26821–26830.

93. Rubin, E.M.; Ishida, B.Y.; Clift, S.M.; Krauss, R.M. Expression of human apolipoprotein A-I in transgenic mice results in reduced plasma levels of murine apolipoprotein A-I and the appearance of two new high density lipoprotein size subclasses. Proc. Natl. Acad. Sci. USA **1991**, *88*, 434–438.

94. Mardones, P.; Pilon, A.; Bouly, M.; Duran, D.; Nishimoto, T.; Arai, H.; Kozarsky, K.F.; Altayó, M.; Miquel, J.F.; Luc, G.; Clavey, V.; Staels, B.; Rigotti, A. Fibrates downregulate hepatic scavenger receptor class B type I (SR-BI) protein expression in mice. J. Biol. Chem. **2003**, *278*, 7884–7890.

95. Yarmush, M.L.; Jayaraman, A. Advances in proteomic technologies. Annu. Rev. Biomed. Eng. **2002**, *4*, 349–373.

96. Ashby, J.; Brady, A.; Elcombe, C.R.; Elliott, B.M.; Ishmael, J.; Odum, J.; Tugwood, J.D.; Kettle, S.; Purchase, I.F. Mechanistically-based human hazard assessment of peroxisome proliferator-induced hepatocarcinogenesis. Hum. Exp. Toxicol. **1994**, *13* (suppl 2), S1–S117.

97. Palmer, C.N.; Hsu, M.H.; Griffin, K.J.; Raucy, J.L.; Johnson, E.F. Peroxisome proliferator activated receptor-α expression in human liver. Mol. Pharmacol. **1998**, *53*, 14–22.

98. Gervois, P.; Torra, I.P.; Chinetti, G.; Grötzinger, T.; Dubois, G.; Fruchart, J.C.; Fruchart-Najib, J.; Leitersdorf, E.; Staels, B. A truncated human peroxisome proliferator-activated receptor α splice variant with dominant negative activity. Mol. Endocrinol. **1999**, *13*, 1535–1549.

99. Woodyatt, N.J.; Lambe, K.G.; Myers, K.A.; Tugwood, J.D.; Roberts, R.A. The peroxisome proliferator (PP) response element upstream of the human acyl CoA oxidase gene is inactive among a sample human population: significance for species differences in response to PPs. Carcinogenesis **1999**, *20*, 369–372.

100. Lawrence, J.W.; Li, Y.; Chen, S.; DeLuca, J.G.; Berger, J.P.; Umbenhauer, D.R.; Moller, D.E.; Zhou, G. Differential gene regulation in human *versus* rodent hepatocytes by peroxisome proliferator-activated receptor (PPAR) α. PPARα fails to

induce peroxisome proliferation-associated genes in human cells independently of the level of receptor expresson. J. Biol. Chem. **2001**, *276*, 31521–31527.

101. International Human Genome Sequencing Consortium. Initial sequencing and analysis of the human genome. Nature **2001**, *409*, 860–921.

102. Venter, J.C.; Adams, M.D.; Myers, E.W.; Li, P.W.; Mural, R.J.; Sutton, G.G.; Smith, H.O.; Yandell, M.; Evans, C.A.; Holt, R.A.; Gocayne, J.D.; Amanatides, P.; Ballew, R.M.; Huson, D.H.; Wortman, J.R.; Zhang, Q.; Kodira, C.D.; Zheng, X.H.; Chen, L.; Skupski, M.; Subramanian, G.; Thomas, P.D.; Zhang, J.; Gabor Miklos, G.L.; Nelson, C.; Broder, S.; Clark, A.G.; Nadeau, J.; McKusick, V.A.; Zinder, N.; Levine, A.J.; Roberts, R.J.; Simon, M.; Slayman, C.; Hunkapiller, M.; Bolanos, R.; Delcher, A.; Dew, I.; Fasulo, D.; Flanigan, M.; Florea, L.; Halpern, A.; Hannenhalli, S.; Kravitz, S.; Levy, S.; Mobarry, C.; Reinert, K.; Remington, K.; Abu-Threideh, J.; Beasley, E.; Biddick, K.; Bonazzi, V.; Brandon, R.; Cargill, M.; Chandramouliswaran, I.; Charlab, R.; Chaturvedi, K.; Deng, Z.; Di Francesco, V.; Dunn, P.; Eilbeck, K.; Evangelista, C.; Gabrielian, A.E.; Gan, W.; Ge, W.; Gong, F.; Gu, Z.; Guan, P.; Heiman, T.J.; Higgins, M.E.; Ji, R.R.; Ke, Z.; Ketchum, K.A.; Lai, Z.; Lei, Y.; Li, Z.; Li, J.; Liang, Y.; Lin, X.; Lu, F.; Merkulov, G.V.; Milshina, N.; Moore, H.M.; Naik, A.K.; Narayan, V.A.; Neelam, B.; Nusskern, D.; Rusch, D.B.; Salzberg, S.; Shao, W.; Shue, B.; Sun, J.; Wang, Z.; Wang, A.; Wang, X.; Wang, J.; Wei, M.; Wides, R.; Xiao, C.; Yan, C.; Yao, A.; Ye, J.; Zhan, M.; Zhang, W.; Zhang, H.; Zhao, Q.; Zheng, L.; Zhong, F.; Zhong, W.; Zhu, S.; Zhao, S.; Gilbert, D.; Baumhueter, S.; Spier, G.; Carter, C.; Cravchik, A.; Woodage, T.; Ali, F.; An, H.; Awe, A.; Baldwin, D.; Baden, H.; Barnstead, M.; Barrow, I.; Beeson, K.; Busam, D.; Carver, A.; Center, A.; Cheng, M.L.; Curry, L.; Danaher, S.; Davenport, L.; Desilets, R.; Dietz, S.; Dodson, K.; Doup, L.; Ferriera, S.; Garg, N.; Glucksmann, A.; Hart, B.; Haynes, J.; Haynes, C.; Heiner, C.; Hladun, S.; Hostin, D.; Houck, J.; Howland, T.; Ibegwam, C.; Johnson, J.; Kalush, F.; Kline, L.; Koduru, S.; Love, A.; Mann, F.; May, D.; McCawley, S.; McIntosh, T.; McMullen, I.; Moy, M.; Moy, L.; Murphy, B.; Nelson, K.; Pfannkoch, C.; Pratts, E.; Puri, V.; Qureshi, H.; Reardon, M.; Rodriguez, R.; Rogers, Y.H.; Romblad, D.; Ruhfel, B.; Scott, R.; Sitter, C.; Smallwood, M.; Stewart, E.; Strong, R.; Suh, E.; Thomas, R.; Tint, N.N.; Tse, S.; Vech, C.; Wang, G.; Wetter, J.; Williams, S.; Williams, M.; Windsor, S.; Winn-Deen, E.; Wolfe, K.; Zaveri, J.; Zaveri, K.; Abril, J.F.; Guigó, R.; Campbell, M.J.; Sjolander, K.V.; Karlak, B.; Kejariwal, A.; Mi, H.; Lazareva, B.; Hatton, T.; Narechania, A.; Diemer, K.; Muruganujan, A.; Guo, N.; Sato, S.; Bafna, V.; Istrail, S.; Lippert, R.; Schwartz, R.; Walenz, B.; Yooseph, S.; Allen, D.; Basu, A.; Baxendale, J.; Blick, L.; Caminha, M.; Carnes-Stine, J.; Caulk, P.; Chiang, Y.H.; Coyne, M.; Dahlke, C.; Mays, A.; Dombroski, M.; Donnelly, M.; Ely, D.; Esparham, S.; Fosler, C.; Gire, H.; Glanowski, S.; Glasser, K.; Glodek, A.; Gorokhov, M.; Graham, K.; Gropman, B.; Harris, M.; Heil, J.; Henderson, S.; Hoover, J.; Jennings, D.; Jordan, C.; Jordan, J.; Kasha, J.; Kagan, L.; Kraft, C.; Levitsky, A.; Lewis, M.; Liu, X.; Lopez, J.; Ma, D.; Majoros, W.; McDaniel, J.; Murphy, S.; Newman, M.; Nguyen, T.; Nguyen, N.; Nodell, M.; Pan, S.; Peck, J.; Peterson, M.; Rowe, W.; Sanders, R.; Scott, J.; Simpson, M.; Smith, T.; Sprague, A.; Stockwell, T.; Turner, R.; Venter, E.; Wang, M.; Wen, M.; Wu, D.;

Wu, M.; Xia, A.; Zandieh, A.; Zhu, X. The sequence of the human genome. Science **2001**, *291*, 1304–1351.

103. Roglans, N.; Bellido, A.; Rodríguez, C.C.; Cabrero, À.; Novell, F.; Ros, E.; Zambón, D.; Laguna, J.C. Fibrate treatment does not modify the expression of acyl coenzyme A oxidase in human liver. Clin. Pharmacol. Ther. **2002**, *72*, 692–701.

104. Yu, S.; Matsusue, K.; Kashireddy, P.; Cao, W.Q.; Yeldandi, V.; Yeldandi, A.V.; Rao, M.S.; Gonzalez, F.J.; Reddy, J.K. Adipocyte-specific gene expression and adipogenic steatosis in the mouse liver due to peroxisome proliferator-activated receptor $\gamma 1$ (PPAR$\gamma 1$) overexpression. J. Biol. Chem. **2003**, *278*, 498–505.

105. Reddy, J.K. Nonalcoholic steatosis and steatohepatitis. III. Peroxisomal β-oxidation, PPAR α, and steatohepatitis. Am. J. Physiol. Gastrointest. Liver Physiol. **2001**, *281*, G1333–G1339.

How Dietary Long-Chain Fatty Acids Induce Gene Signals Involved in Carcinogenesis

Pascale Anderle

ISREC, Swiss Institute for Experimental Cancer Research, Epalinges, Lausanne, Switzerland

Pierre Farmer

ISREC, Bioinformatic Core Facility (BCF), Epalinges, Lausanne, Switzerland

Alvin Berger

Paradigm Genetics, Inc., Research Triangle Park, USA

Matthew A. Roberts

Nestle Global Strategic Business Unit and R&D, St. Louis, USA

INTRODUCTION

Lipids are components of most foods. Besides their principal dietary role of delivering energy to the body, dietary lipids are involved in normal and aberrant mitogenic processes, and have been linked to the development and progression of colon, liver, and mammary carcinogenesis. Lipids influence multiple signaling pathways involved in disease progression, most of which are difficult to study with a single-gene/protein/biomarker approach. These disease states are inherently multiparametric. Therefore, the highly parallel measurement approach central to nutrigenomics is needed to further scientific understanding. These tools have been largely unavailable in the past; thus the mechanisms by which lipids influence cancer have been mostly unknown.

Table 13.1 summarizes the effects of all types of lipids on carcinogenesis. This review will, however, focus on long-chain polyunsaturated fatty acids (LC-PUFA), which are probably the most commonly studied dietary lipids with respect to their effects on carcinogenesis. The effects of diets high in LC-PUFA have been studied for decades.

Bougnoux[19] extensively reviewed the influence of LC-PUFAs on cancer and carcinogenesis. The inhibitory effects of fish oil, which is rich in n-3 LC-PUFA, on carcinogenesis, have been shown in various models including an axoxymethane-induced rat colon carcinogenesis model and an HT29 colon tumor growth and metastasis nude mouse model.[20,21] These studies did not, however, fully unravel the mechanism of actions of n-3 LC-PUFA in ameliorating colon cancer. Though cell culture and animal studies support a preventive action of LC-PUFA, there is divided opinion on the association between LC-PUFA intake and cancer risk in humans. For example, case–control, international comparisons, and animal studies suggest an association between fat intake and breast cancer incidence, but cohort studies do not show this association.[22]

A number of studies have recently applied microarray and bioinformatic technologies to elucidate the complex cell regulation processes affected by LC-PUFAs.[23–25] Several of these studies are explored in depth by the original authors in other chapters of this book. Herein, we will review selected mechanisms through which LC-PUFA affect carcinogenesis, and studies that utilized "omic approaches" to study LC-PUFA effects on carcinogenesis. This chapter is an extension and expansion of a shortened article by our group on the same topic.[26]

MECHANISTIC EFFECTS OF VARIOUS LIPIDS ON CELL REGULATION

Biological and Carcinogenic Effects of LC-PUFA

Processes affected by dietary LC-PUFA consumption include growth; neurological development; lean and fat mass accretion; reproduction; innate and acquired immunity; infectious pathologies of viruses, bacteria, and parasites; and the incidence and severity of virtually all chronic and degenerative diseases including cancer, atherosclerosis, stroke, arthritis, diabetes, and osteoporosis and neurodegenerative, inflammatory, and skin diseases.[27–30] There is a distinct difference between the effects of n-3 and n-6 LC-PUFA on carcinogenesis. For example, Kim et al.[31] compared the effects of fish oils which are rich in n-3 LC-PUFA to corn oil which is rich in n-6 LC-PUFA (i.e., 18:2n-6 or linoleic acid). They found that colon cell proliferation, an intermediate marker for colon carcinogenesis, was lower in rats that had been fed with fish oil when

Table 13.1 Selected Examples of the Roles of Lipids in Etiology of Cancer

Disease	Lipid class	Specific lipid	Biological role	Drug treatment, or pathway to target to restore profile	Reference
Cancer	FA	DHA	Nonmitogenic		[1, 2]
Cancer	LOX products	15-HETrE	Nonmitogenic		[3–5]
Cancer	LOX products	13S HODE, esterified to DAG, PI	Apoptotic, affects MAPK signaling	15-LOX1 activators, depends on cancer	[6]
Cancer	Antioxidants	Antioxidants	Less free-radical induced genetic damage, prevent metabolic activation of carcinogens		[7]
Cancer	Sphingolipid	CER	Normal keratinocyte differentiation/barrier function/apoptosis	Inhibit ceramidase	[8, 9]
Glial tumors	Sphingolipid	CER	CER associated with less cancer progression, better chemotherapy	CER	[10–12]
Cancer	Inositide	3,4,5 PIP_3	Mitogenic, cytoskeletal remodeling, Ca signaling	Wortmannin, PI3K inhibitor	[13]
Cancer	Inositide	3 PIP	3,4,5 PIP_3 precursor	Wortmannin, PI3K inhibitor	[13]
Neuroblastoma	GS	GD_2			[14]
Dalton' T cell lymphoma	GS	GS	Impair NO production by macrophages		[15]
Melanoma	GS	Shed GM_3, GD_3, GD_{1a}	Apoptotic/melanoma cells escape destruction due to apoptosis of dendritic cells		[16, 17]
Renal cell Carcinoma	GS	GS	Apoptotic/cell death of T lymphocytes		[18]

Note: CER, ceramide; COX, cyclooxygenase; DAG, diacylglycerol; DHA, docosahexaenoic acid; FA, fatty acid; GD_2, GM_3, GD_3, GD_{1a}, classes of gangliosides; GS, ganglioside; HETrE, hydroxyeicosatrienoic acid (from dihomogammalinolenic acid); HODE, hydroxyoctadecadienoic acid, from linoleic acid; LOX, lipoxygenase; MAPK, mitogen activated protein kinase; NO, nitrous oxide; NSAIDs, non steroidal anti-inflammatory drug(s); PG, prostaglandin; PGE_2, class of prostaglandin derived from arachidonic acid; PI, phosphatidylinositol; PI3K, phosphatidylinositol 3 kinase; PIP, phosphatidylinositolphosphate; PIP_3, phosphatidylinositoltriphosphate; SPC, sphingosylphosphorylcholine; SPH, sphingosine; SPN, sphingomyelin; SPNase, sphingomyelinase; SPP, sphingosine 1-phosphate; SPT, serine palmitoyl transferase.

compared with rats fed with corn oil. Cremonezzi et al.[32] suggested that dietary LC-PUFA modulate normal and preneoplastic urothelial proliferation when induced by the tumorigenic agent melamine. They concluded that the frequency of simple urothelial hyperplasias and dysplasia/carcinoma *in situ* was significantly lower in the fish oil group than in the corn oil (rich in n-6 LC-PUFA) and the olein (18:1n-9 enriched) group.

LC-PUFA Conversion to Classical Eicosanoids and Effects on Carcinogenesis

Fatty acids (FAs) on the sn-2 position of phospholipids are cleaved in response to specific stimuli, following a signaling cascade, via the action of phospholipase A_2 to generate free fatty acids, that are then substrates for eicosanoid metabolism. A general pathway describing the subsequent flow of fatty acids to active eicosanoids is shown in Fig. 13.1.

The FAs are substrates for cyclooxygenase (COX), lipoxygenase (LOX), and epoxygenase. COX derivatives from arachidonic acid (AA) include prostaglandin D_2 (PGD_2), prostaglandin $F_2\alpha$ ($PGF_{2\alpha}$), prostacyclin (PGI_2), prostaglandin E_2 (PGE_2), and thromboxane A_2 (TXA_2). 5-LOX derivates from AA include four series leukotrienes and hydroxylated eicosatetraenoate (5-HETE). Similarly, 12 and 15 LOX generate 12 and 15 HETE products from AA. Analogous compounds are formed from released eicosapentaenoic acid (EPA; a component of fish oil), except with an additional double bond. Generally, the eicosanoid derivatives formed from EPA are less potent in affecting carcinogenesis and other processes, than the products derived from AA.[24] This partly explains why EPA has anticarcinogenic properties relative to AA.

Two COX isoforms have been described in the literature, COX-1 and COX-2. The COX-1 enzyme is constitutively expressed in nearly all cell types; the COX-2 isoform is a proinflammatory enzyme whose expression is initiated only after proinflammatory or mitogenic stimulations.[33]

There has been a renewed interest in how COX products affect colon cancer. This interest stems from the observation that 70–80% of human colorectal tumors have altered COX expression.[34,35] Additionally, aspirin and other COX inhibitors have been shown in recent clinical trials to protect against cancer development.[36,37]

Prostaglandins (PGs) may directly affect carcinogenesis as pharmacological inhibition/stimulation and genetic disruption of PG selective receptors reduced carcinogenesis in mice.[38–40] Both synthesis of PGE_2, and signaling through the PGE_2 selective receptor (EP_2) are likely influenced by LC-PUFA.[41] However, the precise subsequent molecular actions of PGs on carcinogenesis are not fully understood. Transactivation of the mitogenic signal of the epithelium growth factor receptor (EGFR) signaling pathway[42] and increased expression of the antiapoptotic Bcl-2 protein following PGE_2 treatment in colon cells have been shown to be essential for carcinogenesis.[43,44]

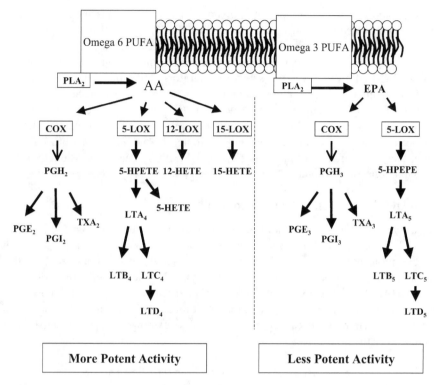

Figure 13.1 Metabolism of n-3 and n-6 LC-PUFA via cycloxygenase and lipoxygenases. Synthesis of diene-series prostaglandins are initiated by the cleavage of an ester bound linking arachidonic acid (AA, an n-6 LC-PUFA; left side of scheme) to phospholipids located in cell membranes through the catalytic activity of phospholipase A_2. AA may then be metabolized by the cyclooxygenase (COX) into one of many structurally related and biologically active prostanoids such as prostacyclin (PGI_2), prostaglandin E2 (PGE_2), and thromboxane A_2 (TXA_2), by the activity of specific cellular PG or TX synthase. Alternatively, AA may be converted into LTA_4 by the enzymatic activity of the 5 lipoxygenase (5-LOX) and further metabolized into LTB_4 or the peptido leukotrienes (LTC_4 and LTD_4; LTE_4 not shown). Eicosapentanoic acid (EPA), an n-3 LC-PUFA competes with AA for access to both oxidative enzymes COX and 5-LOX. The EPA metabolites of both enzymes either the triene-prostaglandins (PGE_3, PGI_2, TXA_2) or 5 series leukotrienes (LTB_5, LTC_5 and LTD_5) are generally less active than the corresponding AA-derived compounds.

Consistent with the earlier molecular mechanisms, Narayanan et al.[25] found decreased expression of Bcl-2 following docosahexaenoic acid (DHA; a component of fish oil) exposure; discussed in greater depth in a later chapter.

The influence of PGs on carcinogenesis might not be strictly limited to tumor cells but also to surrounding tissues as PGs can influence angiogenesis *in vitro* and *in vivo*,[45] and potentially suppress synthesis of proapoptotic cytokines such as TNFα by surrounding immune cells.[46]

Much more work is still needed to fully understand the transcription events initiated by n-3 LC-PUFAs in preventing cancer. A comparison of the efficiency of n-3 LC-PUFAs with COX-2 selective inhibitors would be valuable, particularly using microarray, proteomics, or metabolomic approaches. Indeed, selective inhibition of COX-2 might lead to accumulation of AA for other metabolic pathways such as the 5-LOX pathway (see Fig. 13.1), and LOX metabolites also affect carcinogenic processes. For example, the 5-LOX metabolite LTD$_4$, enhances proliferation of intestinal epithelial cells[47] and expression of the antiapoptotic Bcl-2 protein.[48]

LC-PUFA Conversion to Novel Eicosanoids

In the last few years, a plethora of novel eicosanoids derived from LC-PUFA have been discovered and some of these have potent anticancer properties. These molecules include endocannabinoids, *N*-acyl linked-amino acids and dopamine, and various oxygenated and epoxylated bioactive derivatives of AA and DHA.[24,49–57] A review of these novel eicosanoids is beyond the scope of the present review. Identifying the precise biological roles of these new derivatives and their roles in cancer prevention is an extremely important area for future investigation.

Affects of Dietary Fatty Acids on Various Transcription Factors

LC-PUFA, either as released free fatty acids or eicosanoid derivatives, or via membrane structural changes, are now known to affect the transcription of genes that can largely account for the previously observed phenotypic changes seen after LC-PUFA consumption. LC-PUFA and their derivatives affect transcription by activating nuclear receptors which in turn bind to promoters containing specific consensus sequences.[58–61]

LC-PUFA were believed to act on a single subfamily of orphan nuclear receptors—peroxisome proliferator activated receptors (PPARs); however, they are now known to affect many genes either via direct interactions or indirectly through additional transcription factors including hepatic nuclear-4α (HNF-4α), nuclear factor $\kappa\beta$ (NF-$\kappa\beta$), retinoid X receptor α (RXRα), sterol regulatory element binding protein-1c (SREBP-1c), and liver X receptors (LXRα and LXRβ).[24,60,62] Additional transcriptional factors that LC-PUFA may signal through include Spot14 (Thrsp), transcription factor 7 T-cell specific (Tcf7), transcription factor 2 (Tcf2), nuclear transcription factor-Y β (NF-Y β; Nfyb transcript), D site albumin promoter binding protein (Dbp), general transcription factor II (Gtf2I), activating transcription factor 5 (Atf5), Msx-interacting-zinc finger (Miz1), and pre B-cell leukemia transcription factor 1 (Pbx1).[24] Of the many affected transcription factors the predominant ones discussed in the literature are PPARs and SREBPs. These two families of transcription factors will be discussed in relation to nutrigenomic effects of LC-PUFA on metabolism and carcinogenesis.

The reader should recognize that it is difficult to interpret transcriptional profiling data alone, since dimerization, posttranslational modifications, relative quantities of coactivators and corepressors, and the formation of multiple transcriptional complexes binding to common promoter regions ultimately determines the binding activation potential of transcription factors. Nevertheless, clustering analysis of differentially regulated genes can be used to identify coregulation and thereby specific transcription factors signatures. These signatures then provide clues concerning the various pathways through which FAs may effect gene transcription, and are described subsequently.

Peroxisome Proliferator-Activated Receptors

PPARs are activated by various FAs and FA analogs, such as fibrates and thiazolidinediones.[24,63] The exact ligand for PPAR activation is debatable. Preferred ligands for PPAR activation are thought to be 18 and 20C FAs, as FAs shorter than 14C or longer than 20C will not fit correctly into the PPAR binding pocket and stabilize the AF-2 helix.[60] The net effect would be that PPAR coactivators are not recruited to relax nucleosomal DNA and allow transcription machinery to interact with various promoter elements. Therefore DHA (22:6n-3) may require retroconversion via β-oxidation to an 18 or 20C LC-PUFA prior to activating PPARs.[60,64]

The PPAR subfamily can be divided into three isotypes, designated PPARα, PPARδ, and PPARγ, each with tissue-specific expression. PPARs form heterodimers with retinoid X receptor (RXR), and interact with the peroxisome proliferator response element (PPREs) in gene promoters. PPREs are direct repeats (DR) of a hexanucleotide sequence AGGTCA separated by one nucleotide and are referred to as a DR-1 response element.[65]

PPAR activation has been linked to carcinogenesis for many years as classical PPARα ligands are established rodent carcinogens. PPARα in rodent liver is associated with peroxisome proliferation and with suppression of apoptosis and induction of cell proliferation.[66] An interaction between PPARα and estrogen carcinogenesis has also been elucidated.[67]

PPARγ is involved in induction of cell growth arrest during fibroblast differentiation to adipocytes, and in growth inhibition in human mammary cancer cells.[68] Conversely, PPARγ has been implicated in the regulation of cell cycle and cell proliferation in colon cancer models.[66,69,70] Thus, PPARγ can have different effects depending on the form of cancer. The downstream genes activated in response to PPARγ agonists have now been studied using microarrays.[71] Ligands for PPARγ are currently being discovered. PPARγ is activated by LC-PUFAs, eicosanoids, and antidiabetic agents such as troglitazone, and PPARγ effects are enhanced by ligands of retinoic acid receptor (RAR) and RXR.[68] This has led to the suggestion that combined

LC-PUFA/retinoid therapies may be desirable for treating cancer. The AA 15-LOX metabolite, 15S-HETE, is an example of a specific eicosanoid known to activate PPARγ and inhibit proliferation in PC3 prostate carcinoma cells.[72]

PPARβ was identified as a downstream target gene for APC/β-catenin/ T cell factor-4 tumor suppressor pathway, involved in regulation of c-*myc* and cyclin D1 growth promoting genes.[66] PPARβ protein and its mRNA are known to be increased in colon tumors[73] but PPARβ overexpression was not an inherent property of breast cancer cell lines evaluated. However, the dynamic changes in PPARβ mRNA expression and the ability of PPARβ in MCF-7 cells to respond to ligand indicate PPARβ may play a role in mammary gland carcinogenesis through activation of downstream genes via endogenous fatty acid ligands or exogenous agonists.[73] Berger et al.[24] found that AA rich fungal oil, fish oil, and the combination of the two, increased expression of genes containing a PPRE, and that many of these genes had roles in β-oxidation of fatty acids. RT-PCR was used to validate changes to selected genes. There were dramatic differences in the extent of upregulation in a variety of PPAR-responsive genes following LC-PUFA consumption. Interestingly, PPAR gene expression itself was not found to be affected by dietary LC-PUFA.[24,74] This constancy of PPAR expression indicates that differences in transcription of genes containing a PPRE are likely because of direct activation of PPARs themselves. PPARs could become activated via posttranslational modifications or by increased dimerization to activated RXR (or other activated receptors). In this regard, it has been shown that DHA can activate RXR by fitting into an RXR ligand binding pocket.[75,76]

One gene with a PPRE of particular interest is Cd36. Berger et al.[24] found that AA rich fungal oil, fish oil, and the combination, all reduced Cd36 transcripts. Cd36 has been implicated not only in fat transport, but also in cell proliferation, angiogenesis, and tumor metastasis, thus offering one explanation of how LC-PUFA may signal through PPRE to affect carcinogenesis.[77] In colon cancer cells, Lee and Hwang[78] found that DHA decreased transactivation of PPRE and DNA binding of PPARs in colon tumor cells. The suppression of PPAR transactivation led to reduced expression of Cd36 in HCT116 cells and human monocytic cells (THP-1).

Hepatic SREBPs

SREBPs are critically important basic helix–loop–helix (bHLH) transcription factors activated in response to intracellular cholesterol levels that induce transcription of genes with sterol regulatory element (SRE) and E-box-containing promoters. Within the SREBP promoter element, the liver X receptor (LXR)-responsive elements (LXRE) may have an important role.[79] LC-PUFA may affect SREBP-1c-mediated transcription via several mechanisms. LC-PUFA can antagonize activation of the LXR receptor leading to fewer LXR/RXR heterodimer complexes to bind the LXRE in the SREBP-1c promoter and less SREBP-1c-mediated transcription of SRE-containing genes.[79] LC-PUFA can also prevent maturation of SREBP protein.[80] The PPARα signaling cascade intertwines with SREBP pathways as LC-PUFA also activate PPARα which can induce LXRα expression.[60]

SREBP activation of the SRE promoter is known to increase fatty acid synthesis to meet normal metabolic needs.[24] However, during *in vitro* transformation of human mammary epithelial cells and breast cancer cells, increased MAP kinase and PI 3-kinase signaling can increase SREBP-1c (and other SREBP subtypes) and fatty acid synthetase (FAS) levels for support of new membrane synthesis during cell growth.[81] Likewise, in prostate cancer biopsies, FAS was found to be increased in 94% of the tumors examined and the mechanism of action involves the SREBP and PI 3-kinase pathways.[82] In transformed cells, SREBPs are involved in the activation of FAS, but the signal transduction cascade leading to the increase in SREBP may be different than in nontransformed cells.[83]

Berger et al.[24] found that hepatic SREBP-1 was downregulated after feeding fish oil and AA-rich fungal oil plus fish oil, but importantly, not affected with AA-rich fungal oil alone. It is not clear how AA is handled differently in the liver to prevent its interaction with SREBP. It was recently shown that relatively low levels of n-3 LC-PUFA, such as administered by Berger et al.[24] might affect SREBP-1 mRNA, but another important effect is to prevent maturation of SREBP-1 protein.[80]

Hepatic Nuclear Factors

Hepatic nuclear factors (HNFs) have been recognized to be transcription factors mediating the transcription of a variety of genes putatively involved in energy metabolism. LC-PUFA bind to HNF as CoA derivatives. HNFs may compete with PPARs for binding to DR-1 elements, and could thus affect carcinogenesis via activation of similar target genes, as described for PPARs. Taken together, LC-PUFA can activate PPARα/RXRα, leading to displacement of HNF4α from PPRE; or be converted to CoA derivatives, which in turn can suppress HNF4α transcriptional activity. Alternatively, these transcription factors may respond differently to chronic versus acute administration of LC-PUFA. HNF4α was not differentially regulated in the study by Berger et al.[24] The related transcription factor HNF3γ (winged helix protein, fork head, Foxa3) was, however, downregulated with the mixture of AA-rich fungal oil and fish oil, but not with the oils alone. HNF3γ protein may be involved in etiology of obesity, hyperlipidemia, and diabetes, by regulating glucagon transcription, insulin resistance, and pancreatic cell function.[84,85] HNF3γ signaling may regulate transcription of Pck1 (upregulated twofold with each LC-PUFA evaluated by Berger et al.[24] and G6pc (differentially regulated in the experiments of Berger.[24,86] HNF3γ is also reported to increase transcription of several genes that were not found to be differentially regulated by Berger et al.,[24] including, transferrin, tyrosine amino transferase,[87] and IPF-1/PDX-1.[84]

The effects of HNF3γ signaling on carcinogenesis have not been examined to our knowledge.

Hepatic cAMP Signaling

cAMP is an important signaling molecule in specific cellular domains.[88] Levels are controlled through synthesis via adenylate cyclase, and degradation via phosphodiesterases (PDEs). Calmodulin PDEs (CaM-PDEs) are sensitive to calmodulin and calcium upregulation.[89] cAMP can then affect cell signaling via interactions with regulatory subunits of cAMP activated kinases, including protein kinase A, leading to their subsequent activation, and via binding of cAMP response element binding proteins (CREBs) to cAMP response elements (CREs) or cAMP response units (CRU) consisting of CREs. In liver, cAMP regulates Pck1 transcription via binding of CREB, as well as CCAAT/enhancer binding protein (C/EBP), to the CRU. Berger et al.[24] found that liver Cam-Pde1c was not regulated by AA-rich fungal oil feeding and fish oil feeding alone, but was significantly upregulated sixfold with the combination of both oils. Berger et al.[24] found that a calcium/calmodulin-dependent protein kinase 2 (CKII; Camk2b) was not affected with AA-rich fungal oil feeding or the combination diet, but was downregulated with fish oil feeding alone. CKII is a calmodulin target involved in brain synaptic plasticity. In the liver, CKII may phosphorylate SAG protein (sensitive to apoptosis gene), which in turn has a role in preventing oxidation by inhibiting cytochrome *c* release and caspase activation. Further, CKII may activate CPT-1 via cytoskeletal phosphorylation of cytokeratin. Thus, activation of CKII may lead to increased β-oxidation, antioxidant protection, and possible effects on carcinogenesis. The overall extent to which LC-PUFA may affect carcinogenesis via cAMP signaling is an important area for future investigation.

In summary, PPAR and SREBP transcription factors can be activated by LC-PUFA in nontransformed cells, having normal effects on carbohydrate and lipid metabolism. It is fascinating that activation of these same transcription factors, perhaps via a different signaling cascade, is associated with carcinogenesis in transformed cells. This leads to a chicken and egg type question. Once a cell becomes carcinogenic, PPAR and SREBP signaling cascades may be turned on to increase the need for membrane lipid and for other processes. Alternatively, a cell may become carcinogenic because of aberrant activation of PPAR and SREBP, and other transcription factors.

RELEVANCE OF MICROARRAYS TO ASSESS THE INFLUENCE OF LIPIDS ON CELL REGULATION

Genomic, proteomic, and metabolomic technologies are fundamentally changing the nature of lipid research. It is now possible to simultaneously measure thousands of molecular events surrounding the perturbation of various aspects of lipid metabolism.[90] Scientists are now extending the analysis of gene expression and its global perspective to examine other layers of biological organization (i.e., proteins and metabolites).[91] Unfortunately, lipid metabolic pathway information has been discouragingly underdeveloped relative to genes, proteins, and regulatory elements.

The primary causative factors in disease are often the altered biochemical composition of cells and tissues. Thus, the link between the gene regulatory control and biochemical mechanisms will be crucial for understanding biological control of lipid metabolism. The identification of relationships among genes, transcripts, proteins, and metabolites are essential components of integrative metabolism.[24,90] Whereas proteomics and metabolomics technologies are in the developmental phase, genomics approaches, in particular DNA microarrays, have become standard procedures when studying global changes in a biological system.

As mentioned earlier, the mechanisms of action that lipids have on cell regulation are complex. Thus, the use of microarrays, which allows the measurement of mRNA levels of thousands of different genes simultaneously, is a useful tool to elucidate such complex systems. To date only a couple of studies have used microarrays to study the effects of PUFA on cell regulation, and these are reviewed here.

Narayanan et al.[23] investigated whether the n-3 LC-PUFA and fish oil component DHA, could affect a specific set of cell cycle regulatory genes with roles in cell death induction; and modulate several sets of membrane proteins involved in lipid signaling. Using a dual-labeling system containing 3800 80-oligomers they compared nonstimulated Caco-2 cells and Caco-2 cells that had been stimulated for 48 h with DHA. Out of 3800 genes 504 were identified as being differently expressed between treated and untreated cells when using a threshold of twofold change. It has to be noted that the true number of verified differentially regulated genes may be significantly lower and there could be false positives. When plotting the fluorescence signals of treated vs. untreated cells a bias vs. the Cy5 labeled sample (treated cells) is possible (see Fig. 2 in Narayanan et al.[23]). Some, but not all, of the genes changed in microarray experiments were verified by RT-PCR. Standardization with housekeeping genes was not, however, performed to assess true differences.

A principal conclusion of the study was that exposure to DHA decreased COX-2 expression. Interestingly, PGE_2 influenced expression of COX-2,[39] suggesting DHA decreased COX-2 synthesis by inhibiting PGE_2 synthesis. This would strongly support one of the most cited mechanisms explaining the protective activity of n-3 LC-PUFA on carcinogenesis (i.e., interference with PG synthesis) (Fig. 13.1).

They also reported a coregulatory pattern of $PPAR\alpha$, $PPAR\gamma$, and $RXR\alpha$ in DHA treated cells. Activation of $PPAR\alpha$ and $PPAR\gamma$ corresponded with changes in expression of $RXR\alpha$, a ligand activated transcription factor, a cellular retinoic acid-binding protein, and insulin-like growth factor binding protein 1. Thus, they concluded that PPAR activation seems to induce genes encoding proteins involved in lipid transport and oxidation processes.

In a follow-up study with the same dataset, Narayanan et al.[25] focused on the effects of DHA on modulation of inducible nitric oxide synthase (iNOS) and related proinflammatory genes using microarrays. Besides microarray analysis of mRNA levels these authors also studied the cellular localization of iNOS and the apoptotic effect of DHA in Caco-2. They observed a downregulation of iNOS expression using RT-PCR as iNOS was not present on the chip in DHA treated cells and a ~54% inhibition of cell growth, partly by inducing apoptosis. More information on this study is provided in a chapter by the original authors.

Similarly to Narayanan et al.,[23,25] Kitajka et al.,[61] Puskas et al.,[92] and Berger et al.[24] studied the effects of dietary lipids on gene regulation using microarrays. These studies did not focus on aspects related to carcinogenesis, but provide some insight into the effects of dietary lipids on gene regulation in cells. The studies by Puskas et al. are discussed in greater depth in Chapter 5.

As mentioned previously, Berger et al.[24] published a study in mice using the murine 11k GeneChip whose target sequences are known. They studied the influence of four different types of diets (control diet, diet enriched in AA-rich fungal oil, diet enriched in fish oil, and diet enriched in both fungal and fish oils) in mice on the liver and the hippocampus. Using the so-called limit fold change approach[93] as a selection criterion for differential expression they identified 239 and 356 genes, respectively, in liver and hippocampus. With respect to the expression profile in the liver, they reported a dramatic difference in the expression of a variety of PPAR-responsive genes, but no significant change in expression of the PPAR genes themselves. Thus, the authors suggested that the constancy of PPAR expression indicates that differences in the transcription of genes containing a PPRE are likely due to direct interaction with PPAR proteins. The study did not focus on carcinogenesis per se, although numerous oncogenes were affected in both organs at the expression level. It should be noted that all of the raw data from Berger et al.[24] are available to be mined for additional information on affected oncogenes genes involved in cellular signaling cascades, and genes involved in the cell cycle, at the gene expression omnibus.[94] A fresh look at this study using updated bioinformatic analysis is presented in this book in Chapters 2–3.

UTILIZING OTHER "OMIC" APPROACHES TO ASSESS THE INFLUENCE OF LIPIDS ON CELL REGULATION

The emphasis on the gene, with the human genome project directed at sequencing the entire genome of humans and other species, has shown that the information obtained has distinct limitations, in addition to offering considerable opportunities. The central limitation is that the genome provides a blueprint and reflects the potential of an organism, but the genome itself does actually represent the phenotype. Thus, all cells contain the complete genome of a given organism, but not all genes are expressed within each cell type. The omic approaches, including transcriptomics, proteomics, and metabolomics, are now being combined with novel whole body reporter type assays as well as systemic gene knock out and knock in approaches to address this problem.

Use of Proteomics in Cancer Studies

Proteomics is the systematic evaluation of changes in the protein constituency of a cell. The ultimate goal is to characterize the information flow through protein pathways that interconnect the extracellular microenvironment with the control of gene transcription.

Proteomics[95,96] and phosphoproteomics[97] are new tools for cancer researchers. Proteomics provide confirmation of the changes to metabolites and enhanced understanding of the mechanism of action of drugs and the signaling cascade leading to changes in metabolites; and the proteomic endpoints may themselves become novel biomarkers. The importance of examining changes in the posttranslational state of proteins is important since it is not the absolute level of a protein, but its posttranslational state, that determines its biological activity. Changes to proteins can be monitored with modern proteomic approaches.[97–102]

Various proteomics technologies are being applied. Low-throughput technologies such as 2D-PAGE coupled with mass spectrometry for protein identification have been proven useful for cancer biomarker discovery. Other newer technologies such as multidimensional separation systems directly coupled to mass spectrometry have been shown to increase the sensitivity and throughput when compared with the traditional 2D-PAGE analysis.[103]

Metabolomics and Lipidomics

Metabolomics is the latest of the omic based sciences that is beginning to garner academic, governmental, and industrial interest worldwide.[104–110]

Human metabolism is mainly the expression of a steady state in the biosynthesis and degradation of proteins (turnover) that function either as enzymes, receptors, transporters, channels, hormones, and other signaling molecules or that provide structural elements for cells, organs, or the skeleton. Between the proteins there is a variable flow of metabolic intermediates that serve as building blocks for proteins, carbohydrates, lipids, or hetero-oligomers and provide the fuel for ATP synthesis. Whereas in the past the expression of metabolism was principally measured through the concentrations and/or changes of metabolic intermediates, the new molecular tools allow us now to determine every step in the flow of biological information from DNA, to mRNA, to proteins, to metabolites, to function.[111]

Very important levels of biological regulation beyond gene and protein expression control are the protein–protein interactions and alterations of the protein activity by metabolic intermediates in terms of kinetic effects and/or allosteric modulation of function. So, whatever information is gathered at mRNA and protein expression levels will not sufficiently allow predictions to be made on all metabolic consequences. The final stage along the line from gene to mRNA to protein to function is therefore the analysis of the pattern and the concentrations of the metabolites that flow between the proteins, organelles, cells, and organs. Analysis of the entire metabolome as the sum of all detectable low and intermediate molecular weight compounds rather than individual metabolites will be performed by the metabolomics approaches. To fully apply the concept of individual metabolic assessment, two things are necessary: analytical platforms that can assess the broad range of metabolites in an individual, and the knowledge of metabolism that is necessary to interpret the analytical data.

Using lipids as an example, before major predictive value can be ascribed to these individual variations in metabolism, a more detailed knowledge of the implications of variations in lipid metabolism must be developed. It might come as a surprise to most lipid biochemists to learn that the technologies necessary to measure lipid metabolites as a fraction of the overall metabolome are quite mature. Analytics is no longer the limiting step. What limits the implementation of metabolomics for assessing human health and nutrition is an incomplete knowledge of metabolism and the ramifications of altered metabolism. When a more complete human metabolic network is known, including the network of reactions that link tissue metabolism to the plasma metabolome, a complete metabolomic analysis of plasma will reveal the status of many biochemical reactions, the perturbations of metabolism linked to human health and, importantly, the full effects of nutrition on modulating these metabolisms and the risk for disease.[112]

Metabolomics provides an accurate estimate of phenotypic changes to an organism, and is now being applied to understanding carcinogenesis.[113,114] Lipidomics is a component of metabolomics focusing on global changes to lipids.[115,116] Several academic groups are now working on global lipidomics such as the US-based "MAPS Consortium", comprising 30 researchers at 16 universities (http://www.lipidmaps.org); the US-based "Omega-3 Research Institute" (http://www.omega3ri.org); the "Austrian Genomics of Lipid-associated Disorders" consortium (GEN-AU Project GOLD; http://www.bmt.tugraz.at/research/Gen.htm); and the Europe-based "Dietary Lipids as Determinants of Cell Function and Future Health Initiative" consortium (http://eoi.cordis.lu/dsp_details.cfm?ID=28372). Metabolomics, and specifically lipidomics, can be used to discover novel lipid biomarkers for diseases such as cancer. As one example, it was recently demonstrated that specific lipid ratios (increased ceramide/sphingomyelin; decreased lysophosphatidylcholine/phosphatidylcholine) showed strong predictive power for sepsis-related mortality.[117] The cholesterol-lowering, commercially important, statin class of drugs have suffered setbacks owing to serious side effects; for example, Baycol ingestion can cause rhabdomyolysis.[118] Lipid profiling can be used to identify drug toxicity and side effects of statins, by monitoring changes to lipid metabolites peripheral to the cholesterol synthetic pathway.[119] By analogy, lipidomics could be used to develop safer drugs for treating cancer.

The three technologies used currently for metabolomic analyses are nuclear magnetic resonance (NMR) spectroscopy, mass spectrometry, and classical chromatographic techniques. Each of these technologies meets some of the criteria listed earlier, and each has limitations that preclude it from becoming a universal assessment technology.[112]

Novel Whole Body Reporter Type Assays and Systemic Gene Knock Out and Knock in Approaches

Other new tools for studying the effects of lipids on carcinogenesis are rapidly emerging. One approach is to systematically knock out genes one at a time using RNAi (siRNA) technologies to assess how lipids mediate carcinogenesis in the absence of specific genes.[120,121] Genes can then be knocked back in using semiautomated approaches. Semiautomated approaches for assessing how lipids affect genes involved in carcinogensis are also possible using whole body reporter assays in organisms such as zebra fish. Zebra fish can also be fed fluorescent lipids to rapidly examine the kinetics of distribution within the body and potentially the ability of lipophilic drugs to reach a cancer target (reviewed in Chapter 19).

OUTLOOK

Continued use of omic approaches, namely genomics, proteomics, and metabolomics will dramatically increase our understanding of how dietary lipids affect carcinogenesis, leading to healthier foods. Robust transcriptome-wide technology platforms are now in routine use enabling us to at least partly understand how nutrients such as dietary LC-PUFA activate genes through common promoter binding sites and subsequently affect carcinogenesis. This perspective points to a new era for nutrition, where as a field it can take the lead in putting the appropriate technologies together in order to understand truly integrative metabolism and its links to cancer. It points to a field of nutrition that will offer new recommendations on health well-being founded by sound molecular science.

Role of Bioinformatic Tools

Driven by the rapid progress in genomics, the enormous task of cataloging, cross-referencing, and database organization for massive biological datasets is now being tackled by both academia and the private sector. These advances in information management have enabled a conceptual shift in the scientific study, from single object (e.g., a gene, a protein) to system biology studies, aimed at capturing the true complexity of biological systems through global analysis.[122] Though extremely important within this context, lipid metabolic pathway information has been discouragingly underdeveloped relative to genes, proteins, and regulatory elements. What is profoundly needed are innovative user friendly, intuitive, and informative informatic solutions to enable biologists to interpret the "data deluge" generated.

Although the primary causative factors in disease are altered protein activities or altered biochemical composition of cells and tissues, changes at the genetic level might be the ultimate cause for the disease. Thus, the link between the gene regulatory control and the primary causative factors will be crucial for application in drug development, medicine, nutrition, and other therapeutic courses of action. The identification of relationships among genes, transcripts, proteins, and metabolites, are essential components to understand integrative metabolism.[24,90] The annotation of genes with such information has been attempted by a number of public efforts, most notably from KEGG (Kyoto Encyclopedia of Genes and Genomes).[123,124] KEGG is a knowledge base for systematic analysis of gene functions in terms of the networks of genes and molecules having as three major goals (i) the computerizing of the current knowledge of genetics, biochemistry, and molecular and cellular biology in terms of the pathway of interacting molecules or genes, (ii) the collection of gene catalogs for all organisms with completely sequenced genomes and selected organisms with partial genomics (consistent annotation), and (iii) the cataloging of chemical elements, compounds, and other substances in living cells.

The manipulation of this data and visualization within the context of flexible and dynamic pathway information is incomplete. Further work in this area will be of immense benefit to the study of lipids. Software is now available to superimpose gene expression data onto said pathways, providing a powerful means to identify biological regulation of metabolism through the coexpression of gene data obtained from microarrays. One tool that assists in the identification of important biological processes is GenMAPP (gene microarray pathway profiler), a program for viewing and analyzing microarray data on microarray pathway profiles (MAPPs) representing biological pathways or any other functional grouping of genes. GenMAPP currently has over 50 MAPP files depicting various biological pathways and gene families. Several other pathway programs such as KEGG, EcoCyc/MetaCyc, Pathway Processor (which uses KEGG), and ViMAc are available for integration with microarray data analysis, but these programs focus on well-defined metabolic pathways, and like GenMAPP, would benefit from a broader base of pathway information.[125,126] The association of differently regulated genes in a given set of experiments of metabolic pathways is certainly an essential step when trying to elucidate the mechanisms involved. However, even more essential will be the understanding of how these genes are regulated (i.e., the identification of common transcription binding sites).

As mentioned earlier, PPARs represent an important point of metabolic control and are a good example to illustrate the relevance of nutrigenomics to gain insight into food–disease interactions. With nutrigenomic data in hand one can determine if any of the differentially expressed genes have similar promoter elements (i.e., have common PPAR binding sites), which will link their transcriptional regulation in a functional sense.[24,127] Although the data exists to perform such *in silico* investigations of common promoter elements, many links are still missing including, a basic understanding of how complex multi-promoter sites operate, the importance of gene-networks, and how complex interactions change functional roles of genes and protein. Furthermore, the link between basic understanding in molecular biology, to biochemical pathways, and understanding of metabolic flux, is beginning only now.

Role of Nutrigenomics Data Accessibility and Knowledge Exchange

Of great importance is public access of microarray data from studies on cancer. This is important not only for corroboration and additional analyses, but also as a starting point for the developers of bioinformatic procedures and tools that can be tailored to understand the biological complexity of carcinogenesis better. Fortunately, there is a growing trend to publish raw data, which enables future researchers to perform additional analysis against endpoints that were not originally intended by the original researchers. Nutrigenomics will be empowered as more authors make these results available for the community to study lipid metabolism *in silico*. Finally, there should be more effort among lipid and nutrition researchers to share frozen tissue samples, rather than each individual researcher largely repeating similar resource intensive experiments. Since lipids and mRNA are degradable materials, we can also imagine that there could be a set protocol that lipid and nutrition researchers should follow when performing lipid-microarray experiments to assure that the biological samples collected are not degraded. It is due to such degradation issues, that each researcher currently feels compelled to perform their own experiments. If there were agreement amongst lipid researchers as to the levels of fatty acids to test, for example, standardized "low, medium, and high" levels, this would also be of great benefit.

ACKNOWLEDGMENTS

We thank David Mutch for his contribution to the literature search and to Dr. Mark Waldron for his careful review of our manuscript. We would also like to thank Prof. Bruce German for his insight and inspiration in the field of lipid nutrition. This work was supported by Medic Foundation and Nestlé.

REFERENCES

1. Serhan, C.N.; Hong, S.; Gronert, K.; Colgan, S.P.; Devchand, P.R.; Mirick, G.; Moussignac, R.L. Resolvins: a family of bioactive products of omega-3 fatty acid transformation circuits initiated by aspirin treatment that counter proinflammation signals. J. Exp. Med. **2002**, *196*, 1025–1037.

2. Berger, A.; Mutch, D.M.; Bruce German, J.; Roberts, M.A. Dietary effects of arachidonate-rich fungal oil and fish oil on murine hepatic and hippocampal gene expression. Lipids Health Dis. **2002**, *1*, 2.

3. Kelavkar, U.; Glasgow, W.; Eling, T.E. The effect of 15-lipoxygenase-1 expression on cancer cells. Curr. Urol. Rep. **2002**, *3*, 207–214.

4. Ziboh, V.A.; Miller, C.C.; Cho, Y. Significance of lipoxygenase-derived mono-hydroxy fatty acids in cutaneous biology. Prostaglandins Other Lipid Mediat. **2000**, *63*, 3–13.

5. Ziboh, V.A.; Cho, Y.; Mani, I.; Xi, S. Biological significance of essential fatty acids/prostanoids/lipoxygenase-derived monohydroxy fatty acids in the skin. Arch. Pharm. Res. **2002**, *25*, 747–758.

6. Hsi, L.C.; Wilson, L.C.; Eling, T.E. Opposing effects of 15-lipoxygenase-1 and -2 metabolites on MAPK signaling in prostate. Alteration in peroxisome proliferator-activated receptor gamma. J. Biol. Chem. **2002**, *277*, 40549–40556.

7. Shureiqi, I.; Chen, D.; Lee, J.J.; Yang, P.; Newman, R.A.; Brenner, D.E.; Lotan, R.; Fischer, S.M.; Lippman, S.M. 15-LOX-1: a novel molecular target of nonsteroidal anti-inflammatory drug-induced apoptosis in colorectal cancer cells. J. Natl. Cancer Inst. **2000**, *92*, 1136–1142.

8. Okamoto, R.; Arikawa, J.; Ishibashi, M.; Kawashima, M.; Takagi, Y.; Imokawa, G. Sphingosylphosphorylcholine is upregulated in the stratum corneum of patients with atopic dermatitis. J. Lipid Res. **2003**, *44*, 93–102.

9. Ohnishi, Y.; Okino, N.; Ito, M.; Imayama, S. Ceramidase activity in bacterial skin flora as a possible cause of ceramide deficiency in atopic dermatitis. Clin. Diagn. Lab. Immunol. **1999**, *6*, 101–104.

10. Riboni, L.; Campanella, R.; Bassi, R.; Villani, R.; Gaini, S.M.; Martinelli-Boneschi, F.; Viani, P.; Tettamanti, G. Ceramide levels are inversely associated with malignant progression of human glial tumors. Glia **2002**, *39*, 105–113.

11. Bleicher, R.J.; Cabot, M.C. Glucosylceramide synthase and apoptosis. Biochim. Biophys. Acta **2002**, *1585*, 172–178.

12. Liour, S.S.; Yu, R.K. Differential effects of three inhibitors of glycosphingolipid biosynthesis on neuronal differentiation of embryonal carcinoma stem cells. Neuro-chem. Res. **2002**, *27*, 1507–1512.

13. Gascard, P.; Tran, D.; Sauvage, M.; Sulpice, J.C.; Fukami, K.; Takenawa, T.; Claret, M.; Giraud, F. Asymmetric distribution of phosphoinositides and phos-phatidic acid in the human erythrocyte membrane. Biochim. Biophys. Acta **1991**, *1069*, 27–36.

14. Pastorino, F.; Brignole, C.; Marimpietri, D.; Sapra, P.; Moase, E.H.; Allen, T.M.; Ponzoni, M. Doxorubicin-loaded Fab' fragments of anti-disialoganglioside immuno-liposomes selectively inhibit the growth and dissemination of human neuroblastoma in nude mice. Cancer Res. **2003**, *63*, 86–92.

15. Bharti, A.C.; Singh, S.M. Inhibition of macrophage nitric oxide production by gangliosides derived from a spontaneous T cell lymphoma: the involved mech-anisms. Nitric Oxide **2003**, *8*, 75–82.

16. Peguet-Navarro, J.; Sportouch, M.; Popa, I.; Berthier, O.; Schmitt, D.; Portoukalian, J. Gangliosides from human melanoma tumors impair dendritic cell differentiation from monocytes and induce their apoptosis. J. Immunol. **2003**, *170*, 3488–3494.

17. Shen, W.; Ladisch, S. Ganglioside GD1a impedes lipopolysaccharide-induced maturation of human dendritic cells. Cell. Immunol. **2002**, *220*, 125–133.

18. Kudo, D.; Rayman, P.; Horton, C.; Cathcart, M.K.; Bukowski, R.M.; Thornton, M.; Tannenbaum, C.; Finke, J.H. Gangliosides expressed by the renal cell carcinoma cell line SK-RC-45 are involved in tumor-induced apoptosis of T cells. Cancer Res. **2003**, *63*, 1676–1683.

19. Bougnoux, P. Polyunsaturated fatty acids and cancer. Rev. Prat. **2000**, *50*, 1513–1515.

20. Calder, P.C.; Davis, J.; Yaqoob, P.; Pala, H.; Thies, F.; Newsholme, E.A. Dietary fish oil suppresses human colon tumour growth in athymic mice. Clin. Sci. (Lond.) **1998**, *94*, 303–311.

21. Chang, W.L.; Chapkin, R.S.; Lupton, J.R. Fish oil blocks azoxymethane-induced rat colon tumorigenesis by increasing cell differentiation and apoptosis rather than decreasing cell proliferation. J. Nutr. **1998**, *128*, 491–497.

22. Willett, W.C. Diet and breast cancer. J. Intern. Med. **2001**, *249*, 395–411.

23. Narayanan, B.A.; Narayanan, N.K.; Reddy, B.S. Docosahexaenoic acid regulated genes and transcription factors inducing apoptosis in human colon cancer cells. Int. J. Oncol. **2001**, *19*, 1255–1262.

24. Berger, A.; Mutch, D.M.; Bruce German, J.; Roberts, M.A. Dietary effects of arachidonate-rich fungal oil and fish oil on murine hepatic and hippocampal gene expression. Lipids Health Dis. **2002**, *1*, 2.

25. Narayanan, B.A.; Narayanan, N.K.; Simi, B.; Reddy, B.S. Modulation of inducible nitric oxide synthase and related proinflammatory genes by the omega-3 fatty acid docosahexaenoic acid in human colon cancer cells. Cancer Res. **2003**, *63*, 972–979.

26. Anderle, P.; Farmer, P.; Berger, A.; Roberts, M.-A. A nutrigenomic approach to understanding the mechanisms by which dietary long chain fatty acids induce gene signals and control mechanisms involved in carcinogenesis. Nutrition: Int. J. Appl. Basic Nutr. Sci. **2003**, *20*(1), 103–108.

27. Martinez, M.; Vazquez, E.; Garcia-Silva, M.T.; Manzanares, J.; Bertran, J.M.; Castello, F.; Mougan, I. Therapeutic effects of docosahexaenoic acid ethyl ester in patients with generalized peroxisomal disorders. Am. J. Clin. Nutr. **2000**, *71*, 376S–385S.

28. Volker, D.; Fitzgerald, P.; Major, G.; Garg, M. Efficacy of fish oil concentrate in the treatment of rheumatoid arthritis. J. Rheumatol. **2000**, *27*, 2343–2346.

29. Ziboh, V.A.; Miller, C.C.; Cho, Y. Metabolism of polyunsaturated fatty acids by skin epidermal enzymes: generation of antiinflammatory and antiproliferative metabolites. Am. J. Clin. Nutr. **2000**, *71*, 361S–366S.

30. Hu, F.B.; Bronner, L.; Willett, W.C.; Stampfer, M.J.; Rexrode, K.M.; Albert, C.M.; Hunter, D.; Manson, J.E. Fish and omega-3 fatty acid intake and risk of coronary heart disease in women. J. Am. Med. Assoc. **2002**, *287*, 1815–1821.

31. Kim, D.Y.; Chung, K.H.; Lee, J.H. Stimulatory effects of high-fat diets on colon cell proliferation depend on the type of dietary fat and site of the colon. Nutr. Cancer **1998**, *30*, 118–123.

32. Cremonezzi, D.C.; Silva, R.A.; del Pilar Diaz, M.; Valentich, M.A.; Eynard, A.R. Dietary polyunsatured fatty acids (PUFA) differentially modulate melamine-induced preneoplastic urothelial proliferation and apoptosis in mice. Prostaglandins Leukot. Essent. Fatty Acids **2001**, *64*, 151–159.

33. Vane, J.R.; Botting, R.M. A better understanding of anti-inflammatory drugs based on isoforms of cyclooxygenase (COX-1 and COX-2). Adv. Prostaglandin Thromboxane Leukot. Res. **1995**, *23*, 41–48.

34. Eberhart, C.E.; Coffey, R.J.; Radhika, A.; Giardiello, F.M.; Ferrenbach, S.; DuBois, R.N. Up-regulation of cyclooxygenase 2 gene expression in human color-ectal adenomas and adenocarcinomas. Gastroenterology **1994**, *107*, 1183–1188.
35. Chapple, K.S.; Cartwright, E.J.; Hawcroft, G.; Tisbury, A.; Bonifer, C.; Scott, N.; Windsor, A.C.; Guillou, P.J.; Markham, A.F.; Coletta, P.L.; Hull, M.A. Localization of cyclooxygenase-2 in human sporadic colorectal adenomas. Am. J. Pathol. **2000**, *156*, 545–553.
36. Giardiello, F.M.; Hamilton, S.R.; Krush, A.J.; Piantadosi, S.; Hylind, L.M.; Celano, P.; Booker, S.V.; Robinson, C.R.; Offerhaus, G.J. Treatment of colonic and rectal adenomas with sulindac in familial adenomatous polyposis. N. Engl. J. Med. **1993**, *328*, 1313–1316.
37. Williams, C.S.; Mann, M.; DuBois, R.N. The role of cyclooxygenases in inflam-mation, cancer, and development. Oncogene **1999**, *18*, 7908–7916.
38. Watanabe, K.; Kawamori, T.; Nakatsugi, S.; Ohta, T.; Ohuchida, S.; Yamamoto, H.; Maruyama, T.; Kondo, K.; Narumiya, S.; Sugimura, T.; Wakabayashi, K. Inhibitory effect of a prostaglandin E receptor subtype EP(1) selective antagonist, ONO-8713, on development of azoxymethane-induced aberrant crypt foci in mice. Cancer Lett. **2000**, *156*, 57–61.
39. Sonoshita, M.; Takaku, K.; Sasaki, N.; Sugimoto, Y.; Ushikubi, F.; Narumiya, S.; Oshima, M.; Taketo, M.M. Acceleration of intestinal polyposis through prostaglandin receptor EP2 in Apc(Delta 716) knockout mice. Nat. Med. **2001**, *7*, 1048–1051.
40. Hansen-Petrik, M.B.; McEntee, M.F.; Jull, B.; Shi, H.; Zemel, M.B.; Whelan, J. Prostaglandin E(2) protects intestinal tumors from nonsteroidal anti-inflammatory drug-induced regression in Apc(Min/+) mice. Cancer Res. **2002**, *62*, 403–408.
41. Bagga, D.; Wang, L.; Farias-Eisner, R.; Glaspy, J.A.; Reddy, S.T. Differential effects of prostaglandin derived from omega-6 and omega-3 polyunsaturated fatty acids on COX-2 expression and IL-6 secretion. Proc. Natl. Acad. Sci. USA **2003**, *100*, 1751–1756.
42. Pai, R.; Soreghan, B.; Szabo, I.L.; Pavelka, M.; Baatar, D.; Tarnawski, A.S. Prostaglandin E2 transactivates EGF receptor: a novel mechanism for pro-moting colon cancer growth and gastrointestinal hypertrophy. Nat. Med. **2002**, *8*, 289–293.
43. Sun, Y.; Tang, X.M.; Half, E.; Kuo, M.T.; Sinicrope, F.A. Cyclooxygenase-2 overexpression reduces apoptotic susceptibility by inhibiting the cytochrome *c*-dependent apoptotic pathway in human colon cancer cells. Cancer Res. **2002**, *62*, 6323–6328.
44. Tang, X.; Sun, Y.J.; Half, E.; Kuo, M.T.; Sinicrope, F. Cyclooxygenase-2 over-expression inhibits death receptor 5 expression and confers resistance to tumor necrosis factor-related apoptosis-inducing ligand-induced apoptosis in human colon cancer cells. Cancer Res. **2002**, *62*, 4903–4908.
45. Dormond, O.; Foletti, A.; Paroz, C.; Ruegg, C. NSAIDs inhibit alpha V beta 3 integrin-mediated and Cdc42/Rac-dependent endothelial-cell spreading, migration and angiogenesis. Nat. Med. **2001**, *7*, 1041–1047.
46. Farmer, P.; Pugin, J. β-Adrenergic agonists exert their "anti-inflammatory" effects in monocytic cells through the IκB/NF-κB pathway. Am. J. Physiol. Lung Cell. Mol. Physiol. **2000**, *279*, L675–L682.

47. Paruchuri, S.; Hallberg, B.; Juhas, M.; Larsson, C.; Sjolander, A. Leukotriene D$_4$ activates MAPK through a Ras-independent but PKCε-dependent pathway in intestinal epithelial cells. J. Cell Sci. **2002**, *115*, 1883–1893.
48. Wikstrom, K.; Ohd, J.F.; Sjolander, A. Regulation of leukotriene-dependent induction of cyclooxygenase-2 and Bcl-2. Biochem. Biophys. Res. Commun. **2003**, *302*, 330–335.
49. Burstein, S.H.; Huang, S.M.; Petros, T.J.; Rossetti, R.G.; Walker, J.M.; Zurier, R.B. Regulation of anandamide tissue levels by *N*-arachidonylglycine. Biochem. Pharmacol. **2002**, *64*, 1147–1150.
50. Cowart, L.A.; Wei, S.; Hsu, M.H.; Johnson, E.F.; Krishna, M.U.; Falck, J.R.; Capdevila, J.H. The CYP4A isoforms hydroxylate epoxyeicosatrienoic acids to form high affinity peroxisome proliferator-activated receptor ligands. J. Biol. Chem. **2002**, *277*, 35105–35112.
51. Huang, S.M.; Bisogno, T.; Trevisani, M.; Al-Hayani, A.; De Petrocellis, L.; Fezza, F.; Tognetto, M.; Petros, T.J.; Krey, J.F.; Chu, C.J.; Miller, J.D.; Davies, S.N.; Geppetti, P.; Walker, J.M.; Di Marzo, V. An endogenous capsaicin-like substance with high potency at recombinant and native vanilloid VR1 receptors. Proc. Natl. Acad. Sci. USA **2002**, *99*, 8400–8405.
52. Serhan, C.N.; Hong, S.; Gronert, K.; Colgan, S.P.; Devchand, P.R.; Mirick, G.; Moussignac, R.L. Resolvins: a family of bioactive products of omega-3 fatty acid transformation circuits initiated by aspirin treatment that counter proinflammation signals. J. Exp. Med. **2002**, *196*, 1025–1037.
53. Walker, J.M.; Huang, S.M. Endocannabinoids in pain modulation. Prostaglandins Leukot. Essent. Fatty Acids **2002**, *66*, 235–242.
54. Amer, R.K.; Pace-Asciak, C.R.; Mills, L.R. A lipoxygenase product, hepoxilin A(3), enhances nerve growth factor-dependent neurite regeneration post-axotomy in rat superior cervical ganglion neurons *in vitro*. Neuroscience **2003**, *116*, 935–946.
55. Capdevila, J.H.; Nakagawa, K.; Holla, V. The CYP P450 arachidonate monooxygenases: enzymatic relays for the control of kidney function and blood pressure. Adv. Exp. Med. Biol. **2003**, *525*, 39–46.
56. Chu, C.J.; Huang, S.M.; De Petrocellis, L.; Bisogno, T.; Ewing, S.A.; Miller, J.D.; Zipkin, R.E.; Daddario, N.; Appendino, G.; Di Marzo, V.; Walker, J.M. *N*-oleoyldopamine, a novel endogenous capsaicin-like lipid that produces hyperalgesia. J. Biol. Chem. **2003**, *278*, 13633–13639.
57. Hong, M.Y.; Chapkin, R.S.; Barhoumi, R.; Burghardt, R.C.; Turner, N.D.; Henderson, C.E.; Sanders, L.M.; Fan, Y.Y.; Davidson, L.A.; Murphy, M.E.; Spinka, C.M.; Carroll, R.J.; Lupton, J.R. Fish oil increases mitochondrial phospholipid unsaturation, upregulating reactive oxygen species and apoptosis in rat colonocytes. Carcinogenesis **2002**, *23*, 1919–1925.
58. Abeywardena, M.Y.; Head, R.J. Long chain n-3 polyunsaturated fatty acids and blood vessel function. Cardiovasc. Res. **2001**, *52*, 361–371.
59. Nakamura, M.T.; Cho, H.P.; Xu, J.; Tang, Z.; Clarke, S.D. Metabolism and functions of highly unsaturated fatty acids: an update. Lipids **2001**, *36*, 961–964.
60. Jump, D.B. Dietary polyunsaturated fatty acids and regulation of gene transcription. Curr. Opin. Lipidol. **2002**, *13*, 155–164.
61. Kitajka, K.; Puskas, L.G.; Zvara, A.; Hackler, L., Jr.; Barcelo-Coblijn, G.; Yeo, Y.K.; Farkas, T. The role of n-3 polyunsaturated fatty acids in brain: modulation of rat

brain gene expression by dietary n-3 fatty acids. Proc. Natl. Acad. Sci. USA **2002**, *99*, 2619–2624.

62. Joseph, S.B.; Castrillo, A.; Laffitte, B.A.; Mangelsdorf, D.J.; Tontonoz, P. Reciprocal regulation of inflammation and lipid metabolism by liver X receptors. Nat. Med. **2003**, *9*, 213–219.

63. Kliewer, S.A.; Willson, T.M. The nuclear receptor PPARγ- bigger than fat. Curr. Opin. Genet. Dev. **1998**, *8*, 576–581.

64. Sprecher, H. Metabolism of highly unsaturated n-3 and n-6 fatty acids. Biochimiophys. Acta **2000**, *1486*, 219–231.

65. Cherkaoui-Malki, M.; Meyer, K.; Cao, W.Q.; Latruffe, N.; Yeldandi, A.V.; Rao, M.S.; Bradfield, C.A.; Reddy, J.K. Identification of novel peroxisome proliferator-activated receptor alpha (PPARα) target genes in mouse liver using cDNA microarray analysis. Gene Expr. **2001**, *9*, 291–304.

66. Boitier, E.; Gautier, J.C.; Roberts, R. Advances in understanding the regulation of apoptosis and mitosis by peroxisome-proliferator activated receptors in preclinical models: relevance for human health and disease. Comp. Hepatol. **2003**, *2*, 3.

67. Gonzalez, F.J. The peroxisome proliferator-activated receptor alpha (PPARα): role in hepatocarcinogenesis. Mol. Cell. Endocrinol. **2002**, *193*, 71–79.

68. Stoll, B.A. Linkage between retinoid and fatty acid receptors: implications for breast cancer prevention. Eur. J. Cancer Prev. **2002**, *11*, 319–325.

69. Bastie, C. PPARδ and PPARγ: roles in fatty acids signalling, implication in tumorigenesis. Bull. Cancer **2002**, *89*, 23–28.

70. Houseknecht, K.L.; Cole, B.M.; Steele, P.J. Peroxisome proliferator-activated receptor gamma (PPARγ) and its ligands: a review. Domest. Anim. Endocrinol. **2002**, *22*, 1–23.

71. Gupta, R.A.; Brockman, J.A.; Sarraf, P.; Willson, T.M.; DuBois, R.N. Target genes of peroxisome proliferator-activated receptor gamma in colorectal cancer cells. J. Biol. Chem. **2001**, *276*, 29681–29687.

72. Shappell, S.B.; Gupta, R.A.; Manning, S.; Whitehead, R.; Boeglin, W.E.; Schneider, C.; Case, T.; Price, J.; Jack, G.S.; Wheeler, T.M.; Matusik, R.J.; Brash, A.R.; Dubois, R.N. 15S-Hydroxyeicosatetraenoic acid activates peroxisome proliferator-activated receptor gamma and inhibits proliferation in PC3 prostate carcinoma cells. Cancer Res. **2001**, *61*, 497–503.

73. Suchanek, K.M.; May, F.J.; Lee, W.J.; Holman, N.A.; Roberts-Thomson, S.J. Peroxisome proliferator-activated receptor beta expression in human breast epithelial cell lines of tumorigenic and non-tumorigenic origin. Int. J. Biochem. Cell. Biol. **2002**, *34*, 1051–1058.

74. Carlsson, L.; Linden, D.; Jalouli, M.; Oscarsson, J. Effects of fatty acids and growth hormone on liver fatty acid binding protein and PPARα in rat liver. Am. J. Physiol. Endocrinol. Metab. **2001**, *281*, E772–E781.

75. Ahuja, H.S.; Szanto, A.; Nagy, L.; Davies, P.J. The retinoid X receptor and its ligands: versatile regulators of metabolic function, cell differentiation and cell death. J. Biol. Regul. Homeost. Agents **2003**, *17*, 29–45.

76. Fan, Y.Y.; McMurray, D.N.; Ly, L.H.; Chapkin, R.S. Dietary (n-3) polyunsaturated fatty acids remodel mouse T-cell lipid rafts. J. Nutr. **2003**, *133*, 1913–1920.

77. Chen, M.; Pych, E.; Corpron, C.; Harmon, C.M. Regulation of CD36 expression in human melanoma cells. Adv. Exp. Med. Biol. **2002**, *507*, 337–342.

78. Lee, J.Y.; Hwang, D.H. Docosahexaenoic acid suppresses the activity of peroxisome proliferator-activated receptors in a colon tumor cell line. Biochem. Biophys. Res. Commun. **2002**, *298*, 667–674.

79. Yoshizawa, K.; Rimm, E.B.; Morris, J.S.; Spate, V.L.; Hsieh, C.C.; Spiegelman, D.; Stampfer, M.J.; Willett, W.C. Mercury and the risk of coronary heart disease in men. N. Engl. J. Med. **2002**, *347*, 1755–1760.

80. Nakatani, T.; Kim, H.J.; Kaburagi, Y.; Yasuda, K.; Ezaki, O. A low fish oil inhibits SREBP-1 proteolytic cascade, while a high-fish-oil feeding decreases SREBP-1 mRNA in mice liver: relationship to anti-obesity. J. Lipid Res. **2003**, *44*, 369–379.

81. Yang, Y.A.; Morin, P.J.; Han, W.F.; Chen, T.; Bornman, D.M.; Gabrielson, E.W.; Pizer, E.S. Regulation of fatty acid synthase expression in breast cancer by sterol regulatory element binding protein-1c. Exp. Cell Res. **2003**, *282*, 132–137.

82. Verhoeven, G. Androgens and increased lipogenesis in prostate cancer. Cell biologic and clinical perspectives. Verh. K. Acad. Geneeskd. Belg. **2002**, *64*, 189–195 (discussion 195-196).

83. Yang, Y.A.; Han, W.F.; Morin, P.J.; Chrest, F.J.; Pizer, E.S. Activation of fatty acid synthesis during neoplastic transformation: role of mitogen-activated protein kinase and phosphatidylinositol 3-kinase. Exp. Cell Res. **2002**, *279*, 80–90.

84. Navas, M.A.; Vaisse, C.; Boger, S.; Heimesaat, M.; Kollee, L.A.; Stoffel, M. The human HNF-3 genes: cloning, partial sequence and mutation screening in patients with impaired glucose homeostasis. Hum. Hered. **2000**, *50*, 370–381.

85. Cederberg, A.; Gronning, L.M.; Ahren, B.; Tasken, K.; Carlsson, P.; Enerback, S. FOXC2 is a winged helix gene that counteracts obesity, hypertriglyceridemia, and diet-induced insulin resistance. Cell **2001**, *106*, 563–573.

86. Lin, B.; Morris, D.W.; Chou, J.Y. The role of HNF1α, HNF3γ, and cyclic AMP in glucose-6-phosphatase gene activation. Biochemistry **1997**, *36*, 14096–14106.

87. Kaestner, K.H.; Hiemisch, H.; Luckow, B.; Schutz, G. The HNF-3 gene family of transcription factors in mice: gene structure, cDNA sequence, and mRNA distribution. Genomics **1994**, *20*, 377–385.

88. Schwartz, J.H. The many dimensions of cAMP signaling. Proc. Natl. Acad. Sci. USA **2001**, *98*, 13482–13484.

89. Cooper, H.J.; Marshall, A.G. Electrospray ionization Fourier transform mass spectrometric analysis of wine. J. Agric. Food Chem. **2001**, *49*, 5710–5718.

90. Roberts, M.A.; Mutch, D.M.; German, J.B. Genomics: food and nutrition. Curr. Opin. Biotech. **2001**, *12*, 516–522.

91. German, J.B.; Roberts, M.A.; Fay, L.B.; Watkins, S.M. Metabolomics and individual metabolic assessment: the next great challenge for nutrition. J. Nutr. **2002**, *132*, 2486–2487.

92. Puskas, L.G.; Kitajka, K.; Nyakas, C.; Barcelo-Coblijn, G.; Farkas, T. Short-term administration of omega 3 fatty acids from fish oil results in increased transthyretin transcription in old rat hippocampus. Proc. Natl. Acad. Sci. USA **2003**, *100*, 1580–1585.

93. Mutch, D.M.; Berger, A.; Mansourian, R.; Rytz, A.; Roberts, M.A. The limit fold change model: a practical approach for selecting differentially expressed genes from microarray data. BMC Bioinformatics **2002**, *3*, 17.

94. Gene Expression Omnibus. http://www.ncbi.nlm.nih.gov/geo/query/acc.cgi?acc=GSE91. 2003.

95. Hanash, S. Disease proteomics. Nature **2003**, *422*, 226–232.

96. Wulfkuhle, J.D.; Liotta, L.A.; Petricoin, E.F. Proteomic applications for the early detection of cancer. Nat. Rev. Cancer **2003**, *3*, 267–275.
97. Mann, M.; Jensen, O.N. Proteomic analysis of post-translational modifications. Nat. Biotechnol. **2003**, *21*, 255–261.
98. Smith, R.W. Proteomics: the role of mass spectrometry. Pharmaceut. Can. **2002**, *3*, 6–9.
99. Aebersold, R.; Mann, M. Mass spectrometry-based proteomics. Nature **2003**, *422*, 198–207.
100. Langbein, W. Mass spec meets oncology. Genome Tech. **2003**, *02.03*, 42–46.
101. Taylor, C.F.; Paton, N.W.; Garwood, K.L.; Kirby, P.D.; Stead, D.A.; Yin, Z.; Deutsch, E.W.; Selway, L.; Walker, J.; Riba-Garcia, I.; Mohammed, S.; Deery, M.J.; Howard, J.A.; Dunkley, T.; Aebersold, R.; Kell, D.B.; Lilley, K.S.; Roepstorff, P.; Yates, J.R.; Brass, A.; Brown, A.J.; Cash, P.; Gaskell, S.J.; Hubbard, S.J.; Oliver, S.G. A systematic approach to modeling, capturing, and disseminating proteomics experimental data. Nat. Biotechnol. **2003**, *21*, 247–254.
102. Wu, C.C.; Yates, J.R. The application of mass spectrometry to membrane proteomics. Nat. Biotechnol. **2003**, *21*, 262–267.
103. Wulfkuhle, J.D.; Liotta, L.A.; Petricoin, E.F. Proteomic applications for the early detection of cancer. Nat. Rev. Cancer **2003**, *3*, 267–275.
104. Fiehn, O. Metabolomics–the link between genotypes and phenotypes. Plant Mol. Biol. **2002**, *48*, 155–171.
105. Phelps, T.J.; Palumbo, A.V.; Beliaev, A.S. Metabolomics and microarrays for improved understanding of phenotypic characteristics controlled by both genomics and environmental constraints. Curr. Opin. Biotechnol. **2002**, *13*, 20–24.
106. Sumner, S.C.J.; Liu, G. Metabolomics holds key to intelligent discovery efforts. Genetic Engineering News **2002**, *22*, 32–33.
107. Varnau, M.; Singhania, A. Dynamic metabolomics industry emerges. Genetic Engineering News **2002**, *22*, 15-17; 93.
108. Watkins, S.M.; German, J.B. Metabolomics and biochemical profiling in drug discovery and development. Curr. Opin. Mol. Ther. **2002**, *4*, 224–228.
109. Weckwerth, W.; Fiehn, O. Can we discover novel pathways using metabolomic analysis? Curr. Opin. Biotechnol. **2002**, *13*, 156–160.
110. Adams, A. Metabolomics: small-molecule 'omics. Scientist **2003**, *17*, 38.
111. Daniel, H. Genomics and proteomics: importance for the future of nutrition research. Br. J. Nutr. **2002**, *87* (suppl 2), S305–S311.
112. Watkins, S.M.; German, J.B. Toward the implementation of metabolomic assessments of human health and nutrition. Curr. Opin. Biotechnol. **2002**, *13*, 512–516.
113. Goodenowe, D.B. Metabolic Profiling: Why and how the comprehensive analysis of small molecules (metabolites) in biological systems can put the "function" back in functional genomics. Presented at Cambridge Healthtech Institute's 2nd Annual Metabolic Profiling, Pathways in Discovery, December 2–3, Durham, North Carolina, 2002.
114. Ritchie, S.; Heath, D.; Goodenowe, D. An integrated approach to studying colon cancer: correlating comprehensive non-targeted metabolomic and gene expression array data in human cell lines. ICSB 2002 Third International Conference on Systems Biology, December 13–15, 2002; Karolinska Institutet: Stockholm, Sweden, 2002.

115. Fisher-Wilson, J. Long-suffering lipids gain respect: technical advances and enhanced understanding of lipid biology fuel a trend toward lipidomics. Scientist **2003**, *17*, 5.

116. Han, X.; Gross, R.W. Global analyses of cellular lipidomes directly from crude extracts of biological samples by ESI mass spectrometry: a bridge to lipidomics. J. Lipid Res. **2003**, *44*, 1071–1079.

117. Drobnik, W.; Liebisch, G.; Audebert, F.X.; Frohlich, D.; Gluck, T.; Vogel, P.; Rothe, G.; Schmitz, G. Plasma ceramide and lysophosphatidylcholine inversely correlate with mortality in sepsis patients. J. Lipid Res. **2003**, *44*, 754–761.

118. Schmitz, G.; Drobnik, W. Pharmacogenomics and pharmacogenetics of cholesterol-lowering therapy. Clin. Chem. Lab. Med. **2003**, *41*, 581–589.

119. He, F. Measuring metabolic responses of hepatocytes to drug treatment using FTMS. Presented at Cambridge Healthtech Institute's 2nd Annual Metabolic Profiling, Pathways in Discovery, December 2–3, Durham, North Carolina, 2002.

120. Kamath, R.S.; Fraser, A.G.; Dong, Y.; Poulin, G.; Durbin, R.; Gotta, M.; Kanapin, A.; Le Bot, N.; Moreno, S.; Sohrmann, M.; Welchman, D.P.; Zipperlen, P.; Ahringer, J. Systematic functional analysis of the *Caenorhabditis elegans* genome using RNAi. Nature **2003**, *421*, 231–237.

121. Kamath, R.S.; Ahringer, J. Genome-wide RNAi screening in *Caenorhabditis elegans*. Methods **2003**, *30*, 313–321.

122. Anderle, P.; Duval, M.; Kuklin, A.; Dragarichi, S.; Medrano, J.; Littlejohn, T.; Vilanova, D.; Roberts, M.A. Gene expression databases and data mining. Biotechniques **2003**, *34*, S36–S44.

123. Ogata, H.; Goto, S.; Sato, K.; Fujibuchi, W.; Bono, H.; Kanehisa, M. KEGG: Kyoto encyclopedia of genes and genomes. Nucl. Acids Res. **1999**, *27*, 29–34.

124. Kanehisa, M.; Goto, S. KEGG: kyoto encyclopedia of genes and genomes. Nucl. Acids Res. **2000**, *28*, 27–30.

125. Dahlquist, K.D.; Salomonis, N.; Vranizan, K.; Lawlor, S.C.; Conklin, B.R. GenMAPP, a new tool for viewing and analyzing microarray data on biological pathways. Nat. Genet. **2002**, *31*, 19–20.

126. Doniger, S.W.; Salomonis, N.; Dahlquist, K.D.; Vranizan, K.; Lawlor, S.C.; Conklin, B.R. MAPPFinder: using Gene Ontology and GenMAPP to create a global gene-expression profile from microarray data. Genome Biol. **2003**, *4*, R7.

127. Tavazoie, S.; Hughes, J.D.; Campbell, M.J.; Cho, R.J.; Church, G.M. Systematic determination of genetic network architecture. Nat. Genet. **1999**, *22*, 281–285.

14

Omega-3 Polyunsaturated Fatty Acid Against Colon Cancer: DNA Microarray Analysis of Gene–Nutrient Interactions

Bhagavathi A. Narayanan, Narayanan K. Narayanan, and Bandaru S. Reddy

Institute for Cancer Prevention,
Valhalla, NY, USA

INTRODUCTION

Colorectal cancer, which is the fourth most common cancer in the world, is the second most common malignancy in the western world including the USA.[1] Dietary factors are important determinants of colorectal cancer in different populations worldwide. Epidemiological studies indicate a positive association between the diets rich in fish and fish oil that contains *n*-PUFA (polyunsaturated fatty acid) including docosahexaenoic (DHA) and eicosapentaenoic acid (EPA) and colorectal cancer development.[2] Animal model studies have unequivocally provided evidence that the colon-tumor promoting effect of dietary fat depends on its fatty acid composition and that high dietary n-3 PUFAs lack colon-tumor promoting effects when compared with diets high in n-6 PUFAs and saturated fats.[3] Although the mechanisms by which diets high in n-3 PUFAs reduce the risk of colorectal cancer are not fully known, studies conducted thus far indicate that modulation of ras-p21 activity, eicosanoids production via the influence on COX-2 activity, and the induction of apoptosis by the types of dietary fat especially n-3 PUFAs may play a key role in colon carcinogenesis.[4]

This chapter does not aim to provide a comprehensive review on mechanisms of action of n-3 PUFAs against colon carcinogenesis; the purpose of this review, however, is to provide a brief overview of the effect of DHA, an n-3 PUFA, on global changes in the gene expression pattern.

DOCOSAHEXAENOIC ACID

The composition of the human brain is dominated by arachidonic acid (AA) and DHA, and are preferentially taken up by the developing brain from the early phases of its growth with adult levels reaching soon after birth.[5] AA and DHA are essential structural and functional constituents of cell membranes. They are especially required for the growth and function of the brain and vascular systems, which are the primary biofocus of human fetal growth. As lipids of the cell membrane are important determinants of several biochemical and physiological effects, cell membrane abnormalities are predicted due to deficiency in DHA. Low levels of DHA in the brain are associated with the onset of sporadic Alzheimer's disease (AD) and cognitive decline during aging.[6] In addition, molecular dynamics and experimental evidence suggest that DHA is also a ligand for the retinoid X receptor (RXR), in neural tissue, related to the function of retina.[7–10] Overall, several theories exist regarding the biological and molecular dynamics of DHA and its function in normal and diseased conditions. Various mechanisms have been suggested to account for these physiological changes in the brain and retina, as reviewed by Kurlack and Stephenson.[11–13] DHA plays a crucial role in controlling major cellular functions, such as influencing the membrane receptor functions of rhodopsin;[14,15] regulation of membrane bound enzymes such as Na/K-dependent ATPase;[16] dopaminergic and serotoninergic neurotransmission;[17] signal transduction via effects on inositol phosphates, diacylglycerol, and protein kinase C;[18] and regulation of the synthesis of eicosanoids derived from AA.[11]

Epidemiological and preclinical studies demonstrate that consumption of diets high in omega-3 PUFAs such as DHA reduces the risk of human cancer, particularly of the colorectum.[2,19–27] This cancer preventive effect of DHA is attributed to its role in the regulation of membrane bound enzymes; effects on signal transduction pathways via inositol phosphatases, diacylglycerol, and protein kinases; and synthesis of eicosanoids and differentiation.

The most important aspect of our ongoing interest in using DHA as a chemopreventive agent against colon cancer has focused more on the gene–nutrient interactions, where dietary DHA can induce global changes in the gene expression pattern and thus reprogram the genes involved in anti-inflammatory effects, lipid metabolism, cell cycle regulation, and apoptosis (Fig. 14.1).

Genomic Interactions of DHA

Gene expression is a central concept in molecular biology particularly with regard to the transcription of potential gene clones and their response to the

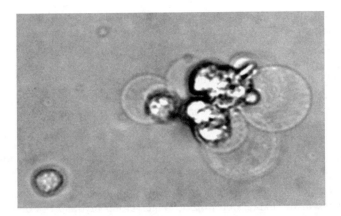

Figure 14.1 CaCo-2 cells treated with DHA—apoptotic cells.

influence of external factors, such as environment, drugs, and nutrients. Control of gene expression, frequently exquisite in terms of cell specificity and timing, forms most part of the explanation for biological process.[28] The importance of gene expression analysis with respect to progression of cancer research is to describe the coordination in time and space of numerous cellular activities related to carcinogenesis such as cell proliferation, migration, inflammation, metabolism, differentiation, and apoptosis. Understanding the role of relevant genes at the transcription level in these processes thus necessitates the use of methods to determine patterns of transcription and functional relevance of hundreds of genes simultaneously. Application of DNA microarray technology to understand genomewide expression profiling in normal and cancer cells or tissues has more potential value in early detection and prevention of several types of cancer. The gradual accumulation of aberrant gene function also can explain the progression of pathologic states seen in the esophagus from early dysplasia through mild to severe dysplasia and finally to cancer.[29] Therefore, high-throughput analysis of gene expression via SAGE and microarray technologies, such as cDNA and oligonucleotide arrays, promises to revolutionize cancer research and to further our understanding of several biological processes related to cancer. Our long-term goals, in evaluating the utility of DHA, against colon cancer lead us to examine the effects of this fatty acid DHA on different groups of genes and transcription factors that are presumed to be altered in human colon cancer at the transcription level.

Global Gene Expression Profile in DHA-Treated Human Colon Cancer Cells

Using human DNA oligonucleotide arrays with 3800 gene spots, we hybridized 20 μg of total RNA from human colon cancer cells treated with a very low dose of DHA (5 μg/mL). Profiling of global gene expression in response to DHA after 48 h was found to inhibit several clusters of genes involved in the human colon carcinogenesis. Reversal of some key metabolic effects due to changes in expression level of a few proinflammatory genes at the transcription level added interesting findings to unmask the functional mechanism of DHA.[4,30] The expression profiles of genes indicated a reprograming pattern of previously known and unknown genes at the transcription level that provided clues to the possible functional mechanism of DHA. Of 3800, about 504 (average) genes were expressed after 48 h of DHA treatment.

Overall, altered expression of the transcription factors includes downregulation of nine members of the RNA II polymerases, transcription corepressor-associated proteins, and enhancer binding proteins such as AP2, in addition to changes in the expression of zinc finger group of transcription factors. With respect to triggering apoptosis mechanisms, activation of cytochrome c, which in turn triggers caspases, was associated with the elevated expression of proapoptotic caspases 10, 13, 8, 5, and 9 in DHA-treated cells. Similarly, modulation of cell cycle regulatory machinery by DHA via activation of cyclin-dependent kinase inhibitors, such as p21, p27, p57, and p19, and growth arrest specific proteins by more than twofold is consistent with the induction of apoptosis and inactivation of antiapototic Bcl2 family of genes. Inactivation of prostaglandin (PG) family of genes, lipoxygenases, and altered expression of peroxisome proliferation activated receptors (PPARα and -γ) by DHA indicates a lipid peroxidation-induced apoptosis in addition to the effect reflected on the modification of cell cycle regulatory genes. These findings support the concept that a genomewide expression profiling of human colon cancer precursor genes and transcription factors provides a set of novel regulatory mechanism(s) to determine the chemopreventive benefits of n-3 PUFAs including DHA, and thus to prevent the inflammation and neoplasia.

Measurements of the intensities of the expressed genes are represented in Fig. 14.2, as simple bivariate scatter plots comparing the profiles of DHA-treated CaCo-2 colon cancer cells with the vehicle-treated control CaCo-2 cells. After 48 h of treatment with DHA, changes in the gene expression pattern was observed as indicated by the shifts of the data points in the scatter plot.

As shown in the pie chart, the overall expression profile indicated no change in the transcription level of almost 87% of the genes, with upregulation in 10% and downregulation in 3% of the totally expressed genes (Fig. 14.3) Biochemical functions of these genes in the expression profiles are diverse and include oncogenes, transcription factors, cell cycle regulators, proteases, pro- and antiapoptotic genes, target genes of cyclooxygenases, membrane bound enzymes, effects on signal transduction pathways via inositol phosphatases, diacylglycerol, and protein kinases, synthesis of eicosanoids, differentiation of cytokines and growth factors, G-protein receptors, map kinases, and other metabolic genes as listed in Table 14.1

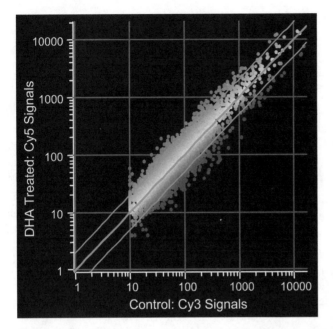

Figure 14.2 Global genomic changes induced by DHA scatter graph. Scatter plot view of gene expression. Expression intensity Cy5/Cy3 ratios of untreated vs. DHA-treated CaCo-2 cells. The ratios (Cy5/Cy3) of genes that have a twofold or higher expression are considered as induced and those with below 0.5-fold or less are considered as repressed. Approximately 504 out of 3800 genes (13%) were expressed in DHA-treated cells (Narayanan et al.[4]).

To confirm the changes in the level of expression observed in the micro-array, we measured the expression level of transcripts of several representative genes by quantitative RT–PCR (Fig. 14.4) using sequence-specific primers. In repeated experiments, we found that more than 70% (7 out of 10) of the expressed genes from microarray analysis could be confirmed by RT–PCR.

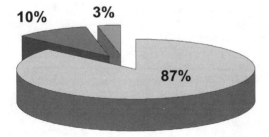

Figure 14.3 The pie chart shows the overall expression profile. Almost 87% of the genes indicated no change in the transcription level, 10% of the genes were upregulated, and 3% of the genes were downregulated, out of the 504 genes totally expressed.

Table 14.1   Differentially Expressed Groups of Genes Altered by DHA (Narayanan et al.[4,30])

GenBank ID	Coordinate	Gene name	Mean CY5/CY3 ratio
Lipoxygenase			
NM_000697	A2d5	Arachidonate 12-lipoxygenase	1.55 ± 0.49
NM_001139	A2d6	Arachidonate 12-lipoxygenase, 12R type	0.43 ± 0.33
NM_001140	A2d7	Arachidonate 15-lipoxygenase	1.65 ± 0.59
NM_001141	A2d8	Arachidonate 15-lipoxygenase, second type	0.33 ± 0.32
NM_000698	A2d9	Arachidonate 5-lipoxygenase	0.48 ± 0.55
NM_001629	A2e1	Arachidonate 5-lipoxygenase-activating protein	0.46 ± 0.09
RNA polymerase			
NM_000937	H1i3	Polymerase (RNA) II polypeptide A	0.47 ± 0.29
NM_000938	H1i4	Polymerase (RNA) II polypeptide B	0.56 ± 0.24
NM_002694	H3g1	Polymerase (RNA) II polypeptide C	0.81 ± 0.47
NM_004805	I1h3	Polymerase (RNA) II polypeptide D	0.58 ± 0.23
NM_002695	H3g2	Polymerase (RNA) II polypeptide E	0.75 ± 0.25
NM_002696	H3g3	Polymerase (RNA) II polypeptide G	0.55 ± 0.36
NM_006233	I3c1	Polymerase (RNA) II polypeptide I	0.51 ± 0.14
NM_006234	I3c2	Polymerase (RNA) II polypeptide J	0.50 ± 0.34
NM_005034	I2b4	Polymerase (RNA) II polypeptide K	0.50 ± 0.31
Transcription activators and repressors			
NM_005171	D1a6	Activating transcription factor 1	0.52 ± 0.13
NM_004083	D1e3	DNA-damage-inducible transcript 3	2.30 ± 0.35
NM_001938	E4a2	TBP binding (negative cofactor 2)	2.13 ± 0.38
NM_002586	H2i3	Pre-B-cell leukemia transcription factor 2	0.95 ± 0.17
NM_001030	K3g1	Ribosomal protein S27 (metallopanstimulin 1)	0.41 ± 0.12
NM_003442	L1c2	Zinc finger protein 143 (clone *pHZ-1*)	0.39 ± 0.09
NM_003443	L1c3	Zinc finger protein 151 (*pHZ-67*)	0.25 ± 0.18
NM_003450	L1c7	Zinc finger protein 174	0.45 ± 0.27
Cell cycle-regulating kinases and proteins			
NM_005255	D4h3	Cyclin G-associated kinase	0.83 ± 0.14
NM_001258	C2d8	Cyclin-dependent kinase 3	0.55 ± 0.11
NM_001274	C2f1	CHK1 (checkpoint, *Schizosaccharomyces pombe*) homolog	2.26 ± 0.78
NM_006716	B1a8	Activator of S phase kinase	1.27 ± 0.40

(continued)

Table 14.1 Continued

GenBank ID	Coordinate	Gene name	Mean CY5/CY3 ratio
cdk inhibitors			
NM_000076	B3b2	Cyclin-dependent kinase inhibitor 1C (*p57, Kip2*)	2.47 ± 0.52
NM_001262	C2e1	Cyclin-dependent kinase inhibitor 2C (*p18*, inhibits CDK4)	2.81 ± 0.15
NM_004064	D2b9	Cyclin-dependent kinase inhibitor 1B (*p27, Kip1*)	2.51 ± 0.25
NM_001800	D1i7	Cyclin-dependent kinase inhibitor 2D (*p19*, inhibits *CDK4*)	2.45 ± 0.21
Oncogenes and tumor suppressors			
NM_005343	G2f5	v-Ha-ras Harvey rat sarcoma viral oncogene homolog	1.94 ± 0.69
NM_004985	G2e1	v-Ki-ras2 Kirsten rat sarcoma 2 viral oncogene homolog	0.54 ± 0.13
NM_004585	I1g7	Retinoic acid receptor responder (tazarotene induced) 3	2.11 ± 0.32
Bcl2 family proteins			
NM_006538	B2d1	BCL2-like 11 (apoptosis facilitator)	0.62 ± 0.11
NM_001196	B2g1	BH3 interacting domain death agonist	1.13 ± 0.23
NM_001188	C2a4	BCL2-antagonist/killer 1	0.97 ± 0.18
Simple lipid metabolism			
NM_000134	B3f6	Fatty acid binding protein 2, intestinal	2.44 ± 0.40
NM_001443	C3h1	Fatty acid binding protein 1, liver	2.44 ± 0.09
NM_001442	C3g9	Fatty acid binding protein 4, adipocyte	2.17 ± 0.19
NM_001444	C3h2	Fatty acid binding protein 5 (psoriasis-associated)	2.70 ± 0.35
NM_000300	H1c2	Phospholipase A2, group IIA (platelets, synovial fluid)	0.47 ± 0.33
NM_005336	E4f4	High density lipoprotein binding protein (vigilin)	0.48 ± 0.13
NM_004457	D3f4	Fatty-acid-Coenzyme A ligase, long-chain 3	1.18 ± 0.20
NM_004458	D3f9	Fatty-acid-Coenzyme A ligase, long-chain 4	0.53 ± 0.28
NM_000780	C1c7	Cytochrome P450, subfamily VIIA, polypeptide 1	1.06 ± 0.19

(*continued*)

Table 14.1 Continued

GenBank ID	Coordinate	Gene name	Mean CY5/CY3 ratio
NM_001608	A1b2	Acyl-Coenzyme A dehydrogenase, long chain	0.35 ± 0.06
Caspases			
NM_001230	C2c4	Caspase 10, apoptosis-related cysteine protease	3.18 ± 0.86
NM_003723	L3d4	Caspase 13, apoptosis-related cysteine protease	6.54 ± 0.18
NM_001228	C2c2	Caspase 8, apoptosis-related cysteine protease	2.13 ± 0.82
NM_004347	D1a9	Caspase 5, apoptosis-related cysteine protease	3.26 ± 0.65
NM_001229	C2c3	Caspase 9, apoptosis-related cysteine protease	3.11 ± 0.23
Phospholipases and phosphoinositol kinases			
NM_002663	H3e4	Phospholipase D2	0.65 ± 0.07
NM_005028	I2a8	Phosphatidylinositol-4-phosphate 5-kinase, type II, alpha	2.78 ± 0.35
NM_006219	I3b4	Phosphoinositide-3-kinase, catalytic, beta polypeptide	1.60 ± 0.39
NM_002649	H3d5	Phosphoinositide-3-kinase, catalytic, gamma polypeptide	1.14 ± 0.33
NM_002647	H3d3	Phosphoinositide-3-kinase, class 3	1.80 ± 0.80
NM_003706	L3b9	Phospholipase A2, group IVC	1.04 ± 0.23
Cytoskeleton/motility proteins			
X00351	E3b7	Actin, beta	1.04 ± 0.17
NM_001614	F1c2	Actin, gamma 1	0.96 ± 0.39
NM_001615	F1c3	Actin, gamma 2, smooth muscle, enteric	3.07 ± 0.72
NM_002373	F4i2	Microtubule-associated protein 1A	0.66 ± 0.11
NM_005909	G3f7	Microtubule-associated protein 1B	0.41 ± 0.21
NM_002374	F4i3	Microtubule-associated protein 2	1.10 ± 0.34
NM_002375	F4i4	Microtubule-associated protein 4	0.60 ± 0.22
Hormone receptors			
NM_000955	H2a7	Prostaglandin E receptor 1 (subtype EP1), 42 kDa	0.44 ± 0.11
NM_000956	H2a8	Prostaglandin E receptor 2 (subtype EP2), 53 kDa	0.49 ± 0.16
NM_000959	H2a9	Prostaglandin F receptor (FP)	0.48 ± 0.09

(*continued*)

Table 14.1 Continued

GenBank ID	Coordinate	Gene name	Mean CY5/CY3 ratio
NM_000960	H2b1	Prostaglandin I2 (prostacyclin) receptor (IP)	0.28 ± 0.13
NM_006917	K2e4	Retinoid X receptor, gamma	0.41 ± 0.09
NM_000964	H2b3	Retinoic acid receptor, alpha	0.44 ± 0.15
Nuclear receptors			
NM_005036	I2b6	Peroxisome proliferative activated receptor, alpha	3.71 ± 0.57
NM_005037	I2b7	Peroxisome proliferative activated receptor, gamma	1.85 ± 0.28
NM_000901	F2h4	Nuclear receptor subfamily 3, group C, member 2	0.38 ± 0.19
NM_002135	F3c5	Nuclear receptor subfamily 4, group A, member 1	0.44 ± 0.30
Growth factor and chemokine receptors			
NM_003327	K4c4	Tumor necrosis factor receptor superfamily, member 4	2.12 ± 0.51
NM_003820	L4b5	Tumor necrosis factor receptor superfamily, member 14	0.92 ± 0.16
NM_004195	L4f9	Tumor necrosis factor receptor superfamily, member 18	2.83 ± 1.10

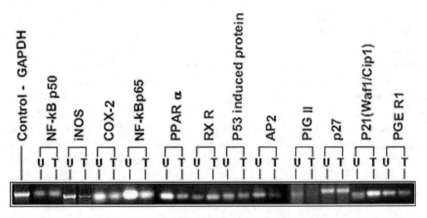

Figure 14.4 RT–PCR analysis. Differential expression of potential molecular targets modulated by DHA in CaCo-2 cells. Agarose gel (2%) showing the RT–PCR products for gene transcripts with differential expression (Narayanan et al.[4,30]).

Effects of DHA on Cell Cycle Regulatory Genes

Altered expression of transcription factors includes downregulation of several RNA polymerases, transcription corepressor-associated proteins, and enhancer binding proteins such as AP-2. These changes are consistent with shutdown of transcription in response to DHA treatment. Similarly, changes in the expression of three of the zinc finger group of transcription factors were also noted. Increased expression of cytochrome c, which triggers apoptosis by activating caspases and downregulating Bcl2, an antiapoptotic factor, was consistent with the cytotoxic effect of DHA in maintaining colonic cell homeostasis. Activation of cytochrome c and several of the cell cycle-related genes such as cyclin-dependent kinase inhibitors p21 and p27, growth arrest specific protein, S100 calcium binding protein, and caspase 10 were associated with the inactivation of Bcl2 family of genes suggesting the effect of DHA on colonic cell cycle regulation and its possible role in apoptosis and cell cycle regulation.

DHA Activates Retinoic Acid Receptor (RXR)

The underlying mechanisms by which n-3 PUFAs exert a chemopreventive effect involve modulation of nuclear receptors. RXRs are a family of nuclear receptors implicated in cancer chemoprevention. As DHA enriched in fish oil reduces colonocyte proliferation and enhances apoptosis relative to n-6 PUFA-treated cells, we examined whether DHA can serve as a specific ligand for RXRα activation and other genes related to its function in differentiation. First of all, using microarray analysis we demonstrated the activation of RXR and the cdk inhibitor p21$^{(wafl/cipl)}$ by DHA, both of which are known to play a key role in the differentiation associated with inhibition of cell proliferation, thus supporting its role as a potential chemopreventive agent. Results from functional analysis of gene expression data from microarray consistently indicated activation of several genes involved in cellular differentiation (Fig. 14.5) by DHA. PPARα is an orphan receptor that is the member of the nuclear hormone receptors, which include retinoic acid receptors. Using microarray analysis for the first time, we could report on the coregulatory pattern of PPARα and -γ and retinoic acid receptor (RXRα) in DHA-treated cells. Activation PPARα and -γ by more than twofold corresponds with a change in the expression of retinoic acid receptor (RXRα), a ligand activated transcription factor, cellular retinoic acid binding protein 1, and insulin-like growth factor binding proteins (Fig. 14.5). PPAR activation leads to induction of genes encoding proteins involved in lipid transport and oxidation processes as could be reviewed from the list of overall expressed genes (Table 14.1)

DHA Effect on Proinflammatory Genes

DHA Effect on COX-2 Target Genes

In-depth analysis of the microarray data indicated changes in the expression of COX-2 and the gene transcripts that are perhaps controlled by COX-2, such as

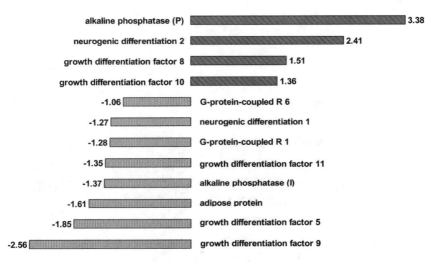

Figure 14.5 Effect of DHA on levels of expression of genes related to differentiation. Functional analysis of genomics from microarray data demonstrated activation of several genes involved in the cellular differentiation (Narayanan et al.[4]).

inducible nitric oxide synthase (iNOS). Major changes are found to be associated with the inactivation of a cluster of PG family of genes and receptors, although very little effect was observed in COX-2 expression (Fig. 14.6). These results suggest that a direct effect of DHA on a cascade of biochemical events resulted in the inactivation of PGs and their receptors. However, constitutive overexpression of COX-2 and its impact on the cluster of PGs in colon carcinogenesis need further investigation. Impact of DHA on the inactivation of a family of PGs and activation of a downstream transcriptional mediator PPARα indicates the role of DHA in lipid peroxidation-induced apoptosis by PUFA. Normally, COX-2 overexpression is tightly regulated in colonic epithelial cells. Microarray analysis indicated changes in the level of COX-2 transcripts in parallel with the inactivation of at least five members of PGs family at the steady state level of mRNA. An increase in the level of fatty acid binding proteins T2 and T6 in conjunction with the downregulation of lipoxygenases, PGs, and phospholipase A2 was observed. Changes in the expression of intestinal fatty acid binding proteins due to DHA treatment in CaCo-2 cells may reveal yet another functional significance related to cellular lipid transport enhancement. Further studies on the status of the fatty acid binding proteins and mRNA may provide insight into the mechanism by which diet influences colonic physiology.

DHA Effect on Lipoxygenases

Microarray data analysis further indicated changes in the expression and inactivation of arachidonate 5-lipoxygenase T1, arachidonate 12-lipoxygenase,

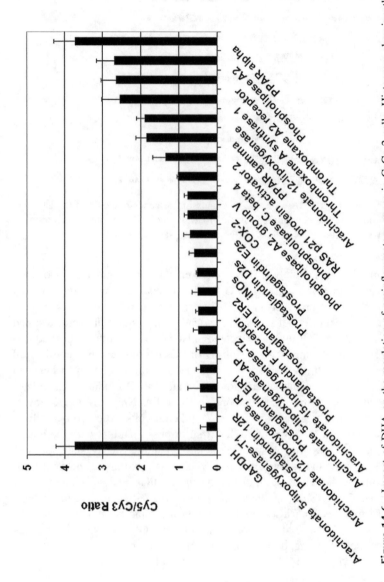

Figure 14.6 Impact of DHA on the regulation of proinflammatory genes in CaCo-2 cells. Histogram showing the relative intensity of expressed genes (the data plotted here is an average of three independent experiments) on the basis of the Cy5/Cy3 ratios of DHA (Narayanan et al.[4]).

arachidonate 12-lipoxygenase R, arachidonate 15-lipoxygenase T2, and arachidonate 5-lipoxygenase AP. The functional significance of DHA in regulating endogenous tumor promoters such as COX-2 derived PGs and lipoxygenases in parallel with the regulation of other proinflammatory gene transcripts reflects the possible recovery of "genetic instability" in colon cancer due to DHA.

Downregulation of iNOS Expression

Immunofluorescence detection based on nuclear positive staining for iNOS expression demonstrated that the number of cells that were positive for iNOS were lower in CaCo-2 cells treated with DHA for 48 h than in the controls (Fig. 14.5). Importantly, the lower number of iNOS-positive cells observed in this study is consistent with an increase in the number of apoptotic cells as determined by DAPI staining. To relate the level of iNOS expression to the number of apoptotic cells, iNOS-positive cells and apoptotic cells were quantified. As shown in Fig. 14.7, an inverse relation between iNOS expression and rate of apoptosis is clearly evident. Notably, the results from Western blot analysis indicate a dose-dependent inhibition of iNOS expression by DHA.

Effect of DHA on Interferons and NF-κB

A major inhibitory effect of DHA was observed on expression of interferons alpha 5, 10, 21, 16, 14, 13, and gamma and beta isoforms. However, 25% of the interferons did not show any remarkable change (either activation or inhibition) as summarized in Table 14.1. RT–PCR analyses indicated a twofold inactivation of NF-κB p65, although not much change could be observed in the transcripts of NF-κB p50 in DHA-treated CaCo-2 cells. Repeated

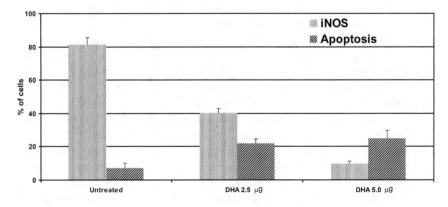

Figure 14.7 Effect of DHA on iNOS expression. Densitometric analysis of iNOS protein bands detected by Western blot analysis in CaCo-2 cells.

experiments using RT–PCR analyses revealed inactivation of iNOS and NF-κB p65. It is noteworthy that major changes associated with CaCo-2 cell growth inhibition, inactivation of iNOS, and induction of apoptosis by DHA are consistent with the downregulation of COX-2, NF-κB p65, the family of interferons, and several lypoxygenases as shown in Table 14.1. Further studies are in progress to demonstrate the biological significance of differential expression patterns of iNOS, NF-κB, and TNF isoforms by DHA.

NF-κB Regulation in DHA-Treated Cell

Results from microarray and RT–PCR analysis indicated inactivation of NF-κB p65 by twofold, although not much change could be observed in the transcripts of NF-κB p50 in DHA-treated CaCo-2 cells. Further analysis on the expression of lipoxygenases indicated downregulation of several isoforms by more than twofold. In addition to the alterations reflected in the transcripts of lipoxygenases, impact of DHA on the level of expression of thromboxane A synthase 1 and thromboxane A receptor 2 is not significant. Similarly, a differential expression of phospholiapse A2 and phospholipase C β4 reflects a coregulatory pattern along with the other PGs. As shown in Fig. 14.6, another set of genes found to be inactivated by more than twofold iNOs are NF-κB p65. Repeated experiments using microarray and RT–PCR analysis revealed the inactivation of iNOS and NF-κB p65. However, NF-κB p50 did not show any change in the level of expression.

DISCUSSION

DNA microarray technology was used to determine the mechanism by which DHA inhibits colon cancer cell growth, an important step in rational use of omega-3 fatty acids as chemopreventive agents. Biochemical functions of the genes in the expression profiles are diverse and include oncogenes, transcription factors, cell cycle regulators, proteases, pro- and antiapoptotic genes, target genes of cyclooxygenases, cytokines and growth factors, G-protein receptors, map kinases, and other metabolic genes. This study altered about 1% of all the genes investigated, a percentage almost identical to earlier reports: 1%[31] and 1.2%[32] using cDNA rays and SAGE analysis, respectively. The accuracy of the results is determined by a higher percentage (7 out of 10) of correct positives when compared with other analysis such as SAGE.[33] Both DHA-treated and -untreated CaCo-2 cells have normal expression of several categories of gene clusters, although alterations in the number of tumor-associated genes and transcription factors outweigh in the DHA-treated cells than that in the untreated cells. Furthermore, alterations in gene expression due to DHA treatment in colon cancer cells was observed to be in the multiple signaling pathways, and is best demonstrated in the present work by the comparison of alterations in cell cycle regulatory genes, and COX-2 target genes, as well as an impact on proinflammatory genes involved in the downstream lipid signaling pathways such as ERK-5/p44 and p38 signaling. Expression of 5% of transcription factors required for transcription of several structural genes were inactivated; however, a complete inactivation could not have taken place, as this would be lethal.[31] Because genetic or epigenetic lesions during colon carcinogenesis leads to over-expression, suppression, or gene mutation, altered gene expression due to chemo-preventive agents such as DHA may result from modifications or regulation of nonmutated genes. It is noteworthy that DHA inhibits CaCo-2 cell proliferation and induces apoptosis, which involves the synergy in the regulation of cell cycle regulatory genes and COX-2 target genes as shown in the results from microarray analysis.

The inhibitory effects of DHA on cell cycle progression and induction of apoptosis were directly paralleled by an increase in the activation of several pro-apoptotic caspases and genes, such as p21, waf1/cip1, and p27, in addition to the changes observed in the expression several other cdk inhibitors as listed in the results. Increased expression of cytochrome c, which triggers apoptosis by acti-vating caspases and downregulation of Bcl2, an antiapoptotic factor, is consistent with the cytotoxic effect of DHA as observed in the present study in maintaining colonic cell homeostasis.

DHA has been shown to inhibit the growth of metastatic melanoma cells and the growth inhibition was correlated with a quantitative increase in the hyper-phosphorylated pRb, p27, and another cdk inhibitor in selected melanoma cells.[34] Decreased expression of Bcl2 level by DHA might increase the sensi-tivity of cells to lipid perxoidation and then to programed cell death as observed in HT29 colon cells.[35] Our data further indicate the effect of DHA, a potential modulator of lipid signaling molecules,[36] on the genetic abnormalities in the transcriptional pathways that induces the activation of cyclooxygenases-2, PGs, and PPAR gene family. Impact of DHA on the inactivation of a family of PGs and activation of a downstream transcriptional mediator PPARα (Fig. 14.4) indicates the role of DHA in lipid peroxidation-induced apoptosis. PPARα activation leads to the induction of genes encoding proteins involved in lipid transport and oxidation including peroxisomal acyl-CoA oxidase.[36,37] In the current study, PPARα showed more than twofold increase in the expression, suggesting the possible role of DHA in lipid signaling pathways in human colon cancer. Metabolites of n-3 fatty acids may have one or two magni-tude greater affinity for PPARα and are consequently far more potent transcrip-tional activators of PPARα-dependent genes.[38] Activation of growth arrest specific protein (fourfold increase) thus suggests the potential role of DHA in activating proapoptotic genes, possibly the consequence of several cascade of molecular events induced by DHA promotes cell growth inhibition.

Differential expression of NF-κB isoforms, for instance, activation of NF-κB p65 in CaCo-2 cells by DHA without affecting NF-κB p50 in agreement with earlier studies with n-3 alpha-linolenic and EPAs, indicates accumulation of PUFA from n-6 and n-3 series elicited an intracellular oxidative stress, resulting in the activation of oxidative stress-responsive transcription factors such as AP1 and NF-κB.[39] The biological significance of DHA-induced modulation of oxidative stress genes in colon cancer needs further investigation. DHA has been shown to reduce the ras p21 localization to the membrane without affecting posttranslational lipidation and lowers GTP binding and downstream p42/44 (ERK)-dependent signaling.[40] Extracellular signal-related kinase (ERK-5) is found to be inactivated in DHA-treated cells, indicating an impact on the downstream lipid signaling pathways. Results from our present global gene expression analysis indicated an orchestered regulation in the level of expression (activation or inactivation) of a family of developmentally regulated G-protein receptors, GTP binding proteins associated with a downregulatory pattern of Ras p21.

CONCLUSION

In summary, here we report the generation of a vast array of DHA-responsive signaling genes and molecules representing the possibilities of more than one signaling pathway involved in colon cancer growth inhibition. The main

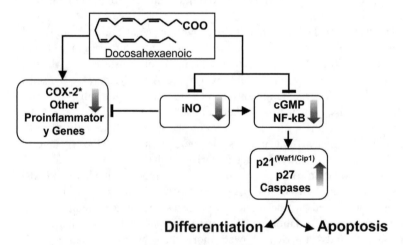

Figure 14.8   Schematic diagram of potential molecular mechanisms of DHA: the illustration presented here depicts the key molecular and cellular events mediated by DHA in inhibiting COX-2 and iNOS target genes. Altered expression of these genes at the mRNA and protein levels in CaCo-2 cells after 48 h of DHA treatment were evident from the DNA microarray, RT–PCR analysis, and Western blot analysis. A simultaneous reprograming of genes at the transcription level involved in differentiation such as p21 (waf1/Cip1), and p27 and apoptosis by activating caspases listed in Table 14.1 is evident from the present study. Because iNOS inhibition and p21 expression can be both p53-dependent and -independent pathways, there may be multiple pathways for DHA for its chemopreventive action. The cascade of molecular events regulated by DHA shows a unique relationship between proinflammatory genes including COX-2, iNOS, and differentiation initiating factors that result in the maintenance of colonic tissue homeostasis (Narayanan et al.[4]).

advantage of such a comprehensive approach on genomewide expression profiling in response to DHA on human colon cancer precursor genes and transcription factors is that it provides a set of novel regulatory mechanism(s) to determine the chemopreventive benefits of DHA, and thus to prevent colon cancer (Fig. 14.8).

ACKNOWLEDGMENTS

This study was supported in part by USPHS grants CA-37663 and CA-17613 from National Cancer Institute. We thank Ilse Hoffmann for editing the chapter.

REFERENCES

1. Greenlee, R.T.; Murray, T.; Bolden, S.; Wingo, P.A. Cancer statistics 2000. CA Cancer J. Clin. **2000**, *50*, 7–33.
2. Caygill, C.P.J.; Charland, S.L.; Lippin, J.A. Fat, fish, fish oil, and cancer. Br. J. Cancer **1996**, *74*, 159–164.
3. World Cancer Research Fund and American Institute for Cancer research. Panel on food, nutrition and the prevention of cancer. In *Food, Nutrition and the Prevention of Cancer: A Global Perspective*; American Institute for Cancer Research: Washington, DC, 1997; 216–251.
4. Narayanan, B.A.; Narayanan, N.K.; Simi, B.; Reddy, B.S. Modulation of inducible nitric oxide synthase and related pro-inflammatory genes by the omega-3 fatty acid docosahexaenoic acid in human colon cancer cells. Cancer Res. **2003**, *63*, 972–979.
5. Sinclair, A.J.; Crawford, M.A. The accumulation of arachidonate and docosahexaenoate in the developing rat brain. J. Neurochem. **1972**, *19*, 1753–1758.
6. Farooqui, A.A.; Yi Ong, W.; Lu, X.R.; Halliwell, B.; Horrocks, L.A. Neurochemical consequences of kainate-induced toxicity in brain: involvement of arachidonic acid release and prevention of toxicity by phospholipase A(2) inhibitors. Brain Res. Brain Res. Rev. **2001**, *38*, 61–78.
7. de Urquiza, A.M.; Liu, S.; Sjoberg, M.; Zetterstrom, R.H.; Griffiths, W.; Sjovall, J.; Perlmann, T. Docosahexaenoic acid, a ligand for the retinoid X receptor in mouse brain. Science **2000**, *290*, 2140–2144.
8. Xi, Z.P.; Wang, J.Y. Effect of dietary n-3 fatty acids on the composition of long- and very-long-chain polyenoic fatty acid in rat retina. J. Nutr. Sci. Vitaminol. (Tokyo) **2003**, *49*, 210–213.
9. Nishizawa, C.; Wang, J.Y.; Sekine, S.; Saito, M. Effect of dietary DHA on DHA levels in retinal rod outer segments in young versus mature rats. Int. J. Vitam. Nutr. Res. **2003**, *73*, 259–265.
10. Moriguchi, K.; Yuri, T.; Yoshizawa, K.; Kiuchi, K.; Takada, H.; Inoue, Y.; Hada, T.; Matsumura, M.; Tsubura, A. Dietary docosahexaenoic acid protects against *N*-methyl-*N*-nitrosourea-induced retinal degeneration in rats. Exp. Eye Res. **2003**, *77*, 167–173.
11. Kurlak, L.O.; Stephenson, T.J. Plausible explanations for effects of long chain polyunsaturated fatty acids (LCPUFA) on neonates. Arch. Dis. Child Fetal Neonatal. Ed. **1999**, *80*, F148–54.
12. Lauritzen, L.; Hansen, H.S.; Jorgensen, M.H.; Michaelsen, K.F. The essentiality of long chain n-3 fatty acids in relation to development and function of the brain and retina. Prog. Lipid Res. **2001**, *40*, 1–94.

13. Salem, N., Jr.; Litman, B.; Kim, H.Y.; Gawrisch, K. Mechanisms of action of docosahexaenoic acid in the nervous system. Lipids **2001**, *36*, 945–959.
14. Litman, B.J.; Niu, S.L.; Polozova, A.; Mitchell, D.C. The role of docosahexaenoic acid containing phospholipids in modulating G protein-coupled signaling pathways: visual transduction. J. Mol. Neurosci. **2001**, *16*, 237–242.
15. Feller, S.E.; Gawrisch, K.; MacKerell, A.D., Jr. Polyunsaturated fatty acids in lipid bilayers: intrinsic and environmental contributions to their unique physical properties. J. Am. Chem. Soc. **2002**, *124*, 318–326.
16. Bourre, J.M. Nature, origin and role of fatty acids of the nervous system: an essential fatty acid, an alpha-linolenic acid, changing the structure and the cerebral function. Bull. Acad. Natl. Med. **1989**, *173*, 1137–1148.
17. Zimmer, L.; Delion-Vancassel, S.; Durand, G.; Guilloteau, D.; Bodard, S.; Besnard, J.C.; Chalon, S. Modification of dopamine neurotransmission in the nucleus accumbens of rats deficient in n-3 polyunsaturated fatty acids. J. Lipid Res. **2000**, *41*, 32–40.
18. Vaidyanathan, V.V.; Rao, K.V.; Sastry, P.S. Regulation of diacylglycerol kinase in rat brain membranes by docosahexaenoic acid. Neurosci. Lett. **1994**, *179*, 171–174.
19. Anti, M.F.; Armelao, G.; Marra, A.; Percesepe, G.M.; Bartoli, P.; Palozz, P.; Parrella, C.; Canetta, N.; Gentiloni, N.; De Vitis, I.; Gasbarrini, G. Effects of different doses of fish oil on rectal cell proliferation in patients with sporadic colonic adenomas. Gastroenterology **1994**, *107*, 1709–1718.
20. Chang, W.C.; Chapkin, R.S.; Lupton, J.R. Predictive value of proliferation, differentiation and apoptosis as intermediate markers for colon tumorigenesis. Carcinogenesis **1997**, *4*, 721–730.
21. Deschner, E.E.; Lytle, J.; Wong, G.; Ruperto, J.; Newmark, H.L. The effect of dietary omega-3 fatty acids (fish oil) on azoxymethanol-induced focal areas of dysplasia and colon tumor incidence. Cancer **1990**, *66*, 2350–2356.
22. Lee, D.Y.; Lupton, J.R.; Aukema, H.M.; Chapkin, R.S. Dietary fat and fiber alter rat colonic mucosal lipid mediators and cell proliferation. J. Nutr. **1993**, *123*, 1808–1817.
23. Minoura, T.; Takata, T.; Sakaguchi, M.; Takada, H.; Yamamura, M.; Yamamoto, M. Effect of dietary eicosapentaenoic acids on azoxymethaneinduced colon carcinogenesis in rat. Cancer Res. **1988**, *48*, 4790–4794.
24. Park, B.H.; Vogelstein, B.; Kinzler, K.W. Genetic disruption of PPARdelta decreases the tumorigenicity of human colon cancer cells. Proc. Natl. Acad. Sci. USA **2001**, *98*, 2598–2603.
25. Reddy, B.S. Dietary fat, calories, and fiber in colon cancer. Prev. Med. **1993**, *22*, 738–749.
26. Reddy, B.S.; Rao, C.V. Novel approaches for colon cancer prevention by cyclooxygenase-2 inhibitors. J. Environ. Pathol. Toxicol. Oncol. **2002**, *21*, 155–164.
27. Wilier, W.C.; Stampfer, M.J.; Colditz, G.A.; Rosner, B.A.; Speizer, F.E. Relations of meat, fat, and fiber intake to the risk of colon cancer in the prospective study among women. New Engl. J. Med. **1990**, *323*, 1664–1672.
28. Smith, L.; Greenfield, A. DNA microarrays and development. Hum. Mol. Genet. **2003**, *12* (suppl 1), R1–8.
29. Uys, P.; van Helden, P.D. On the nature of genetic changes required for the development of esophageal cancer. Mol. Carcinog. **2003**, *36*, 82–89.
30. Narayanan, B.A.; Narayanan, N.K.; Reddy, B.S. Decosahexaenoic acid (DHA) regulated genes and transcription factors inducing apoptosis in human colon cancer cells. Int. J. Oncol. **2001**, *19*, 1255–1262.

31. Backert, S.; Gelos, M.; Kobalz, U.; Hanski, M.L.; Bohm, C.; Mann, B.; Lovin, N.; Gratchev, A.; Mansmann, U.; Moyer, M.P.; Riecken, E.O.; Hanski, C. Differential gene expression in colon carcinoma cells and tissues detected with a cDNA array. Int. J. Cancer **1999**, *82*, 868–874.

32. Zhang, L.; Zhou, W.; Velculescu, V.E.; Kern, S.; Hruban, R.H.; Hamilton, S.R.; Vogelstein, B.; Kinzler, K.W. Gene expression profiles in normal and cancer cells. Science **1997**, *276*, 1268–1272.

33. Gelos, M.; Backert, S.; Klussmann, E.; Gratchev, A.; Kobalz, U.; Mann, B.; Moyer, M.P.; Boehm, C.; Riecken, E.O.; Hanski, C. Detection of differential gene expression in colorectal cancer cell lines: comparison of three methods. J. Mol. Med. **1998**, *76*, B13.

34. Albino, A.P.; Juan, G.; Traganos, F.; Reinhart, L.; Connolly, J.; Rose, D.P.; Darzynkiewicz, Z. Cell cycle arrest and apoptosis of melanoma cells by docosa-hexaenoic acid: association with decreased pRb phosphorylation. Cancer Res. **2000**, *60*, 4139–4145.

35. Chen, Z.Y.; Istfan, N.W. Docosahexaenoic acid is a potent inducer of apoptosis in HT-29 colon cancer cells. Prostaglandins Leukot. Essent. Fatty Acids **2000**, *63*, 301–308.

36. Clarke, S.D. Polyunsaturated fatty acid regulation of gene transcription: a mechanism to improve energy balance and insulin resistance. Br. J. Nutr. **2000**, *83*, S59–S66.

37. Aoyama, T.; Peters, J.M.; Iritani, N.; Nakajima, T.; Furihata, K.; Hashimoto, T.; Gonzales, F.J. Altered constitutive expression of fatty acid metabolizing enzymes in mice lacking the peroxisome proliferator-activated receptor α (PPARα). J. Biol. Chem. **1998**, *273*, 5678–5684.

38. Krey, G.; Braissant, O.; L'Horset, F.; Kalkoven, E.; Perround, M.; Parker, M.G.; Wahli, W. Fatty acids, eicosanoids and hypolipidemic agents identified as ligands of peroxisome proliferator-activated receptors by coactivator-dependent receptor ligand assay. Mol. Endocrinol. **1997**, *11*, 779–791.

39. Maziere, C.; Conte, M.A.; Degonville, J.; Ali, D.; Maziere, J.C. Cellular enrichment with polyunsaturated fatty acids induces an oxidative stress and activates the tran-scription factors AP1 and NF-kappa B. Biochem. Biophys. Res. Commun. **1999**, *265*, 116–122.

40. Collett, E.D.; Davidson, L.A.; Fan, Y.Y.; Lupton, J.R.; Chapkin, R.S. n-6 and n-3 polyunsaturated fatty acids differentially modulate oncogenic Ras activation in colonocytes. Am. J. Physiol. Cell Physiol. **2001**, *280*, C1066–75.

15

Ellagic Acid Modulates Lipid Metabolism in Prostate Cancer: Gene–Nutrient Interactions

Bhagavathi A. Narayanan, Narayanan K. Narayanan, Otto Geoffroy, and Daniel W. Nixon

Institute for Cancer Prevention, Valhalla, NY, USA

INTRODUCTION

Several epidemiological studies support the hypothesis that a diet high in fat and low in carbohydrates, fruits, vegetables, and fiber increases the risk of a number of cancers, including cancer of the colon.[1] In contrast, regular intake of a diet rich in plant foods provides phytochemicals that possess health-protective effects. Polyphenols constitute the most frequently occurring plant metabolites and are an integral part of many types of diets. Interest in phenolics has increased greatly, owing to their antioxidant, antiproliferative, and anti-inflammatory properties that can in part mediate a whole array of beneficial effects to human health, such as prevention and treatment of cancer, reducing the risk for cardiovascular diseases, and correcting immunological disorders.[2–4] Polyphenols occur primarily in conjugated form with one or more sugar residues linked to their hydroxyl groups.[5] The hydrolyzable tannins or ellagitannins are a distinct subset of tannins with a polyphenolic structure primarily composed of esters of glucose with gallic acid and ellagic acid (EA). In the following sections we will discuss the biological and pharmacological importance of EA (Fig. 15.1), particularly with a focus on its role in lipid metabolism at the genomic level.

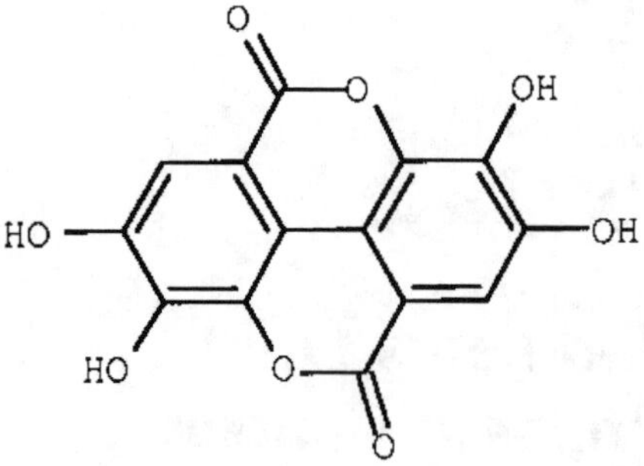

Figure 15.1 Ellagic acid structure.

PHARMACOKINETIC EVALUATION OF EA

EA, a hydrolyzable ellagitannin, is present in a variety of fruits, berries, and nuts, but the highest concentration has been found in raspberries, strawberries, walnuts, and peanuts. EA has exhibited multiple biological activities against cancer and has been demonstrated in a variety of tumor cells and animal models.[6-9]

Absorption of EA by the human gut and pharmacokinetic properties of EA from a dietary source were evaluated on the basis of the absorption in 12 healthy volunteers aged 20-60, who were inpatients for 24 h at the Medical University of South Carolina (MUSC), General Clinical Research Center (P.I.—Dr. Nixon) after fasting for 8 h. Upon collection of baseline serum and urine samples, each volunteer consumed 150 g of raspberry puree, which contained 38 mg of EA. Total urine and serial blood samples were then collected over 24 h. The pharmacokinetic results for 12 volunteers are listed in Table 15.1. Roughly one-quarter (25.4%) of the EA present in the raspberry puree was absorbed and excreted unchanged in the urine over the 24 h study period.

It is apparent that higher levels of EA seen in some 24 h urinary output samples must be because of tannins that are being converted to EA in the stomach or small intestine. Although the oral bioavailability of EA from a single dose of raspberries was low, the elimination half-life was relatively long (8.57 h). This suggests that EA is not rapidly excreted or modified/metabolized once it is in

Table 15.1 Pharmacokinetic Data for EA in Raspberries

	Age	AUC_{24} ($\mu g\, h/L/m^2$)	C_{max} ($\mu g/L$)	Cl_R ($L/h/m^2$)	Ue_{24} (%)	Ue_{24}/BSA (%/m²)
Mean	32.9	1031	258.9	3.018	25.4	14.2
SD	11.1	452	104	1.88	22.1	11.7

Note: AUC_{24}, area under the serum concentration vs. time plot over 24 h divided by BSA; C_{max}, the maximal serum concentration of EA reached; Cl_R, renal clearance; Ue_{24}, 24 h urinary excretion of EA as a percentage of dose; BSA, body surface area.

the blood stream. Food sources of EA may therefore be suitable for human cancer clinical prevention trials.

ANTIPROLIFERATIVE EFFECT OF DIETARY EA ON HUMAN COLON EPITHELIAL CELLS

Another study of volunteers with and without a history of colon cancer or polyps was undertaken with MUSC IRB approval, in which the subjects ingested a weighed amount (150 g) of raspberry puree daily; this puree contained 38–40 mg of EA. The volunteers had undergone sequential colon biopsies that were evaluated by KI-67 staining and TUNEL assays to assess cell proliferation and cell death (apoptosis). Results from biopsies ($n = 26$) from seven subjects to date show a significant effect of EA intake on the rate of inhibition of cell proliferation and cell death. After the raspberry puree intake, biopsies from healthy volunteers did not show a significant change in the total number of KI-positive staining in the crypts, when compared with those taken from volunteers with polyps and with cancer after 12 months (Fig. 15.2), suggesting an antiproliferative effect against colon cancer, seemingly induced by the intake of raspberry puree. The results from the TUNEL assays for cell death revealed increasingly higher number of apoptotic cells per crypt in biopsies taken from polyp-bearing patients and cancer patients. After regular raspberry puree intake, subjects with colon cancer showed a decline in the cell proliferation rate within 12 months, time period certainly indicative of a protective effect. However, information pertaining to EA mechanism of action against cancer at the molecular level is still in its infant stage. Studies conducted in our laboratory provided extensive information

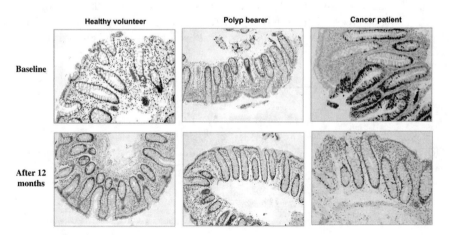

Figure 15.2 Antiproliferative effect attributed to the key component EA present in raspberry puree. Note the change in the total number of KI-positive stained crypts in biopsies taken from volunteers with polyps and with cancer.

on the potential role of EA in regulating cell cycle, apoptosis, and lipid metabolism at the genomic and protein levels.[7–9]

EFFECTS OF EA ON INSULIN GROWTH RECEPTOR AND PEROXISOME PROLIFERATION-ACTIVATED RECEPTOR PPARγ

Peroxisome proliferation-activated receptor (PPARγ) is a nuclear receptor that is activated by PUFAs, eicosanoids, and a few phytochemicals. Recently, we have demonstrated that EA downregulates insulin-like growth factor (IGF)II and activates PPARγ expression in colon and prostate cancer cells.[7,8] PPARγ regulates adipocyte differentiation and is highly expressed in the normal human colonic mucosa, so that the existence of a link with the inverse relation between the levels of IGF and PPARγ may account for the activation of PPAR by specific anticancer agents. Acromegalic patients have an increased prevalence of colonic polyps because of the elevated serum-IGFI level. However, the mechanism underlying this process is poorly understood. There is evidence to indicate synergistic interaction between the nuclear receptors and the nutritional agents. PPAR activation may increase in response to imbalance in the IGFI or IGFII concentrations.[10,11] Microarray results and studies of EA and DHA with Western blot analysis indicated that reduced expression of IFGII mRNA in CaCo-2 cells is associated with an activation of PPAR.[7,8] The EA-mediated inhibitory effect on IGF signaling via activation of PPARγ is evident in our studies with prostate and colon cancer cells. The differential expressions of TGF-β and IGF-like growth factors, receptors, and binding proteins, in conjunction with the coregulation of several related growth factors, suggest their positive role in inducing cell cycle arrest and apoptosis in a p53-independent manner without showing any dramatic effect on the level of p53.[9] Plant phenol-mediated cell cycle arrest in concert with the induction of p21 and downregulation of IGFII in EA-treated cells (Fig. 15.3)[7] is likely connected to the gene expression pattern of growth factors and their orchestrated

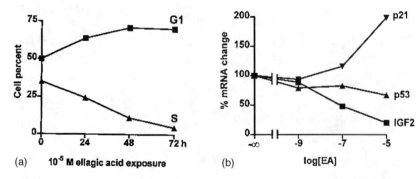

Figure 15.3 Plant phenol-mediated cell cycle arrest in concert with the induction of p21 and downregulation of IGFII in EA-treated cells. (a) Cell cycle arrest and (b) downregulation of IGFII (Narayanan et al.[8]).

activation and downregulation that lead to growth arrest in human prostate cancer cells. Among the 12 hormonal receptors, expressions of androgen receptor-associated protein and TR3 orphan receptors were downregulated significantly by EA.

GENE–NUTRIENT INTERACTIONS

Genomic interactions of functional food ingredients against colon cancer were one of the first research projects in functional genomics. The main focus of our interest in understanding gene–nutrient interactions paved the way to analyze transcriptomics involving measurements of differential expression of genes at the mRNA level. Using high-density cDNA microarrays, we identified genes involved in prostate and colon carcinogenesis. Quantitative RT–PCR was performed to verify and validate a minimum of 10% of the selected and highly expressed genes. In-depth analysis of the data provided insight into alterations on cell cycle regulation, apoptosis, and regulation of lipid metabolism.

EA-Modulated Differential Expression of Genes in Prostate Cancer

Although deeper biological insight is likely to develop from microarray analysis using multiple time points with different cell lines treated independently or in combination, gene expression analysis for two samples: (a) RNA from EA-treated vs. (b) RNA from EA untreated using cDNA microarrays containing 2400 human genes demonstrated alterations in the expression of several clones of genes within 48 h of induction. Genes altered at least by twofold increase with EA were found to be 2.99% of the total expressed genes. Repressed genes amounted to 11.29%, suggesting a significant number of genes are repressed by EA (Fig. 15.4).

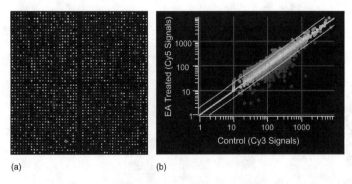

(a)   (b)

Figure 15.4 Scanned image of hybridized MICROMAX human cDNA microarray containing 2400 genes (a) Total RNA from LNCaP cells treated with EA for 48 h was used for microarray analysis. (b) Scatter plot view showing the distribution of differential gene expression pattern.

UNT   TPA   TPA+EA   EA (5uM)   EA (10uM)

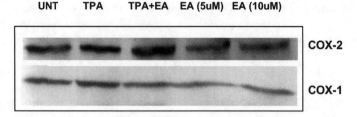

Figure 15.5   Effect of EA on TPA-induced COX-2.

EA-Modulated Gene Clusters Involved in Lipid Metabolism

One of our research interests was in understanding the anti-inflammatory function of EA. With that goal in mind, we analyzed the inhibitory effect of EA on COX-2 and PGE2 levels and the results are presented in Figs. 15.5 and 15.6.

As the expression of PGH synthases are the site of action of nonsteroidal anti-inflammatory drugs such as aspirin and related phytochemicals such as EA, we are focusing on the important regulators of inflammation and mito-genesis. Towards this end, we are examining many of the inflammation-response genes at the transcription level. Transcriptional regulation of the proinflammatory genes coupled to regulators of lipid metabolism such as phospholipases and lipoproteins, responsible for providing fatty acid substrate, were identified and their expression levels are presented in Fig. 15.7 and Table 15.2.

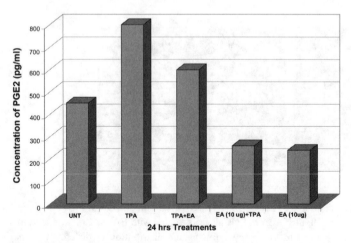

Figure 15.6   Effect of EA on COX-2 and PGE2 levels.

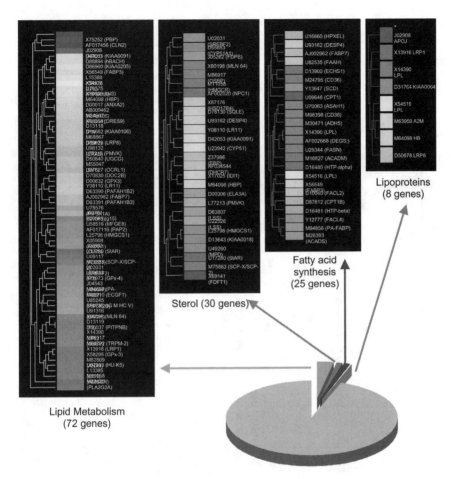

Figure 15.7 Hierarchal clustering analysis. This analysis was carried out using all 2400 genes to classify the gene expression profile in EA-treated LNCaP cells. This hierarchal clustering was carried out using 36 genes to classify them according to similarities and closeness of the association among the genes involved with lipid metabolism (see Table 15.2 for gene description and details).

We identified six major clusters of highly responsive genes that are differentially expressed. Interestingly, a few clusters of genes that come under the category of oncogenes and protein kinases are coregulated with a twofold increase by EA. In contrast, a higher percentage of a similar set of genes from these two categories was downregulated simultaneously by EA, suggesting a selective inducing and repressing effect in these two clusters of genes in the same cell type. Notably, a large number of glycoproteins and transglutaminases showed alterations effected by EA.

Table 15.2 Effects of EA on Gene Clusters Associated with Lipid Metabolism

GenBank ID	Expression ratio	Gene	Gene description	Function
Fatty acid synthesis and lipid metabolism				
M22430	2.52	PLA2G2A	RASF-A PLA2	Lipid catabolism
D16480	2.04	HTP-alpha	Mitochondrial enoyl-CoA hydratase	Fatty acid metabolism
U67963	1.95	HU-K5	Lysophospholipase	Lipid metabolism
L13385	1.93	LIS1	Miller–Dieker lissencephaly protein	Lipid metabolism
X58295	1.89	GPx-3	GPx-3 mRNA for plasma glutathione peroxidase	Response to lipid hydroperoxide
M16827	1.83	ACADM	Medium-chain acyl-CoA dehydrogenase	Fatty acid beta-oxidation
D30037	1.74	PITPNB	Phosphatidylinositol transfer protein (PI-TPbeta)	Transport
D13119	1.74	P2	P2-ATP synthase	Proton transport
U29344	1.72	FASN	Breast carcinoma fatty acid synthase	Fatty acid metabolism
AF002668	1.71	DEGS; MLD	Putative fatty acid desaturase MLD	Fatty acid desaturation
U70063	1.65	ASAH1	Acid ceramidase	Ceramide metabolism
M98398	1.63	CD36	Antigen CD36	Transport
M64722	1.61	TRPM-2	TRPM-2	Complement activation
M30471	1.59	ADH5	Class III alcohol dehydrogenase	Ethanol oxidation
U09813	1.47	P3	Mitochondrial ATP synthase	Proton transport
X71973	1.45	GPx-4	GPx-4 phospholipid hydroperoxide glutathione peroxidase	Phospholipid metabolism
M94856	1.39	PA-FABP	Fatty acid binding protein	Epidermal differentiation
M26393	1.39	ACADS	Short-chain acyl-CoA dehydrogenase	Fatty acid beta-oxidation
Y12777	1.37	FACL4	Acyl-CoA synthetase-like protein	Fatty acid metabolism
M31210	1.35	ECGF1	Endothelial differentiation protein (edg-1) gene	G-protein-coupled receptor protein signaling pathway
U85245	1.3	PIP5K2B	Phosphatidylinositol-4-phosphate 5-kinase type II beta	Cell surface receptor linked signal transduction

1.25	U91316	BACH: ACT	Acyl-CoA thioester hydrolase	Lipid metabolism
1.21	D16481	HTP-beta	Mitochondrial 3-ketoacyl-CoA thiolase beta	Fatty acid beta-oxidation
1.15	U96132	ERAB	Amyloid beta-peptide binding protein	Lipid metabolism
1.14	U57627	OCRL1	Fetal brain oculocerebrorenal syndrome	Lipid metabolism
1.13	D87812	CPT1B	Carnitine palmitoyltransferase I	Fatty acid beta-oxidation
1.13	D50840	UGCG	Ceramide glucosyltransferase	Glucosylceramide biosynthesis
1.12	M55047	SYT	Synaptotagmin	Transport
1.09	M68867	RBP2	Cellular retinoic acid binding protein II (*CRABP*)	Epidermal differentiation
1.08	X98654	DRES9	DRES9 protein	Transport
1.08	D14662	KIAA0106	KIAA0106 gene	Response to oxidative stress
1.08	D13118	P1	P1 ATP synthase	Proton transport
1.06	D10040	FACL2	mRNA for long-chain acyl-CoA synthetase	Fatty acid metabolism
1.05	M34667	PLCG1	Phospholipase C-gamma	Intracellular signaling cascade
1.03	D00017	ANXA2	mRNA for lipocortin II	Skeletal development
1.03	AB009462	hLRp105	LDL receptor related protein 105	Receptor-mediated endocytosis
1.02	M64098	HBP	High-density lipoprotein binding protein	Lipid transport
1.02	X70508	INS	Insulinoma preproinsulin	Glucose metabolism; lipid metabolism
1.02	U78575	PIP5K1A	Phosphatidylinositol-4-phosphate 5-kinase alpha	Glycerophospholipid metabolism
0.95	L15388	GRK5	G-protein-coupled receptor kinase (*GRK5*)	Regulation of G-protein-coupled receptor protein signaling pathway
0.94	D86960	KIAA0205	KIAA0205 gene	Phospholipid biosynthesis
0.94	X56549	FABP3	Muscle fatty-acid binding protein	Transport; lipid binding
0.93	D88894	hBACH	Brain acyl-CoA hydrolase	Lipid metabolism
0.89	D63390	PAFAH1B2	Acetylhydrolase IB beta	Lipid catabolism
0.86	D00632	GPX3	Glutathione peroxidase	Response to lipid hydroperoxide
0.86	D70830	DOC2B	mRNA for Doc2 beta	Transport
0.85	U82535	FAAH	Fatty acid amide hydrolase	Fatty acid metabolism

(*continued*)

Table 15.2 Continued

GenBank ID	Expression ratio	Gene	Gene description	Function
AJ002962	0.83	FABP7	hB-FABP	Fatty acid metabolism
D63391	0.83	PAFAH1B3	Platelet-activating factor acetylhydrolase IB gamma	Neurogenesis; lipid catabolism
U16660	0.79	HPXEL	Peroxisomal enoyl-CoA hydratase-like protein	Fatty acid beta-oxidation
J02761	0.78	SFTP3	Pulmonary surfactant-associated protein B	Sphingolipid metabolism
U78576	0.74	PIP5K1A	Phosphatidylinositol-4-phosphate 5-kinase alpha	Glycerophospholipid metabolism
Y13647	0.69	SCD	Stearoyl-CoA desaturase	Fatty acid biosynthesis
X05908	0.69	ANXA1	Lipocortin	Lipid metabolism
J02883	0.69	CLPS	Colipase	Lipid catabolism
AF017116	0.68	PAP2	Type-2 phosphatidic acid phosphohydrolase	Lipid metabolism
U58516	0.67	MFGE8	Breast epithelial antigen BA46	Cell adhesion; oncogenesis
Y09565	0.65	g15	1-Acylglycerol-3-phosphate O-acyltransferase	Phosphatidic acid biosynthesis
M24795	0.6	CD36	CD36 antigen	Fatty acid metabolism
D13900	0.58	ECHS1	Mitochondrial short-chain enoyl-CoA hydratase	Fatty acid beta-oxidation
U09117	0.56	PLCD1	Phospholipase c delta 1	Phospholipid metabolism
U09648	0.48	CPT1	Carnitine palmitoyltransferase II precursor	Fatty acid beta-oxidation
AF017456	0.33	CLN2	Lysosomal pepstatin insensitive protease	Proteolysis and peptidolysis
U46689	0.27	ALDH10	Microsomal aldehyde dehydrogenase (*ALD10*)	Epidermal differentiation
U05598	0.12	AKR1C2	Dihydrodiol dehydrogenase	Canalicular bile acid transport; lipid metabolism; digestion
Sterol				
AF002020	2.05	NPC1	Niemann-Pick C disease protein	Cholesterol transport
M11058	1.86	HMGCR	3-Hydroxy-3-methylglutaryl coenzyme A reductase	Cholesterol biosynthesis
X80198	1.73	MLN 64	MLN64	Cholesterol metabolism
M86917	1.66	OSBP	Oxysterol binding protein	Steroid metabolism

D55653	1.53	CYP51A1	Lanosterol 14-demethylase	Cholesterol biosynthesis
U02031	1.48	SREBF2	Sterol regulatory element binding protein-2	Cholesterol metabolism
J05262	1.39	FDPS	Farnesyl pyrophosphate synthetase	Cholesterol biosynthesis
D00306	1.25	ELA3A	Pancreatic protease E precursor	Cholesterol metabolism
L77213	1.17	PMVK	Phosphomevalonate kinase	Cholesterol biosynthesis
Z37986	1.1	EBP	Phenylalkylamine binding protein	Cholesterol biosynthesis
AF034544	1.08	DHCR7	Delta7-sterol reductase	Cholesterol biosynthesis
X17025	1.03	IDI1	Homolog of yeast IPP isomerase	Cholesterol biosynthesis
M64098	1.02	HBP	High-density lipoprotein binding protein	Lipid transport
U23942	0.95	CYP51	Lanosterol 14-demethylase cytochrome P450	Cholesterol biosynthesis
D42053	0.93	KIAA0091	KIAA0091 gene	Cholesterol metabolism
U93162	0.89	DESP4	C4-sterol methyl oxidase	Sterol biosynthesis
D78130	0.89	SQLE	Squalene epoxidase	Sterol biosynthesis
Y08110	0.88	LR11	Mosaic protein LR11	Lipid transport
X87176	0.85	HSD17B4	17-Beta-hydroxysteroid dehydrogenase	Steroid biosynthesis
U22526	0.77	LSS	2,3-Oxidosqualene-lanosterol cyclase	Steroid biosynthesis
D63807	0.74	LSS	Lanosterol synthase	Steroid biosynthesis
D13643	0.68	KIAA0018	KIAA0018 gene	Cholesterol biosynthesis
L25798	0.68	HMGCS1	3-Hydroxy-3-methylglutaryl coenzyme A synthase	Cholesterol biosynthesis
U49260	0.67	MPD	Mevalonate pyrophosphate decarboxylase	Cholesterol biosynthesis
M75883	0.6	SCP-X/SCP-2	Sterol carrier protein X/sterol carrier protein 2	Steroid biosynthesis
U17280	0.55	StAR	Steroidogenic acute regulatory protein	C21-steroid hormone biosynthesis
X69141	0.41	FDFT1	Squalene synthase	Cholesterol biosynthesis
D85181		SC5DL	Sterol-C5-desaturase	Sterol biosynthesis

(*continued*)

Table 15.2 Continued

GenBank ID	Expression ratio	Gene	Gene description	Function
Lipoprotein				
X13916	1.9	LRP1	LDL-receptor related protein	Lipid metabolism
X14390	1.7	LPL	Lipoprotein lipase	Fatty acid metabolism
D31764	1.45	KIAA0064	KIAA0064 gene	Low-density lipoprotein catabolism
D50678	1.1	LRP8	Apolipoprotein E receptor 2	Lipid metabolism
M63959	1.01	A2M	Alpha-2-macroglobulin receptor-associated protein	Vesicle-mediated transport
X54516	0.98	LPL	Lipoprotein lipase	Fatty acid metabolism
J02908	0.43	APOJ	Apolipoprotein J	Lipid metabolism

Effects on Proinflammatory Genes

Phenolic antioxidants have both anticancer and anti-inflammatory effects. However, their mechanisms of action remain unclear. Generally, in inflammation, cyclooxygenases (COX-1 and COX-2), prostaglandins, NF-κB, and tumor necrosis factor-alpha (TNF-α) play a pivotal role. Although nuclear transcription factor NF-κB controls the expression of many genes, including the TNF-α gene, most of the anti-inflammatory agents inhibit the COX enzymes first. Without exception, EA inhibited TPA-induced COX-2 protein expression and altered the PGE2 level in prostate cancer cells within 24 h at a very low dose of 10 μg. However, the mechanisms involved in COX-2 inhibition by EA either via NF-κB or TNF-α-mediated reprograming of genes remain to be investigated. Among the transcription factors, a higher percentage of clones that are upregulated by EA are involved in cell cycle regulation and apoptosis, including p53 target genes and cdk inhibitors such as p21. Among a few families of genes, both nuclear factors NF-κBp50 and NF-κBp65 are upregulated at least twofold. As we know from other studies, the anti-inflammatory mechanism of green tea polyphenols is mediated at least in part through the downregulation of TNF-α gene expression by blocking NF-κB activation.[12–14]

CONCLUSION

The concept of cancer chemoprevention has matured greatly. Significant reversal or suppression of premalignancy in several sites by chemopreventive agents appears possible using natural agents. Tumor initiation and progression in the colonic mucosa is a multistep, multigene process involving the activation of some genes and the loss of expression of others. The latter category includes the tumor suppressor genes APC and p53. Studies directed towards understanding the genetic link among loss or inactivation of tumor suppressor genes, colorectal tumorigenesis, and chemoprevention through natural nontoxic agents will enable the mechanism-based identification of molecular targets. Hyperproliferation in normal mucosa, leading to early adenomas and cellular proliferation, growth, and progression of neoplasia are also aspects of colorectal carcinogenesis that can be controlled by nontoxic natural chemopreventive agents. Simple phenolic compounds and related antioxidants have been reported to play a potential role in interfering with both mutagenicity and unregulated proliferation of normal cells. Studies directed towards understanding the role of plant-derived chemopreventive agents in preventing tumor cell growth, detoxification processes, downregulation of carcinogen-activating enzymes, and activation of phase II-detoxifying enzyme activity, which protects cells from a wide variety of toxicants, are very much needed at this point. In addition, the mechanistic principles that function through cellular and molecular targets apply to the identification of potential dietary chemopreventive agents against specific cancers including colorectal cancer. Selective and differential effects of phytochemicals on colon tumorigenesis and the reprograming of genes involved in different cascade of events related to tumor growth regression in preclinical model will help to identify potential molecular targets for EA and compounds with similar functions.

ACKNOWLEDGMENTS

This study was supported in part by USPHS grant CA-17613 (NCI Cancer Center Grant) and AICR Grant No. 01A015. We thank Ilse Hoffmann for editing the chapter.

REFERENCES

1. Sandler, R.S. Cholecystectomy and colorectal cancer. Gastroenterology **1993**, *105*, 286–288.
2. Sun, J.; Chu, Y.F.; Wu, X.; Liu, R.H. Antioxidant and antiproliferative activities of common fruits. J. Agric. Food Chem. **2002**, *50*, 7449–7454.
3. Liu, R.H. Health benefits of fruit and vegetables are from additive and synergistic combinations of phytochemicals. Am. J. Clin. Nutr. **2003**, *78* (3 Suppl), 517S–520S.
4. Lopez-Velez, M.; Martinez-Martinez, F.; Del Valle-Ribes, C. The study of phenolic compounds as natural antioxidants in wine. Crit. Rev. Food Sci. Nutr. **2003**, *43*, 233–244.
5. Bach Knudsen, K.E.; Munck, L.; Eggum, B.O. Effect of cooking, pH and polyphenol level on carbohydrate composition and nutritional quality of a sorghum (*Sorghum bicolor* (L.) *Moench*) food, ugali. Br. J. Nutr. **1988**, *59*, 31–47.
6. Stoner, G.D.; Morse, M.A. Isothyocyanates and plant polyphenols as inhibitors of lung and esophageal cancer. Cancer Lett. **1997**, *114*, 113–119.
7. Narayanan, B.A.; Geoffroy, O.; Willingham, C.M.; Re, G.G.; Nixon, D.W. p53/ p21($^{WAF1/CIP1}$) expression and its possible role in G1 arrest and apoptosis in ellagic acid treated cancer cells. Cancer Lett. **1999**, *136*, 215–221.
8. Narayanan, B.A.; Re, G.G. IGF II down regulation associated cell cycle arrest in colon cancer cells exposed to phenolic antioxidant ellagic acid. Anticancer Res. **2001**, *21*, 359–364.
9. Narayanan, B.A.; Narayanan, N.K.; Stoner, G.D.; Bullock, B.P. Interactive gene expression pattern in prostate cancer cells exposed to phenolic antioxidants. Life Sci. **2002**, *70*, 1–19.
10. Bogazzi, F.; Ultimieri, F.; Raggi, F.; Costa, A.; Gasperi, M.; Cecconi, E.; Mosca, F.; Bartalena, L.; Martino, E. Peroxisome proliferator activated receptor gamma expression is reduced in the colonic mucosa of acromegalic patients. J. Clin. Endocrinol. Metab. **2002**, *87*, 2403–2406.
11. Bogazzi, F.; Ultimieri, F.; Raggi, F.; Russo, D.; Viacava, P.; Cecchetti, D.; Costa, A.; Brogioni, S.; Cosci, C.; Gasperi, M.; Bartalena, L.; Martino, E. Changes in the expression of the peroxisome proliferator-activated receptor gamma gene in the colonic polyps and colonic mucosa of acromegalic patients. J. Clin. Endocrinol. Metab. **2003**, *88*, 3938–3942.

12. Yang, F.; de Villiers, W.J.; McClain, C.J.; Varilek, G.W. Green tea polyphenols block endotoxin-induced tumor necrosis factor-production and lethality in a murine model. J. Nutr. **1998**, *128*, 2334–2340.
13. Pan, M.H.; Lin-Shiau, S.Y.; Ho, C.T.; Lin, J.H.; Lin, J.K. Suppression of lipopolysaccharide-induced nuclear factor-kappaB activity by theaflavin-3,3'-digallate from black tea and other polyphenols through down-regulation of IkappaB kinase activity in macrophages. Biochem. Pharmacol. **2000**, *59*, 357–367.
14. Chen, P.C.; Wheeler, D.S.; Malhotra, V.; Odoms, K.; Denenberg, A.G.; Wong, H.R. A green tea-derived polyphenol, epigallocatechin-3-gallate, inhibits IkappaB kinase activation and IL-8 gene expression in respiratory epithelium. Inflammation **2002**, *26*, 233–241.

16

Changes of Gene Expression in INS-1 cells: The implication of palmitate in lipotoxicity

Jianzhong Xiao,* Per Bendix Jeppesen, and Abudula Reziwanggu
*Department of Endocrinology and Metabolism C, Sygehus THG,
Aarhus University Hospital, Aarhus, Denmark*

Mogens Kruhøffer, Torben Ørntoft, and Kjeld Hermansen**
*Molecular Diagnostic Laboratory, Department of Clinical Biochemistry,
Skejby Hospital, Aarhus University Hospital, Aarhus, Denmark*

INSULIN RESISTANCE AND β-CELL DYSFUNCTION PLAY AN IMPORTANT ROLE IN DEVELOPMENT OF TYPE 2 DIABETES

The pathogenesis of type 2 diabetes is complex and usually requires defects both in β-cell function and in insulin sensitivity. The molecular defects behind diabetes are not fully understood, and have led to heated scientific debates. Nevertheless, there are some points of agreement on the pathology of type 2 diabetes: (a) genetic factors determine the risk of developing type 2 diabetes; (b) the presence of insulin resistance predicts future development of type 2 diabetes; and (c) the failure of β-cells to compensate the insulin resistance leads to diabetes.

---

*Department of Endocrinology, China–Japan Friendship Hospital, Beijing, China.
**Corresponding Author.

INTERACTION OF GENETIC AND ENVIRONMENTAL FACTORS PLAYS IMPORTANT ROLE IN DEVELOPMENT OF TYPE 2 DIABETES

Positional cloning holds great promise for identifying genes that cause the common form of type 2 diabetes and for elucidating the molecular mechanisms that cause insulin deficiency and insulin resistance. Although the common form of type 2 diabetes is polygenic, most of the progress until today has been in identifying genes that cause several uncommon forms of diabetes with monogenic inheritance, which consist of less 5% diabetic patients.[1] Genetic susceptibility of diabetes in a stable population without many immigrants should not be changed within decades or centuries. As we know, however, the incidence of type 2 diabetes has increased tremendously during the last century. These evidences suggest that environmental factors are at least of equal importance to genetic factors in the development of type 2 diabetes. Epidemiological studies have shown that overnutrition, sedentary work, stress, and lack of exercise are associated with diabetes.[2,3] In contrast, diet and exercise intervention have been shown to prevent or delay the development of diabetes.[4-6] Studies concerning the interaction between susceptible genes and environmental factors may be promising tools to unravel the pathogenesis of diabetes. Interestingly, the association of obesity/insulin resistance and the common variant Pro12Ala of PPARγ2 is only present in populations whose diets are rich in saturated fatty acids.[7] This underlines the importance of the interaction between genetic and environmental factors in the development of diabetes.

IMPORTANCE OF β-CELL FUNCTION FOR DEVELOPMENT AND PROGRESS OF TYPE 2 DIABETES

Insulin resistance *per se* does not cause diabetes. This statement has been confirmed in studies demonstrating that, even in absence of skeletal insulin receptors in the mouse, diabetes does not develop.[8] In order to elicit hyperglycemia and type 2 diabetes, a β-cell dysfunction is required. The dysfunction of pancreatic β-cells also plays a key role in the progressive nature of type 2 diabetes. At diagnosis of diabetes, ~50% the β-cells are functionally destroyed. Using the homeostasis model assessment (HOMA), it was demonstrated in the UKPDS that the progressive nature of diabetes in individuals with recently diagnosed type 2 diabetes consists of an ongoing decline in β-cell function without a change in insulin sensitivity.[9,10]

LIPOTOXICITY, AN IMPORTANT FACTOR IN PATHOGENESIS OF TYPE 2 DIABETES

Type 2 diabetes is associated with elevated levels of triglycerides (TG) and non-esterified fatty acids (NEFA). Interestingly, increased TG is an independent risk factor for the development of type 2 diabetes. In male Zucker diabetic fatty

(ZDF) rats, a marked increase in TGs and NEFA appears just before the onset of diabetes. Series of studies confirm that increased TG and NEFA in plasma lead to an impairment of insulin sensitivity and secretion.

Fatty Acids and Insulin Secretion

Fatty acids exert a stimulatory effect on insulin secretion in the short-term and they are important for postprandial insulin secretion. McGarry and Dobbins[11] demonstrated that insulin secretion is correlated to the plasma level of NEFA in rats. Insulin secretion is suppressed when lipolysis is suppressed with nicotinic acid and the NEFA level is subsequently decreased.[11] In contrast, elevated plasma NEFA concentrations elicited by infusion of Intralipid® can reconstitute the nicotinic acid-induced inhibition of insulin secretion. NEFA in basal (fasting) state appears to be permissive for the insulin secretion during nutrient challenge.[11] Efforts to elucidate how fatty acids influence β-cell function have led to a number of important findings. Thus (a) methods that inhibit CPT-1 activity concomitantly increase insulin secretion. This takes place indirectly (e.g., by increasing malonyl-CoA by glucose or other secretagogues[12,13] or by adding direct inhibitors of CPT-1 such as 2-bromopalmitate and etomoxir;[14] (b) exogenous long-chain fatty acids significantly potentiate glucose stimulated insulin secretion (GSIS) from rat islets and clonal β-cell lines;[15–17] (c) the length, saturation, and geometrical configuration (cis/trans) influence the insulinotropic potency of a fatty acid;[15,17] and (d) the AMPK stimulator, AICAR, deactivates acetyl-CoA carboxylase (ACC) and ablates GSIS.

Taken together, these findings led to the proposal that the stimulus–secretion coupling within the β-cell might use an element of glucose–fatty acid cross-talk.[11] An increase in the glucose concentration is sensed by glucokinase (GK), allowing metabolism of the hexose through glycolysis to pyruvate. In the β-cell, the C3 unit can be converted into citrate in the mitochondria. Some fraction of citrate is oxidized in the tricarboxylic acid (TCA) cycle, generating ATP and CO_2, whereas the remainder possibly leaves the mitochondria and is converted in the cytosolic compartment into malonyl-CoA via the sequential action of ATP-citrate lyase (ATP-CL) and ACC. The increase in malonyl-CoA concentration is expected to suppress the CPT-1 activity, and consequently the fatty acid oxidation, resulting in an increase in the cytosolic concentration of long-chain acyl-CoA (LC-CoA). LC-CoA then acts as a signaling molecule for insulin secretion,[18] working in concert with the rise in Ca^{++} caused by alterations in the K_{ATP}^+ and $[Ca^{++}]_i$ channel activities. To summarize, the expansion of the LC-CoA pool in the presence of elevated exogenous fatty acid levels may explain the insulinotropic effect of NEFA.

Lipotoxicity and Type 2 Diabetes

The effect of NEFA to sustain β-cell function in the fasting state and to assure efficient nutrient-stimulated insulin secretion is considered physiological phenomena. Even in the obese state, the hypersecretion of insulin in the presence of increased NEFA reflects a beneficial adaptation protecting against glucose intolerance [11] Increased β-cell mass and/or secretion capacity is responsible for the adaptation to the oversupply of energy and insulin resistance. However, as a result of long-term exposure to fatty acids pancreatic islets become over-worked. On the one hand, it alters insulin secretion by causing an enhanced basal insulin secretion (BIS) in the presence of low glucose as well as a blunted insulin response to high glucose, thus resembling the typical insulin secretion profile in type 2 diabetes.[11,19,20] On the other hand, it may lead to a reduction of β-cell mass by inducing apoptosis or inhibiting proliferation and neogenesis. These alterations in insulin secretion and the reduction of the β-cell mass following chronic augmentation of TG and NEFA represent the characteristic pathogenic changes in lipotoxicity.

Gene Expression Changes During Lipotoxicity

Generally, the immediate biological response of cells to changes in the external milieu is regulated (within seconds or minutes) by modification of enzyme activity. In contrast, the adaptation to more prolonged changes depends on the regulation of gene transcription. Abnormal insulin secretion secondary to chronic exposure of fatty acids may involve the regulation of gene expression, protein translation, processing, modification, and secretion. Studies on fatty acid-induced changes of gene expression in islets or cultured β-cells have focused on a few candidate genes [i.e., *CPT-1*, acetyl-CoA carboxylase (*ACC*), *proinsulin*, *IDX-1*, and *GLUT2* gene], which are thought to play an important role in the adaptation and insulin secretion.[21-23] The maintenance of β-cell mass involves cell neogenesis, replication, and apoptosis. Naturally, it is likely that the chronic exposure to fatty acids induces changes in pancreatic β-cell function resulting not just from changes in one dimension. Accordingly, the expression is modified in a multitude of genes. Consequently, it is necessary to obtain a global view of the gene expression (or profile) for elucidating the mechanisms of lipotoxicity.

GENE CHIP METHOD FOR PROFILING GENE EXPRESSION IN THE BETA CELL

DNA microarrays have been used extensively to monitor mRNA expressions. DNA microarrays determine the expression of many genes in parallel, which helps us to understand, for example, the transcriptional mechanisms underlying the alteration in β-cell function during lipotoxicity.[24-29] Taking advantage of the high density, 15–20 pairs of probes are used for determining the amount of each transcript in high-density oligonucleotide microarrays. One pair of probes consists of one perfect match probe (positive signal) and one mismatch probe (one nucleotide difference in the middle of the chain, used as background signal).

FATTY ACIDS, BETA CELLS, AND THE AFFYMETRIX GENECHIP TECHNOLOGY

To understand the changes in the gene expression profile in beta cells in response to fatty acids, we applied the clonal insulin-secreting cell line INS-1. In the presence of 6.6 mM glucose, cells were cultured at 0, 50, and 200 μM palmitate (P0, P50, P200) from 2 days up to 10 weeks.[28,30] Apart from the determination of insulin secretion capacity, we investigated the expression profiles using the Affymetrix GeneChip technology at days 2 and 42.[28] In addition, we have investigated cell proliferation, glucose oxidation, IRS-1, and IRS-2 expression at the protein level.

Our studies showed that palmitate acutely enhanced the insulin secretion corroborating previous reports.[17] Even at low glucose levels of 1.0 and 3.3 mM, which represents BIS, a similar pattern was observed. BIS reached the highest level after 6 weeks (i.e., being increased 3× and 5× after culture in P50 and P200 for 42 days, respectively). GSIS increased insulin secretion in cells cultured in P200 after 2 days. No significant decrease in total insulin release occurred after prolonged culture in the presence of P200 even after 10 weeks. Calculating the fold increase above basal level, we found that the β-cell responsiveness to glucose or palmitate exposure was blunted after 42 days.[30] The increased BIS and blunted responsiveness to glucose reflect the pathogenesis of diabetes. Hyperinsulinemia compensates for the insulin resistance in the early stage of insulin resistance in type 2 diabetes. Studies in animals indicate that early hyperinsulinemia seems in itself to be capable of causing insulin resistance and subsequently dysfunction of beta cells.[31] Exposure of β-cells to fatty acids for 2 days is usually considered to result in "chronic" effects and previous studies have shown that it suppresses insulin secretion.[32] Although increased BIS is a characteristic feature of lipotoxicity, we failed to detect a decline in the blunted GSIS. The discrepancy may be explained by differences in experimental conditions [e.g., the higher concentrations of fatty acids (0.4–1 mM), glucose (11 mM), or BSA (which determines the level of active unbound fatty acids) used by others.[33–36] The different sensitivity to lipotoxicity between islets and cell lines may also contribute to the differences.

We have chosen lower palmitate concentrations and a borine serum albumin (BSA) concentration of 0.5% in the culture media to prevent direct cytotoxicity of unbound fatty acids. Furthermore, it was observed that the inhibitory action of fatty acids on insulin release was only found in the presence of stimulatory glucose concentrations. The level of 6.6 mM glucose used in present studies may account for the lack of suppression in absolute levels of insulin secretion. The increased BIS and reduced responsiveness to nutrients (in relative but not in absolute terms) may represent the early stage of lipotoxicity in both humans and animals. The so-called "lipotoxicity" in different diabetic models (e.g., the male ZDF rats) are characterized by increased BIS and decreased

β-cell responsiveness (i.e., with less fold increase in stimulated insulin secretion when compared with lean controls). Concomitantly, the absolute plasma insulin level is much higher than in the lean nondiabetic controls in the early stage although insufficient to compensate for the metabolic aberrations.[37] Furthermore, it was found that even the obese female Zucker nondiabetic rat had a higher glucose stimulated insulin secretion than the lean male nondiabetic rats (118 $\pm$ 19 vs. 94 $\pm$ 12 μU/mL) as well as a higher plasma glucose concentration (9.15 $\pm$ 0.02 vs. 5.70 $\pm$ 0.17 mM).[37] After calculating the fold increase of insulin secretion, it is no surprise that the former group had a much lower responsiveness (0.8- vs. 12-fold).[37] At a later stage of diabetes the absolute insulin concentration also declines. Consequently, we propose that the increased BIS and the declined responsiveness (although not the absolute insulin secreted) to nutrient stimulation are the earlier characteristics of lipotoxicity.

General Pattern of Gene Expression and Changes in its Expression

By using the Affymetrix GeneChip, we measured 8740 genes or expression sequence tags (ESTs). Among the 8740 transcripts, 3751 (44%) transcripts were not detected as present in any preparation after day 2 or day 42. In contrast, 2960 transcripts were present in every preparation.[28] The present and absent concordance at different time points was ~88% regardless of the palmitate concentration (Table 16.1). Prolonged culturing did not change the gene expression profiles significantly (i.e., comparing the profiles of day 2 and day 42, 52% transcripts were absent and 40% were present in cells cultured with 0.2 mM palmitate). The concordance was 93%. Similarly, 52% of all 8740 transcripts were absent and 41% were present in cells cultured without palmitate, the concordance was 93%. We believe that the INS-1 cell line is quite stable when it comes to gene expression profiles and the selection effect is negligible, because the concordance is greater under the same condition rather than at same time point.

As shown in Table 16.2 it is obvious that the number of genes that changes in expression depends on the cut-off values set (significant level), which are arbitrary. The larger the numerical "sort score", (SS, a value based on both "fold change" and "average difference change"), the more reliable is the measured difference in expression of genes between the experimental and the control

Table 16.1 The Percentage Concordance in Gene Expression in INS-1 Cells at Various Conditions and Time Points

	Day 2	Day 42	All	0.2 mM palmitate	6.6 mM glucose
Absent	50.0	48.3	44.3	52.3	51.7
Present	37.8	39.6	34.9	40.3	41.4
Concordance	87.8	88.0	79.2	92.6	93.1

Table 16.2 Influence of Cut-Off Value on the Corresponding Number of Genes Showing Changes (Among 8740 Genes or ESTs) in INS-1 Cells in Response to 0.05 mM (P50) and 0.2 mM palmitate (P200)

Cut-off value		Day 2		Day 44		Total
FC	SS	P50	P200	P50	P200	
≥ 2	≥ 0.5	22	121	165	199	507
≥ 2	≥ 0.65	11	99	134	159	403
≥ 3	≥ 0.5	18	90	124	156	388
≥ 3	≥ 0.65	10	84	111	140	345

Note: SS, sort score; FC, fold change. All difference calls are increase or decrease.
Source: From Xiao et al.,[28] permission granted by The Endocrine Society.

groups. If fold change were set to 2 and SS to 0.5, the number of genes that changed was 507. However, changing sort score to 0.65 and fold change to 3, the number of affected genes was reduced to 345. In the present study, genes or ESTs with a sort score value ≥ 0.65, "difference call" being "increase" or "decrease", and showing twofold or higher change were considered to be significant changes.[38,39]

In general, downregulated genes were more common than upregulated genes. The changes to palmitate exposure occurred to be dose and time dependent.[28]

Randle's Cycle also Operates at Transcription Level

Glucose, lipid, and amino acids are major respiratory fuels in animals. It is well documented that fatty acids and glucose compete for respiration.[40] In other words, fatty acids inhibit glucose utilization and oxidation and vice versa. This operation of fatty acid–glucose cycle in muscle tissue is widely accepted. However, in pancreatic islets or β-cells this is a controversial topic. It has been reported that chronic exposure of islets to fatty acids decreases pyruvate dehydrogenase (PDH) activity and inhibits glucose oxidation.[35,36] In contrast, others have reported that chronic exposure to fatty acids might decrease the PDH activity but does not inhibit the glucose oxidation at high glucose concentration,[35,41] whereas fatty acids increase glucose oxidation in the presence of low glucose.[42]

After exposure to fatty acids, the changes in gene expression reflect the adaptation of energy source variation (i.e., genes involved in fatty acids catabolism were upregulated). This was quite similar to the peroxisome proliferator-activated receptor agonists (PPARs). PPARα and -γ play important roles in regulating genes related to fatty acid oxidation and lipogenesis, respectively. Palmitate is a PPARα agonist, inducing expression of genes involved in fatty acid oxidation.[43] During culture of INS-1 cells with 0.2 mM palmitate for 2 days and 42 days, we found that 12 transcripts were increased among the 9 known genes, of which 6 were related to fatty acid catabolism (i.e., lipoprotein lipase, fatty acids translocase/CD36, CPT-1, LCAD, 3-oxoacyl-CoA thiolase, 2,4-dienoyl-CoA reductase) (Xiao et al.[28] and Table 16.3). The rest were glucagon, fibronectin, and tissue factor protein.

The downregulation of the c-erb-A thyroid hormone receptor and the retinoic acid receptor (RAR), which are transcription factors that inhibit the expression of genes involved in fatty acid oxidation, may contribute to increased oxidation of fatty acids. In addition, the downregulation of gastrin-binding protein (gastrin-BP) may lead to enhanced fatty acid oxidation. On the other hand, palmitate may suppress the effect of PPARγ by inhibiting the expression of adipocyte differentiate and determinator 1 and result in suppression of fatty acid synthetase (FAS) and ACC, as seen in the adipocytes. These changes in gene expression may reflect the adaptation of cells to fatty acid exposure (i.e., increased expression of genes regulating fatty acid oxidation and suppressed expression of genes regulating fatty acid synthesis.[21,23,31,44] The alteration of expression of genes involved in fatty acid metabolism was similar between days 2 and 44 in cells cultured with 200 μM palmitate (Xiao et al.[28] and Table 16.3). Nevertheless, a downregulation of ACC was only found in the latter. In contrast, an increased expression of peroxisomal enoyl hydratase-like protein (PEXL), 2,4-dienoyl-CoA reductase was found only at day 2 in cells cultured with 200 μM palmitate.[28] Concomitantly, we detected a more profound suppression of FAS (which may result in more malonyl-CoA accumulation). Whether these differences contributed to the functional differences between days 2 and 44 remain to be elucidated.

Palmitate was shown to suppress the expression of stearoyl-CoA desaturase 2 (SCD2) and the SCD2 homolog gene in INS-1 cells.[28] SCD2 can transform the saturated fatty acid into unsaturated fatty acid, thereby changing the plasma membrane fluidity. The physiological importance of our observation is yet unclear.

Glucose transporter 2 (GLUT2) and glucokinase are two key elements in glucose metabolism and insulin secretion in the β-cells. We did not find any differences in glucokinase and GLUT2 mRNA levels between groups (Xiao et al.[28] and Table 16.3). Type 1 hexokinase (HK1) is supposed to become upregulated during lipotoxicity, supported by the increased glucose oxidation at low concentration. This increment was believed to be associated with increased BIS. Considering the principle of Randle's cycle, the increased glucose oxidation is questionable after exposure to fat. The fact that increased BIS is not accompanied by increased expression of HK1 in this study (Xiao et al.[28] and Table 16.3) suggests that HK1, at least in our model, may not play a major role in the enhanced BIS during lipotoxicity. In the cascade of glycolysis, aldolase A, PFK, and ATP-CL were suppressed by long-term exposure and/or the higher palmitate.[28] The downregulation of PFK is consistent with results from studies on the heart exposed to fatty acids. Because the expression of genes involved in glycolysis were changed in a similar manner in cells exposed to 200 μM palmitate at

Table 16.3 Selected Gene Expression Changes in INS-1 Cells Exposed to 50 (P50) or 200 μM (P200) Palmitate (Fold Compared with Control)

Gene	Genebank accession no.	Function	Day 2		Day 44	
			P50	P200	P50	P200
Genes involved in fatty acid metabolism						
RAR *alpha*	AJ002940	Transcription factor related lipid metabolism	NC	−9.7	−13	−15
FAS	X13527	Fatty acid synthesis	NC	−33.4	−6.3	−8.8
Gastrin-binding protein	X98225	Inhibit mitochondrial fatty acid oxidation	NC	−7.6	−7.9	−4.9
SCD2	AF036761	Saturated fatty acid desaturation	NC	−19.7	−32.5	−49.3
3-Oxoacyl-CoA thiolase	X05341	Fatty acid oxidation	NC	5.7	NC	3.2
CPT-1	L07736	LCFA oxidation	NC	7	NC	2.8
LCAD	J05029	LCFA oxidation	NC	4.5	NC	4.7
EpH	X60328	LCFA oxidation	NC	5.2	NC	18
PEXL	U08976	Peroxisome β oxidation	NC	2.4	NC	NC
c-erb-A TR	X12744	Nuclear receptor	NC	−21.9	NC	−6.4
2,4-dienoyl-CoA reductase	D00569	Unsaturated fatty acid oxidation	NC	2.5	NC	NC
ACC	J03808	Malonyl-CoA synthesis	NC	NC	−8	−6.4
SCD2 homolog	S75730	?	NC	NC	−3.1	−2.8
ADD1	L16995	Transcription factor related lipid metabolism	NC	NC	−27.1	−5.1
Genes involved in glucose metabolism						
ATP-CL	L27075	Glycolysis	NC	−21.5	−47.8	−15.5
HNF 4	X57133	Transcription factors related to glycolysis	NC	−6.8	−6.7	−22
PFK-M	D21869	Glycolysis	NC	−15.1	−11.0	−10.3
Aldolase A	U20643	Glycolysis	NC	−6.1	−4.1	−2.9

(continued)

Table 16.3 Continued

Gene	Genebank accession no.	Function	Day 2		Day 44	
			P50	P200	P50	P200
Fructose-1,6 biophostase	M86240	Counter the action of PFK	NC	5.6	NC	NC
GP	M27726	Glycogen degradation	NC	NC	−17.2	−16.8
Genes involved signal transduction						
IR	M29104	Insulin signal transduction	NC	NC	NC	−4.7
IRS-2	AF083418	Insulin signal transduction, cell growth	NC	NC	−12.6	−12.8
SH2P	AF065161	Cell growth signal transduction	NC	−11	−10	−18
Housekeeping gene						
GAPDH	M17701		NC	NC	NC	NC
beta-actin	V01217		NC	NC	NC	NC
Total			0	17	14	20

Note: NC, no change. Numbers refer to fold change where prefix (−) refer to decrease and no prefix to increase. RAR, retinoic acid receptor alpha 1; FAS, fatty acid synthetase; SCD2, stearoyl-CoA desaturase 2; CPT-1, carnitine palmitoyl transferase-1; LCAD, long-chain acyl-CoA dehydrogenase; EpH, cytosolic epoxide hydrolase; PEXL, peroxisomal enoyl hydratase-like protein; c-erb-A TR, c-erb-A thyroid hormone receptor; ACC, acetyl-CoA carboxylase; ADD1, adipocyte determination and differentiation factors1; ATP-CL, ATP-citrate lyase; PFK, phosphofructokinase-M; HNF4, hepatocyte nuclear factor 4; GP, glycogen phosphorylase; IR, insulin receptor; IRS-1, insulin receptor substrate-1; IRS-2, insulin receptor substrate-2; SH₂P, cytokine-inducible src-homology-2-containing protein.

Source: From Xiao et al.,[28] with permission of The Endocrine Society.

day 2 as in day 42. This modification could not entirely explain the blunted insulin secretion.

Interestingly, we found that hepatic nuclear factor 4 (*HNF4*) gene expression was inhibited after palmitate exposure, whereas higher palmitate concentrations seem to cause an earlier onset and a more marked inhibition.[28] A mutation in the *HNF4α* gene is responsible for the autosomal dominant, early-onset form of type 2 diabetes, the so-called *MODY1*. It has been shown that the MODY1 mutant protein has lost its transcriptional transactivation activity,[45] which affects several genes whose expression is dependent upon HNF4α, including GLUT2, the glycolytic enzymes aldolase B, glyceraldehyde 3-phosphate dehydrogenase, and the liver pyruvate kinase.[44] We found that aldolase A but not B was suppressed accompanying the downregulation of HNF4.[28] The downregulation of genes involved in glycolysis and upregulation of fatty acid oxidation in cells cultured in fatty acid suggest that the fatty acid–glucose reciprocal cycle (Randle cycle)[40] is indeed operative at the gene expression level in INS-1 cells.

To see if the change in gene expression influences glucose oxidation, we have determined the glucose oxidation after 42 days exposure of palmitate (Fig. 16.1).

We did not find that glucose oxidation was impaired after 6 weeks incubation at 16.7 mM glucose. However, the glucose oxidation was inhibited by 400 μM palmitate in the presence of 16.7 mM glucose (Fig. 16.1). This may reflect the situation *in vivo*, where elevated levels of glucose and NEFA are coexistent in diabetes.[46] Consequently, the inhibition of glucose oxidation may contribute to the decline in GSIS.

The Insulin Gene Expression

We did not find any decrease in the expression of insulin after long-term exposure to palmitate with the GeneChip method.[28] However, we do not know, taking into account the enhanced BIS, if a lack of a corresponding degree of increase in insulin content also reflects a relative insufficient function in cells cultured

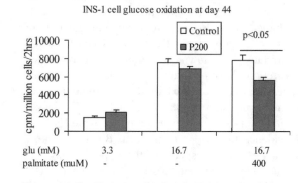

Figure 16.1 Glucose oxidation in INS-1 cells after exposure to 200 μM palmitate for 6 weeks.

in P200.[22] The finding of unchanged type 2 proinsulin processing endopeptidase gene expression suggests that the proinsulin processing might be normal during long-term exposure to palmitate in this study.

Apparently, a change in the cell functions is not only determined by one or a few genes but seems to be ascribed to a series of genes. We found that the genes altered ranged from ion channel genes, to enzymes involved in glucose, fatty acid, and amino acid metabolism, to proteins involved in the signal transduction, cell growth, replication, and apoptosis.[28]

In Table 16.4, downregulated genes that are involved in signal transduction are presented. These genes are downregulated after both 2 days and 42 days exposure to 200 μM palmitate. The accumulation or synergistic effects of different genes may play a critical role in the alteration of cell function.

As mentioned previously, the plasma levels of insulin—both in the fasting and feeding state—are higher in ZDF rats than in their lean controls as diabetes progresses. However, the response to stimulation is almost lost. This indicates that the responsiveness of insulin secretion to nutrient stimulation is an even more sensitive parameter to evaluate the β-cell function than the absolute amounts.[47] Despite a variety of computational methods to ensure the validity of microarray data, concerns still exist regarding intrachip and chip-to-chip variation. Positive and negative data found in microarray should be validated with other methods. Conventional Northern blot analysis requires relatively large amounts of starting RNA and specific probes. In contrast, real-time RT-PCR has several advantages to Northern blot (e.g., the starting material can be relatively less in quantity). To confirm our findings, we have used RT-PCR to determine a few selected genes. This is a more sensitive method than the GeneChip method to detect small differences to various experimental conditions. Our data obtained using RT-PCR corroborated our findings with the Affymetrix GeneChip. Thus, high concentrations of palmitate did not reduce the expression of *GLUT2* and *insulin 1* gene. In contrast, the expression was to some extent increased by palmitate in long-term studies. We found that the increased gene expression of *insulin 1* was correlated to the insulin content of the β-cell. This is consistent with the finding that the absolute amount of insulin secreted is increased during long-term studies. We also detected a downregulation of *insulin 1* and *GLUT2* in INS-1 cells cultured for 2 days at ≥ 10 mM glucose with and without 200 μM palmitate. This down-regulation was accompanied by an inhibition of insulin secretion. Taken together, these data suggest that INS-1 cells may be more vulnerable to glucose toxicity than to lipotoxicity, as judged by insulin secretion and expression of *insulin 1* and *GLUT2* genes. Another possibility is that a stimulatory glucose level is a prerequisite for lipotoxicity. There are lines of evidence that support this idea. By transplanting islets from ZDF rats to normal rats and transplanting the opposite way, the expression of *GLUT2* and *insulin 1* gene was suppressed in the diabetic state and enhanced in the normal state. It suggests that the internal diabetic environment (elevated NEFA and glucose levels in plasma) plays a major role in the

Table 16.4 Changes in the Expression of Genes in INS-1 Cells that Involve Signal Transduction and Cell Proliferation/apoptosis in Response to Exposure to Palmitate (200 μM) for 2 and 44 Days

GenBank accession no.	Name	Day 2			Day 44		
		Avg diff change	FC	SS	Avg diff change	FC	SS
L12382	ADP-ribosylation factor 3	−135	−3.3	−0.97	−107	−2.7	−0.61
L19933	Protein tyrosine phosphatase	−588	−3.6	−2.29	−426	−2.0	−0.62
X12535	Ras-related protein p23	−258	−1.8	−0.37	−242	−1.7	−0.29
AB001452	Sck	−347	−39.1	−12.07	−348	−8.4	−4.87
AF055065	Signal regulatory protein alpha	−63	−4.5	−1.01	−108	−5.8	−1.86
D85760	G alpha 12	−108	−2.5	−0.52	−133	−2.7	−0.67
D30041	Protein kinase beta	−62	−5.5	−1.29	−54	−2.2	−0.28
X13905	Ras-related rab1B	−325	−4.8	−2.54	−387	−4.7	−2.71
X13905	Ras-related rab1B	−214	−24.5	−7.32	−226	−35.4	−9.24
S50461	Signal-transducing G protein alpha 12 subunit	−61	−7.7	−1.80	−47	−4.1	−0.78
S80345	Von Hippel-Lindau tumor suppressor gene homolog	−66	−8.2	−1.69	−74	−12.3	−2.93
D16308	Cyclin D2	−474	−7.1	−4.89	−666	−12.8	−9.47
S78284	bcl-xshort = apoptosis inducer	−189	−4.0	−1.55	−198	−4.1	−1.64
U53486	Corticotropin releasing factor receptor(CRFR)	−382	−5.1	−3.03	−202	−2.9	−0.96
U53500	CRFR variant form C	−196	−22.5	−6.45	−120	−19.2	−3.18
AF065161	Cytokine-inducible SH2-containing protein	−94	−11.3	−3.08	−116	−18.6	−4.92
L48060	Lactogen receptor	−149	−2.6	−0.65	−161	−3.0	−0.90
Z35654	Ost oncogene	−248	−3.3	−1.26	−234	−2.2	−0.58
U34963	Programmed cell death repressor BCL-X-Long	−193	−2.2	−0.51	−235	−3.0	−1.05

Note: SS, sort score; Avg diff change, average difference change; FC, fold change.

change of the expression of *GLUT2* and *insulin 1* gene.[48] Furthermore, the fatty acid inhibition of the expression of *insulin* and *GLUT2* genes was only found in the presence of stimulatory glucose concentrations (16.7 or 30 mM),[34,48,49] suggesting that fatty acids *per se* may not be sufficient to suppress the expression of *GLUT2* and *insulin*. A relatively low glucose concentration (6.6 mM) was used our study, which may, at least in part, explain the difference between our data and other reports. The discrepancies may also be attributed to the different experimental conditions between the *in vivo* and *in vitro* studies, between islets and clonal β cells, and/or different concentrations of fatty acids.[28,32,48]

Lipotoxicity and Cell Proliferation

The pancreatic β-cell mass is determined by the balance between β-cell formation, including proliferation and neogenesis, and the loss of β-cells through necrosis and apoptosis.[50–53] Chronic exposure to fatty acids induces apoptosis and inhibits the proliferation of pancreatic β-cell, as does chronic exposure to high glucose.[54,55] Consistently, we have found that the inhibition of proliferation of INS-1 E cells depends on the concentration of the fatty acid (Fig. 16.2).

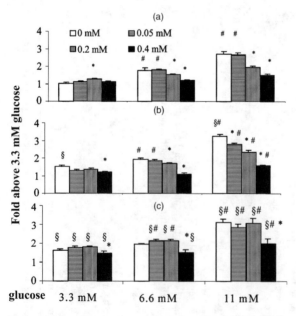

Figure 16.2 Effect of palmitate on the INS-1E cell proliferation in the presence of 3.3, 6.6, and 11 mM glucose (fold above 3.3 mM glucose without IGF-1 and palmitate, $n = 16$ for each bar) and 0, 2, and 10 nM IGF-1 (a, b, and c). *$P < 0.05$ compared with group at same glucose concentration in the absence of palmitate, #$P < 0.05$ compared with group at same concentration of palmitate in the presence of 3.3 mM glucose, §$P < 0.05$ compared with group at same glucose and palmitate concentration in the absence of IGF-1.

In addition, this inhibition was observed only in the presence of a stimulatory glucose level (i.e., the concentration of glucose should be ≥6.6 mM). In contrast, palmitate at low concentrations could increase cell proliferation at low glucose levels (3.3 mM).

The inhibition of proliferation was associated with a downregulation of the *IRS-1* and *IRS-2* gene expression.[56] The question arises: Which of the intermediate fatty acid metabolites is responsible for the inhibitory effect on the β-cell proliferation? Why is a stimulatory glucose concentration required to elicit an inhibitory effect on proliferation (and insulin secretion)? Accumulation of lipids within cells seems to play a critical role. Accumulation of TG is associated with insulin resistance in muscles. An increased TG accumulation within islets has also been shown to correlate with the suppression of insulin secretion.[37] Using ³H-palmitate incorporation technique, we have demonstrated that the accumulation of lipids depends on the concentrations of exogenous fatty acids (i.e., the higher the concentration of fatty acid the more the lipid accumulation has occurred). The prevailing state of fatty acid metabolism also influences the accumulation of fatty acids (Fig. 16.3). In the presence of certain concentrations of exogenous fatty acid, glucose elicits a dose-dependent increase in the accumulation of lipids (Fig. 16.3).

This corroborates many previous studies, which have shown that elevated glucose inhibits fatty acid oxidation and consequently increases lipid

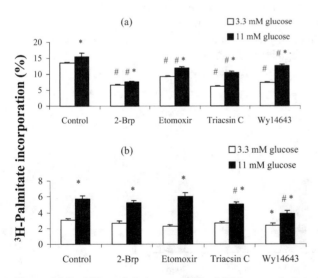

Figure 16.3 Effect of glucose and 200 μM 2-bromopalmitate (2-Brp), 5 μM Etomoxir, 10 μM Triacsin C, and 10 μM Wy14,643 on the incorporation of ³H-palmitate during 24 h in the absence (a) or presence 400 μM palmitate (b) (*n* = 16 for each bar). *$P < 0.05$ compared with group with 3.3 mM glucose in the presence identical reagents. #$P < 0.05$ vs. controls (3.3 or 11 mM glucose only).

accumulation. In the current study, we have found that glucose also causes a dose-dependent inhibition of the *CPT-1* gene expression, which commits long-chain acyl-CoA dehydrogenase (LACD) to β-oxidation in mitochondria (Fig. 16.4). The acyl-CoA synthetase inhibitor triacsin C, causes a decrease in the accumulation of lipids probably because of inhibition of formation of acyl CoA esters. In contrast, the PPARα agonist, Wy14,643, which increases the expression of CPT-1 and also the extramitochondrial fatty acid oxidation reduces the accumulation of lipids. This reduction is correlated with an alleviation of the inhibition of cell proliferation and a decline in insulin secretion.

It is noteworthy, that during long-term cell culture the upregulated expression of genes that are involved in fatty acid oxidation is more pronounced at day 2 than at day 44 (Table 16.3). Thus CPT-1 was induced about sevenfold at day 2 whereas it was induced only about threefold at day 44. During 2 days exposure the expression of CPT-1 did not increase proportionally when the concentration of palmitate was elevated from 400 to 1000 μM.[30] This may reflect that the cells fail to adapt to the long-term/or very high concentration of palmitate. It may lead to lipid accumulation within cells.

We know that corticotropin releasing factor (CRF) and prolactin regulate the β-cell proliferation. Interestingly, the CRF receptor (CRFR) and lactogen receptor expressed in the β-cell lines were found to be downregulated in our study (Table 16.4). Cyclin D2 is an important protein in the cell cycle, the downregulation of this type cyclin may also contribute to the inhibited cell proliferation. We have not focused on apoptosis in our experiments. However, as seen from Table 16.4 we found that the programmed cell death repressor BCL-X long was down regulated, but at the same time the BCL-X short (apoptosis inducer) was also downregulated. Since these two gene products normally work in opposite direction, our finding is puzzling.

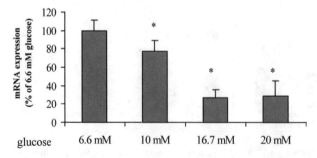

Figure 16.4 The expression of carnitine palmitoyltransferase I (*CPT-1*) gene in INS-1E cells after culture in different concentrations of glucose in presence of 200 μM palmitate for 24 h (*n* = 6). Data presented as ratio to beta actin and normalized with value at 6.6 mM glucose (control). *$P < 0.05$ vs. control.

IR, IRS-1, IRS-2, Sck, IP3 Binding Protein

Obese rats and other diabetic animals have been found to have a low expression of IR in adipose and liver tissues.[31,57]

Interestingly, our study showed a suppression of insulin receptor gene expression after 44 days culture at P200, which is accompanied by the suppression of IRS-2 and cytokine inducible src homology-2 containing protein (Table 16.4 and Xiao et al.[28]). No change was found in the expression of IR during acute exposure to palmitate. The INS-1 cells released more insulin and caused relatively higher insulin levels in the culture medium which may have contributed to the downregulation of the *IR* gene expression. The supplement of insulin receptor antibody in the incubation medium partly inhibits GSIS, indicating that insulin apparently exerts autocrine effects. It is reasonable to speculate that the suppressed IR expression in β-cells may be involved in the abnormal insulin secretion during chronic exposure to fatty acids. We have found that palmitate induces a dose-dependent increase in lipid accumulation, a decrease in the expression of IRS-1 and IRS-2 (Fig. 16.5), a suppression of the insulin secretion, and an inhibition of β-cell proliferation. The IR and the IRS molecules play pleiotropic roles in cell growth, differentiation, and metabolism. Since the discovery of the insulin receptor in insulin-secreting β-cells,[58] a rapidly growing body of evidence indicates that the insulin signal pathway is active and important for the development, replication, and function of β-cells.[59–62] Consequently, the changes in the expression of key proteins in the insulin signal transduction pathway (IR and IRS-1, -2) may contribute to the suppressed insulin secretion and declined cell proliferation during lipotoxicity. The suppression of IRS-2 may contribute to the growth suppression of cells during acute lipotoxicity; however, we have not observed that the suppression of IRS-2 alone was correlated to a decline in insulin secretion. The combined suppression of *IRS-1* and *IRS-2* genes may result in impairment in both insulin and growth factor signal transduction and related functions during acute lipotoxicity. By generating mice with combined heterozygous null mutations in *IR*, *IRS-1*, and *IRS-2*, it

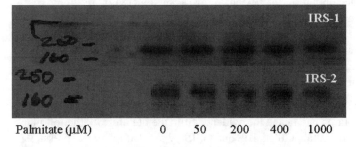

Figure 16.5 The expression of IRS-1 and IRS-2 (Western blotting) in INS-1E cells after 48 h incubation in the presence of 0–1000 µM palmitate.

was reported that diabetes developed in 40% of the $IR/IRS\text{-}1/IRS\text{-}2^{+/-}$, 20% of the $IR/IRS\text{-}1^{+/-}$, 17% of the $IR/IRS\text{-}2^{+/-}$, and 5% of the $IR^{+/-}$ genotypes.[62] The finding that the incidence of diabetes decreases when the defects are reduced corroborates with our results (i.e., the more the defects present in the signal pathway, the more the deterioration of the β-cell function). A concomitant suppression of *IRS-1* and *IRS-2* gene expressions may play an important role in the abnormalities of insulin secretion and cell proliferation during lipotoxicity.

Other Genes Involved in Signal Transduction

Besides the genes mentioned earlier, we also found a few other genes were down-regulated by both short- and long-term fatty acid exposure (Table 16.4). For example, ADP-ribosylation factor 3, protein tyrosine phosphatase, ras-related protein p23, Sck, signal regulatory protein alpha, G alpha 12, protein kinase beta, ras-related rab1B, and signal-transducing G protein alpha 12 subunit were modified. Further studies are required to clarify their roles in lipotoxicity.

CONCLUSION AND PERSPECTIVE

On the basis of our experiments, it seems prudent to conclude, that:

1. Palmitate induces a change in insulin secretion and cell proliferation in the clonal β-cell line INS-1. High concentrations of palmitate suppress insulin secretion in absolute terms whereas low concentrations only suppress the insulin secretion response to nutrient stimulation after long-term exposure.
2. Palmitate modifies a series of gene expression depending on dose and exposure time during lipotoxicity, supporting our hypothesis that many genes are involved in the pathogenesis of lipotoxicity.[28]
3. Palmitate upregulates the expression of genes that are involved in fatty acid oxidation and downregulates expression of genes that are involved in glucose oxidation, suggesting that the fatty acid–glucose cycle was operative at the gene transcription level in the β-cell.
4. An accumulation of β-cell lipids is associated with lipotoxicity. Glucose inhibited the expression of the *CPT-1* gene and increased the accumulation of lipids, explaining why a stimulatory glucose concentration is required to induce lipotoxicity.
5. The fatty acid-induced downregulation of IRS-1 and IRS-2 may contribute to the pathogenesis of lipotoxicity, including the suppression of cell proliferation and insulin secretion.

On the basis of our own data and the evidence from the literature[21] the following scenario on how palmitate affects BIS and GSIS by regulating gene expression in INS-1 cells could be hypothesised: After exposure of INS-1 cells to palmitate, the ATP concentration is augmented from fatty acid oxidation (the major energy source in the presence of low glucose concentration) in the presence of low glucose. Consequently, the increased ATP/ADP ratio closes the ATP-sensitive K-channel and opens the voltage-dependent calcium channels, causing increased insulin release from INS-1 cells (BIS). In contrast, the amount of ATP coming from glucose oxidation is reduced (the major energy source in the presence of high glucose concentration) because of impaired glycolysis, especially in the presence of fatty acids. In accordance with the upregulation of CPT-1, LCAD, and so on, a suppressed ACC may reduce the production of malonyl-CoA (from glycolysis) and uninhibit the activity of CPT-1. Subsequently, more fatty acids will enter the mitochondria for oxidation. The impaired coupling between ATP/ADP and LC-CoA may contribute to the abnormal responsiveness of GSIS.

Our experiments with GeneChip have only been carried out in the clonal β-cell line INS-1. Although cell lines may not be suitable for determining gene expression patterns associated with diseases, for example, diabetes, they are ideal for the identification of downstream targets of a particular gene or signaling pathway. Since palmitate is the only fatty acid used, caution should be exercised when interpreting our results. However, the recently published data on gene expression in Min-6 cells treated with palmitate and oleate[29] confirmed some of our results[28] (e.g., the fatty acids caused changes of the expression of more than 100 genes). Thus the changes in expression of CPT-1 and fructose-1,6 biphophatase show the same pattern. However, discrepancy also exists between the results of Busch et al.[29] and ours.[28] Different cell lines and culture conditions may account for the differences observed. To validate our findings and get more insight, experiments comparing the effect on gene expression, insulin secretion, and cell proliferation of fatty acids with different length, saturation, and spatial configuration of fatty acids are needed. It will be necessary to add studies on pancreatic islets and animals *in vivo* to obtain further insight and provide confidence regarding the biology. To develop the strategy for preventing lipotoxicity, we need to decrease the fatty acid availability (restrict fat intake, increase their oxidation in liver and muscle tissues, and decrease the lipolysis in adipose tissue), and stimulate glucose oxidation in pancreatic islets. This strategy seems to be effective, since the diabetes in ZDF rats, an adipogenic diabetes model, can be prevented or treated with energy restriction, exercise, or troglitazone/rosiglitazone administration.[50,63–65] It will be of interest to test whether the large number of new putative drugs do possess antilipotoxic effects in the β-cell. Interestingly, we found that the diterpene glycoside, stevioside, that exerts antihyperglycemic and blood pressure lowering effects in the diabetic Goto-Kakizaki (GK) rat[65] concomitantly alters the gene profile in the INS-1 cells in the opposite direction of that seen in response to lipotoxicity in a number of genes involved in fatty acid metabolism. Thus stevioside increased FAS and ACC, whereas it suppressed 3-oxoacyl-CoA dehydrogenase, long-chain Acyl CoA dehydrogenase, and CPT-1[65] Fibrates, hypolipidemic agents, that upregulate fatty acid oxidation in the liver have been found to reduce the angiographic progress of coronary-artery disease in type 2 diabetes.[66]

These substances may also prove to be useful in the prevention of lipotoxicity and progressive pancreatic β-cell failure. Lipotoxicity is a complicated phenomenon. GeneChip technology is a suitable method "to find a needle in a haystack". It has been used in relation to diabetes[24] pointing to the fact that the underlying mechanisms of diabetes are not simple. Taking advantage of GeneChip technology many potentially interesting mechanisms have been demonstrated, for example, SCD2 has experienced relatively the most pronounced change in gene expression in our study (Table 16.3), whereas the biological importance of this is unclear.[50] In addition to the alterations in genes involved in glucose and fatty acid metabolism, palmitate induced changes in the gene expression also including transcription factors, ion channel proteins, protein involved in growth, differentiation, and signal transduction. To fully understand the role they play during lipotoxicity more studies are needed.

ACKNOWLEDGMENT

This study was supported by the Danish Medical Research Council, Novo Nordic Foundation, Poul and Erna Sehested Hansens Foundation, the Danish Diabetes Association, and Institute of Experimental Clinical Research, Aarhus University. We thank Lisbet Blak, Kirsten Eriksen, Tove Skrumsager, Dorthe Rasmussen, and Hanne Steen for skilled technical assistance.

REFERENCES

1. Fajans, S.S.; Bell, G.I.; Polonsky, K.S. Molecular mechanisms and clinical pathophysiology of maturity-onset diabetes of the young. N. Engl. J. Med. **2001**, *345*, 971–980.
2. Zimmet, P.; Alberti, K.G.; Shaw, J. Global and societal implications of the diabetes epidemic. Nature **2001**, *414*, 782–787.
3. Zimmet, P.Z. Diabetes epidemiology as a tool to trigger diabetes research and care. Diabetologia **1999**, *42*, 499–518.
4. Pan, X.R.; Li, G.W.; Hu, Y.H.; Wang, J.X.; Yang, W.Y.; An, Z.X.; Hu, Z.X.; Lin, J.; Xiao, J.Z.; Cao, H.B.; Liu, P.A.; Jiang, X.G.; Jiang, Y.Y.; Wang, J.P.; Zheng, H.; Zhang, H.; Bennett, P.H.; Howard, B.V. Effects of diet and exercise in preventing NIDDM in people with impaired glucose tolerance. The Da Qing IGT and Diabetes Study. Diab. Care **1997**, *20*, 537–544.
5. Tuomilehto, J.; Lindstrom, J.; Eriksson, J.G.; Valle, T.T.; Hamalainen, H.; Ilanne-Parikka, P.; Keinanen-Kiukaanniemi, S.; Laakso, M.; Louheranta, A.; Rastas, M.; Salminen, V.; Uusitupa, M. Prevention of type 2 diabetes mellitus by changes in lifestyle among subjects with impaired glucose tolerance. N. Engl. J. Med. **2001**, *344*, 1343–1350.
6. Hu, F.B.; Manson, J.E.; Stampfer, M.J.; Colditz, G.; Liu, S.; Solomon, C.G.; Willett, W.C. Diet, lifestyle, and the risk of type 2 diabetes mellitus in women. N. Engl. J. Med. **2001**, *345*, 790–797.

7. Luan, J.; Browne, P.O.; Harding, A.H.; Halsall, D.J.; O'Rahilly, S.; Chatterjee, V.K.; Wareham, N.J. Evidence for gene-nutrient interaction at the PPARgamma locus. Diabetes **2001**, *50*, 686–689.

8. Kahn, S.E. Clinical review: the importance of beta-cell failure in the development and progression of type 2 diabetes. J. Clin. Endocrinol. Metab. **2001**, *86*, 4047–4058.

9. Matthews, D.R.; Cull, C.A.; Stratton, I.M.; Holman, R.R.; Turner, R.C. UKPDS 26: Sulphonylurea failure in non-insulin-dependent diabetic patients over six years. UK Prospective Diabetes Study (UKPDS) Group. Diab. Med. **1998**, *15*, 297–303.

10. Holman, R.R. Assessing the potential for alpha-glucosidase inhibitors in prediabetic states. Diab. Res. Clin. Pract. **1998** *40* (suppl): S21–S25.

11. McGarry, J.D.; Dobbins, R.L. Fatty acids, lipotoxicity and insulin secretion. Diabetologia **1999**, *42*, 128–138.

12. Tamarit-Rodriguez, J.; Vara, E.; Tamarit, J. Starvation-induced changes of palmitate metabolism and insulin secretion in isolated rat islets stimulated by glucose. Biochem. J. **1984**, *221*, 317–324.

13. Brun, T.; Roche, E.; Assimacopoulos-Jeannet, F.; Corkey, B.E.; Kim, K.H.; Prentki, M. Evidence for an anaplerotic/malonyl-CoA pathway in pancreatic beta-cell nutrient signaling. Diabetes **1996**, *45*, 190–198.

14. McGarry, J.D.; Brown, N.F. The mitochondrial carnitine palmitoyltransferase system. From concept to molecular analysis. Eur. J. Biochem. **1997**, *244*, 1–14.

15. Stein, D.T.; Stevenson, B.E.; Chester, M.W.; Basit, M.; Daniels, M.B.; Turley, S.D.; McGarry, J.D. The insulinotropic potency of fatty acids is influenced profoundly by their chain length and degree of saturation. J. Clin. Invest. **1997**, *100*, 398–403.

16. Elks, M.L. Chronic perifusion of rat islets with palmitate suppresses glucose-stimulated insulin release. Endocrinology **1993**, *133*, 208–214.

17. Alstrup, K.K.; Gregersen, S.; Jensen, H.M.; Thomsen, J.L.; Hermansen, K. Differential effects of cis and trans fatty acids on insulin release from isolated mouse islets. Metabolism **1999**, *48*, 22–29.

18. Prentki, M.; Corkey, B.E. Are the beta-cell signaling molecules malonyl-CoA and cystolic long- chain acyl-CoA implicated in multiple tissue defects of obesity and NIDDM? Diabetes **1996**, *45*, 273–283.

19. Milburn, J.L., Jr.; Hirose, H.; Lee, Y.H.; Nagasawa, Y.; Ogawa, A.; Ohneda, M.; BeltrandelRio, H.; Newgard, C.B.; Johnson, J.H.; Unger, R.H. Pancreatic beta-cells in obesity. Evidence for induction of functional, morphologic, and metabolic abnormalities by increased long chain fatty acids. J. Biol. Chem. **1995**, *270*, 1295–1299.

20. Unger, R.H.; Zhou, Y.T. Lipotoxicity of beta-cells in obesity and in other causes of fatty acid spillover. Diabetes **2001**, *50* (suppl 1), S118–S121.

21. Assimacopoulos-Jeannet, F.; Thumelin, S.; Roche, E.; Esser, V.; McGarry, J.D.; Prentki, M. Fatty acids rapidly induce the carnitine palmitoyltransferase I gene in the pancreatic beta-cell line INS-1. J. Biol. Chem. **1997**, *272*, 1659–1664.

22. Bollheimer, L.C.; Skelly, R.H.; Chester, M.W.; McGarry, J.D.; Rhodes, C.J. Chronic exposure to free fatty acid reduces pancreatic beta cell insulin content by increasing basal insulin secretion that is not compensated for by a corresponding increase in proinsulin biosynthesis translation. J. Clin. Invest. **1998**, *101*, 1094–1101.

23. Zhou, Y.T.; Shimabukuro, M.; Lee, Y.; Koyama, K.; Higa, M.; Ferguson, T.; Unger, R.H. Enhanced de novo lipogenesis in the leptin-unresponsive pancreatic

islets of prediabetic Zucker diabetic fatty rats: role in the pathogenesis of lipotoxic diabetes. Diabetes **1998**, *47*, 1904–1908.

24. Nadler, S.T.; Attie, A.D. Please pass the chips: genomic insights into obesity and diabetes. J. Nutr. **2001**, *131*, 2078–2081.

25. Soukas, A.; Cohen, P.; Socci, N.D.; Friedman, J.M. Leptin-specific patterns of gene expression in white adipose tissue. Genes Dev. **2000**, *14*, 963–980.

26. Nadler, S.T.; Stoehr, J.P.; Schueler, K.L.; Tanimoto, G.; Yandell, B.S.; Attie, A.D. The expression of adipogenic genes is decreased in obesity and diabetes mellitus. Proc. Natl. Acad. Sci. USA **2000**, *97*, 11371–11376.

27. Celis, J.E.; Kruhoffer, M.; Gromova, I.; Frederiksen, C.; Ostergaard, M.; Thykjaer, T.; Gromov, P.; Yu, J.; Palsdottir, H.; Magnusson, N.; Orntoft, T.F. Gene expression profiling: monitoring transcription and translation products using DNA microarrays and proteomics. FEBS Lett. **2000**, *480*, 2–16.

28. Xiao, J.; Gregersen, S.; Kruhøffer, M.; Pedersen, S.B.; Ørntoft, T.F.; Hermansen, K. The effect of chronic exposure to fatty acids on gene expression in clonal insulin-producin cell: studies using high density oligonucleotide microarray. Endocrinology **2001**, *142*, 4777–4784.

29. Busch, A.K.; Cordery, D.; Denyer, G.S.; Biden, T.J. Expression profiling of palmitate- and oleate-regulated genes provides novel insights into effects of chronic lipid exposure on pancreatic beta-cell function. Diabetes **2002**, *51*, 977–987.

30. Xiao, J.; Gregersen, S.; Pedersen, S.B.; Hermansen, K. Differential impact of acute and chronic lipotoxicity on gene expression in INS-1 cells. Metabolism **2002**, *51*, 155–162.

31. Shafrir, E. Development and consequences of insulin resistance: lessons from animals with hyperinsulinaemia. Diab. Metab. **1996**, *22*, 122–131.

32. Unger, R.H. Lipotoxicity in the pathogenesis of obesity-dependent NIDDM. Genetic and clinical implications. Diabetes **1995**, *44*, 863–870.

33. Segall, L.; Lameloise, N.; Assimacopoulos-Jeannet, F.; Roche, E.; Corkey, P.; Thumelin, S.; Corkey, B.E.; Prentki, M. Lipid rather than glucose metabolism is implicated in altered insulin secretion caused by oleate in INS-1 cells. Am. J. Physiol. **1999**, *277*, E521–E528.

34. Gremlich, S.; Bonny, C.; Waeber, G.; Thorens, B. Fatty acids decrease IDX-1 expression in rat pancreatic islets and reduce GLUT2, glucokinase, insulin, and somatostatin levels. J. Biol. Chem. **1997**, *272*, 30261–30269.

35. Zhou, Y.P.; Berggren, P.O.; Grill, V. A fatty acid-induced decrease in pyruvate dehydrogenase activity is an important determinant of beta-cell dysfunction in the obese diabetic db/db mouse. Diabetes **1996**, *45*, 580–586.

36. Zhou, Y.P.; Ling, Z.C.; Grill, V.E. Inhibitory effects of fatty acids on glucose-regulated B-cell function: association with increased islet triglyceride stores and altered effect of fatty acid oxidation on glucose metabolism. Metabolism **1996**, *45*, 981–986.

37. Lee, Y.; Hirose, H.; Ohneda, M.; Johnson, J.H.; McGarry, J.D.; Unger, R.H. Beta-cell lipotoxicity in the pathogenesis of non-insulin-dependent diabetes mellitus of obese rats: impairment in adipocyte-beta-cell relationships. Proc. Natl. Acad. Sci. USA **1994**, *91*, 10878–10882.

38. Lee, C.K.; Klopp, R.G.; Weindruch, R.; Prolla, T.A. Gene expression profile of aging and its retardation by caloric restriction. Science **1999**, *285*, 1390–1393.

39. Lee, C.K.; Weindruch, R.; Prolla, T.A. Gene-expression profile of the ageing brain in mice. Nat. Genet. **2000**, *25*, 294–297.
40. Randle, P.J. Regulatory interactions between lipids and carbohydrates: the glucose fatty acid cycle after 35 years. Diab. Metab. Rev. **1998**, *14*, 263–283.
41. Liu, Y.Q.; Tornheim, K.; Leahy, J.L. Glucose-fatty acid cycle to inhibit glucose utilization and oxidation is not operative in fatty acid-cultured islets. Diabetes **1999**, *48*, 1747–1753.
42. Liang, Y.; Buettger, C.; Berner, D.K.; Matschinsky, F.M. Chronic effect of fatty acids on insulin release is not through the alteration of glucose metabolism in a pancreatic beta-cell line (beta HC9). Diabetologia **1997**, *40*, 1018–1027.
43. Yoshikawa, H.; Tajiri, Y.; Sako, Y.; Hashimoto, T.; Umeda, F.; Nawata, H. Effects of free fatty acids on beta-cell function: a possible involvement of peroxisome proliferator-activated receptors alpha or pancreatic/duodenal homeobox. Metabolism **2001**, *50* (5), 613–618.
44. Jump, D.B.; Clarke, S.D. Regulation of gene expression by dietary fat. Annu. Rev. Nutr. **1999**, *19*, 63–90.
45. Stoffel, M.; Duncan, S.A. The maturity-onset diabetes of the young (MODY1) transcription factor HNF4alpha regulates expression of genes required for glucose transport and metabolism. Proc. Natl. Acad. Sci. USA **1997**, *94*, 13209–13214.
46. Jequier, E. Effect of lipid oxidation on glucose utilization in humans. Am. J. Clin. Nutr. **1998**, *67*, 527S–530S.
47. Carpentier, A.; Mittelman, S.D.; Lamarche, B.; Bergman, R.N.; Giacca, A.; Lewis, G.F. Acute enhancement of insulin secretion by FFA in humans is lost with prolonged FFA elevation. Am. J. Physiol. **1999**, *276*, E1055–E1066.
48. Thorens, B.; Wu, Y.J.; Leahy, J.L.; Weir, G.C. The loss of GLUT2 expression by glucose-unresponsive beta cells of db/db mice is reversible and is induced by the diabetic environment. J. Clin. Invest. **1992**, *90*, 77–85.
49. Jacqueminet, S.; Briaud, I.; Rouault, C.; Reach, G.; Poitout, V. Inhibition of insulin gene expression by long-term exposure of pancreatic beta cells to palmitate is dependent on the presence of a stimulatory glucose concentration. Metabolism **2000**, *49*, 532–536.
50. Finegood, D.T.; McArthur, M.D.; Kojwang, D.; Thomas, M.J.; Topp, B.G.; Leonard, T.; Buckingham, R.E. Beta-cell mass dynamics in Zucker diabetic fatty rats. Rosiglitazone prevents the rise in net cell death. Diabetes **2001**, *50*, 1021–1029.
51. Steil, G.M.; Trivedi, N.; Jonas, J.C.; Hasenkamp, W.M.; Sharma, A.; Bonner-Weir, S.; Weir, G.C. Adaptation of beta-cell mass to substrate oversupply: enhanced function with normal gene expression. Am. J. Physiol. Endocrinol. Metab. **2001**, *280*, E788–E796.
52. Bernard, C.; Berthault, M.F.; Saulnier, C.; Ktorza, A. Neogenesis vs. apoptosis As main components of pancreatic beta cell mass changes in glucose-infused normal and mildly diabetic adult rats. FASEB J. **1999**, *13*, 1195–1205.
53. Kloppel, G.; Lohr, M.; Habich, K.; Oberholzer, M.; Heitz, P.U. Islet pathology and the pathogenesis of type 1 and type 2 diabetes mellitus revisited. Surv. Synth. Pathol. Res. **1985**, *4*, 110–125.
54. Shimabukuro, M.; Zhou, Y.T.; Levi, M.; Unger, R.H. Fatty acid-induced beta cell apoptosis: a link between obesity and diabetes. Proc. Natl. Acad. Sci. USA **1998**, *95*, 2498–2502.

55. Cousin, S.P.; Hugl, S.R.; Wrede, C.E.; Kajio, H.; Myers, M.G, Jr.; Rhodes, C.J. Free fatty acid-induced inhibition of glucose and insulin-like growth factor I-induced deoxyribonucleic acid synthesis in the pancreatic beta-cell line INS-1. Endocrinology **2001**, *142*, 229–240.

56. Iritani, N.; Sugimoto, T.; Fukuda, H.; Komiya, M.; Ikeda, H. Dietary soybean protein increases insulin receptor gene expression in Wistar fatty rats when dietary poly-unsaturated fatty acid level is low. J. Nutr. **1997**, *127*, 1077–1083.

57. Aspinwall, C.A.; Lakey, J.R.; Kennedy, R.T. Insulin-stimulated insulin secretion in single pancreatic beta cells. J. Biol. Chem. **1999**, *274*, 6360–6365.

58. Harbeck, M.C.; Louie, D.C.; Howland, J.; Wolf, B.A.; Rothenberg, P.L. Expression of insulin receptor mRNA and insulin receptor substrate 1 in pancreatic islet beta-cells. Diabetes **1996**, *45*, 711–717.

59. Kulkarni, R.N.; Winnay, J.N.; Daniels, M.; Bruning, J.C.; Flier, S.N.; Hanahan, D.; Kahn, C.R. Altered function of insulin receptor substrate-1-deficient mouse islets and cultured beta-cell lines. J. Clin. Invest. **1999**, *104*, R69–R75.

60. Kulkarni, R.N.; Bruning, J.C.; Winnay, J.N.; Postic, C.; Magnuson, M.A.; Kahn, C.R. Tissue-specific knockout of the insulin receptor in pancreatic beta cells creates an insulin secretory defect similar to that in type 2 diabetes. Cell **1999**, *96*, 329–339.

61. Withers, D.J.; Burks, D.J.; Towery, H.H.; Altamuro, S.L.; Flint, C.L.; White, M.F. IRS-2 coordinates IGF-1 receptor-mediated beta-cell development and peripheral insulin signaling. Nat. Genet. **1999**, *23*, 32–40.

62. Kido, Y.; Burks, D.J.; Withers, D.; Bruning, J.C.; Kahn, C.R.; White, M.F.; Accili, D. Tissue-specific insulin resistance in mice with mutations in the insulin receptor, IRS-1, and IRS-2. J. Clin. Invest. **2000**, *105*, 199–205.

63. Ohtani, K.I.; Shimizu, H.; Sato, N.; Mori, M. Troglitazone (CS-045) inhibits beta-cell proliferation rate following stimulation of insulin secretion in HIT-T 15 cells. Endocrinology **1998**, *139*, 172–178.

64. Kakuma, T.; Lee, Y.; Higa, M.; Wang, Z.; Pan, W.; Shimomura, I.; Unger, R.H. Leptin, troglitazone, and the expression of sterol regulatory element binding proteins in liver and pancreatic islets. Proc. Natl. Acad. Sci. USA **2000**, *97*, 8536–8541.

65. Jeppesen, P.B.; Gregersen, S.; Rolfsen, S.E.D.; Jepsen, M.; Colombo, M.; Agger, A.; Xiao, J.; Kruhøffer, M.; Ørntoft, T.; Hermansen, K. Antihyperglycemic and blood pressure-reducing effects of stevioside in the diabetic Goto-Kakizaki (GK) rat. Metabolism **2003**, *52*, 372–378.

66. Diabetes Atherosclerosis Intervention Study Investigators. Effect of fenofibrate on progression of coronary-artery disease in type 2 diabetes: the Diabetes Atherosclerosis Intervention Study, a randomised study. Lancet **2001**, *357*, 905–910.

17

Use of Microarrays to Investigate the Transcriptional Effects of Protein Tyrosine Phosphatase IB (PTP1B): Implications for Diabetes and Obesity

Cristina M. Rondinone and Jeffrey F. Waring

Metabolic Diseases Research and Department of Cellular and Molecular Toxicology, Global Pharmaceutical Research and Development, Abbott Laboratories, Abbott Park, IL, USA

INTRODUCTION

Type 2 diabetes is a polygenic disease affecting over 100 million people worldwide. Affected patients manifest insulin resistance, hyperinsulinemia, and hyperglycemia.[1] Insulin resistance in peripheral tissues including liver, fat, and skeletal muscle characterizes type 2 diabetes and is also associated with obesity, hypertriglyceridemia, and hypertension. The molecular mechanism underlying insulin resistance is not well understood but appears to involve a defect in the postinsulin receptor (IR) signal transduction pathway.[2]

Insulin is a hormone secreted from pancreatic β-cells in response to increasing glucose concentrations in the blood. Insulin binds to its receptor, a tetrameric complex composed of two α- and two β-subunits (for a review on insulin signaling see Saltiel and Kahn[3]), triggering a conformational change that activates the intrinsic tyrosine kinase activity of the intracellular β-subunit via autophosphorylation of at least six tyrosine residues in the activation loop. Autophosphorylation of tyrosine residues 1146, 1150, and 1151 in the kinase domain activates the insulin receptor kinase and causes the phosphorylation of other protein substrates, including IR substrates (IRS-1 to -4) and other adapter proteins (Gab1 and Shc) that mediate the biological effects of insulin[4,5] (Fig. 17.1). The phosphorylated IRSs serve as adaptor proteins and recruit phosphatidylinositol 3-kinase (PI3K) via the regulatory subunit. PI3K then catalyzes the conversion of phosphatidylinositol to the 3,4-bis- and 3,4,5-trisphosphates that stimulate the activity of phosphoinositide-dependent kinase-1 and -2 (PDK1 and PDK2).[5,6] These phosphoinositide-dependent kinases phosphorylate and activate protein kinase B (PKB) leading to the translocation of the glucose transporter GLUT-4 and to the uptake of glucose into cells (Fig. 17.1).[7–10] In addition, PKB appears to participate in the pathway by phosphorylating glycogen synthase kinase 3 (GSK3) to promote glycogen synthesis via glycogen synthase (GS). GSK3 is constitutively active and phosphorylates GS to inactivate this enzyme, which is required for the incorporation of glucose. Phosphorylation of GSK3 by PKB inactivates the kinase and relieves its block on GS.[11–13]

Protein tyrosine kinases and protein tyrosine phosphatases (PTPs) are known to be important regulators of insulin signal transduction. Much attention has been focused on the dephosphorylation of key tyrosine residues in the activation loop of the receptor and of IRS proteins by PTPs. These phosphatases are represented by a diverse family of ∼100 members, which encompasses both receptor-linked and nontransmembrane enzymes.[14] At least three different phosphatases have been implicated as negative regulators of insulin signaling. These include the transmembrane receptor type phosphatases RPTP1α, LAR, and PTP1B.[15] PTP1B is expressed in all insulin-responsive tissues and it is

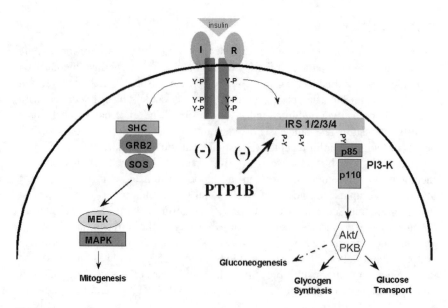

Figure 17.1   Inhibitory effect of protein tyrosine phosphatase 1B (PTP1B) on insulin signaling.

localized primarily on intracellular membranes. PTP1B expression in Rat-1 cells decreased insulin receptor autophosphorylation and glucose incorporation into glycogen.[16] Transient transfection of PTP1B into primary adipocytes impairs insulin-stimulated GLUT-4 translocation.[17] Osmotic loading of PTP1B antibodies into cells enhances insulin-stimulated receptor autophosphorylation and IRS-1 phosphorylation resulting in enhanced insulin signaling.[18] A selective phosphotyrosyl mimetic peptide with an IC50 of 180 nM for PTP1B has been demonstrated to reverse the impairment of insulin-stimulated translocation of GLUT-4 caused by overexpression of PTP1B in rat adipocytes.[19] In addition, overexpression of PTP1B in L6 myocytes and FAO cells inhibited insulin-stimulated tyrosine phosphorylation of IR and IRS-1 and resulted in a significant inhibition of PKB phosphorylation.[20] PTP1B reduction by PTP1B antisense oligonucleotide (ASO) markedly increased insulin-stimulated tyrosine phosphorylation of IR and IRS-1 and resulted in a significant stimulation of PKB in FAO rat hepatoma cells.[21] PTP1B has been also demonstrated to directly dephosphorylate and inactivate IRS-1 suggesting another mechanism to negatively regulate insulin signaling.[22]

The loss of PTP1B potentiates insulin's activity, which would suggest that PTP1B is a negative regulator of insulin signaling (Fig. 17.1). This would place PTP1B downstream of the IR, and presumably it functions to dephosphorylate and inactivate the IR.[23] Alternatively, or in addition to its activity on the IR, PTP1B potentially may attenuate insulin signaling by dephosphorylating IRSs or possibly other phosphotyrosyl insulin-dependent signaling molecules yet to be identified. Although proof that PTP1B directly interacts with the IR in a cellular or *in vivo* context is equivocal, there is a significant amount of evidence to suggest that this is probably the case. The dynamics of the interaction of the insulin receptor with a substrate-trapping mutant of PTP1B has been monitored in living human embryonic kidney cells using bioluminescence resonance energy transfer and insulin dose-dependently stimulates this interaction suggesting that in intact cells the tyrosine-kinase activity of the insulin receptor is tightly controlled by PTP1B.[24]

The involvement of PTP1B in human insulin resistance and diabetes is supported in several studies.[25−27] Increased PTP1B expression and activity has been observed in tissues of obese insulin-resistant patients and was reduced in parallel with the increased insulin sensitivity that accompanied weight loss.[28]

PTP1B AND INSULIN RESISTANCE

The most convincing evidence that PTP1B is involved in the insulin-signaling pathway originates from the phenotype of the PTP1B knockout (KO) mouse[29,30] and, more recently, from results of PTP1B ASO treatments in diabetic rodents.[31−33] PTP1B deficiency in mice results in enhanced insulin sensitivity, as demonstrated by a significant reduction in fed glucose levels that is maintained with one-half the circulating insulin levels.[29] Additionally, there is increased insulin-stimulated phosphorylation of the IR in muscle and liver and an improved glucose clearance in glucose and insulin tolerance tests.

Mice lacking a functional PTP1B gene exhibit increased insulin sensitivity in liver and skeletal muscle but fail to show increased insulin sensitivity in fat.[29,30] They are resistant to weight gain on a high-fat diet[29] and are reported to have low adiposity owing to a marked reduction in fat cell mass without a decrease in adipocyte number.[30] The reason for the obesity resistance observed in PTP1B$^{-/-}$ mice is unclear. Increased leptin sensitivity in PTP1B$^{-/-}$ mice has been suggested as a mechanism for increased energy expenditure.[30] Although these reports demonstrate a major role of PTP1B in the modulation of insulin sensitivity in liver and muscle, they fail to address the effect of reducing PTP1B expression in adipose tissue in obese diabetic animals.

PTP1B AND OBESITY

An unexpected phenotype of the PTP1B KO mouse was its resistance to diet-induced obesity. Because insulin promotes the storage of glucose and fat, it was expected that PTP1B KO mice would be rather more susceptible to obesity. Several factors appear to contribute to this phenotype. For instance, PTP1B KO mice have been reported to exhibit enhanced leptin sensitivity.[34,35] It has been suggested that this may be due to PTP1B acting as a negative regulator of leptin signaling by dephosphorylating the leptin receptor-associated kinase Jak2.[34,36,37] Although a role for PTP1B in leptin signaling seems possible, the studies reported with the PTP1B KO mouse do not conclusively implicate a role for the phosphatase in leptin signaling.

An additional factor of the PTP1B KO mice that may be influencing their obesity resistance is that these animals display tissue-specific insulin sensitivity. Liver and muscle are sensitive to insulin-stimulated phosphorylation of the IR, whereas adipose tissue sensitivity is not any different from that of wild-type littermates.[29,30] The fact that the PTP1B KO mice fail or have a decreased ability to store fat suggests a different role for the phosphatase in fat tissue. In fact, recent results in *ob/ob* mice treated with the PTP1B ASO suggest that it is more likely that PTP1B has a different role in adipose tissue[33,38] (Fig. 17.2). Adipose tissue of *ob/ob* mice that were treated with the PTP1B ASO had a significant decrease in adiposity that was associated with a downregulation of genes involved in lipogenesis; insulin sensitivity in this tissue was not changed relative to *ob/ob* control mice. Therefore, a reduction in PTP1B levels in adipose tissue by genetic or antisense methods affects fat storage but does not enhance insulin sensitivity.

MICROARRAY STUDIES IN FAT

Microarray analysis allows monitoring the expression of thousands of genes at the same time by quantitating the expression levels of messenger RNA (mRNA).[39]

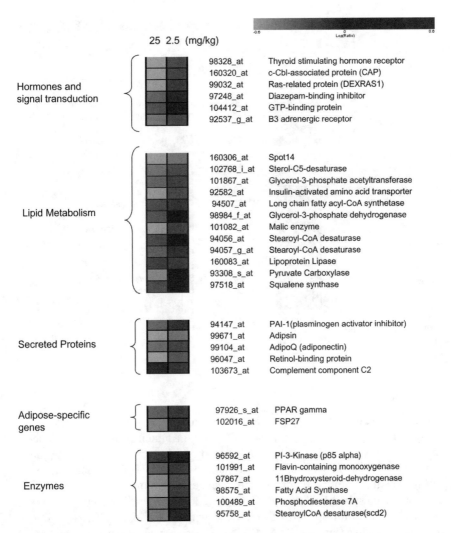

Figure 17.2 Gene expression changes for some of the genes regulated by PTP1B ASO treatment in *ob/ob* adipose tissue. The change in gene expression is shown relative to saline-treated control mice. The 25 mg/kg samples were assayed by microarray twice for confirmation. (Reprinted from Waring et al.,[38] with permission from Elsevier.)

Microarray technology has been applied to a number of different biological systems. Studies have applied microarray analysis to learn about entire biological pathways, to classify tissue types, to identify the function of previously unknown genes, and, in the case of drug discovery, to determine both intended (on-target) and untoward (off-target) effects.[40–43]

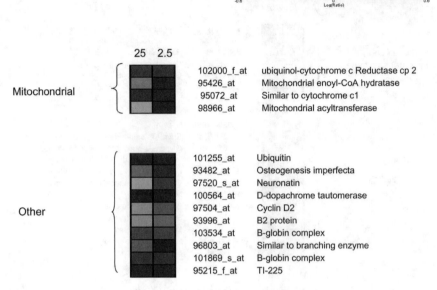

Figure 17.2   Continued.

Lipogenesis

Recently, we have investigated the effects of PTP1B antisense treatment on insulin resistance and the regulation of fat metabolism in *ob/ob* diabetic mice.[31–33] Antisense technology offers a powerful method to study gene function in an organism without the potential developmental issues inherent with using KO mice. Interestingly, hyperglycemic *ob/ob* mice as well as *db/db* mice treated with the PTP1B ASO displayed a significant reduction in blood glucose and insulin levels that correlated with a decrease in PTP1B protein expression in liver and fat.[31] A cDNA microarray approach was used to study gene expression in fat from *ob/ob* mice treated with the PTP1B ASO (Fig. 17.2). Genes involved in lipid metabolism and lipogenesis were influenced by the reduction of PTP1B in fat. In addition, a small but significant difference on growth rate was found with antisense treatment, and epididymal fat weight was significantly reduced by 42%.[33] Furthermore, results of magnetic resonance image analysis have demonstrated that abdominal fat and subcutaneous fat storage were reduced by 27% and 16%, respectively, with PTP1B ASO treatment in Zucker fatty rats.[44]

In rodents, elevated lipogenesis plays a crucial role in the overproduction of fat. It is well known that hypertrophied adipose depots of animals in which obesity results from the disruption of the leptin signaling pathway (such as

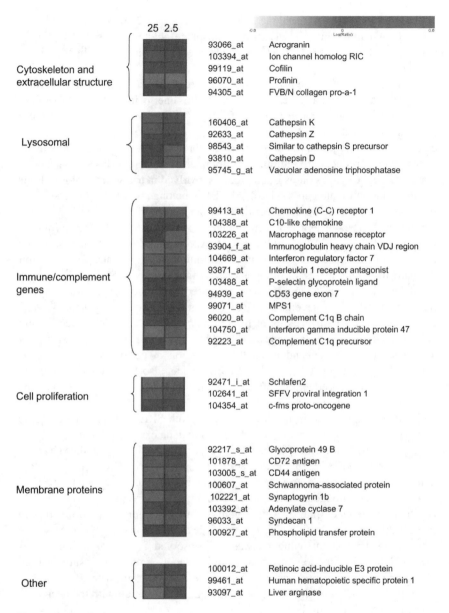

Figure 17.2 Continued.

ob/ob and *db/db* mice and *fa/fa* rats) exhibit disregulated lipogenesis.[45] One of the transcription factors that has been shown to regulate the expression of several key genes of fatty acid and triglyceride metabolism is sterol regulatory element-binding protein (SREBP)-1.[46] The family of SREBP-responsive genes, include several genes that control lipid metabolism, such as fatty acid synthase (FAS),[47] acetyl-CoA carboxylase,[48] glycerol-3-phosphate acyltransferase,[49] leptin,[50] and lipoprotein lipase (LPL),[51] which are important for adipose tissue metabolism.

Adipose tissue is the primary site of energy storage, building triglycerides in response to nutritional excess and releasing this energy in the form of free fatty acids (FFA) and glycerol in times of fasting.[52] The ability to carry out these functions efficiently is dependent on changes in the expression of genes that carry out the lipogenesis and the lipolytic program. Among these are many genes whose products are required for lipogenesis, including FAS, ATP citrate lyase, stearoyl CoA desaturase, and proteins involved in triglyceride biosynthesis (PEPCK and pyruvate carboxylase). SREBP-1 belongs to a family of transcription factors involved in trygliceride and fatty acid metabolism,[49,50,53] and a number of these genes have been shown to be regulated at the transcriptional level by SREBP-1.[46,50] In addition, overexpression of FAS in adipose tissue of obese animals, such as *ob/ob* and *db/db* mice and *fa/fa* rats, which exhibit dramatically high levels of lipogenic activities, involves SREBP-1.[47,53–55] Interestingly, PTP1B ASO treatment decreased expression of the transcription factor SREBP-1 and several SREBP-1 target genes, including spot14, ATP citrate lyase, and FAS in adipose tissue of *ob/ob* mice.[33] This effect would be expected to reduce lipogenesis. In addition, SREBP-1 and its expression are, at least in part, dependent on the expression of peroxisome proliferator-activated receptor-γ (PPARγ).[56,57] SREBP-1 has also been found to activate PPARγ through production of an endogenous ligand[58] or through the PPARγ-1 and -3 promoters[57] and to stimulate expression of leptin.[50] Interestingly, PTP1B reduction in fat downregulated both SREBP-1 and PPARγ gene expression, resulting in a reduction of lipogenic gene expression that contributes to adipocyte hypertrophy. This result is in agreement with other studies showing that moderate reduction of PPARγ activity observed in heterozygous PPARγ-deficient mice decreases triglyceride content in fat, and direct antagonism of PPARγ reduces lipogenesis in fat and prevents adipocyte hypertrophy.[59] In addition, the reduced rate of weight gain and reduced adiposity observed with PTP1B ASO treatment is consistent with previous reports in KO mice where whole-body fat content reduction was also observed.[29,30] In addition, it was expected that a reduction in obesity might lead to improved insulin action. Consistent with these findings the triglyceride content of fat from PTP1B antisense-treated mice was significantly decreased, in correlation with the decreased SREBP-1 expression in these mice.[33]

Obesity generally involves an increase in both cell size and cell number of adipocytes. SREBP-1 may represent a mechanism unifying both.[50] It can increase the percentage of cells that differentiate into adipocytes[51] by increasing the transcriptional activity of PPARγ. Hence, chronic overfeeding and its attending elevation of insulin would be expected to result in increased expression and activity of SREBP-1. This is likely to affect not only lipid accumulation in fat cells through increased expression of FAS, LPL, and other lipogenic genes but also increased differentiation of adipocytes from precursor cells.[50] Although SREBP-1 gene expression is upregulated by insulin and glucose *in vitro*,[50,60] the *in vivo* significance of these factors in SREBP-1 regulation is a matter of controversy. It is likely, however, that in the intact animal, hyperinsulinemia and altered nutritional conditions characteristic of the obese state might induce SREBP-1, additionally contributing to altered adipocyte metabolism so that a decrease in plasma insulin levels might lead to a reduction in SREBP-1 expression. Yet, PPARγ activators induce an increase in the expression of lipogenic genes and fat mass, while simultaneously acting as insulin sensitizers, decreasing insulin and glucose levels in the animals.[61] This is supported by clinical data showing that patients who take synthetic PPARγ activators frequently gain weight.[62] Furthermore, heterozygous PPARγ mutant mice exhibit smaller fat stores on a high-fat diet.[56,63]

Our results are similar to a study of leptin treatment in which a reduction was observed in the expression of genes encoding lipogenic genes such as SREBP-1, spot14, and FAS.[64] Thus, mRNA levels of genes involved in fatty acid and cholesterol synthesis are rapidly repressed by leptin administration, in association with an acute decrease in plasma insulin levels and decreased SREBP-1 expression.[65]

Recently, microarray technology has been applied to study the regulation of genes in diabetic mouse models.[64,66] In two separate cases, researchers showed that in adipose tissue from *ob/ob* mice, which have a missense mutation in the gene encoding leptin, the pattern of gene expression was the reverse of the pattern observed during adipocyte differentiation. This suggests that adipocytes from *ob/ob* mice have decreased lipogenic capacity, similar to preadipocytes.[67] The lack of lipogenic adipocytes in *ob/ob* mice results in a shift of lipogenesis from adipose tissue to the liver, resulting in hepatic steatosis and the upregulation of genes in the liver such as SREBP-1, FAS, ATP citrate lyase, and malic enzyme.[68]

In summary, our results demonstrated that PTP1B reduction decreased fat depots and triglyceride levels in fat by downregulating important genes known to be involved in lipogenesis.[33,38] Our observation provides the first evidence of a PTP1B effect in adipose tissue, with a possible implication of PTP1B in the enlargement of adipocyte energy stores and obesity.

Adipogenesis

The first hallmark of the adipogenesis process is the dramatic alteration in cell shape as the cells convert from fibroblastic to spherical shape. These morphological modifications are paralleled by changes in the level and type of extracellular matrix (ECM) components and the level of cytoskeletal components.[69] Recent findings indicate that these events are key for regulating adipogenesis as they may promote expression of critical adipogenic transcription factors, including

CCAAT/enhancer binding protein-α (C/EBPα) and PPARγ. In addition, genes differentially regulated during adipogenesis have been categorized into early, intermediate, and late mRNA/protein markers.[69–72] Moreover, progressive acquisition of the adipocyte phenotype is associated with changes in the expression of over 2000 genes, as highlighted in a recent study using microarray technology to monitor global changes in gene expression profiles during 3T3-L1 differentiation.[73] Several transcription factor families that exhibit diverse modes of activation and function are key regulators of the adipogenesis process. Members of the C/EBP family and PPARγ-2 are involved in terminal differentiation by their subsequent transactivation of adipocyte-specific genes.[71,72] Exposure of confluent preadipocytes to the adipogenic cocktail induces expression of C/EBPβ and C/EBPδ, which in turn activate PPARγ-2 and C/EBPα. Production of the appropriate ligand of PPARγ-2 by the differentiating preadipocyte is likely a limiting step in this transcriptional cascade. SREBP-1 is another key transcription factor known to modulate transcription of numerous liver genes involved in lipid metabolism as well as adipocyte differentiation.[46,72] SREBP-1 and more recently C/EBPβ and C/EBPδ were proposed to play roles in regulation of PPARγ-2 ligand production.[72,74] In addition to C/EBPs, PPARγ-2, and SREBP-1, several other transcription factors, including GATA-binding transcription factors GATA-2 and GATA-3, and cAMP response element binding protein (CREB), play a critical role in the molecular control of the preadipocyte–adipocyte transition.[75,76] GATA-2 and GATA-3 are specifically expressed in white preadipocytes, and their mRNAs are downregulated during adipocyte differentiation. Constitutive expression of GATA-2 and GATA-3 suppresses adipocyte differentiation and traps cells at the preadipocyte stage. This effect is mediated, at least in part, through direct suppression of the activity of the PPARγ-2 promoter. The transcription factor CREB is constitutively expressed prior to and during adipogenesis, and is upregulated by differentiation. Overexpression of a constitutively active CREB in 3T3-L1 preadipocytes is necessary and sufficient to initiate adipogenesis, whereas overexpression of a dominant-negative CREB alone blocks adipogenesis in cells treated with conventional differentiation-inducing agents.[75]

During the terminal phase of differentiation, activation of the transcriptional cascade leads to increased activity, protein, and mRNA levels for enzymes involved in triacylglycerol synthesis and degradation. Glucose transporters, insulin receptor number, and insulin sensitivity also increase. Synthesis of adipocyte-secreted products including leptin, adipsin, resistin, and adipocyte-complement-related protein (adiponectin) begins, producing a highly specialized endocrine cell that will play key roles in various physiological processes.[77]

Microarray results from fat of animals treated with PTP1B ASO showed that a number of genes shown to be upregulated during adipocyte differentiation were downregulated with PTP1B ASO treatment, including spot14, adipsin, retinal-binding protein, c-Cbl-associating protein, PAI-1, and adiponectin.[33,38] Interestingly, many of these genes are regulated by SREBP-1 and PPARγ and

PTP1B ASO induces the downregulation of these transcription factors. In contrast, some other genes such as C/EBPβ, C/EBPδ, CHOP-10, and GLUT-4, that have been shown to be highly regulated during adipocyte differentiation, were not regulated by PTP1B ASO treatment. Previous studies on adipocyte differentiation *in vitro* and *in vivo* have shown that there are numerous transcriptional programs and phases necessary for adipogenesis.[78] Soukas et al.[78] obtained similar results concerning the regulation of genes involved in lipid metabolism of adipocytes from *ob/ob* vs. lean mice. The results also showed a downregulation of genes that have been shown to be upregulated in adipocyte differentiation. In addition, gene expression changes were characterized when *ob/ob* mice were treated with leptin, a hormone produced in adipocytes which functions in satiety and regulation of adipose tissue mass.[79] Microarray results showed that treatment with leptin caused a further downregulation of some of the genes, such as FAS, involved in lipid metabolism, while normalizing others.[59] Previous studies have suggested that leptin activates a program that reverses differentiation in adipocytes.[80]

Many of the genes involved in the adipocyte differentiation, such as spot14, adipsin, retinol-binding protein, and malic enzyme, were found to be differentially expressed in adipose tissue between *ob/ob* and lean mice.[66] These genes have also been shown to be upregulated during adipocyte differentiation.[81] In fact, the microarray results show that approximately half of the genes shown to be regulated during adipocyte differentiation were regulated in the opposite manner with PTP1B ASO treatment. This is in contrast to treatment with thiazolidinediones (TZDs), which have been shown to act by causing adipocyte differentiation and have been shown to upregulate genes indicative of mature adipocytes.[82,83] For instance, genes such as adipsin, c-Cbl-associating protein, and plasminogen activator inhibitor-1 (PAI-1), which are expressed in highly differentiated adipocytes, have been shown to be upregulated by treatment with rosiglitazone or other TZDs but they are downregulated with treatment with PTP1B ASO.[84,85] These insulin-sensitizing drugs are ligands for PPARγ and have been shown to cause adipocyte differentiation. This adipocyte differentiation, in the absence of increased energy storage, would produce more fat cells of smaller average size, which are more sensitive to insulin-dependent glucose uptake.[86]

Since PTP1B ASO treatment does not lead to adipocyte differentiation, it is interesting to speculate that treatment with an antagonist to PTP1B would not lead to the weight gain seen with TZD treatment. In fact, as opposed to what is seen with TZD-treated mice, *ob/ob* mice treated with PTP1B ASO more closely resemble what is seen in mice with decreased PPARγ expression. Mice treated with PTP1B ASO also showed a downregulation of PPARγ gene expression. Yamauchi et al.[87] showed that heterozygous PPARγ mice displayed decreased lipogenesis in white adipose tissue, liver, and muscle were resistant to increased adipose mass on high-fat diets, and showed increased serum leptin levels and improved insulin resistance. Possibly, decreasing PTP1B expression may improve insulin sensitivity through some of the same mechanisms as seen with decreased PPARγ expression.

GENES AFFECTING INSULIN SENSITIVITY

Another gene that was shown to be downregulated in adipose tissue by treatment with PTP1B ASO, and which might be important for the normalization of glucose levels and improvement of insulin sensitivity, is 11β-hydroxysteroid dehydrogenase type 1 (11β-HSD1). 11β-HSD1 is widely expressed in numerous tissues, including liver, adipose, and the central nervous system. Recent studies have suggested that 11β-HSD1 amplifies glucocorticoid action in the liver, adipose tissue, and brain.[88] Glucocorticoid excess leads to insulin resistance/type II diabetes and dislipidemia. Mice deficient for 11β-HSD1 showed increased lipid catabolism with *ad lib* feeding, reduced intracellular glucocorticoid concentrations during fasting, and increased hepatic insulin sensitivity after refeeding.[89] Finally, recent studies have shown that 11β-HSD1 is increased in adipose tissue from obese patients when compared with lean patients.[90] The fact that lowering PTP1B results in decreased levels of 11β-HSD1 is intriguing. Whether this is due to secondary effects or possibly a direct effect of PTP1B is unclear at this time.

Tumor necrosis factor alpha (TNF-alpha) has been purported to play an important role in insulin resistance, potentially through PPARγ.[85,91] No change in RNA expression of TNF-alpha in either fat or liver was observed,[38] suggesting that PTP1B ASO treatment restores insulin sensitivity through a mechanism other than reducing TNF-alpha expression.

MICROARRAY STUDIES IN LIVER

Disregulation of fat metabolism, resulting in adipose tissue becoming refractory to suppression of fat mobilization by insulin, occurs early in the development of insulin resistance. This results in an excess of circulating lipid metabolites that would normally have been absorbed by adipose tissue and leads to fat deposition in other tissues, such as skeletal muscle and liver.[92] Therefore a crucial element in any treatment for type 2 diabetes is to reverse the flux of FFAs from the adipose tissues to the liver.

In order to better understand the function of PTP1B in liver steatosis, we have utilized microarrays to characterize the response to PTP1B ASO treatment in *ob/ob* mice. Histopathological examination of livers from *ob/ob* mice treated with PTP1B ASO showed a marked reduction in lipid accumulation and resulted in a downregulation of genes involved in lipogenesis in liver.[38] These data suggest PTP1B improved insulin sensitivity in *ob/ob* mice without causing an increase in adipocyte mass or increased liver steatosis, as seen in some cases with TZD therapy, and suggest that therapeutic modalities targeting PTP1B inhibition may have clinical benefit in type 2 diabetes.

In genetically obese animals, the expression of many lipogenic genes in liver and muscle increase as a result of lipid storage being shifted to the liver.[67] In agreement with this, histopathology findings showed reduced levels of lipid in livers from *ob/ob* mice treated with PTP1B ASO. Figure 17.3 depicts the difference in histologically detectable hepatocellular lipid accumulation in the livers of *ob/ob* mice treated with PTP1B ASO compared with saline control mice. Saline-treated mice consistently exhibited marked diffuse hepatocellular lipid accumulation, whereas PTP1B ASO-treated mice exhibited mild or occasionally moderate focal to multifocal hepatocellular lipid accumulation. Although it has been shown that lipogenesis can also shift to the muscle in genetically obese animals, similar gene changes were not seen in the muscle.

Figure 17.4 shows a heat map of gene expression changes in the liver and muscle from *ob/ob* mice treated with PTP1B ASO compared with saline-treated controls. PTP1B ASO treatment resulted in a decrease in genes involved in lipogenesis in liver including SREBP-1, ATP citrate lyase, spot14, and malic enzyme, among others.

Lipogenic enzymes which are involved in energy storage through synthesis of fatty acids and triglycerides are coordinately regulated at the transcriptional level during different metabolic states.[93,94] Recent *in vivo* studies suggest that

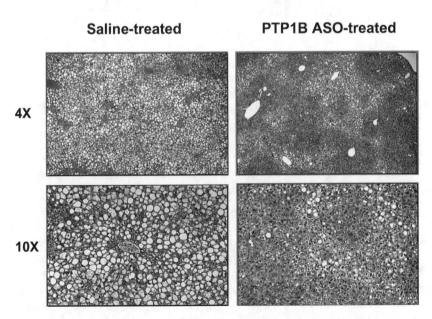

Saline-treated   **PTP1B ASO-treated**

4X

10X

Figure 17.3 Histopathology slide showing sections of liver from an *ob/ob* mouse treated with saline or with PTP1B ASO at 25 mg/kg. The top sections are at 4× magnification, the bottom sections are at 10× magnification. (Reprinted from Waring et al.,[38] with permission from Elsevier.)

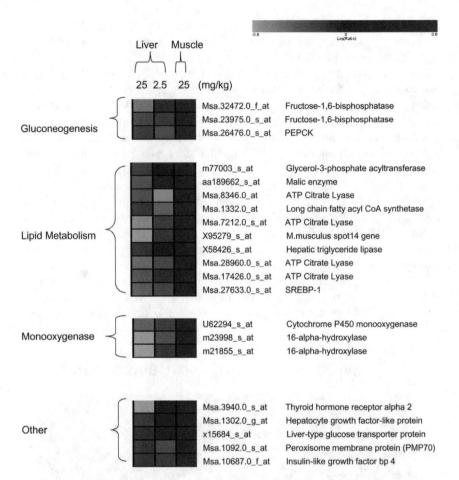

Figure 17.4 Gene expression changes in liver and muscle from *ob/ob* mice treated with PTP1B ASO relative to saline-treated control mice. The results in liver are from mice treated at 25 and 2.5 mg/kg and the results in muscle are from mice treated at 25 mg/kg. (Reprinted from Waring et al.,[38] with permission from Elsevier.)

SREBP-1 plays a crucial role in the dietary regulation of most hepatic lipogenic genes. These include studies of the effects of the overexpression of SREBP-1 on hepatic lipogenic gene expression in transgenic mice,[95,96] as well as physiological changes of SREBP-1 protein in normal mice after dietary manipulation such as placement on fasting–refeeding regimens.[97] The similarly coordinated changes in SREBP-1 and lipogenic gene expression at fasting and refeeding were also observed in adipose tissue.[98]

Absence of SREBP-1 severely impaired the marked induction of hepatic mRNAs of fatty acid synthetic genes such as acetyl-CoA carboxylase, FAS, and stearoyl-CoA desaturase-1 (SCD-1), that was observed upon refeeding in the wild-type mice.[99] Furthermore, the refeeding responses of other lipogenic enzymes (glycerol-3-phosphate acyltransferase, ATP citrate lyase, malic enzyme, glucose-6-phosphate dehydrogenase, and spot14 mRNAs were completely abolished in SREBP-1 KO mice. These data basically confirmed that SREBP-1 plays a crucial role in the induction of lipogenesis, but not in cholesterol biosynthesis, in liver when excess energy by carbohydrates is consumed. However, the extent to which the gene regulation depends upon SREBP-1 considerably differs among the lipogenic genes and organs. For instance, glycolytic enzymes such as pyruvate kinase and glucokinase are sometimes classified as lipogenic enzymes, but are not highly regulated by SREBP-1. Gene expression of FAS and SCD-1 is highly controlled by SREBP-1, but other factors for FAS regulation also seem to be involved. Another interesting observation was made in adipose tissue of refed SREBP-1 KO mice.[99] The absence of SREBP-1 did not severely affect overall gene expression of lipogenic enzymes in fat as observed in the livers, suggesting that control of lipogenic gene expression might show some tissue specificity.

Interestingly, a recent paper showed that SREBP-1 was induced in livers from mice deficient for IRS-2.[100] As PTP1B ASO-treated mice show an increase in IRS-2 expression, this may account for the downregulation of SREBP-1 in the liver.[31,33]

The downregulation of genes involved in lipogenesis in the liver is in contrast to what has been seen previously with TZD treatment, where an increase in hepatic lipogenic gene expression was seen.[61] Previous research in Zucker diabetic rats has shown that TZDs generally have little effect on hepatic lipogenesis[101] or actually result in an increase in lipogenic gene expression.[61] Lipid accumulation in the liver has been shown to play a key role in insulin resistance, and the alleviation of this condition may be one mechanism whereby PTP1B ASO improves insulin sensitivity.[102]

PTP1B ASO treatment resulted in a decrease in genes involved in gluconeogenesis such as PEPCK, fructose-1,6-bisphosphatase, and glucose-6-phosphatase. Previous research has shown that mice deficient in leptin levels display increased expression of genes involved in gluconeogenesis.[103] Additionally, treatment of perfused livers with leptin resulted in decreased gluconeogenesis.[104] Whether treatment with PTP1B ASO results in a direct or indirect effect on gluconeogenesis is unclear. It has been suggested that excessive hepatic fatty acid synthesis might be the cause of insulin resistance.[105] Further research is necessary to determine if the primary target for PTP1B ASO treatment is in adipocytes or in liver.

None of the genes involved in adipogenesis or gluconeogenesis were regulated in muscle. Previous studies have shown that antisense treatment in animals shows high distribution in fat and liver, with little distribution in muscle.[106] Thus, it is very possible the gene changes seen in this tissue were due to secondary effects.

CONCLUSIONS

Treatment of *ob/ob* mice with PTP1B antisense results in a downregulation of genes involved in lipogenesis, adipogenesis, and gluconeogenesis. PTP1B ASO may be acting in an opposite manner to PPARγ activation and possibly may be acting downstream of the leptin pathway. On the basis of our results we would postulate that mice expressing decreased levels of PTP1B would display increased leptin sensitivity. Our results strongly suggest that therapies targeting PTP1B should result in lower blood glucose and insulin sensitivity without leading to increased weight gain seen with other treatments.

REFERENCES

1. Reaven, G.M. Role of insulin resistance in human disease. Diabetes **1988**, *37*, 1595–1607.
2. Olefsky, J.M.; Garvey, W.T.; Henry, R.R.; Brillon, D.; Matthaei, S.; Freidenberg, G.R. Cellular mechanisms of insulin resistance in non-insulin-dependent (type II) diabetes. Am. J. Med. **1988**, *85*, 86–105.
3. Saltiel, A.R.; Kahn, C.R. Insulin signaling and the regulation of glucose and lipid metabolism. Nature **2001**, *414*, 799–806.
4. Czech, M.P.; Corvera, S. Signaling mechanisms that regulate glucose transport. J. Biol. Chem. **1999**, *274*, 1865–1868.
5. White, M.F. The IRS-signalling system: a network of docking proteins that mediate insulin action. Mol. Cell. Biochem. **1998**, *182*, 3–11.
6. Cohen, P.; Alessi, D.R.; Cross, D.A.E. PDK1, one of the missing links in insulin signal transduction. FEBS Lett. **1997**, *410*, 3–10.
7. Hill, M.M.; Clark, S.F.; Tucker, D.F.; Birnbaum, M.J.; James, D.E.; Macaulay, S.L. A role for protein kinase Bbeta/Akt2 in insulin-stimulated GLUT4 translocation in adipocytes. Mol. Cell. Biol. **1999**, *19*, 7771–7781.
8. Kupriyanova, T.A.; Kandror, K.V. Akt-2 binds to Glut4-containing vesicles and phosphorylates their component proteins in response to insulin. J. Biol. Chem. **1999**, *274*, 1458–1464.
9. Wang, Q.; Somwar, R.; Bilan, P.J.; Liu, Z.; Jin, J.; Woodgett, J.R.; Klip, A. Protein kinase B/Akt participates in GLUT4 translocation by insulin in L6 myoblasts. Mol. Cell. Biol. **1999**, *19*, 4008–4018.
10. Ueki, K.; Yamamoto-Honda, R.; Kaburagi, Y.; Yamauchi, T.; Tobe, K.; Burgering, B.M.; Coffer, P.J.; Komuro, I.; Akanuma, Y.; Yazaki, Y.; Kadowaki, T. Potential role of protein kinase B in insulin-induced glucose transport, glycogen synthesis, and protein synthesis. J. Biol. Chem. **1998**, *273*, 5315–5322.
11. Moule, S.K.; Welsh, G.I.; Edgell, N.J.; Foulstone, E.J.; Proud, C.G.; Denton, R.M. Regulation of protein kinase B and glycogen synthase kinase-3 by insulin and beta-adrenergic agonists in rat epididymal fat cells. Activation of protein kinase B by wortmannin-sensitive and -insensitive mechanisms. J. Biol. Chem. **1997**, *272*, 7713–7719.
12. Hurel, S.J.; Rochford, J.J.; Borthwick, A.C.; Wells, A.M.; Vandenheede, J.R.; Turnbull, D.M.; Yeaman, S.J. Insulin action in cultured human myoblasts:

contribution of different signalling pathways to regulation of glycogen synthesis. Biochem. J. **1996**, *320*, 871–877.

13. Cross, D.A.; Watt, P.W.; Shaw, M.; van der Kaay, J.; Downes, C.P.; Holder, J.C.; Cohen, P. Insulin activates protein kinase B, inhibits glycogen synthase kinase-3 and activates glycogen synthase by rapamycin-insensitive pathways in skeletal muscle and adipose tissue. FEBS Lett. **1997**, *406*, 211–215.

14. Andersen, J.N.; Mortensen, O.H.; Peters, G.H.; Drake, P.G.; Iversen, L.F.; Olsen, O.H.; Jansen, P.G.; Andersen, H.S.; Tonks, N.K.; Moller, N.P. Structural and evolutionary relationships among protein tyrosine phosphatase domains. Mol. Cell. Biol. **2001**, *21*, 7117–7136.

15. Goldstein, B.J. Protein-tyrosine phosphatases and the regulation of insulin action. In *Diabetes mellitus: A Fundamental and Clinical Text*; LeRoith, D., Olefsky, J.M., Taylor, S.I., Eds.; Lippincott Publishing: New York, 2000; 206–217.

16. Dadke, S.; Kusari, A.; Kusari, J. Phosphorylation and activation of protein tyrosine phosphatase (PTP)1B by insulin receptor. Mol. Cell. Biochem. **2001**, *221*, 147–154.

17. Chen, H.; Wertheimer, S.J.; Lin, C.H.; Katz, S.L.; Amrein, K.E.; Burn, P.; Quon, M.J. Protein-tyrosine phosphatases PTP1B and syp are modulators of insulin-stimulated translocation of GLUT4 in transfected rat adipose cells. J. Biol. Chem. **1997**, *272*, 8026–8031.

18. Ahmad, F.; Li, P.M.; Meyerovitch, J.; Goldstein, B.J. Osmotic loading of neutralizing antibodies defines a role for protein-tyrosine phosphatase 1B in negative regulation of the insulin action pathway. J. Biol. Chem. **1995**, *270*, 20503–20508.

19. Chen, H.; Cong, L.N.; Li, Y.; Yao, Z.J.; Wu, L.L.; Zhang, Z.Y.; Burke, T.R., Jr.; Quon, M.J. A phosphotyrosyl mimetic peptide reverses impairment of insulin-stimulated translocation of GLUT4 caused by overexpression of PTP1B in rat adipose cells. Biochemistry **1999**, *38*, 384–389.

20. Egawa, K.; Maegawa, H.; Shimizu, S.; Morino, K.; Nishio, Y.; Bryer-Ash, M.; Cheung, A.T.; Kolls, J.K.; Kikkawa, R.; Kashiwagi, A. Protein-tyrosine phosphatase-1B negatively regulates insulin signaling in 16 myocytes and Fao hepatoma cells. J. Biol. Chem. **2001**, *276*, 10207–10211.

21. Clampit, J.E.; Meuth, J.L.; Smith, H.T.; Reilly, R.M.; Jirousek, M.R.; Trevillyan, J.M.; Rondinone, C.M. Reduction of protein-tyrosine phosphatase-1B increases insulin signaling in FAO hepatoma cells. Biochem. Biophys. Res. Commun. **2003**, *300*, 261–267.

22. Goldstein, B.J.; Bittner-Kowalczyk, A.; White, M.F.; Harbeck, M. Tyrosine dephosphorylation and deactivation of insulin receptor substrate-1 by protein-tyrosine phosphatase 1B. J. Biol. Chem. **2000**, *275*, 4283–4289.

23. Seely, B.L.; Staub, P.A.; Reichart, D.R.; Berhanu, P.; Milarski, K.L.; Raltiel, A.R.; Kusari, J.; Olefsky, J.M. Protein tyrosine phosphatase 1B interacts with the activated insulin receptor. Diabetes **1996**, *45*, 1379–1385.

24. Boute, N.; Boubekeur, S.; Lacasa, D.; Issad, T. Dynamics of the interaction between the insulin receptor and protein tyrosine-phosphatase 1B in living cells. EMBO Reports **2003**, *4*, 313–319.

25. Di Paola, R.; Frittitta, L.; Miscio, G.; Bozzali, M.; Baratta, R.; Centra, M.; Spampinato, D.; Santagati, M.G.; Ercolino, T.; Cisternino, C.; Soccio, T.; Mastroianno, S.; Tassi, V.; Almgren, P.; Pizzuti, A.; Vigneri, R.; Trischitta, V. A

variation in 3′ UTR of hPTP1B increases specific gene expression and associates with insulin resistance. Am. J. Hum. Genet. **2002**, *70*, 806–812.

26. Mok, A.; Cao, H.; Zinman, B.; Hanley, A.J.; Harris, S.B.; Kennedy, B.P.; Hegele, R.A. A single nucleotide polymorphism in protein tyrosine phosphatase PTP-1B is associated with protection from diabetes or impaired glucose tolerance in Oji-Cree. J. Clin. Endocrinol. Metab. **2002**, *87*, 724–727.

27. Wu, X.; Hoffstedt, J.; Deeb, W.; Singh, R.; Sedkova, N.; Zilbering, A.; Zhu, L.; Park, P.K.; Arner, P.; Goldstein, B.J. Depot-specific variation in protein-tyrosine phosphatase activities in human omental and subcutaneous adipose tissue: a potential contribution to differential insulin sensitivity. J. Clin. Endocrinol. Metab. **2001**, *86*, 5973–5980.

28. Ahmad, F.; Considine, R.V.; Bauer, T.L.; Ohannesian, J.P.; Marco, C.C.; Goldstein, J. Improved sensitivity to insulin in obese subjects following weight loss is accompanied by reduced protein-tyrosine phosphatases in adipose tissue. Metabolism **1997**, *46*, 1140–1145.

29. Elchebly, M.; Payette, P.; Michaliszyn, E.; Cromlish, W.; Collins, S.; Loy, A.L.; Normandin, D.; Cheng, A.; Himms-Hagen, J.; Chan, C.C.; Ramachandran, C.; Gresser, M.J.; Tremblay, M.L.; Kennedy, B.P. Increased insulin sensitivity and obesity resistance in mice lacking the protein tyrosine phosphatase-1B gene. Science **1999**, *283*, 1544–1548.

30. Klaman, L.D.; Boss, O.; Peroni, O.D.; Kim, J.K.; Martino, J.L.; Zabolotny, J.M.; Moghal, N.; Lubkin, M.; Kim, Y.B.; Sharpe, H.A.; Stricker-Krongrad, A.; Shulman, G.I.; Neel, B.G.; Kahn, B.B. Increased energy expenditure, decreased adiposity, and tissue-specific insulin sensitivity in protein-tyrosine phosphatase 1B-deficient mice. Mol. Cell. Biol. **2000**, *20*, 5479–5489.

31. Zinker, B.A.; Rondinone, C.M.; Trevillyan, J.M.; Gum, R.J.; Clampit, J.E.; Waring, J.F.; Xie, N.; Wilcox, D.; Jacobson, P.; Frost, L.; Kroeger, P.E.; Reilly, R.M.; Koterski, S.; Opgenorth, T.J.; Ulrich, R.G.; Crosby, S.; Butler, M.; Murray, S.F.; McKay, R.A.; Bhanot, S.; Monia, B.P.; Jirousek, M.R. PTP1B antisense oligonucleotide lowers PTP1B protein, normalizes blood glucose, and improves insulin sensitivity in diabetic mice. Proc. Natl. Acad. Sci. USA **2002**, *99*, 11357–11362.

32. Gum, R.J.; Gaede, L.L.; Koterski, S.L.; Heindel, M.; Clampit, J.E.; Zinker, B.A.; Trevillyan, J.M.; Ulrich, R.G.; Jirousek, M.R.; Rondinone, C.M. Reduction of protein tyrosine phosphatase 1B increases insulin-dependent signaling in ob/ob mice. Diabetes **2003**, *52*, 21–28.

33. Rondinone, C.M.; Trevillyan, J.M.; Clampit, J.; Gum, R.J.; Berg, C.; Kroeger, P.; Frost, L.; Zinker, B.A.; Reilly, R.; Ulrich, R.; Butler, M.; Monia, B.P.; Jirousek, M.R.; Waring, J.F. Protein tyrosine phosphatase 1B reduction regulates adiposity and expression of genes involved in lipogenesis. Diabetes **2002**, *51*, 2405–2411.

34. Cheng, A.; Uetani, N.; Simoncic, P.D.; Chaubey, V.P.; Lee-Loy, A.; McGlade, C.J.; Kennedy, B.P.; Tremblay, M.L. Attenuation of leptin action and regulation of obesity by protein tyrosine phosphatase 1B. Dev. Cell **2002**, *2*, 497–503.

35. Zabolotny, J.M.; Bence-Hanulec, K.K.; Stricker-Krongrad, A.; Haj, F.; Wang, Y.; Minokoshi, Y.; Kim, Y.B.; Elmquist, J.K.; Tartaglia, L.A.; Kahn, B.B.; Neel, B.G. PTP1B regulates leptin signal transduction *in vivo*. Dev. Cell **2002**, *2*, 489–495.

36. Myers, M.P.; Andersen, J.N.; Cheng, A.; Tremblay, M.L.; Horvath, C.M.; Parisien, J.P.; Salmeen, A.; Barford, D.; Tonks, N.K. TYK2 and JAK2 are substrates of protein-tyrosine phosphatase 1B. J. Biol. Chem. **2001**, *276*, 47771–47774.
37. Kaszubska, W.; Falls, H.D.; Schaefer, V.G.; Haasch, D.; Frost, L.; Hessler, P.; Kroeger, P.E.; White, D.W.; Jirousek, M.R.; Trevillyan, J.M. Protein tyrosine phosphatase 1B negatively regulates leptin signaling in a hypothalamic cell line. Mol. Cell. Endocrinol. **2002**, *195*, 109–118.
38. Waring, J.F.; Ciurlionis, R.; Clampit, J.E.; Morgan, S.; Gum, R.J.; Jolly, R.A.; Kroeger, P.; Frost, L.; Trevillyan, J.; Zinker, B.A.; Jirousek, M.; Ulrich, R.G.; Rondinone, C.M. PTP1B antisense-treated mice show regulation of genes involved in lipogenesis in liver and fat. Mol. Cell. Endocrinol. **2003**, *203*, 155–168.
39. Waring, J.F.; Gum, R.; Morfitt, D.; Jolly, R.A.; Ciurlionis, R.; Heindel, M.; Gallenberg, L.; Buratto, B.; Ulrich, R.G. Identifying toxic mechanisms using DNA microarrays: evidence that an experimental inhibitor of cell adhesion molecule expression signals through the aryl hydrocarbon nuclear receptor. Toxicology **2002**, *181*, 537–550.
40. Golub, T.R.; Slonim, D.K.; Tamayo, P.; Huard, C.; Gaasenbeek, J.R.; Mesirov, J.P.; Coller, H.; Loh, M.L.; Downing, J.R.; Caligiuri, M.A.; Bloomfield, C.D.; Lander, E.S. Molecular classification of cancer: class discovery and class prediction by gene expression monitoring. Science **1999**, *286*, 531–537.
41. Hughes, T.R.; Marton, M.J.; Jones, A.R.; Roberts, C.J.; Stoughton, R.; Armour, C.D.; Bennett, H.A.; Coffey, E.; Dai, H.; He, Y.D.; Kidd, M.J.; King, A.M.; Meyer, M.R.; Slade, D.; Lum, P.Y.; Stepaniants, S.B.; Shoemaker, D.D.; Gachotte, D.; Chakraburtty, K.; Simon, J.; Bard, M.; Friend, S.H. Functional discovery via a compendium of expression profiles. Cell **2000**, *102*, 109–126.
42. Scherf, U.; Ross, D.T.; Waltham, M.; Smith, L.H.; Lee, J.K.; Tanabe, L.; Kohn, K.W.; Reinhold, W.C.; Myers, T.G.; Andrews, D.T.; Scudiero, D.A.; Eisen, M.B.; Sausville, E.A.; Pommier, Y.; Botstein, D.; Brown, P.O.; Weinstein, J.N. A gene expression database for the molecular pharmacology of cancer. Nat. Gene. **2000**, *24*, 236–244.
43. Waring, J.F.; Halbert, D.N. The promise of toxicogenomics. Curr. Opin. Mol. Ther. **2002**, *4*, 229–235.
44. Nuss, M.; Hradil, V.; Shapiro, R.; Adler, A.; Mika, A.; Clampit, J.; Bush, E.; Xie, N.; Jacobson, P.; Jirousek, M.; Zinker, B. Effects of protein phosphatase 1B (PTP1B) antisense oligonucleotide (ASO) treatment on fat volume using MRI in Zucker fatty rats (Abstract). Diabetes **2001**, *50*, A377.
45. Bray, G.A.; York, D.A. The MONA LISA hypothesis in the time of leptin. Recent Prog. Horm. Res. **1998**, *53*, 95–117.
46. Osborne, T.F. Sterol regulatory element-binding proteins (SREBPs): key regulators of nutritional homeostasis and insulin action. J. Biol. Chem. **2000**, *275*, 32379–32382.
47. Boizard, M.; Le Liepvre, X.; Lemarchand, P.; Foufelle, F.; Ferré, P.; Dugail, I. Obesity-related overexpression of fatty-acid synthase gene in adipose tissue involves sterol regulatory element-binding protein transcription factors. J. Biol. Chem. **1998**, *273*, 29164–29171.
48. Lopez, J.M.; Bennett, M.K.; Sanchez, H.B.; Rosenfeld, J.M.; Osborne, T.F. Sterol regulation of acetyl coenzyme A carboxylase: a mechanism for coordinate control of cellular lipid. Proc. Natl. Acad. Sci. USA **1996**, *93*, 1049–1053.

49. Ericsson, J.; Jackson, S.M.; Kim, J.B.; Spiegelman, B.M.; Edwards, P.A. Identification of glycerol-3-phosphate acyltransferase as an adipocyte determination and differentiation factor 1- and sterol regulatory element-binding protein-responsive gene. J. Biol. Chem. **1997**, *272*, 7298–7305.

50. Kim, J.B.; Sarraf, P.; Wright, M.; Yao, K.M.; Mueller, E.; Solanes, G.; Lowell, B.B.; Spiegelman, B.M. Nutritional and insulin regulation of fatty acid synthetase and leptin gene expression through ADD1/SREBP-1. J. Clin. Invest. **1998**, *101*, 1–9.

51. Kim, J.B.; Spiegelman, B.M. ADD1/SREBP-1 promotes adipocyte differentiation and gene expression linked to fatty acid metabolism. Genes Dev. **1996**, *10*, 1096–1107.

52. Mora, S.; Pessin, J.E. An adipocentric view of signaling and intracellular trafficking. Diab. Metab. Res. Rev. **2002**, *18*, 345–356.

53. Shimano, H.; Horton, J.D.; Hammer, R.E.; Shimomura, I.; Brown, M.S.; Goldstein, J.L. Overproduction of cholesterol and fatty acids causes massive liver enlargement in transgenic mice expressing truncated SREBP-1a. J. Clin. Invest. **1996**, *98*, 1575–1584.

54. Shimomura, I.; Shimano, H.; Horton, J.D.; Goldstein, J.L.; Brown, M.S. Differential expression of exons 1a and 1c in mRNAs for sterol regulatory element binding protein-1 in human and mouse organs and cultured cells. J. Clin. Invest. **1997**, *99*, 838–845.

55. Bennett, M.K.; Lopez, J.M.; Sanchez, H.B.; Osborne, T.F. Sterol regulation of fatty acid synthase promoter: coordinate feedback regulation of two major lipid pathways. J. Biol. Chem. **1995**, *270*, 25578–25583.

56. Kubota, N.; Terauchi, Y.; Miki, H.; Tamemoto, H.; Yamauchi, T.; Komeda, K.; Satoh, S.; Nakano, R.; Ishii, C.; Sugiyama, T.; Eto, K.; Tsubamoto, Y.; Okuno, A.; Murakami, K.; Sekihara, H.; Hasegawa, G.; Naito, M.; Toyoshima, Y.; Tanaka, S.; Shiota, K.; Kitamura, T.; Fujita, T.; Ezaki, O.; Aizawa, S.; Nagai, R.; Tobe, K.; Kimura, S.; Kadowaki, T. PPAR-gamma mediates high-fat diet-induced adipocyte hypertrophy and insulin resistance. Mol. Cell. **1999**, *4*, 597–609.

57. Fajas, L.; Schoonjans, K.; Gelman, L.; Kim, J.B.; Najib, J.; Martin, G.; Fruchart, J.C.; Briggs, M.; Spiegelman, B.M.; Auwerx, J. Regulation of peroxisome proliferator-activated receptor gamma expression by adipocyte differentiation and determination factor 1/sterol regulatory element binding protein 1: implications for adipocyte differentiation and metabolism. Mol. Cell. Biol. **1999**, *19*, 5495–5503.

58. Kim, J.B.; Wright, H.M.; Wright, M.; Spiegelman, B.M. ADD1/SREBP-1 activates PPARgamma through the production of endogenous ligand. Proc. Natl. Acad. Sci. USA **1998**, *95*, 4333–4337.

59. Yamauchi, T.; Kamon, J.; Waki, H.; Murakami, K.; Motojima, K.; Komeda, K.; Ide, T.; Kubota, N.; Terauchi, Y.; Tobe, K.; Miki, H.; Tsuchida, A.; Akanuma, Y.; Nagai, R.; Kimura, S.; Kadowaki, T. The mechanisms by which both heterozygous peroxisome proliferator-activated receptor gamma (PPARgamma) deficiency and PPARgamma agonist improve insulin resistance. J. Biol. Chem. **2001**, *276*, 41245–41254.

60. Hasty, A.H.; Shimano, H.; Yahagi, N.; Amemiya-Kudo, M.; Perrey, S.; Yoshikawa, T.; Osuga, J.; Okazaki, H.; Tamura, Y.; Iizuka, Y.; Shionoiri, F.; Ohashi, K.; Harada, K.; Gotoda, T.; Nagai, R.; Ishibashi, S.; Yamada, N. Sterol regulatory element-binding protein-1 is regulated by glucose at the transcriptional level. J. Biol. Chem. **2000**, *275*, 31069–31077.

61. Way, J.M.; Harrington, W.W.; Brown, K.K.; Gottschalk, W.K.; Sundseth, S.S.; Mansfield, T.A.; Ramachandran, R.K.; Willson, T.M.; Kliewer, S.A. Comprehensive messenger ribonucleic acid profiling reveals that peroxisome proliferator-activated receptor gamma activation has coordinate effects on gene expression in multiple insulin-sensitive tissues. Endocrinology **2001**, *142*, 1269–1277.

62. Fuchtenbusch, M.; Standl, E.; Schatz, H. Clinical efficacy of new thiazolidinediones and glinides in the treatment of type 2 diabetes mellitus. Exp. Clin. Endocrinol. Diab. **2000**, *108*, 151–163.

63. Miles, P.D.; Barak, Y.; He, W.; Evans, R.M.; Olefsky, J.M. Improved insulin-sensitivity in mice heterozygous for PPAR-gamma deficiency. J. Clin. Invest. **2000**, *105*, 287–292.

64. Soukas, A.; Cohen, P.; Socci, N.D.; Friedman, J.M. Leptin-specific patterns of gene expression in white adipose tissue. Genes Dev. **2000**, *14*, 963–980.

65. Liang, C.P.; Tall, A.R. Transcriptional profiling reveals global defects in energy metabolism, lipoprotein and bile acid synthesis and transport with reversal by leptin treatment in ob/ob mouse liver. J. Biol. Chem. **2001**, *276*, 49066–49076.

66. Nadler, S.T.; Stoehr, J.P.; Schueler, K.L.; Tanimoto, G.; Yandell, B.S.; Attie, A.D. The expression of adipogenic genes is decreased in obesity and diabetes mellitus. Proc. Natl. Acad. Sci. USA **2000**, *97*, 11371–11376.

67. Nadller, S.T.; Attie, A.D. Please pass the chips: genomic insights into obesity and diabetes. J. Nutr. **2001**, *138*, 2078–2081.

68. Shimomura, I.; Matsuda, M.; Hammer, R.E.; Bashmakov, Y.; Brown, M.S.; Goldstein, J.L. Decreased IRS-2 and increased SREBP-1c lead to mixed insulin resistance and sensitivity in livers of lipodystrophic and *ob/ob* mice. Mol. Cell. **2000**, *6*, 77–86.

69. Gregoire, F.M.; Sma, C.M.; Sul, H.S. Understanding adipocyte differentiation. Physiol. Rev. **1998**, *78*, 783–809.

70. Morrison, R.F.; Farmer, S.R. Hormonal signaling and transcriptional control of adipocyte differentiation. J. Nutr. **2000**, *130*, 3116S–3121S.

71. Rangwala, S.M.; Lazar, M.A. Transcriptional control of adipogenesis. Annu. Rev. Nutr. **2000**, *20*, 535–559.

72. Rosen, E.D.; Spiegelman, B.M. Molecular regulation of adipogenesis. Annu. Rev. Cell. Dev. Biol. **2000**, *16*, 145–171.

73. Guo, X.; Liao, K. Analysis of gene expression profile during 3T3-L1 preadipocyte differentiation. Gene **2000**, *251*, 45–53.

74. Hamm, J.K.; Park, B.H.; Farmer, S.R. A role for C/EBPß in regulating peroxisome proliferator-activated receptor-γ activity during adipogenesis in 3T3-L1 preadipocytes. J. Biol. Chem. **2001**, *276*, 18464–18471.

75. Reusch, J.E.; Colton, L.A.; Klemm, D.J. CREB activation induces adipogenesis in 3T3-L1 cells. Mol. Cell. Biol. **2000**, *20*, 1008–1020.

76. Tong, Q.; Dalgin, G.; Xu, H.; Ting, C.N.; Leiden, J.M.; Hotamisligil, G.S. Function of GATA transcription factors in preadipocyte-adipocyte transition. Science **2000**, *290*, 134–138.

77. Kim, S.; Moustaid-Moussa, N. Secretory, endocrine and autocrine/paracrine function of the adipocyte. J. Nutr. **2000**, *130*, 3110S–3115S.

78. Soukas, A.; Socci, N.D.; Saatkamp, B.D.; Novelli, S.; Friedman, J.M. Distinct transcriptional profiles of adipogenesis *in vivo* and *in vitro*. J. Biol. Chem. **2001**, *276*, 34167–34174.

79. Friedman, J.M.; Halaas, J.L. Leptin and the regulation of body weight in mammals. Nature **1998**, *395*, 763–770.

80. Zhou, Y.T.; Wang, Z.W.; Higa, M.; Newgard, C.B.; Unger, R.H. Reversing adipocyte differentiation: implications for treatment of obesity. Proc. Natl. Acad. Sci. USA **1999**, *96*, 2391–2395.

81. Cornelius, P.; MacDougald, O.A.; Lane, M.D. Regulation of adipocyte development. Annu. Rev. Nutr. **1994**, *14*, 99–129.

82. Kletzien, R.F.; Clarke, S.D.; Ulrich, R.G. Enhancement of adipocyte differentiation by an insulin-sensitizing agent. Mol. Pharmacol. **1992**, *41*, 393–398.

83. Hallakou, S.; Doare, L.; Foufelle, F.; Kergoat, M.; Guerre-Millo, M.; Berthault, M.F.; Dugail, I.; Morin, J.; Auwerx, J.; Ferre, P. Pioglitazone induces *in vivo* adipocyte differentiation in the obese Zucker fa/fa rat. Diabetes **1997**, *46*, 1393–1399.

84. Ihara, H.; Urano, T.; Takada, A.; Loskutoff, D.J. Induction of plasminogen activator inhibitor 1 gene expression in adipocytes by thiazolidinediones. FASEB J. **2001**, *15*, 1233–1235.

85. Baumann, C.; Chokshi, N.; Saltiel, A.R.; Ribon, V. Cloning and characterization of a functional peroxisome proliferator activator receptor-gamma-responsive element in the promoter of the CAP gene. J. Biol. Chem. **2000**, *275*, 9131–9135.

86. Spiegelman, B.M. PPAR-γ adipogenic regulator and thiazolidinedione receptor. Diabetes **1998**, *47*, 507–514.

87. Yamauchi, T.; Kamon, J.; Waki, H.; Murakami, K.; Motojima, K.; Komeda, K.; Ide, T.; Kubota, N.; Terauchi, Y.; Tobe, K.; Miki, H.; Tsuchida, A.; Akanuma, Y.; Nagai, R.; Kimura, S.; Kadowaki, T. The mechanisms by which both heterozygous PPARgamma deficiency and PPARgamma agonist improve insulin resistance. J. Biol. Chem. **2001**, *276*, 41245–41254.

88. Seckl, J.R.; Walker, B.R. 11beta-hydroxysteroid dehydrogenase type 1- a tissue-specific amplifier of glucocorticoid action. Endocrinology **2001**, *142*, 1371–1376.

89. Morton, N.M.; Holmes, M.C.; Fievet, C.; Staels, B.; Tailleux, A.; Mullins, J.J.; Seckl, J.R. Improved lipid and lipoprotein profile, hepatic insulin sensitivity, and glucose tolerence in 11-hydroxysteroid dehydrogenase type 1 null mice. J. Biol. Chem. **2001**, *276*, 41293–41300.

90. Paulmyer-Lacroix, O.; Boullu, S.; Oliver, C.; Alessi, M.C.; Grino, M. Expression of the mRNA coding for 11beta-hydroxysteroid dehydrogenase type 1 in adipose tissue from obese patients: an *in situ* hybridization study. J. Clin. Endocrinol. Metab. **2002**, *87*, 2701–2705.

91. Moller, D.E. Potential role of TNF-alpha in the pathogenesis of insulin resistance and type 2 diabetes. Trends Endocrinol. Metab. **2000**, *11*, 212–217.

92. Lewis, G.F.; Carpentier, A.; Adeli, K.; Giacca, A. Disordered fat storage and mobilization in the pathogenesis of insulin resistance and type 2 diabetes. Endocr. Rev. **2002**, *23*, 201–229.

93. Goodridge, A.G. Dietary regulation of gene expression: enzymes involved in carbohydrate and lipid metabolism. Annu. Rev. Nutr. **1987**, *7*, 157–185.

94. Hillgartner, F.B.; Salati, L.M.; Goodridge, A.G. Physiological and molecular mechanisms involved in nutritional regulation of fatty acid synthesis. Physiol. Rev. **1995**, *75*, 47–76.

95. Shimano, H.; Horton, J.D.; Hammer, R.E.; Shimomura, I.; Brown, M.S.; Goldstein, J.L. Overproduction of cholesterol and fatty acids causes massive liver enlargement in transgenic mice expressing truncated SREBP-1a. J. Clin. Invest. **1996**, *7*, 1575–1584.

96. Shimano, H.; Horton, J.D.; Shimomura, I.; Hammer, R.E.; Brown, M.S.; Goldstein, J.L. Isoform 1c of sterol regulatory element binding protein is less active than isoform 1a in livers of transgenic mice and in cultured cells. J. Clin. Invest. **1997**, *99*, 846–854.

97. Horton, J.D.; Bashmakov, Y.; Shimomura, I.; Shimano, H. Regulation of sterol regulatory element binding proteins in livers of fasted and refed mice. Proc. Natl. Acad. Sci. USA **1998**, *95*, 5987–5992.

98. Kim, J.B.; Sarraf, P.; Wright, M.; Yao, K.M.; Mueller, E.; Solanes, G.; Lowell, B.B.; Spiegelman, B.M. Nutritional and insulin regulation of fatty acid synthetase and leptin gene expression through ADD1/SREBP-1. J. Clin. Invest. **1998**, *101*, 1–9.

99. Shimano, H.; Yahagi, N.; Amemiya-Kudo, M.; Hasty, A.H.; Osuga, J.; Tamura, Y.; Shionoiri, F.; Iizuka, Y.; Ohashi, K.; Harada, K.; Gotoda, T.; Ishibashi, S.; Yamada, N. Sterol regulatory element-binding protein-1 as a key transcription factor for nutritional induction of lipogenic enzyme genes. J. Biol. Chem. **1999**, *274*, 35832–35839.

100. Tobe, K.; Suzuki, R.; Aoyama, M.; Yamauchi, T.; Kamon, J.; Kubota, N.; Terauchi, Y.; Matsui, J.; Akanuma, Y.; Kimura, S.; Tanaka, J.; Abe, M.; Ohsumi, J.; Nagai, R.; Kadowaki, T. Increased expression of the sterol regulatory element-binding protein-1 gene in insulin receptor substrate-2($-/-$) mouse liver. J. Biol. Chem. **2001**, *276*, 38337–38340.

101. Murakami, K.; Tobe, K.; Ide, T.; Mochizuki, T.; Ohashi, M.; Akanuma, Y.; Yazaki, Y.; Kadowaki, T. A novel insulin sensitizer acts as a coligand for peroxisome proliferator-activated receptor-alpha (PPAR-alpha) and PPAR-gamma: effect of PPAR-alpha activation on abnormal lipid metabolism in liver of Zucker fatty rats. Diabetes **1998**, *47*, 1841–1847.

102. Kim, J.K.; Fillmore, J.J.; Chen, Y.; Yu, C.; Moore, I.K.; Pypaert, M.; Lutz, E.P.; Kako, Y.; Velez-Carrasco, W.; Goldberg, I.J.; Breslow, J.L.; Shulman, G.I. Tissue-specific overexpression of lipoprotein lipase causes tissue-specific insulin resistance. Proc. Natl. Acad. Sci. USA **2001**, *98*, 7522–7577.

103. Shimomura, I.; Hammer, R.E.; Ikemoto, S.; Brown, M.S.; Goldstein, J.L. Leptin reverses insulin resistance and diabetes mellitus in mice with congenital lipodystrophy. Nature **1999**, *401*, 73–76.

104. Ceddia, R.B.; Lopes, G.; Souza, H.M.; Borba-Murad, G.R.; William, W.N.; Bazotte, R.B.; Curi, R. Acute effects of leptin on glucose metabolism of *in situ* rat perfused livers and isolated hepatocytes. Int. J. Obes. Relat. Metab. Disord. **1999**, *23*, 1207–1212.

105. McGarry, J.D. What if Minkowski had been ageusic? An alternative angle on diabetes. Science **1992**, *258* (5083), 766–770.

106. Levin, A.A. A review of the issues in the pharmacokinetics and toxicology of phosphorothioate antisense oligonucleotides. Biochim. Biophys. Acta **1999**, *1489*, 69–84.

18

Metabonomics as a Tool for Understanding Lipid Metabolism

David J. Grainger and David E. Mosedale

Department of Medicine, Cambridge University, Addenbrooke's Hospital, Cambridge, UK

Translational Research Unit, Papworth Hospital NHS Trust, Papworth Everard, Cambridge, UK

Elaine Holmes and Jeremy K. Nicholson

Division of Biomedical Sciences, Imperial College of Science, Technology, and Medicine, London, UK

INTRODUCTION

Understanding lipid metabolism is important for both academic and practical purposes: metabolic pathways involving lipids form a large part of the web of intermediary metabolism, and a knowledge of lipid handling is an essential part of understanding the organism as a whole. Equally, defects in lipid metabolism have been associated with a wide range of human diseases, including coronary heart disease (CHD),[1] which remains the single biggest cause of premature death in the developed world.

However, gaining such an understanding has proved extremely difficult. Mammalian metabolism typically involves thousands of intermediates, products, and metabolites that fall into the broad structural definition of lipids. Worse still, many thousands more make their way into mammalian metabolic pathways through dietary sources, yielding a bewildering array of exogenous metabolic products that may or may not affect endogenous metabolic pathways in some way. Gaining any kind of understanding of such a complex, dynamic system with so many players depends on being able to monitor hundreds, or preferably thousands, of the participants at the same time.

Fortunately, over the last decade, we have seen tremendous advances in the techniques that permit such system-wide analyses to be carried out. Genomic techniques examining variations among arrays of genes simultaneously has pioneered the way, spawning fairly robust techniques to assess genetic variation between individuals at many disparate loci in a single assay, as well as methods to measure the levels of RNA transcripts from different genes as a first estimate of the likely activity of their gene products. Though technically still more demanding, we have also increasingly seen the development of proteomic approaches that can estimate the levels of hundreds of different proteins in a biological sample. These system-wide analytical tools allow the investigator, for the first time, to obtain a handle on the levels of many of the key effectors of lipid metabolism—the enzymes, transport proteins, and their receptors.

By measuring the levels of different RNA transcripts or proteins at the same time, it is possible to build a much more refined picture of lipid metabolism than has previously been possible. Focusing largely or exclusively on a single player in such a complex system can yield surprisingly informative results, but usually only about a tiny fraction of population. For example, Goldstein and Brown's[2,3] pioneering work on the LDL receptor demonstrated the essential role of this protein in cholesterol transport, and how its absence led to one particular type of familial hypercholesterolemia. We now know, however, that variations in LDL receptor levels probably do not contribute to much of the normal variation in lipoprotein levels among healthy individuals,[4] although it is clearly of central importance among the small number of familial hypercholesterolemia sufferers. A much more detailed view of the factors controlling, for example, plasma lipoprotein levels emerges only when multiple enzymes, transport proteins, and receptors are considered simultaneously; the current status of such models of lipid metabolism is reviewed extensively elsewhere in this volume.[5]

Unfortunately, approaches for estimating the levels of the key gene products regulating lipid metabolism give only part of the picture.[6] It will never be possible to gain a full understanding of lipid metabolism without examining the levels of the lipid intermediates, products, and metabolites themselves. A good analogy would be a factory producing widgets through a complex process involving different machines (Fig. 18.1): it would be very difficult to understand what factors controlled the amount of finished widgets accumulating in the stock room by simply counting the number of each type of machine. Instead, process engineers usually require information about the flux of partly finished widgets through each step of the process. There are a number of reasons for this: Firstly, unless all of the properties of all of the machines are completely understood, it will be impossible to convert the information about machine numbers into a model that predicts the stock of finished widgets. Secondly, it

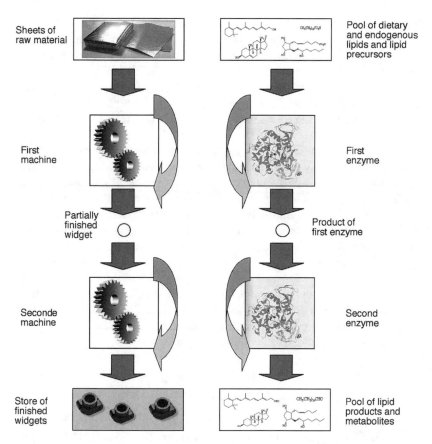

Figure 18.1 Comparison of process engineering and lipid biochemistry. The left panel is a model of a simple engineering process in which two machines acting serially on sheets of raw material produce widgets. The right panel shows an analogous biochemical pathway in which lipid precursors are converted through the serial action of two enzymes into lipid products. Process engineers study the widget production process by counting the flux of partially finished widgets through the system because they do not have a full mathematical description of the properties of each of the machines (the curved arrows). In contrast, biologists have tended to study lipid metabolism by measuring the levels of enzymes, rather than by studying the levels of hundreds or thousands of lipid precursors, intermediates, and metabolites. However, in the absence of useful mathematical models for the properties of most enzymes (the curved arrows), the process engineering approach is likely to be more informative. Metabonomics represents the biochemical equivalent of the process engineering approach to analyze such systems.

would be impossible to understand how the process would react to all possible variations in the nature of the starting materials. If the raw material were slightly too thick, for example, it might occasionally jam one of the machines with partly finished widgets. In lipid biochemistry, the parallels are obvious: our application of genomic and proteomic technologies has progressed far ahead of our knowledge of enzymology. Fully understanding the properties of a single enzyme remains a painstaking biochemical exercise, and such a detailed picture is known for only a tiny fraction of the proteins that participate in lipid metabolism. Worse still, even if we had such a picture of the contributing enzymes, we could not predict how they would function in the presence of the diverse, and varying, array of dietary lipids and their metabolites that are the "raw materials" for the lipid metabolic process (Fig. 18.1).

The only solution, therefore, is to adopt the paradigm of the process engineer and measure not only the levels of the enzymes, transport proteins, and receptors, but also of as many of the lipid intermediates, products, and metabolites as possible. In the past decade, great strides have been made in the techniques required for such a system-wide analysis of small molecule metabolites. Although it is yet to realize its full potential, this field of metabonomics is already able to provide novel insights into lipid metabolism, both in terms of an academic understanding of the underlying biochemistry and in practical terms, providing a potential, new clinical diagnostic platform. The purpose of this review is to examine the application of metabonomics to lipid metabolism, and highlight how it can, in parallel with genomic and proteomic approaches, provide a deeper and more refined understanding of this beautiful, dynamic, and complex system than has previously been possible.

ANALYTICAL TOOLS FOR METABONOMICS

Two analytical tools have been used in almost all of the metabonomics studies of lipid metabolism published to date: nuclear magnetic resonance (NMR) spectroscopy and liquid chromatography followed by mass spectrometry (LC–MS). There are advantages and disadvantages to each, and it is likely that application of both approaches to the same samples would be the ideal solution for assembling a complete system-wide description of a metabolic system.

NMR spectroscopy, which utilizes the context-specific response of nuclei placed in a magnetic field to identify the molecules present in a sample, has the advantages of being a highly reproducible, robust analytical method. We have found that variation between replicate aliquots of the same sample is typically less than 1% if the repeated measurements are performed in a single batch on two separate occasions several days apart (Fig. 18.2). Few techniques, including LC–MS, can match this level of quantitative reproducibility. In contrast, estimates of a metabolite concentration by enzyme assay may have replicate reproducibility of 5–10% depending on the operator, and estimates of protein concentration by proteomics rarely achieve coefficients of variance less than 25%.

Using proton NMR spectroscopy also has the advantage of low bias: in principle, every biological molecule will be represented in the resulting spectrum because all of them contain at least one proton. In marked contrast, different

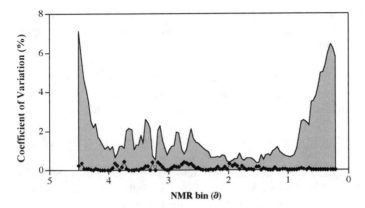

Figure 18.2 Replicate reproducibility of NMR spectroscopy applied to human serum. A single sample of human serum (33-year-old male) was divided into 16 aliquots and frozen. On two separate occasions, 14 days apart, eight aliquots were thawed and ^{1}H-NMR spectra were obtained exactly as described previously. After phase and baseline correction and normalization to total spectral area, the spectra were data reduced to 256 integral bins over the region 9.98–0.00 p.p.m. and the coefficient of variation was calculated for all 16 samples for each bin (gray area). The average replicate reproducibility was 1.8% in the range 0.2–4.5 p.p.m. and 1.1% in the range 0.9–4.0 p.p.m., where most of the signal resides. The net day-to-day variation is also shown (black line) and was negligible (<0.5%) across the whole spectral region studied.

molecules differ dramatically in their ionization potential and certain classes of molecules may be invisible to an LC–MS-based analysis.

However, NMR is inherently an insensitive technique. Consequently, only metabolites that are present at relatively high molar concentrations in the sample are likely to be detected, whereas LC–MS is considerably more sensitive (at least, for those molecules that ionize under the selected experimental conditions). Typically, metabolites present at concentrations less than 1 μM are not readily detected by NMR spectroscopy of complex biofluids. Although sensitivity may not be a major issue for studying lipoprotein transport (as many of the key participants are present at high levels), it may severely impede studies of rare lipid metabolites, such as lipid mediators involved in cell signaling.

LC–MS offers a major advantage that is particularly applicable to the study of lipoprotein metabolism: under the right conditions, it has superior resolving power to distinguish closely related lipid intermediates. Much of the signal derived from, for example, saturated fatty acids is compressed to a single small region of the NMR spectrum and when the sample is a complex mixture of metabolites it rapidly becomes difficult, or indeed impossible, to resolve fatty acid chain length. In marked contrast, both the LC and the MS steps in LC–MS can aid resolution of related molecular structures. When the class of molecules, which is the focus of the study, is known in advance (e.g., when specifically studying lipid metabolism), it is possible to select chromatographic columns that maximize the resolution of that molecular class, favoring, for instance, hydrophobic solid phases for analyzing lipids. Similarly, MS provides unequivocal assignment of chain length information, although it is poorly able, or often unable, to distinguish isomers (such as double bond positions in unsaturated fatty acids) when analyzing complex biofluids. These advantages have led to proponents of the LC–MS approach coining the term "lipidomics" to describe a system-wide description of lipid metabolites obtained in this way.[7,8]

Both LC–MS and NMR spectroscopy yield what is essentially a molecular fingerprint of the biofluid under investigation. In the simplest experiments, most of the information present is related to the chemical structure of the component molecules. Although this has obvious advantages in terms of advancing our understanding of lipid metabolic pathways, it has the disadvantage of being difficult to correlate with older, well-established methods of describing lipoprotein classes. Classically, lipoproteins have been categorized on the basis of their buoyant density (or less frequently on particle size), but such physical descriptions only correlate partially with chemical composition measures.[9] In crude terms, HDL particles are relatively cholesterol rich and triglyceride poor compared with larger, more buoyant LDL particles but this simple rule of thumb hides a great deal of molecular complexity. The extensive work of Otvos and colleagues,[10–14] one of the pioneers of the application of metabonomics to lipid metabolism, has begun to unravel some of these complex relationships. In a recent study, they have shown that the molecular composition of the LDL particle is considerably more important in determining the risk of CHD than the simple LDL-cholesterol (LDL-C) concentration determined by conventional methods.[13,14]

Although simple NMR and LC–MS approaches can yield useful, though rather crude, metabolic fingerprints, there exist numerous refinements to the techniques, which can increase the utility of the approach for specific applications. For example, Nicholson, Lindon and colleagues have extensively described NMR pulse sequences that yield much more information than a simple one-dimensional NMR spectrum.[15–18] In the context of understanding lipid metabolism, diffusion editing may be the most relevant: by exploiting the fact that molecular aggregates (such as lipoprotein particles) tumble in solution more slowly than individual small molecules, it is possible to obtain NMR spectra that specifically highlight the signals from aggregates tumbling above or below a specific rate.[17]

An optimum metabonomic dataset would include reproducible information about a wide variety of metabolites, without bias for particular structural classes, abundancies, or interactions with supramolecular assemblies. Even with the considerable technical advances of the last few years, real world metabonomic datasets still fall short of this ideal, and for the most demanding of the applications the best solution is the use of several complementary techniques to assemble a composite dataset that best approximates the currently unattainable optimum.

EXTRACTING A PEARL OF UNDERSTANDING FROM AN OCEAN OF DATA

The obvious advantages of obtaining system-wide datasets have already been set out. The major disadvantage is the problem of coping with the flood of data that it generates. This bioinformatic challenge remains a significant hurdle to fully exploiting genomic, proteomic, and metabonomic datasets. Even a fairly small experiment might readily generate a million or more datapoints, and the task of extracting any comprehensible patterns from such a sea of information is daunting.[19,20]

Unfortunately, classical mathematical approaches to such multivariate problems (such as multiple linear regression or MLR) are rarely applicable in the real world of "omics" datasets. Several key assumptions of MLR, such as the requirements for the number of observations to exceed the number of variables and for the measured variables to be fully independent, are violated to such an extent as to render regression approaches valueless. A typical omics dataset may have 10,000 measurements (variables) on only 100 individuals (observations), whereas the intercorrelations among the variables are likely to be highly significant (an average pairwise correlation coefficient above 0.5 would not be unusual for data of spectroscopic origin). Mathematical approaches distinct from MLR are clearly required.

A range of alternative methods have been developed for identifying interesting patterns in very large datasets, in many cases driven by the specific problems of analyzing datasets obtained from genechip analytical approaches. Examples range from very simple algorithms (such as hierachichal cluster analysis[21]) to complex solutions (such as genetic computing[22] or neural networks[23]). However, the most extensively used approaches in metabonomics are the projection-based methods refined and popularized by Wold and colleagues.[20]

The principle of the projection methods such as principal component analysis (PCA) or partial least squares (PLS) is to reduce the complexity of the dataset by extracting principle components that are simple linear combinations of the original variables, which describe as much of the variance of the original matrix as possible. An analogy of this process is the projection of a two-dimensional shadow by a three-dimensional object as shown in Fig. 18.3. In the left panel, the shadow of the key projected along the long axis of the shaft is not recognizable as a key: this particular projection has lost most of the information encoded in the original object. In the right panel is shown an optimum 2D projection of the key, which is instantly recognizable. The mathematical algorithms in PCA are designed to select the most descriptive low-dimensional projections of the original multidimensional dataset.

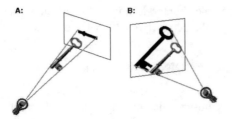

Figure 18.3 The impact of choosing the right projection. A projection is a lower-dimensional representation of a complex higher-dimensional structure. Here, the two-dimensional projection of a three-dimensional key is shown. In panel A, a projection is chosen that poorly reflects the variance of the original three-dimensional object, and the resulting two-dimensional representation is not even identifiable as a key. In panel B, the optimum projection is chosen that maximizes the retention of the information in the original three-dimensional object in the two-dimensional representation, which is now clearly recognizable. PCA is a mathematical method to select the optimum projection of an X-matrix of K variables and N observations into A principal components (dimensions) where $A \ll K$. Reproduced with permission from http://www.graingerlab.org.

Projection approaches are robust tools for extracting patterns from very large datasets. They can cope with many of the problems that prevented the use of MLR on omics datasets: there can be many more variables than observations, the variables can be intercorrelated and the data can be noisy and contain many missing values.[20]

There are many refinements of the simple PCA approach that have found utility in analyzing metabonomic datasets. For example, it is possible to select the projection not only to describe the matrix of measured observations, but also to correlate with some response (e.g., to maximally distinguish individuals with disease from those without). This is the basis for PLS, which is an example of a supervised projection. Such methods can be tremendously powerful if used carefully, but there are a number of pitfalls to trap the unwary,[24] and considerable care should be taken in interpreting the resulting models. The application of projection methods to very large multivariate datasets has been reviewed extensively elsewhere,[20] and is beyond the scope of this review.

METABONOMIC VIEW OF LIPOPROTEIN METABOLISM

Lipid metabolism is a very large field, with thousands of enzymes interconverting tens of thousands of intermediates. As a result, one specific aspect of lipid metabolism, namely, lipoprotein transport, has been the focus for most of the early attempts to apply metabonomics to understanding lipid metabolism. Lipoprotein transport has a number of attractive facets that make it particularly appealing for early metabonomic studies: Firstly, lipoproteins are abundant in blood, and sample preparation and analytical methodologies have been extensively optimized for blood. Studying other tissues, such as tumor cells, is possible but more challenging.[25,26] Secondly, lipoproteins are abundant, ameliorating any concerns about analytical sensitivity. Thirdly, lipoprotein transport has already been implicated in various clinically relevant phenotypes making improved understanding of this area a priority. Finally, lipoprotein transport has been extensively studied for decades using classical approaches and is already fairly well understood, allowing many of the conclusions from early metabonomic investigations to be validated against known models of lipoprotein trafficking.

Otvos and colleagues[10–14] have studied lipoprotein particle composition extensively using NMR spectroscopy. In some sense their approach is not truly a metabonomic one, as they interpret their data by reference to classical measurements of lipoprotein particle classes rather than building a model from scratch using only the molecular composition data in their NMR spectra. Nevertheless, this approach has proved very valuable. Not only does NMR-based lipoprotein analysis provide a swift and accurate way to estimate lipoprotein class distributions for many hundreds or thousands of samples, which has been very useful in large epidemiological studies such as the Framingham cohort,[27] but it also provides additional information not obtained through classical analyses. For example, they have obtained evidence that the cholesterol content of each individual particle may be an important additional indicator of future risk of CHD.[28]

We have also begun to investigate the relationship between the metabolic profile, determined by proton NMR spectroscopy and lipoprotein transport variables. In our study, 600 MHz proton-NMR spectra of human serum collected from individuals with either normal coronary arteries (defined by invasive coronary angiography) or severe coronary artery disease were obtained.[29] Following phasing and baseline correction, the spectra were reduced to 256 integral segments ("bins") and normalized to total spectral area. For each bin, the Pearson's correlation coefficient between the bin intensity and the serum concentration of triglyceride (Trig), LDL-C, HDL-cholesterol (HDL-C), and total cholesterol (Total-C) were computed. In each case, the lipoprotein measurements were made using classical methods, as employed in clinical chemistry laboratories. The results are shown in Fig. 18.4.

Each of the classical lipoprotein measures correlated strongly with particular regions of the spectrum. The strongest correlations we observed are shown in Table 18.1, and agree well with those assignments published previously.[10] Strikingly, we found strong correlations between unique bins in the spectrum and each of the four lipoprotein variables, provided that we used the classical lipoprotein measurements made on the same blood sample as the metabonomic analysis. If we used existing clinical measurements of lipoprotein levels taken a few days prior to the blood sample used for metabonomic analysis, then only the correlations between HDL-C and the spectrum remained. This illustrates

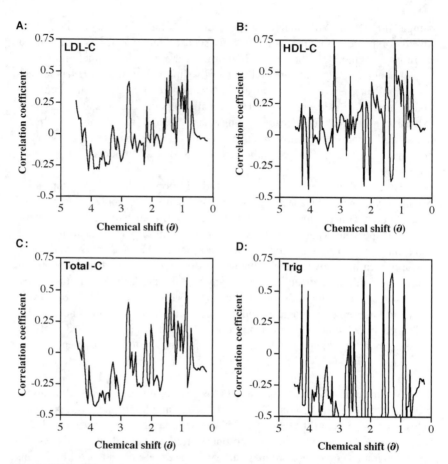

Figure 18.4 Eigenspectra illustrating the association between NMR spectra of human sera and classical lipoprotein measurements. The correlation coefficient between the signal in each bin and the four classical lipoprotein measures (a: LDL-C; b: HDL-C; c: total-C and d: total triglyceride) have been plotted to generate eigenspectra, as described in the legend to Table 18.1. Note the strong inverse relationship between HDL-C and total triglyceride.

an important concept when applying omics technology to obtain a snapshot of a dynamic system: parameters that are more temporally stable are more likely to correlate with stable phenotypes than are noisy parameters.

It is important to stress that even for the strongest correlations (e.g., between bin $\partial 3.22$ and the serum concentration of HDL-C), only about 50% of the variance in the bin can be attributed to classical HDL-C measures. The remainder of the variance is presumably due to variations in the molecular composition of HDL between individuals, as well as to signals that happen to fall in the bin centered on $\partial 3.22$, which do not originate from HDL components. For the

Table 18.1 Correlations Between NMR Chemical Shifts and Classical Lipoprotein Measurements Made on the Same Sample

	NMR bin (∂)	Correlation	Notes
Triglyceride	2.22	0.669	Negatively correlated with age
	1.58	0.648	Negatively correlated with age
	1.30	0.642	
	0.90	0.596	
	1.34	0.582	
	2.26	0.571	Negatively correlated with age
	2.02	0.556	Negatively correlated with age
	1.26	0.553	
	4.26	0.548	
	4.06	0.493	
Total cholesterol	0.86	0.598	
	1.38	0.476	Best specificity over LDL
	1.54	0.467	Associated with gender
LDL-cholesterol	0.86	0.546	
	1.42	0.527	Best specificity over total cholesterol
	1.54	0.447	Associated with gender
HDL-cholesterol	3.22	0.746	
	1.22	0.737	
	0.82	0.509	
	1.50	0.495	
	2.70	0.466	
	0.66	0.457	

Note: NMR spectra were collected for 78 individuals as described previously[29]. After phase and baseline correction and normalization to total spectral area, the Pearson's R correlation statistic was calculated between each bin and the classical lipoprotein measurements were made on the same serum samples. Each statistically significant ($P < 0.05$) correlation is shown, with chemical shift denoting the center of the bin. Each bin was also tested for independent association with age or gender or smoking status by MLR modeling, and the significant results from these models are also noted.

other classes, which correlate even less well with NMR bin intensities, the variance contribution is less than 20%. Consequently, it is entirely unsurprising that a model computed using the NMR bin intensities was able to predict disease susceptibility better than a model based on classical lipoprotein measures.[29]

Probing the association between NMR bin intensity and classical lipoprotein measurements still further, it is possible to establish that the majority of the signal in bin $\partial 3.22$ that correlates with classical HDL-C measurements is because of the signal from choline, possibly in phosphatidyl choline esters. Consequently, the residual following regression of the $\partial 3.22$ intensity onto HDL-C levels is an estimate of the variation in choline concentration within HDL particles from different individuals. This residual is powerfully associated with the presence of coronary artery disease (unpublished data), underlining the additional diagnostic information likely to be contained in a structurally based metabolic profile compared with a classically derived lipoprotein profile.[10,29]

Whereas our studies[29] and those of Otvos and colleagues[13,28] have demonstrated the potential for metabolic profiling of lipid metabolism as diagnostic tool in humans, others have adopted similar approaches to phenotyping lipid metabolism defects in mice. For example, Watkins et al.[30] reported the detailed information on the impact of the PPAR-γ agonist rosiglitazone on the lipid metabolome. Although the dose of drug used was extremely high, nevertheless it demonstrates the power of this technique to uncover unsuspected changes in lipid metabolism that might contribute to the therapeutic (or toxicological) profile of a drug.[31,32]

All of these metabonomic investigations into lipid metabolism are in their early stages. Many of the technical difficulties have been solved, and the tools are now available to address many outstanding questions about mammalian lipid metabolism of both academic and practical value. The next few years are likely to see metabonomics increasingly contribute to the advancement of knowledge in this area.

CLINICAL APPLICATIONS OF METABONOMICS

Gradually, clinical diagnostic applications of metabonomics are making their way into the clinic. In diverse areas, from CHD[29] to cancer, metabonomics-based tests are being evaluated for their diagnostic potential. To date, however, it is in the field of perinatal testing for in-born errors of metabolism that this approach has been developed furthest.[33] Metabonomics is now routinely used for these tests in Germany, where over 500,000 individuals have been screened, proving the utility of the approach.

In the field of lipoprotein metabolism, application of this technology in the clinic has been restricted to the introduction of NMR-based methods for obtaining information about classical lipoprotein particles.[10,11,13] Although this undoubtedly offers some benefits, it does not realize the full potential of metabonomics for disease diagnosis.

Our approach has been to collect metabolic profiles, initially using simple 1D proton NMR spectroscopy of human serum, which encapsulate information about lipoproteins as well as a wide range of other metabolites, and then to apply projection-based analysis methods to determine whether the spectra contain reproducible patterns associated with the presence of disease. Such an approach has the advantage that it makes as few assumptions as possible about what might be associated with a particular disease, and adopts an unbiased modeling approach to identify disease biomarkers.

The results of our first pilot-scale studies in CHD were published recently.[29] We found that subtle patterns do indeed exist in the metabolic profiles from individuals with severe coronary artery disease that allows them to be distinguished, with excellent accuracy, from healthy individuals. Furthermore, these metabolic signatures could provide useful information about the disease severity, among individuals with graded disease progression, established by invasive coronary angiography.

Perhaps unsurprisingly, many of the bins that contributed prominently to the signature defining the coronary artery disease phenotype had signals due to lipid metabolites as major contributors to their variance. Defects in lipoprotein metabolism are already known to be a major contributor to the risk of CHD.[1] In general, the metabolic signature we uncovered was consistent with the expected pattern: bins containing signals from HLD-C were generally of lower intensity among heart disease sufferers, whereas bins containing signals from LDL-C or triglyceride were increased. Importantly, however, the model constructed using the NMR bin intensities was significantly superior in its classification of subjects than any similar model built on the classical lipoprotein measurements. This likely reflects both the important contribution of subtle compositional differences among lipoproteins from different individuals (consistent with earlier studies[28]) but also an important contribution from nonlipoprotein signals.

The studies performed to date demonstrate the clinical potential of such an approach, but they provide only the faintest glimmer of the precise molecular nature of the metabolic defects associated with a particular disease. Refined studies, already underway, which exploit the whole gamut of data capture tools (various 2D-NMR pulse sequences, LC–MS, gene chip data, and limited proteomic data) as well as more advanced implementations of the projection-based model building tools (such as adaptive binning, wavelet transformation, and signal compression filters[20]) will yield a considerably more detailed picture of precisely what metabolic pathways the diagnostic changes in signal intensity are actually reflecting. The outcome from these studies will take a step further than the current clinical diagnostics application, providing valuable pathophysiological insights into lipid (and indeed nonlipid) metabolic pathways, which are instrumental in the cause of human disease.

Diagnosis of an existing medical condition is a potential useful application of metabonomics, particularly in cases where establishing the presence and severity of a disease on the basis of symptoms alone is difficult. However, metabonomics (like genomics and proteomics) offers considerable potential for broader applications. For example, it is now well-established that lipid-lowering drugs of the statin class offer similar protection from heart disease irrespective of the LDL-C level of the subject.[34] It is possible that our ability to target statins effectively to those who will benefit the most could be done better than simply relying on classical lipoprotein measurements. Using metabonomics, it may be possible to identify metabolic profiles associated with a good response to statins, while eliminating others from the treatment program who show only a poor response. Such pharmacometabonomics may offer considerable advantages over the more established pharmacogenomic approaches. It is possible (although completely hypothetical) that response to statins might be modulated as much by the nature of the dietary fatty acid composition as by the genetic status of the

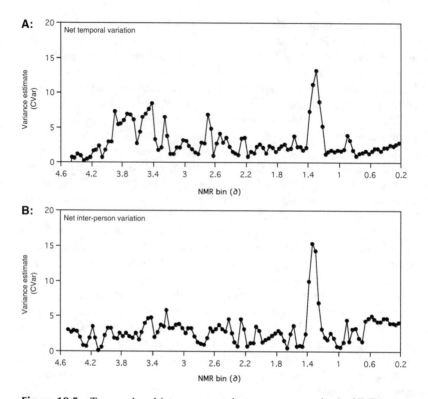

Figure 18.5 Temporal and interperson variance components in the NMR spectrum of human serum. Serum samples were prepared from 17 different individuals on six separate occasions over a 3 month period, and ^{1}H-NMR spectra were collected as previously described. After phase and baseline correction and normalization to total spectral area, the spectra were data reduced to 256 integral bins over the region 9.98–0.00 p.p.m. (a) The coefficient of variation was calculated between the six samples for each individual. The mean coefficient of variation across the 17 individuals is shown (panel A), with the replicate assay variability (Fig. 18.2) subtracted to yield the net temporal variation over a 3 month period. Some regions of the spectra (particularly 3.4–4.0 p.p.m. due to sugars and 1.2–1.4 p.p.m. due to lipids) were considerably more temporally variable than others (e.g., 2.16 or 2.28 p.p.m.). (b) An average value for each bin was calculated for each individual by taking the mean of the six values obtained at different time points, representing an estimate of the time-averaged value in that bin for each individual. The coefficient of variation between 17 individuals was then calculated, and is shown with the replicate assay variability subtracted. Most, but not all, regions of the spectra contain a significant amount of stable variation between individuals (which could, in principle, be used to diagnose the presence of chronic phenotypes such as coronary artery disease), with the bins between 1.0 and 1.4 p.p.m., mainly due to lipids, having approximately 4× more variation between individuals than the rest of the spectrum, on average (even after correction for short-term temporal variation). Note that bins with a larger stable variance component are more likely to contribute to supervised projection models than bins with low stable variance components.

subject. If that were true, then pharmacometabonomics alone would be able to separate responders from nonresponders.

On the other hand, it is commonly assumed that genomics offers considerable advantages over metabonomics for prognostic applications. Assessing the risk of developing a disease in the future (prior to any detectable symptoms) requires measurements of characteristics about an individual, which are temporally stable over a timescale at least as long as the time taken to subsequently develop the disease. The highly stable genomic profile meets this criterion ideally, but is there a component of the metabolic profile that is similarly temporally stable and which could also be used for this purpose?

To address this question, we have carried out a range of simple pilot experiments. For example, we have performed ^{1}H-NMR-based metabonomic analysis of sera collected from 17 different individuals on six separate occasions over a 6 month period, and demonstrated that there is a significant variance component that is stable over this timescale and distinguishes the different individuals (Fig. 18.5). We have also performed a simple analysis of metabolic profiles in pairs of twins. By comparing identical twins (who share 100% of their genetic material) with nonidentical twins (who share only half their genetic material), we have been able to find evidence to suggest that there is a temporally stable component of the metabolic profile that must be unchanged not only for months or even years, but also for decades. Although such studies are in their early stages, it is nevertheless clear that metabonomics can in principle complement genomics in assigning future disease risk in the population.

CONCLUSIONS

Metabonomics is essentially a two-step process: In the first step, an appropriate analytical technique is selected to obtain a broad, and hopefully unbiased, picture of the array of small molecule metabolites present in a sample. In the second step, an appropriate pattern recognition tool is used to mine the vast datasets generated in the first step, to compile generally applicable "rules" from the millions of "examples".

As with genomic and proteomic approaches, metabonomics has tremendous potential for unraveling the complexity of lipid metabolism. Indeed it seems likely that a combination of omics technologies will ultimately deliver the most complete picture. Progress is finally being made, converting this potential into reality, but today we have hardly scratched the surface of what can be achieved. The next 5 years promise to be very exciting indeed.

ACKNOWLEDGMENTS

We are very grateful to Metabometrix Ltd. who funded much of our work on the metabonomics of lipid metabolism. D.J.G. is a British Heart Foundation Senior Research Fellow.

REFERENCES

1. Beisiegel, U. Lipoprotein metabolism. Eur. Heart J. **1998**, *19* (suppl. A), A20–A23.
2. Brown, M.S.; Goldstein, J.L. A receptor-mediated pathway for cholesterol homeostasis. Science **1986**, *232*, 34–47.
3. Tolleshaug, H.; Hobgood, K.K.; Brown, M.S.; Goldstein, J.L. The LDL receptor locus in familial hypercholesterolemia: multiple mutations disrupt transport and processing of a membrane receptor. Cell **1983**, *32*, 941–951.
4. Knoblauch, H.; Busjahn, A.; Munter, S.; Nagy, Z.; Faulhaber, H.D.; Schuster, H.; Luft, F.C. Heritability analysis of lipids and three gene loci in twins link the macrophage scavenger receptor to HDL cholesterol concentrations. Arterioscler. Thromb. Vasc. Biol. **1997**, *17*, 2054–2060.
5. See Chapters 2, 4 and 8.
6. German, J.B.; Roberts, M.A.; Watkins, S.M. Genomics and metabolomics as markers for the interaction of diet and health: lessons from lipids. J. Nutr. **2003**, *133*, 2078S–2083S.
7. Serhan, C.N.; Hong, S.; Gronert, K.; Colgan, S.P.; Devchand, P.R.; Mirick, G.; Moussignac, R.L. Resolvins: a family of bioactive products of omega-3 fatty acid transformation circuits initiated by aspirin treatment that counter proinflammation signals. J. Exp. Med. **2002**, *196*, 1025–1037.
8. Marcheselli, V.L.; Hong, S.; Lukiw, W.J.; Hua Tian, X.; Gronert, K.; Musto, A.; Hardy, M.; Gimenez, J.M.; Chiang, N.; Serhan, C.N.; Bazan, N.G. Novel docosanoids inhibit brain ischemia-reperfusion-mediated leukocyte infiltration and pro-inflammatory gene expression. J. Biol. Chem. **2003**, *278*, 43807–43817.
9. Coresh, J.; Kwiterovich, P.O., Jr.; Smith, H.H.; Bachorik, P.S. Association of plasma triglyceride concentration and LDL particle diameter, density, and chemical composition with premature coronary artery disease in men and women. J. Lipid Res. **1993**, *34*, 1687–1697.
10. Otvos, J.D.; Jeyarajah, E.J.; Bennett, D.W. Quantification of plasma lipoproteins by proton nuclear magnetic resonance spectroscopy. Clin. Chem. **1991**, *37*, 377–386.
11. Otvos, J.D.; Jeyarajah, E.J.; Bennett, D.W.; Krauss, R.M. Development of a proton nuclear magnetic resonance spectroscopic method for determining plasma lipoprotein concentrations and subspecies distributions from a single, rapid measurement. Clin. Chem. **1992**, *38*, 1632–1638.
12. Freedman, D.S.; Bowman, B.A.; Srinivasan, S.R.; Berenson, G.S.; Otvos, J.D. Distribution and correlates of high-density lipoprotein subclasses among children and adolescents. Metabolism **2001**, *50*, 370–376.
13. Otvos, J.D.; Jeyarajah, E.J.; Cromwell, W.C. Measurement issues related to lipoprotein heterogeneity. Am. J. Cardiol. **2002**, *90*, 22i–29i.
14. Kuller, L.; Arnold, A.; Tracy, R.; Otvos, J.; Burke, G.; Psaty, B.; Siscovick, D.; Freedman, D.S.; Kronmal, R. Nuclear magnetic resonance spectroscopy of lipoproteins and risk of coronary heart disease in the cardiovascular health study. Arterioscler. Thromb. Vasc. Biol. **2002**, *22*, 1175–1180.
15. Nicholson, J.K.; Foxall, P.J.; Spraul, M.; Farrant, R.D.; Lindon, J.C. 750 MHz ^{1}H and ^{1}H-^{13}C NMR spectroscopy of human blood plasma. Anal. Chem. **1995**, *67*, 793–811.
16. Liu, M.; Mao, X.A.; Ye, C.; Nicholson, J.K.; Lindon, J.C. Three-dimensional maximum-quantum correlation HMQC NMR spectroscopy (3D MAXY-HMQC). J. Magn. Reson. **1997**, *129*, 67–73.

17. Beckwith-Hall, B.M.; Thompson, N.A.; Nicholson, J.K.; Lindon, J.C.; Holmes, E. A metabonomic investigation of hepatotoxicity using diffusion-edited ^{1}H NMR spectroscopy of blood serum. Analyst **2003**, *128*, 814–818.

18. Spraul, M.; Nicholson, J.K.; Lynch, M.J.; Lindon, J.C. Application of the one-dimensional TOCSY pulse sequence in 750 MHz ^{1}H-NMR spectroscopy for assignment of endogenous metabolite resonances in biofluids. J. Pharm. Biomed. Anal. **1994**, *12*, 613–618.

19. Shaw, A.D.; Winson, M.K.; Woodward, A.M.; McGovern, A.C.; Davey, H.M.; Kaderbhai, N.; Broadhurst, D.; Gilbert, R.J.; Taylor, J.; Timmins, E.M.; Goodacre, R.; Kell, D.B.; Alsberg, B.K.; Rowland, J.J. Rapid analysis of high-dimensional bioprocesses using multivariate spectroscopies and advanced chemometrics. Adv. Biochem. Eng. Biotechnol. **2000**, *66*, 83–113.

20. Eriksson, L.; Johansson, E.; Kettaneh-Wold, N.; Wold, S. *Multi- and Megavariate Data Analysis*; Umetrics Academy: Umea, 2001; 533 pp.

21. Goodacre, R.; McGovern, A.C. Chemometric analyses with self organising feature maps. In *Kohonen Maps*; Oja, E., Kaski, S., Eds.; Elsevier: Amsterdam, 1998.

22. Taylor, J.; Goodacre, R.; Wade, W.G.; Rowland, J.J.; Kell, D.B. The deconvolution of pyrolysis mass spectra using genetic programming: application to the identification of some Eubacterium species. FEMS Microbiol. Lett. **1998**, *160*, 237–246.

23. Holmes, E.; Nicholson, J.K.; Tranter, G. Metabonomic characterization of genetic variations in toxicological and metabolic responses using probabilistic neural networks. Chem. Res. Toxicol. **2001**, *14*, 182–191.

24. Sheridan, R.P.; Nachbar, R.B.; Bush, B.L. Extending the trend vector: the trend matrix and sample-based partial least squares. J. Comput. Aided Mol. Des. **1994**, *8*, 323–340.

25. Shockcor, J.P.; Holmes, E. Metabonomic applications in toxicity screening and disease diagnosis. Curr. Top. Med. Chem. **2002**, *2*, 35–51.

26. Cheng, L.L.; Ma, M.J.; Becerra, L.; Ptak, T.; Tracey, I.; Lackner, A.; Gonzalez, R.G. Quantitative neuropathology by high resolution magic angle spinning proton magnetic resonance spectroscopy. Proc. Natl. Acad. Sci. USA **1997**, *94*, 6408–6413.

27. Ordovas, J.M.; Cupples, L.A.; Corella, D.; Otvos, J.D.; Osgood, D.; Martinez, A.; Lahoz, C.; Coltell, O.; Wilson, P.W.; Schaefer, E.J. Association of cholesteryl ester transfer protein-TaqIB polymorphism with variations in lipoprotein subclasses and coronary heart disease risk: the Framingham study. Arterioscler. Thromb. Vasc. Biol. **2000**, *20*, 1323–1329.

28. Otvos, J. Why cholesterol measurements may be misleading about lipoprotein levels and cardiovascular disease risk—clinical implications of lipoprotein quantification using NMR spectroscopy. J. Lab. Med. **2002**, *26*, 544–550.

29. Brindle, J.T.; Antti, H.; Holmes, E.; Tranter, G.; Nicholson, J.K.; Bethell, H.W.; Clarke, S.; Schofield, P.M.; McKilligin, E.; Mosedale, D.E.; Grainger, D.J. Rapid and noninvasive diagnosis of the presence and severity of coronary heart disease using ^{1}H-NMR-based metabonomics. Nat. Med. **2002**, *8*, 1439–1444.

30. Watkins, S.M.; Reifsnyder, P.R.; Pan, H.J.; German, J.B.; Leiter, E.H. Lipid metabolome-wide effects of the PPARgamma agonist rosiglitazone. J. Lipid Res. **2002**, *43*, 1809–1817.

31. Nicholson, J.K.; Wilson, I.D. Opinion: understanding 'global' systems biology: metabonomics and the continuum of metabolism. Nat. Rev. Drug Discov. **2003**, *2*, 668–676.

32. Nicholson, J.K.; Connelly, J.; Lindon, J.C.; Holmes, E. Metabonomics: a platform for studying drug toxicity and gene function. Nat. Rev. Drug Discov. **2002**, *1*, 153–161.

33. Moolenaar, S.H.; Engelke, U.F.; Wevers, R.A. Proton nuclear magnetic resonance spectroscopy of body fluids in the field of inborn errors of metabolism. Ann. Clin. Biochem. **2003**, *40*, 16–24.

34. Liao, J.K. Beyond lipid lowering: the role of statins in vascular protection. Int. J. Cardiol. **2002**, *86*, 5–18.

19

Zebrafish Model of Lipid Metabolism for Drug Discovery

Amy L. Rubinstein

Zygogen, LLC,
Atlanta, GA, USA

Shiu-Ying Ho and Steven A. Farber

Department of Microbiology and Immunology,
Kimmel Cancer Center,
Thomas Jefferson University,
Philadelphia, PA, USA

The field of lipid biology is benefiting from the numerous genomic resources that have been developed in recent years to study diverse areas of biology. For example, many of the chapters in this book describe the application of micro-arrays to the study of lipid metabolism. Similarly, recent technologies have been developed that allow metabolic processes to be visualized in whole living organisms, such as zebrafish. In particular, zebrafish assays have been developed to examine lipid signaling in embryos and larvae as well as lipid metabolism in the context of the digestive system of larvae.

The zebrafish is a teleost fish that has become a popular model for studies of developmental biology. The advantages of the zebrafish include the large number of embryos produced by adult zebrafish, the development of embryos outside the mother, and the transparent nature of the embryos. Furthermore, genetic and molecular tools, including the ability to perform large-scale forward genetic screens as well as the more recent addition of antisense technologies, have increased the value of the zebrafish model.

Recently, the utility of zebrafish has been expanded to include models of human disease.[1,2] The utility of these models is dependent upon the degree to which zebrafish and human physiology have been conserved. Many comparative studies have been performed that show the similarity between fish and human physiological processes, and for several genes and proteins, functional conservation has been demonstrated. For the purposes of this chapter, discussion will be limited to lipid metabolism, the digestive system, and molecules involved in regulation of lipid levels. As studies in zebrafish are just beginning, much of what is known about lipid metabolism has been gleaned from studies performed in other teleost fish, such as salmon and trout. A brief overview of these studies will also be presented.

LIPID METABOLISM IN FISH

Lipid metabolism has been studied in a number of fish species and many conserved aspects have been described.[3,4] For example, ultrastructural evidence of chylomicrons has been observed following ingestion of a meal. Lipids are transported by lipoproteins of varying sizes in both mammals and fish, including chylomicrons, very low density lipoproteins (VLDL), low density lipoproteins (LDL), high density lipoproteins (HDL), and apolipoproteins.[4] Furthermore, the distribution of total lipids among these lipoprotein classes is comparable. Cholesterol is present in both free and esterified forms. Fish store fat in multiple tissues in addition to adipose tissue, such as in the liver and muscle.[3] Most lipids are stored as triacylglycerol, similar to the situation observed in mammals. The genes encoding apolipoproteins E and A-I were recently cloned and characterized in the zebrafish.[5,6] Both genes were found to be expressed in the yolk syncytial layer of early embryos, an extraembryonic structure that performs many of the functions, such as the metabolism of nutrients, that will later be performed by the liver.[5]

The lipid profile of zebrafish larvae has been analyzed by incorporating radioactively labeled lipid precursors followed by thin layer chromatography (TLC) or by gas chromatography-mass spectrometry (GC-Mass) (S. Ho and S. Farber, unpublished data). This isotopic labeling technique has been applied to examine lipid profiles (lipomics) in a single larva. Briefly, different stages of zebrafish larva were immersed into media containing radioactive lipid precursors following by lipid extraction and TLC separation. Lipid fractions on the TLC plate were then analyzed with a radioactivity scanner. An example of a lipomics study using ^{14}C-oleic acid labeling at different time points for a single 4 days post fertilization (dpf) larva is shown in Fig. 19.1. As the incubation time increased, the fraction represented by free fatty acids decreased, as indicated by the appearance of radioactivity mainly in the phosphatidylcholine (PC) fraction. An increase in the triacylglyeride fraction was also observed. This technique can potentially be applied to screen drugs that alter lipid metabolism by immersing zebrafish larvae in embryo media containing radioactive lipids with or without drugs.

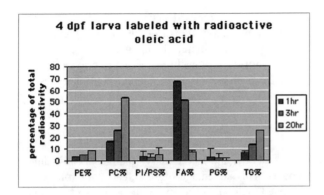

Figure 19.1   Larvae (4 dpf) labeled with radioactive oleic acid. At 1 h (dark grey bars) the majority of radioactivity (66.92%) is present as free fatty acids (FA). After 3 h (red bars), radioactivity in the FA fraction has decreased to 50.91% and radioactivity in the phosphatidylcholine (PC) fraction has increased to 25.52%. By 20 h (light grey bars), little radioactivity is present as FA (8.69%). The majority of FA has been incorporated into PC (52.81%) and triglyceride (TG, 25.53%). Other lipids detected include phosphatidylethanolamine (PE), phosphophatidylserine/inositol (PI/PS), and phosphatidylglyceride (PG).

GC-Mass can analyze not only the lipid class but also the composition of acyl chains. For example, 44 larvae (5 dpf) were pooled together for analysis of their lipid composition using GC-Mass. The major lipid class for 5 dpf larvae was determined to be PC with the major acyl chain being 16:0. This technique can also be applied to study subtle differences between acyl chain composition in zebrafish larvae with or without drug treatment.

Lipid metabolism in the context of the digestive system can also be studied in zebrafish larvae. The zebrafish digestive tract comprises the same essential elements as the mammalian digestive system. It consists of a single intestine that serves the function of the mammalian stomach and intestines, as well as an esophagus, liver, gall bladder, and pancreas. By 5 dpf, all of these structures have developed in the zebrafish larva. Furthermore, the brush border of the intestinal epithelium has developed and intestinal folds are apparent.[7] Peristaltic contractions of the intestinal wall are detectable as early as 4 dpf. Whereas comparative studies have demonstrated that the zebrafish digestive system develops later in embryogenesis than its mammalian counterpart, the molecular determinants of gut formation have been largely conserved.[8]

ATHEROSCLEROSIS IN FISH

Some studies have suggested that a disease similar to human atherosclerosis occurs in fish. For example, members of the salmonid family of teleost fish have been shown to develop atherosclerotic lesions in their coronary arteries. This was initially thought to occur as part of the natural senescence process occuring in migrating salmon; however, later evidence suggested that fish begin to develop atherosclerosis well before maturity.[9] Lesions generally consist of vascular smooth muscle, with no lipid or calcium component. However, lipids do appear to contribute to the severity of lesions in that a diet rich in cholesterol has been shown to be correlated with an increased prevalence of coronary lesions in salmon.[10]

Z-LIPOTRACK

The transparency of zebrafish embryos and larvae makes them especially amenable to fluorescent technologies. This advantage of zebrafish was exploited by Farber et al.[11,12] who developed methods for assaying the activity of phospholipase enzymes. BODIPY-labeled phospholipids were used to determine that a high molecular weight calcium-dependent phospholipid A_2 (cPLA$_2$) constitutes the majority of phospholipase activity during early zebrafish embryogenesis.[11] To assay PLA$_2$ activity in living embryos, a quenched BODIPY-labeled phospholipid encapsulated in liposomes was introduced into zebrafish embryos by soaking. Using this method, cPLA$_2$ activity was localized to perinuclear membranes in embryonic cells and inhibited by a cPLA$_2$-specific inhibitor, methylarachidonyl fluorophosphonate (MAFP). These data confirmed that the majority of phospholipase activity in developing embryos was due to cPLA$_2$ rather than low molecular weight secreted PLA$_2$ isoforms.

By 6 dpf, zebrafish larvae are swimming, swallowing water, and feeding. To assay PLA$_2$ activity in larvae, a water-soluble quenched fluorophore, N-((6-(2,4-dinitrophenyl)amino)hexanoyl)-1-palmitoyl-2-BODIPY-FL-pentanoyl-sn-glycero-3-phosphoethanolamine (PED6) was designed such that cleavage by PLA$_2$ resulted in an increase in fluorescence.[12,13] When larvae are immersed in a solution of PED6, they will swallow it. PED6 is emulsified in the lumen of the intestine and cleaved by phospholipases, resulting in the liberation of fluorescent short chain fatty acids.[12] The by-products of PED6 processing are then absorbed into the intestinal epithelium and transported to the liver where they are incorporated into bile (as in humans, a cholesterol-containing emulsifier), and are eventually concentrated in the gall bladder. A brightly fluorescent gall bladder is observed within 1 h after larvae are fed PED6, indicating the speed of intestinal/hepatobiliary transport. Fluorescent bile is secreted back into the lumen of the intestine, illuminating the intestine to allow clear visualization of many details of intestinal morphology. Similar results are observed when zebrafish are fed other fluorescent lipids, such as NBD cholesterol (Molecular Probes). However, NBD cholesterol is not quenched and fluoresces without the aid of intestinal lipases.

These remarkable observations led to the development of an assay to track lipids *in vivo*, which has been termed Z-Lipotrack. Z-Lipotrack can be used to examine the effect of small molecule compounds on lipid metabolism. For example, fluorescence in the intestine and gall bladder is highly attenuated in

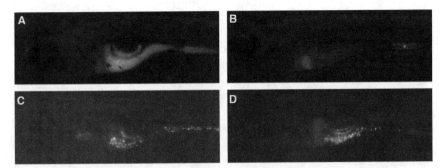

Figure 19.2 Larvae (6 dpf) incubated with PED6 (a, b) or fluorescent beads (c, d). (a, c) Control larvae, (b, d) larvae incubated in 40 μM ezetimibe. Arrows point to the intestine. PED6 fluorescence was significantly diminished by ezetimibe treatment whereas the number of fluorescent beads present in the intestine was not significantly different.

PED6-fed fish if they are simultaneously fed cholesterol synthesis inhibitors, such as atorvastatin.[12] This is likely because cholesterol is a necessary precursor of bile that is required for PED6 to be emulsified in the lumen of the intestine. Emulsification of PED6 is likely to be an important step for efficient intestinal phospholipase activity; thus a lack of emulsification prevents PED6 from becoming fluorescent. Furthermore, unemulsified lipids cannot be properly absorbed by the intestinal epithelium. A similar effect has been observed when larvae are fed fluorescent cholesterol.[12] Thus, this assay can be used to monitor a variety of functional aspects of the zebrafish digestive system, including PLA_2 activity, lipid absorption, and cholesterol synthesis.

We have tested several compounds of known activity for their effect on the Z-Lipotrack assay. For example, ezetimibe, which prevents cholesterol absorption through the brush border cells of the intestine,[14] also reduces the fluorescence that is observed when fish are fed PED6, as shown in Fig. 19.2. Fluorescent beads present in the intestine of drug-treated larvae verified that ezetimibe does not affect the ability of larvae to swallow (Fig. 19.2). Similar to the situation observed with atorvastatin, we hypothesize that a reduction in cholesterol available for emulsification of lipids is likely the cause of this reduced fluorescence (S. Farber, A. White, T. Doan, and A. Rubinstein, unpublished data).

Z-LIPOTRACK FOR GENETIC AND COMPOUND SCREENS

Z-Lipotrack can be used to determine the role of specific genes in lipid metabolism or digestive organ morphology. Forward genetics in the zebrafish is straightforward and large-scale screens for mutations caused by the chemical mutagen ethylnitrosourea have been conducted.[15,16] The first mutagenesis screens focused on mutations that were easily visible because of altered embryonic morphology, including those that affect the digestive system.[3] Mutations that cause degeneration of the pancreas or anterior intestine were shown to result in reduced PED6 fluorescence.[12]

Z-Lipotrack can be especially useful, however, for identifying mutations with no obvious morphological defects. For example, the digestive tract of the recessive lethal *fat free* mutant has a normal appearance, but fails to properly absorb either PED6 or fluorescent cholesterol.[12] Further analysis of *fat free* mutant larvae demonstrated that the *fat free* gene plays an essential role in cholesterol and phospholipid processing. Efforts are currently underway to determine the molecular nature of the *fat free* gene (S. Farber, unpublished data).

Z-Lipotrack may also be used to test the function of genes believed to play a role in lipid metabolism. This can be accomplished by using antisense technology. Two types of antisense molecules have been shown to effectively knockdown gene function in the zebrafish: morpholinos[17] and peptide nucleic acids (gripNAs[18]). Morpholinos can be obtained from Gene Tools (www.gene-tools.com) and gripNAs from Active Motif (www.activemotif.com). To test the function of a specific gene, an antisense molecule is designed to interfere with either the translation or the correct splicing of the gene.[17–19] The molecules are injected into zebrafish embryos at the one cell stage. By 6 dpf, when the Z-Lipotrack assay is performed, the number of antisense molecules per cell is greatly reduced because of increased cell number. Thus, the efficacy of the antisense approach to test late stage gene function may be limited. However, it has been shown to work in some instances (A. Rubinstein, and T. Doan, unpublished data).

We are also exploring the utility of the Z-Lipotrack assay for primary compound screening. To this end, we have screened a 640 compound library using PED6. Preliminary results showed that ~5% of the compounds had a reproducible effect on PED6 fluorescence at concentrations of both 10 and 50 μM (C. Eilertson, A. White, T. Doan, and A. Rubinstein, unpublished data).

Several mechanisms of action could be proposed for how knocking down a gene or treating with a compound results in a decrease in PED6 fluorescence. For example, a phospholipase inhibitor could prevent cleavage of the quenched fluorophore, resulting in a lack of fluorescence. Any compound that affects overall cholesterol levels in the fish could result in a bile acid deficiency, and thus prevent proper emulsification, digestion, and uptake of PED6. Compounds in this category would include cholesterol synthesis inhibitors such as statins as well as cholesterol-lowering fibrates. Compounds that directly prevent uptake of lipids across the brush border of the intestine would also result in a decrease in fluorescence. Compounds that cause significant damage to the intestinal system will prevent lipids from being absorbed and processed properly. Finally, compounds that interfere with larval swallowing would necessarily result in a lack of fluorescent lipid within the intestine.

Since so many possible scenarios can be envisioned for a compound that affects the Z-Lipotrack assay, it was necessary to devise additional assays that can help elucidate the mechanism of action for any particular compound. A very simple assay to determine whether larvae are swallowing normally involves immersion of the larvae in a solution of fluorescent microspheres (2 μm, Polysciences Inc.), which can be observed in the intestine if swallowed[12] (Fig. 19.2). Another assay involves the use of a fluorescent short chain fatty acid (BODIPY-FL-C5, Molecular Probes). As short chain fatty acids are not as dependent upon emulsification as larger lipids, they can be observed in the intestine even if bile acids have been reduced as a result of inhibition of cholesterol synthesis.

To determine whether a compound has phospholipase activity, drug-treated larvae can be fed an unquenched PC analog (such as BODIPY-FL-C5-HPC, Molecular Probes), followed by TLC of larval lipid extracts. If PLA$_2$ function is impaired by the compound, only the unprocessed fluorescent lipid substrate will be visible on the TLC plate. Untreated larvae will show evidence of PLA$_2$ processing in the form of free fatty acids that are used to form cholesteryl ester.

APPLICATIONS OF Z-TAG TECHNOLOGY TO IDENTIFY DRUG TARGETS

Another fluorescent zebrafish technology may be useful for the identification of novel molecular targets for the absorption and transport of lipids. Z-Tag utilizes zebrafish promoters to drive tissue-specific expression of fluorescent proteins.[20] Stable transgenic zebrafish can be homogenized and subjected to fluorescence activated cell sorting (FACS) to obtain relatively pure populations of fluorescent cells. This method may simplify the collection of specific cell types from which RNA can be extracted as the starting material for a cDNA library. Thus, production of a highly tissue-specific library is feasible.

Transgenic lines that express fluorescent proteins in the zebrafish digestive system have already been developed.[21,22] An even more useful Z-Tag transgenic line would express fluorescent proteins in a subset of cells in the digestive system. For example, a cDNA library prepared from the brush border cells of the intestine could represent a valuable resource for genes involved in the transport of cholesterol and other lipids. Whereas cholesterol transport was once thought to be a passive process, many investigators now argue that a specific cholesterol transporter exists in the brush border cells of the intestine.[23–25] Thus, genes expressed in these cells may encode protein components of the transporter and represent important molecular targets for cholesterol absorption inhibitors. Promoters for genes such as the zebrafish *annexin 2b* gene, which is expressed specifically in brush border cells,[26] could be used to create an intestine-specific Z-Tag transgenic line.

Additionally, the zebrafish affords the opportunity to rapidly test genes identified in the Z-Tag intestine-specific library for function by using morpholinos or gripNAs in conjunction with Z-Lipotrack, as described earlier. Knocking down a protein specifically required for cholesterol transport might be expected to yield a result similar to that described for ezetimibe, which specifically inhibits cholesterol transport through the brush border.[27]

SUMMARY

The zebrafish is rapidly emerging as an important player in lipid metabolism research. Though further studies may be required to firmly establish the relevance of zebrafish to human physiology, evidence to date suggests that enough similarities exist to make the zebrafish a valuable model system for lipid research. Furthermore, the unique ability to visualize lipid absorption and transport *in vivo* afforded by the Z-Lipotrack technology in combination with the genetic and molecular tools available for the zebrafish should help to advance our understanding of lipid biology.

REFERENCES

1. Dooley, K.; Zon, L.I. Zebrafish: a model system for the study of human disease. Curr. Opin. Genet. Dev. **2000**, *10*, 252–256.
2. Rubinstein, A.L. Zebrafish: from disease modeling to drug discovery. Curr. Opin. Drug Discov. Dev. **2003**, *6*, 218–223.
3. Sheridan, M.A. Lipid dynamics in fish: aspects of absorption, transportation, deposition and mobilization. Comp. Biochem. Physiol. **1988**, *90B*, 679–690.
4. Babin, P.J.; Vernier, J.M. Plasma lipoproteins in fish. J. Lipid Res. **1989**, *30*, 467–489.
5. Babin, P.J.; Thisse, C.; Durliat, M.; Andre, M.; Akimenko, M.A.; Thisse, B. Both apoliprotein E and A-I genes are present in a nonmammalian vertebrate and are highly expressed during embryonic development. Proc. Natl. Acad. Sci. USA **1997**, *94*, 8622–8627.
6. Monnot, M.J.; Babin, P.J.; Poleo, G.; Andre, M.; Laforest, L.; Ballagny, C.; Akimenko, M.A. Epidermal expression of apolipoprotein E gene during fin and scale development and fin regeneration in zebrafish. Dev. Dyn. **1999**, *214*, 207–215.
7. Pack, M.; Solnica-Krezel, L.; Malicki, J.; Neuhauss, S.C.F.; Schier, A.F.; Stemple, D.L.; Driever, W.; Fishman, M.C. Mutations affecting development of zebrafish digestive organs. Development **1966**, *123*, 321–328.
8. Wallace, K.N.; Pack, M. Unique and conserved aspects of gut development in zebrafish. Dev. Biol. **2003**, *255*, 12–29.
9. Farrell, A.P. Coronary arteriosclerosis in salmon: growing old or growing fast? Comp. Biochem. Physiol. **2002**, *132A*, 723–735.
10. Farrell, A.P.; Saunders, R.L.; Freeman, H.C.; Mommsen, T.P. Arteriosclerosis in Atlantic salmon. Effects of dietary cholesterol and maturation. Arteriosclerosis **1986**, *6*, 453–461.
11. Farber, S.A.; Olson, E.S.; Clark, J.D.; Halpern, M.E. Characterization of Ca^{2+}-dependent phospholipase A_2 activity during zebrafish embryogenesis. J. Biol. Chem. **1999**, *274*, 19338–19346.
12. Farber, S.A.; Pack, M.; Ho, S.Y.; Johnson, I.D.; Wagner, D.S.; Dosch, R.; Mullins, M.C.; Hendrickson, H.S.; Hendrickson, E.K.; Halpern, M.E. Genetic analysis of digestive physiology using fluorescent phospholipid reporters. Science **2001**, *292*, 1385–1388.
13. Hendrickson, H.S.; Hendrickson, E.K.; Johnson, I.D.; Farber, S.A. Intramolecularly quenched BODIPY-labeled phospholipid analogs in phospholipase A_2 and platelet-activating factor acetylhydrolase assays and *in vivo* fluorescence imaging. Anal. Biochem. **1999**, *276*, 27–35.

14. Rosenblum, S.B.; Huynh, T.; Afonso, A.; Davis, H.R.; Yumibe, N.; Clader, J.W.; Burnett, D.A. Discovery of 1-(4-fluorophenyl)-(3R)-[3-(4-fluorophenyl)-(3S)-hydroxypropyl]-(4S)-(4-hydroxyphenyl)-2-azetidinone (SCH 58235): a designed, potent, orally active inhibitor of cholesterol absorption. J. Med. Chem. **1998**, *41*, 973–980.
15. Haffter, P.; Granato, M.; Brand, M.; Mullins, M.C.; Hammerschmidt, M.; Kane, D.A.; Odenthal, J.; van Eeden, F.J.; Jiang, Y.J.; Heisenberg, C.P.; Kelsh, R.N.; Furutani-Seiki, M.; Vogelsang, E.; Beuchle, D.; Schach, U.; Fabian, C.; Nusslein-Volhard, C. The identification of genes with unique and essential functions in the development of the zebrafish, *Danio rerio*. Development **1996**, *123*, 1–36.
16. Driever, W.; Solnica-Krezel, L.; Schier, A.F.; Neuhauss, S.C.; Malicki, J.; Stemple, D.L.; Stainier, D.Y.; Zwartkruis, F.; Abdelilah, S.; Rangini, Z.; Belak, J.; Boggs, C. A genetic screen for mutations affecting embryogenesis in zebrafish. Development **1996**, *123*, 37–46.
17. Nasevicius, A.; Ekker, S.C. Effective targeted gene 'knockdown' in zebrafish. Nat. Genet. **2000**, *26*, 216–220.
18. Urtishak, K.A.; Choob, M.; Tian, X.; Sternheim, N.; Talbot, W.S.; Wickstrom, E.; Farber, S.A. Targeted gene knockdown in zebrafish using negatively charged peptide nucleic acid mimics. Dev. Dyn. **2003**, *228*, 405–413.
19. Draper, B.W.; Morcos, P.A.; Kimmel, C.B. Inhibition of zebrafish fgf8 pre-mRNA splicing with morpholino oligos: a quantifiable method for gene knockdown. Genesis **2001**, *30*, 154–156.
20. Long, Q.; Meng, A.; Wang, H.; Jessen, J.R.; Farrell, M.J.; Lin, S. GATA-1 expression pattern can be recapitulated in living transgenic zebrafish using GFP reporter gene. Development **1997**, *124*, 4105–4111.
21. Field, H.A.; Ober, E.A.; Roeser, T.; Stainier, D.Y. Formation of the digestive system in zebrafish. I. Liver morphogenesis. Dev. Biol. **2003**, *253*, 279–290.
22. Her, G.M.; Chiang, C.C.; Chen, W.Y.; Wu, J.L. *In vivo* studies of liver-type fatty acid binding protein (L-FABP) gene expression in liver of transgenic zebrafish (*Danio rerio*). FEBS Lett. **2003**, *538*, 125–133.
23. Hauser, H.; Dyer, J.H.; Nandy, A.; Vega, M.A.; Werder, M.; Bieliauskaite, E.; Weber, F.E.; Compassi, S.; Gemperli, A.; Bofelli, D.; Wehrli, E.; Schulthess, G.; Phillips, M.C. Identification of a receptor mediating absorption of dietary cholesterol in the intestine. Biochemistry **1998**, *37*, 17843–17850.
24. Detmers, P.A.; Patel, S.; Hernandez, M.; Montenegro, J.; Lisnock, J.; Pikounis, B.; Steiner, M.; Kim, D.; Sparrow, C.; Chao, Y.S.; Wright, S.D. A target for cholesterol absorption inhibitors in the enterocyte brush border membrane. Biochim. Biophys. Acta **2000**, *1486*, 243–252.
25. Kramer, W.; Glombik, H.; Petry, S.; Heuer, H.; Schafer, H.L.; Wendler, W.; Corsiero, D.; Girbig, F.; Weyland, C. Identification of binding proteins for cholesterol absorption inhibitors as components of the intestinal cholesterol transporter. FEBS Lett. **2000**, *487*, 293–297.
26. Farber, S.A.; De Rose, R.A.; Olson, E.S.; Halpern, M.E. The zebrafish annexin gene family. Genome Res. **2003**, *13*, 1082–1096.
27. Van Heek, M.; Farley, C.; Compton, D.S.; Hoos, L.; Davis, H.R. Ezetimibe selectively inhibits intestinal cholesterol absorption in rodents in the presence and absence of exocrine pancreatic function. Br. J. Pharmacol. **2001**, *134*, 409–417.

Index